[数据库技术丛书]

MySQL 8.x 从入门到精通（视频教学版）

李小威 编著

清华大学出版社
北京

内 容 简 介

MySQL 是比较受欢迎的开源关系型数据库之一。本书通过大量实用的操作案例，详细讲解 MySQL 8.x 数据库操作方法和技巧。本书配套资源提供了所有例子的源代码、PPT 课件、培训班形式的同步教学视频、命令速查手册、QQ 群与微信群答疑，以方便读者参考和自学。

本书共分 25 章。主要内容包括 MySQL 8.x 的安装与配置、数据库和数据表基本操作、数据类型和运算符、MySQL 函数、数据的增删改查、索引的设计和使用、存储过程和函数、视图、触发器、存储引擎的选择、分区和事务控制、性能优化、锁定机制、服务器性能优化、性能监控、数据备份与还原、日志、权限与安全管理、高可用架构、复制、MySQL Utilities、MySQL Proxy。最后通过两个案例系统的数据库设计，进一步讲解 MySQL 在系统开发中的应用。本书注重实战操作，帮助读者循序渐进地掌握 MySQL 的各项管理与开发技术。

本书适合 MySQL 数据库初学者、MySQL 数据库开发人员和 MySQL 数据库管理员，同时也能作为高等院校相关专业师生的教学用书。

本书封面贴有清华大学出版社防伪标签，无标签者不得销售。
版权所有，侵权必究。举报：010-62782989，beiqinquan@tup.tsinghua.edu.cn。

图书在版编目（CIP）数据

MySQL 8.x 从入门到精通：视频教学版 / 李小威编著. —北京：清华大学出版社，2022.7（2023.4重印）
（数据库技术丛书）
ISBN 978-7-302-61285-8

Ⅰ. ①M… Ⅱ. ①李… Ⅲ. ①SQL 语言—程序设计 Ⅳ. ①TP311.138

中国版本图书馆 CIP 数据核字（2022）第 122453 号

责任编辑：夏毓彦
封面设计：王　翔
责任校对：闫秀华
责任印制：杨　艳

出版发行：清华大学出版社
　　　　　网　　址：http://www.tup.com.cn，http://www.wqbook.com
　　　　　地　　址：北京清华大学学研大厦 A 座　　邮　　编：100084
　　　　　社 总 机：010-83470000　　邮　　购：010-62786544
　　　　　投稿与读者服务：010-62776969，c-service@tup.tsinghua.edu.cn
　　　　　质 量 反 馈：010-62772015，zhiliang@tup.tsinghua.edu.cn

印 装 者：三河市铭诚印务有限公司
经　　销：全国新华书店
开　　本：190mm×260mm　　印　张：33.5　　字　数：903 千字
版　　次：2022 年 8 月第 1 版　　印　次：2023 年 4 月第 2 次印刷
定　　价：129.00 元

产品编号：096983-01

前　　言

本书是面向 MySQL 数据库管理系统初学者的一本高质量的入门书。目前国内对掌握 MySQL 的人才需求旺盛，各大知名企业高薪招聘技术能力强的 MySQL 开发人员和管理人员。本书根据这样的需求，为初学者量身定做，内容注重实战，通过实例的操作与分析，引领读者快速掌握 MySQL 开发和管理技术。

本书内容

第 1 章主要介绍 MySQL 的安装与配置，包括 MySQL 基本概念、Windows 平台下的安装和配置、如何启动 MySQL 服务、MySQL 常用图形管理工具、Linux 平台下的安装和配置等。

第 2 章介绍数据库和数据表的基本操作，主要包括创建数据库、删除数据库、创建数据表、查看数据表结构、修改数据表和删除数据表。

第 3 章介绍数据类型和运算符，主要包括 MySQL 数据类型介绍、如何选择数据类型和常见运算符介绍。

第 4 章介绍 MySQL 函数，包括数学函数、字符串函数、日期和时间函数、条件判断函数、系统信息函数、加密函数、其他函数和窗口函数。

第 5 章介绍如何查询数据表中的数据，主要包括基本查询语句、单表查询、使用聚合函数查询、连接查询、子查询、合并查询结果、为表和字段取别名以及使用正则表达式查询。

第 6 章介绍如何插入、更新与删除数据，包括插入数据、更新数据、删除数据。

第 7 章介绍 MySQL 中的索引设计和使用，包括索引简介、如何创建各种类型的索引和如何删除索引。

第 8 章介绍 MySQL 中的存储过程和函数，包括存储过程和函数的创建、调用、查看、修改和删除。

第 9 章介绍 MySQL 视图，主要介绍视图的概念、创建视图、查看视图、修改视图、更新视图和删除视图。

第 10 章介绍 MySQL 触发器，包括创建触发器、查看触发器、触发器的使用和删除触发器。

第 11 章介绍存储引擎的选择，包括 MySQL 的架构、MySQL 存储引擎简介、更改数据表的存储引擎、各种存储引擎的特性和选择合适的存储引擎。

第 12 章介绍 MySQL 分区和事务控制，包括合并表、分区表、事务控制、MySQL 分布式事务。

第 13 章介绍如何对 MySQL 进行性能优化，包括优化简介、优化查询、优化数据库结构、临时表性能优化、创建全局通用表空间和隐藏和显示索引。

第 14 章介绍 MySQL 的锁定机制，包括认识 MySQL 的锁定机制、MyISAM 的锁定机制、InnoDB 的锁定机制、跳过锁等待。

第 15 章介绍 MySQL 服务器性能优化，包括优化 MySQL 服务器简介、影响 MySQL 性能的重要参数、MySQL 日志设置优化、MySQL I/O 设置优化、MySQL 并发设置优化、服务器语句超时处理、线程和临时表的优化、增加资源组。

第 16 章介绍 MySQL 性能监控，包括基本监控系统方法、开源监控利器 Nagios 实战、MySQL 监控利器 Cacti 实战。

第 17 章介绍 MySQL 数据库的备份和恢复，主要包括数据备份、数据恢复、数据库的迁移和数据表的导出和导入。

第 18 章介绍 MySQL 日志，主要包括日志简介、二进制日志、错误日志、通用查询日志和慢查询日志。

第 19 章介绍 MySQL 权限与安全管理，主要包括 MySQL 中的各种权限表、账户管理、权限管理、MySQL 的访问控制机制、提升安全性和管理角色。

第 20 章介绍企业中 MySQL 的高可用架构，主要包括 MySQL 高可用的简单介绍、MySQL 主从复制、MySQL+DRBD+HA、LVS+Keepalived+MySQL 单点写入主主同步方案、MMM 高可用架构。

第 21 章介绍 MySQL 复制数据的操作方法和技巧。

第 22 章介绍 MySQL Utilities 管理 MySQL 数据库的方法和技巧。

第 23 章介绍 MySQL Proxy 实现数据库的读写分离的方法和技巧。

第 24 章介绍新闻发布系统的数据库设计过程。

第 25 章介绍论坛管理系统的数据库设计过程。

本书特色

内容全面：涵盖了所有 MySQL 的基础知识点，可让读者由浅入深地掌握 MySQL 数据库开发技术。

图文并茂：在介绍案例的过程中，每一个操作均有对应步骤和过程说明。这种图文结合的方式使读者在学习过程中能够直观、清晰地看到操作的过程以及效果，便于读者更快地理解和掌握。

易学易用：颠覆传统"看"书的观念，变成一本能"操作"的图书。

案例丰富：把知识点融汇于系统的案例实训当中，并且结合综合案例进行讲解和拓展。进而达到"知其然，并知其所以然"的效果。

疑难提示：本书对读者在学习过程中可能会遇到的疑难问题以"提示""注意"等形式进行

了说明,以免读者在学习的过程中走弯路。

答疑支持:本书提供答疑 QQ 群,可直接索要源代码、教学幻灯片和精品教学视频等资源。

课件、源码、教学视频等配套资源下载

本书配套的课件、源码、教学视频、命令速查手册、QQ 群与微信群答疑信息,需要使用微信扫描下面二维码获取,可按扫描后的页面提示,把下载链接转发到自己的邮箱中下载。如果发现问题或疑问,请电子邮件联系 booksaga@163.com,邮件主题为"MySQL 8.x 从入门到精通"。

读者对象

本书适合以下读者学习使用:

- MySQL 数据库初学者。
- 对数据库开发有兴趣,希望快速、全面掌握 MySQL 的人员。
- 对其他数据库有一定的了解,想转到 MySQL 平台上的开发者。
- 高等院校和培训机构相关专业的师生。

鸣谢与技术支持

本书由李小威主编,还有王英英、张工厂、刘增杰、刘玉萍、胡同夫、刘玉红等人参加了编写工作。虽然本书倾注了众多编者的努力,但由于我们水平有限、时间仓促,书中难免存在不足之处,请读者谅解。如果遇到问题或有意见和建议,敬请与我们联系,我们将全力解决。

编 者
2022 年 6 月

目 录

第 1 章 MySQL 的安装与配置 1
1.1 什么是 MySQL 1
1.1.1 客户端/服务器软件 1
1.1.2 MySQL 版本 2
1.2 Windows 平台下安装与配置 MySQL 8.0 2
1.2.1 安装 MySQL 8.0 2
1.2.2 配置 MySQL 8.0 5
1.3 启动服务并登录 MySQL 数据库 9
1.3.1 启动 MySQL 服务 9
1.3.2 登录 MySQL 数据库 10
1.3.3 配置 Path 变量 11
1.4 MySQL 常用图形管理工具 13
1.5 Linux 平台下安装与配置 MySQL 8.0 14
1.5.1 Linux 操作系统下的 MySQL 版本介绍 14
1.5.2 安装和配置 MySQL 的 RPM 包 15
1.5.3 安装和配置 MySQL 的源码包 18

第 2 章 数据库和数据表的基本操作 19
2.1 创建数据库 19
2.2 删除数据库 20
2.3 创建数据表 21
2.3.1 创建表的语法形式 21
2.3.2 使用主键约束 22
2.3.3 使用外键约束 23
2.3.4 使用非空约束 24
2.3.5 使用唯一性约束 25
2.3.6 使用默认约束 25
2.3.7 设置表的属性值自动增加 26
2.4 查看数据表结构 27
2.4.1 查看表基本结构语句 DESCRIBE 27
2.4.2 查看表详细结构语句 SHOW CREATE TABLE 28

2.5 修改数据表 ··· 28
    2.5.1 修改表名 ··· 29
    2.5.2 修改字段的数据类型 ·· 29
    2.5.3 修改字段名 ·· 30
    2.5.4 添加字段 ··· 31
    2.5.5 删除字段 ··· 33
    2.5.6 修改字段的排列位置 ·· 34
    2.5.7 删除表的外键约束 ·· 35
2.6 删除数据表 ··· 36
    2.6.1 删除没有被关联的表 ·· 36
    2.6.2 删除被其他表关联的主表 ·· 37

第3章 数据类型和运算符 39

3.1 MySQL 数据类型介绍 ··· 39
    3.1.1 整数类型 ··· 39
    3.1.2 小数类型 ··· 41
    3.1.3 日期与时间类型 ··· 42
    3.1.4 文本字符串类型 ··· 52
    3.1.5 二进制字符串类型 ·· 56
3.2 如何选择数据类型 ·· 59
3.3 常见运算符介绍 ··· 60
    3.3.1 运算符概述 ·· 60
    3.3.2 算术运算符 ·· 61
    3.3.3 比较运算符 ·· 62
    3.3.4 逻辑运算符 ·· 69
    3.3.5 位运算符 ··· 71
    3.3.6 运算符的优先级 ··· 74

第4章 MySQL 函数 75

4.1 MySQL 函数简介 ·· 75
4.2 数学函数 ·· 75
    4.2.1 绝对值函数 ABS(x)和返回圆周率的函数 PI() ······································· 75
    4.2.2 平方根函数 SQRT(x)和求余函数 MOD(x,y) ·· 76
    4.2.3 获取整数的函数 CEIL(x)、CEILING(x)和 FLOOR(x) ··························· 76
    4.2.4 获取随机数的函数 RAND()和 RAND(x) ··· 77
    4.2.5 函数 ROUND(x)、ROUND(x,y)和 TRUNCATE(x,y) ····························· 78
    4.2.6 符号函数 SIGN(x) ·· 79

目录 | VII

- 4.2.7 幂运算函数 POW(x,y)、POWER(x,y)和 EXP(x) ································ 79
- 4.2.8 对数运算函数 LOG(x)和 LOG10(x) ·· 79
- 4.2.9 角度与弧度相互转换的函数 RADIANS(x)和 DEGREES(x) ············ 80
- 4.2.10 正弦函数 SIN(x)和反正弦函数 ASIN(x) ·· 80
- 4.2.11 余弦函数 COS(x)和反余弦函数 ACOS(x) ······································ 81
- 4.2.12 正切函数、反正切函数和余切函数 ·· 81

4.3 字符串函数 ·· 82

- 4.3.1 计算字符串字符数的函数和字符串长度的函数 ······························ 82
- 4.3.2 合并字符串函数 CONCAT(s1,s2,…)、CONCAT_WS(x,s1,s2,…) ···· 83
- 4.3.3 替换字符串的函数 INSERT(s1,x,len,s2) ·· 83
- 4.3.4 字母大小写转换函数 ·· 84
- 4.3.5 获取指定长度的字符串的函数 LEFT(s,n)和 RIGHT(s,n) ················· 85
- 4.3.6 填充字符串的函数 LPAD(s1,len,s2)和 RPAD(s1,len,s2) ··················· 85
- 4.3.7 删除空格的函数 LTRIM(s)、RTRIM(s)和 TRIM(s) ························· 86
- 4.3.8 删除指定字符串的函数 TRIM(s1 FROM s) ······································· 86
- 4.3.9 重复生成字符串的函数 REPEAT(s,n) ·· 87
- 4.3.10 空格函数 SPACE(n)和替换函数 REPLACE(s,s1,s2) ······················· 87
- 4.3.11 比较字符串大小的函数 STRCMP(s1,s2) ·· 88
- 4.3.12 获取子串的函数 SUBSTRING(s,n,len)和 MID(s,n,len) ··················· 88
- 4.3.13 匹配子串开始位置的函数 ·· 89
- 4.3.14 字符串逆序的函数 REVERSE(s) ·· 89
- 4.3.15 返回指定位置的字符串的函数 ·· 89
- 4.3.16 返回指定字符串位置的函数 FIELD(s,s1,s2,…,sn) ·························· 90
- 4.3.17 返回子串位置的函数 FIND_IN_SET(s1,s2) ····································· 90
- 4.3.18 选取字符串的函数 MAKE_SET(x,s1,s2,…,sn) ································ 90

4.4 日期和时间函数 ··· 91

- 4.4.1 获取当前日期的函数和获取当前时间的函数 ·································· 91
- 4.4.2 获取当前日期和时间的函数 ·· 92
- 4.4.3 UNIX 时间戳函数 ·· 92
- 4.4.4 返回 UTC 日期的函数和返回 UTC 时间的函数 ······························· 92
- 4.4.5 获取月份的函数 MONTH(date)和 MONTHNAME(date) ················· 93
- 4.4.6 获取星期的函数 DAYNAME(d)、DAYOFWEEK(d)和 WEEKDAY(d) ···· 93
- 4.4.7 获取星期数的函数 WEEK(d)和 WEEKOFYEAR(d) ························ 94
- 4.4.8 获取天数的函数 DAYOFYEAR(d)和 DAYOFMONTH(d) ··············· 95
- 4.4.9 获取年份、季度、小时、分钟和秒钟的函数 ·································· 96
- 4.4.10 获取日期的指定值的函数 EXTRACT(type FROM date) ················· 96

4.4.11 时间和秒钟转换的函数 · 97
4.4.12 计算日期和时间的函数 · 97
4.4.13 将日期和时间格式化的函数 · 100
4.5 条件判断函数 · 102
4.5.1 IF(expr,v1,v2)函数 · 102
4.5.2 IFNULL(v1,v2)函数 · 103
4.5.3 CASE 函数 · 103
4.6 系统信息函数 · 104
4.6.1 获取 MySQL 版本号、连接数和数据库名的函数 · 104
4.6.2 获取用户名的函数 · 106
4.6.3 获取字符串的字符集和排序方式的函数 · 106
4.6.4 获取最后一个自动生成的 ID 值的函数 · 107
4.7 加密函数 · 108
4.7.1 加密函数 MD5(str) · 108
4.7.2 加密函数 SHA(str) · 108
4.7.3 加密函数 SHA2(str, hash_length) · 109
4.8 其他函数 · 109
4.8.1 格式化函数 FORMAT(x,n) · 109
4.8.2 不同进制的数字进行转换的函数 · 109
4.8.3 IP 地址与数字相互转换的函数 · 110
4.8.4 加锁函数和解锁函数 · 111
4.8.5 重复执行指定操作的函数 · 111
4.8.6 改变字符集的函数 · 112
4.8.7 改变数据类型的函数 · 112
4.9 窗口函数 · 113

第 5 章 查询数据 · 115
5.1 基本查询语句 · 115
5.2 单表查询 · 117
5.2.1 查询所有字段 · 117
5.2.2 查询指定字段 · 118
5.2.3 查询指定记录 · 120
5.2.4 带 IN 关键字的查询 · 121
5.2.5 带 BETWEEN...AND...的范围查询 · 122
5.2.6 带 LIKE 的字符匹配查询 · 123
5.2.7 查询空值 · 124
5.2.8 带 AND 的多条件查询 · 126

		5.2.9 带 OR 的多条件查询	126
		5.2.10 查询结果不重复	127
		5.2.11 对查询结果排序	128
		5.2.12 分组查询	131
		5.2.13 使用 LIMIT 限制查询结果的数量	136
	5.3	使用集合函数查询	137
		5.3.1 COUNT()函数	137
		5.3.2 SUM()函数	138
		5.3.3 AVG()函数	139
		5.3.4 MAX()函数	140
		5.3.5 MIN()函数	141
	5.4	连接查询	141
		5.4.1 内连接查询	142
		5.4.2 外连接查询	144
		5.4.3 复合条件连接查询	146
	5.5	子查询	147
		5.5.1 带 ANY、SOME 关键字的子查询	147
		5.5.2 带 ALL 关键字的子查询	147
		5.5.3 带 EXISTS 关键字的子查询	148
		5.5.4 带 IN 关键字的子查询	149
		5.5.5 带比较运算符的子查询	151
	5.6	合并查询结果	152
	5.7	为表和字段取别名	154
		5.7.1 为表取别名	154
		5.7.2 为字段取别名	155
	5.8	使用正则表达式查询	157
		5.8.1 查询以特定字符或字符串开头的记录	157
		5.8.2 查询以特定字符或字符串结尾的记录	158
		5.8.3 用符号 "." 来替代字符串中的任意一个字符	158
		5.8.4 使用 "*" 和 "+" 来匹配多个字符	159
		5.8.5 匹配指定字符串	159
		5.8.6 匹配指定字符中的任意一个	160
		5.8.7 匹配指定字符以外的字符	161
		5.8.8 使用 {n,} 或者 {n,m} 来指定字符串连续出现的次数	162
	5.9	通用表表达式	162

第 6 章 插入、更新与删除数据 166

6.1 插入数据 166
6.1.1 为表的所有字段插入数据 166
6.1.2 为表的指定字段插入数据 168
6.1.3 同时插入多条记录 169
6.1.4 将查询结果插入到表中 170
6.2 更新数据 172
6.3 删除数据 173
6.4 为表增加计算列 175
6.5 DDL 的原子化 176

第 7 章 索引的设计和使用 178

7.1 索引简介 178
7.1.1 索引的含义和特点 178
7.1.2 索引的分类 179
7.1.3 索引的设计原则 180
7.2 创建索引 180
7.2.1 创建表的时候创建索引 180
7.2.2 在已经存在的表上创建索引 185
7.3 删除索引 191
7.4 统计直方图 193
7.4.1 直方图的优点 193
7.4.2 直方图的基本操作 193

第 8 章 存储过程和函数 195

8.1 创建存储过程和函数 195
8.1.1 创建存储过程 195
8.1.2 创建存储函数 197
8.1.3 变量的使用 198
8.1.4 定义条件和处理程序 199
8.1.5 光标的使用 202
8.1.6 流程控制的使用 203
8.2 调用存储过程和函数 207
8.2.1 调用存储过程 207
8.2.2 调用存储函数 208
8.3 查看存储过程和函数 208
8.3.1 使用 SHOW STATUS 语句查看存储过程和函数的状态 208

        8.3.2　使用 SHOW CREATE 语句查看存储过程和函数的定义 ·················· 209
        8.3.3　从 information_schema.Routines 表中查看存储过程和函数的信息 ······ 210
　8.4　修改存储过程和函数 ·· 211
　8.5　删除存储过程和函数 ·· 212
　8.6　全局变量的持久化 ··· 213

第 9 章　视图··· 214
　9.1　视图概述 ·· 214
        9.1.1　视图的含义 ·· 214
        9.1.2　视图的作用 ·· 215
　9.2　创建视图 ·· 215
        9.2.1　创建视图的语法形式 ··· 216
        9.2.2　在单表上创建视图 ·· 216
        9.2.3　在多表上创建视图 ·· 217
　9.3　查看视图 ·· 218
        9.3.1　使用 DESCRIBE 语句查看视图基本信息 ······································· 218
        9.3.2　使用 SHOW TABLE STATUS 语句查看视图基本信息 ······················· 218
        9.3.3　使用 SHOW CREATE VIEW 语句查看视图详细信息 ························ 219
        9.3.4　在 views 表中查看视图详细信息 ··· 220
　9.4　修改视图 ·· 221
        9.4.1　使用 CREATE OR REPLACE VIEW 语句修改视图 ·························· 221
        9.4.2　使用 ALTER 语句修改视图 ·· 222
　9.5　更新视图 ·· 222
　9.6　删除视图 ·· 225

第 10 章　MySQL 触发器··· 226
　10.1　创建触发器 ·· 226
         10.1.1　创建只有一个执行语句的触发器 ··· 226
         10.1.2　创建有多个执行语句的触发器 ·· 227
　10.2　查看触发器 ·· 229
         10.2.1　利用 SHOW TRIGGERS 语句查看触发器信息 ······························ 229
         10.2.2　在 triggers 表中查看触发器信息 ··· 231
　10.3　触发器的使用 ··· 232
　10.4　删除触发器 ·· 233

第 11 章　存储引擎的选择··· 234
　11.1　MySQL 的架构 ·· 234
         11.1.1　MySQL 物理文件的组成 ··· 235

11.1.2　MySQL 各逻辑块简介 ·················· 237
　　11.1.3　MySQL 各逻辑块协调工作 ·············· 239
11.2　MySQL 存储引擎简介 ······················ 240
11.3　更改数据表的存储引擎 ······················ 242
11.4　各种存储引擎的特性 ························ 242
　　11.4.1　MyISAM ···························· 243
　　11.4.2　InnoDB 存储引擎 ····················· 245
　　11.4.3　MEMORY ···························· 247
　　11.4.4　MERGE ······························ 248
11.5　选择合适的存储引擎 ························ 250

第 12 章　MySQL 分区和事务控制 ················ 252

12.1　合并表 ······································ 252
12.2　分区表 ······································ 254
　　12.2.1　认识分区表 ·························· 254
　　12.2.2　RANGE 分区 ·························· 254
　　12.2.3　LIST 分区 ···························· 256
　　12.2.4　HASH 分区 ··························· 257
　　12.2.5　线性 HASH 分区 ······················ 257
　　12.2.6　KEY 分区 ····························· 258
　　12.2.7　复合分区 ···························· 259
12.3　事务控制 ···································· 261
12.4　MySQL 分布式事务 ·························· 264
　　12.4.1　分布式事务的原理 ···················· 264
　　12.4.2　分布式事务的语法 ···················· 265

第 13 章　MySQL 性能优化 ······················ 267

13.1　优化简介 ···································· 267
13.2　优化查询 ···································· 268
　　13.2.1　分析查询语句 ························ 268
　　13.2.2　索引对查询速度的影响 ················ 271
　　13.2.3　使用索引查询 ························ 272
　　13.2.4　优化子查询 ·························· 274
13.3　优化数据库结构 ······························ 274
　　13.3.1　将字段很多的表分解成多个表 ·········· 274
　　13.3.2　增加中间表 ·························· 276
　　13.3.3　增加冗余字段 ························ 277

		13.3.4 优化插入记录的速度	277
		13.3.5 分析表、检查表和优化表	279
13.4	临时表性能优化		281
13.5	创建全局通用表空间		282
13.6	隐藏和显示索引		283

第 14 章 MySQL 的锁定机制 ... 285

14.1	认识 MySQL 的锁定机制		285
14.2	MyISAM 的锁定机制		289
	14.2.1	MyISAM 表级锁的锁模式	289
	14.2.2	获取 MyISAM 表级锁的争用情况	291
	14.2.3	MyISAM 表级锁加锁方法	292
	14.2.4	MyISAM Concurrent Insert 的特性	294
	14.2.5	MyISAM 表锁优化建议	295
14.3	InnoDB 的锁定机制		296
	14.3.1	InnoDB 行级锁模式	296
	14.3.2	获取 InnoDB 行级锁的争用情况	300
	14.3.3	InnoDB 行级锁的实现方法	304
	14.3.4	间隙锁（Net-Key 锁）	307
	14.3.5	InnoDB 在不同隔离级别下加锁的差异	309
	14.3.6	InnoDB 存储引擎中的死锁	309
	14.3.7	InnoDB 行级锁优化建议	311
14.4	跳过锁等待		311

第 15 章 MySQL 服务器性能优化 ... 313

15.1	优化 MySQL 服务器简介		313
	15.1.1	优化服务器硬件	313
	15.1.2	优化 MySQL 的参数	314
15.2	影响 MySQL 性能的重要参数		315
	15.2.1	查看性能参数的方法	315
	15.2.2	key_buffer_size 的设置	319
	15.2.3	内存参数的设置	321
	15.2.4	日志和事务参数的设置	322
	15.2.5	存储和 I/O 相关参数的设置	324
	15.2.6	其他重要参数的设置	325
15.3	MySQL 日志设置优化		326
15.4	MySQL I/O 设置优化		328

15.5 MySQL 并发设置优化 330
15.6 服务器语句超时处理 331
15.7 线程和临时表的优化 331
    15.7.1 线程的优化 331
    15.7.2 临时表的优化 332
15.8 增加资源组 333

第 16 章 MySQL 性能监控 335

16.1 基本监控系统方法 335
    16.1.1 ps 命令 335
    16.1.2 top 命令 336
    16.1.3 vmstat 命令 338
    16.1.4 mytop 命令 339
    16.1.5 sysstat 工具 341
16.2 开源监控利器 Nagios 实战 345
    16.2.1 安装 Nagios 之前的准备工作 346
    16.2.2 安装 Nagios 主程序 347
    16.2.3 整合 Nagios 到 Apache 服务 348
    16.2.4 安装 Nagios 插件包 351
    16.2.5 监控服务器的 CPU、负载、磁盘 I/O 使用情况 352
    16.2.6 配置 Nagios 监控 MySQL 服务器 356
16.3 MySQL 监控利器 Cacti 实战 359
    16.3.1 Cacti 工具的安装 359
    16.3.2 Cacti 监控 MySQL 服务器 363

第 17 章 数据备份与恢复 368

17.1 数据备份 368
    17.1.1 使用 mysqldump 命令备份 368
    17.1.2 直接复制整个数据库目录 373
    17.1.3 使用 mysqlhotcopy 工具快速备份 374
17.2 数据恢复 374
    17.2.1 使用 MySQL 命令恢复 374
    17.2.2 直接复制到数据库目录 375
    17.2.3 mysqlhotcopy 快速恢复 375
17.3 数据库迁移 376
    17.3.1 相同版本的 MySQL 数据库之间的迁移 376
    17.3.2 不同版本的 MySQL 数据库之间的迁移 376

17.3.9　不同数据库之间的迁移 ··· 377
　17.4　表的导出和导入 ··· 377
　　　17.4.1　使用 SELECT…INTO OUTFILE 导出文本文件 ··· 377
　　　17.4.2　使用 mysqldump 命令导出文本文件 ·· 381
　　　17.4.3　使用 MySQL 命令导出文本文件 ··· 383
　　　17.4.4　使用 LOAD DATA INFILE 方式导入文本文件 ·· 386
　　　17.4.5　使用 mysqlimport 命令导入文本文件 ··· 388

第 18 章　MySQL 日志 390

　18.1　日志简介 ··· 390
　18.2　二进制日志 ··· 391
　　　18.2.1　启动和设置二进制日志 ··· 391
　　　18.2.2　查看二进制日志 ··· 392
　　　18.2.3　删除二进制日志 ··· 393
　　　18.2.4　使用二进制日志恢复数据库 ·· 395
　　　18.2.5　暂时停止二进制日志功能 ·· 395
　18.3　错误日志 ··· 396
　　　18.3.1　启动和设置错误日志 ··· 396
　　　18.3.2　查看错误日志 ·· 396
　　　18.3.3　删除错误日志 ·· 397
　18.4　通用查询日志 ··· 398
　　　18.4.1　启动通用查询日志 ·· 398
　　　18.4.2　查看通用查询日志 ·· 398
　　　18.4.3　删除通用查询日志 ·· 399
　18.5　慢查询日志 ··· 399
　　　18.5.1　启动和设置慢查询日志 ··· 399
　　　18.5.2　查看慢查询日志 ··· 400
　　　18.5.3　删除慢查询日志 ··· 400

第 19 章　MySQL 权限与安全管理 401

　19.1　权限表 ··· 401
　　　19.1.1　user 表 ··· 401
　　　19.1.2　db 表 ·· 403
　　　19.1.3　tables_priv 表和 columns_priv 表 ·· 405
　　　19.1.4　procs_priv 表 ·· 405
　19.2　账户管理 ··· 406
　　　19.2.1　登录和退出 MySQL 服务器 ·· 406

19.2.2 新建普通用户 ... 407
19.2.3 删除普通用户 ... 409
19.2.4 root 用户修改普通用户密码 ... 411
19.3 权限管理 ... 412
19.3.1 MySQL 的各种权限 ... 412
19.3.2 授权 ... 414
19.3.3 收回权限 ... 415
19.3.4 查看权限 ... 416
19.4 访问控制 ... 417
19.4.1 连接核实阶段 ... 417
19.4.2 请求核实阶段 ... 417
19.5 提升安全性 ... 418
19.5.1 密码到期更换策略 ... 418
19.5.2 安全模式安装 ... 420
19.6 管理角色 ... 420

第 20 章 MySQL 高可用架构 ... 422
20.1 MySQL 高可用简介 ... 422
20.2 MySQL 主从复制架构 ... 423
20.2.1 MySQL 主从架构设计 ... 423
20.2.2 配置环境 ... 423
20.2.3 服务器的安装配置 ... 424
20.2.4 LVS 的安装配置 ... 426
20.3 MySQL+DRBD+HA 主备架构 ... 428
20.3.1 什么是 DRBD ... 428
20.3.2 MySQL+DRBD+HA 架构设计 ... 428
20.3.3 配置环境 ... 429
20.3.4 安装配置 Heartbeat ... 429
20.3.5 安装配置 DRBD ... 431
20.4 LVS+Keepalived+MySQL 单点写入主主同步架构 ... 433
20.4.1 配置环境 ... 434
20.4.2 LVS+Keepalived 的安装 ... 438
20.4.3 LVS+Keepalived 的配置 ... 439
20.4.4 Master 和 Backup 的启动 ... 441
20.5 MMM 高可用架构 ... 442
20.5.1 MMM 高可用架构简介 ... 442
20.5.2 配置环境 ... 442

	20.5.3 MMM 的安装	445
	20.5.4 Monitor 服务器的配置	446
	20.5.5 各个数据库服务器的配置	447
	20.5.6 MMM 的管理	447

第 21 章 MySQL 复制 ... 449

21.1 MySQL 复制概述 ... 449
21.2 Windows 环境下的 MySQL 主从复制 ... 450
- 21.2.1 复制前的准备工作 ... 450
- 21.2.2 Windows 环境下实现主从复制 ... 450
- 21.2.3 Windows 环境下主从复制测试 ... 456

21.3 Linux 环境下的 MySQL 复制 ... 457
- 21.3.1 下载并安装 MySQL 8.0 ... 458
- 21.3.2 单机主从复制前的准备工作 ... 459
- 21.3.3 mysqld_multi 实现单机主从复制 ... 462
- 21.3.4 不同服务器之间实现主从复制 ... 468
- 21.3.5 MySQL 主从复制启动选项 ... 470
- 21.3.6 指定复制的数据库或者表 ... 471

21.4 查看从服务器的复制进度 ... 477
21.5 复制环境的监控和维护 ... 478
- 21.5.1 了解服务器的状态 ... 478
- 21.5.2 服务器复制出错的原因 ... 479

21.6 切换主从服务器 ... 482
21.7 多源复制的改进 ... 485

第 22 章 MySQL Utilities ... 488

22.1 MySQL Utilities 概述 ... 488
22.2 安装与配置 ... 488
- 22.2.1 下载与安装 MySQL Utilities ... 489
- 22.2.2 MySQL Utilities 连接数据库 ... 490

22.3 管理与维护 ... 491
- 22.3.1 使用 mysqldbcompare 比较数据 ... 491
- 22.3.2 使用 mysqldbcopy 复制数据 ... 492
- 22.3.3 使用 mysqldbexport 导出数据 ... 492
- 22.3.4 使用 mysqldbimport 导入数据 ... 493
- 22.3.5 使用 mysqldiff 比较对象的定义 ... 493

第 23 章　MySQL Proxy 494

23.1　概述 494
23.2　安装与配置 495
　　23.2.1　下载与安装 MySQL Proxy 495
　　23.2.2　配置 MySQL Proxy 参数 496
23.3　使用 MySQL Proxy 实现读写分离 498

第 24 章　新闻发布系统数据库设计 499

24.1　系统概述 499
24.2　系统功能 500
24.3　数据库设计和实现 500
　　24.3.1　设计表 501
　　24.3.2　设计索引 505
　　24.3.3　设计视图 506
　　24.3.4　设计触发器 506

第 25 章　论坛管理系统数据库设计 508

25.1　系统概述 508
25.2　系统功能 509
25.3　数据库设计和实现 509
　　25.3.1　设计方案图表 509
　　25.3.2　设计表 511
　　25.3.3　设计索引 514
　　25.3.4　设计视图 515
　　25.3.5　设计触发器 516

第 1 章

MySQL 的安装与配置

MySQL 支持多种系统平台，不同平台下的安装与配置过程也不相同。在 Windows 平台下可以使用二进制的安装软件包或免安装版的软件包进行安装，二进制的安装包提供了图形化的向导安装方式，而免安装版直接解压缩即可使用。Linux 平台下使用命令行安装 MySQL，但由于 Linux 是开源操作系统，有众多的分发版本，因此针对不同的 Linux 平台需要下载相应的 MySQL 安装包。本章将主要讲解 Windows 和 Linux 两个平台下 MySQL 8.0.28（本书也称 MySQL 8.0 或者 MySQL 8）的安装和配置过程。

1.1 什么是 MySQL

MySQL 是一个小型关系数据库管理系统。与其他大型数据库管理系统（例如 Oracle、DB2、SQL Server 等）相比，MySQL 规模小、功能有限，但是它体积小、速度快、成本低，并且提供的功能对稍微复杂的应用来说已经够用，这些特性使得 MySQL 成为世界上最受欢迎的开放源代码数据库。本节将介绍 MySQL 的特点。

1.1.1 客户端/服务器软件

主从式架构（Client-Server Model）或客户端/服务器（Client/Server）结构（简称 C/S 结构），是一种网络架构，通常在该网络架构下的软件可分为客户端（Client）和服务器（Server）。

服务器（也称服务器端）是整个应用系统资源的存储与管理中心，多个客户端则各自处理相应的功能，共同实现完整的应用。在客户端/服务器结构中，客户端的请求被传送到数据库服务器，数据库服务器进行处理后，将结果返回给客户端，从而减少了网络数据传输量。

用户使用应用程序时，首先启动客户端通过有关命令告知服务器进行连接以完成各种操作，而服务器则按照此请示提供相应的服务。每一个客户端软件的实例都可以向一个服务器或应用程序服务器发出请求。

这种系统的特点就是，客户端和服务器程序不需要在同一台计算机上运行，这些客户端和服务器程序通常归属不同的计算机。

主从式架构通过不同的途径应用于很多不同类型的应用程序，比如现在人们最熟悉的、因特网上使用的网页。例如，当顾客想要在当当网站上买书的时候，电脑和网页浏览器就被当作一个客户端，同时组成当当网的电脑、数据库和应用程序就被当作服务器。当顾客的网页浏览器向当当网请求搜寻数据库相关的图书时，当当网服务器从当当网的数据库中找出所有该类型的图书信息，结合成一个网页，再发送回顾客的浏览器。服务器一般使用高性能的计算机，并配合使用不同类型的数据库，比如 Oracle、Sybase 或者是 MySQL 等；客户端需要安装专门的软件，比如专门开发的客户端工具、浏览器等。

1.1.2 MySQL 版本

针对不同用户，MySQL 分为两个不同的版本：

- MySQL Community Server（社区版服务器）：该版本完全免费，但是官方不提供技术支持。
- MySQL Enterprise Server（企业版服务器）：能够以很高的性价比为企业提供数据仓库应用，支持 ACID 事物处理，提供完整的提交、回滚、崩溃恢复和行级锁定功能。但是该版本需付费使用，官方提供电话技术支持。

提示：MySQL Cluster 主要用于架设集群服务器，需要在社区版或企业版基础上使用。

MySQL 版本的命名机制由 3 个数字和 1 个后缀组成，例如：MySQL-8.0.28 版本。

（1）第 1 个数字（8）是主版本号，描述了文件格式，所有版本 8 的发行版都有相同的文件格式。

（2）第 2 个数字（0）是发行级别，主版本号和发行级别组合在一起便构成了发行序列号。

（3）第 3 个数字（28）是在此发行系列的版本号，随每次新分发版本递增。通常选择已经发行的最新版本。

在 MySQL 开发过程中，同时存在多个发布系列，每个发布处在软件成熟度的不同阶段。

（1）MySQL 8.0 是最新开发的稳定（GA）发布系列，是将执行新功能的系列，目前已经可以正常使用。

（2）MySQL 5.7 是比较稳定（GA）发布系列。只针对漏洞修复重新发布，没有增加会影响稳定性的新功能。

1.2 Windows 平台下安装与配置 MySQL 8.0

Windows 平台下安装 MySQL，可以使用图行化的安装包，图形化的安装包提供了详细的安装向导，通过向导，读者可以一步一步地完成 MySQL 的安装。本节将介绍使用图形化安装包安装 MySQL 8.0 的步骤。

1.2.1 安装 MySQL 8.0

要想在 Windows 中运行 MySQL，需要 32 位或 64 位 Windows 操作系统，例如 Windows 7、

Windows 8、Windows 10、Windows Server 2012 等。Windows 可以将 MySQL 服务器作为服务来运行。通常，在安装时需要具有系统的管理员权限。

Windows 平台下提供两种安装方式：MySQL 二进制分发版（.msi 安装文件）和免安装版（.zip 压缩文件）。一般来讲，应当使用二进制分发版，因为该版本比其他的分发版使用起来要简单，不再需要其他工具来启动就可以运行 MySQL。

1. 下载 MySQL 安装文件

下载 MySQL 安装文件的具体操作步骤如下：

步骤 01 打开浏览器，在地址栏中输入网址 "https://dev.mysql.com/downloads/installer/"，单击【转到】按钮，打开 MySQL Community Server 8.0.28 下载页面，选择 Microsoft Windows 平台，选择离线安装版本，单击右侧的【Download】按钮开始下载，如图 1.1 所示。

步骤 02 在弹出的提示页面中开始下载，单击【Login】按钮，如图 1.2 所示。

图 1.1　MySQL 下载页面

图 1.2　开始下载页面

提示：这里 32 位的安装程序有两个版本，分别为 mysql-installer-web-community 和 mysql-installer-community，其中 mysql-installer-web-community 为在线安装版本，mysql-installer-community 为离线安装版本。

步骤 03 弹出用户登录页面，输入用户名和密码后，单击【登录】按钮，如图 1.3 所示。

步骤 04 弹出开始下载页面，单击【Download Now】按钮，即可开始下载，如图 1.4 所示。

图 1.3　用户登录页面

图 1.4　开始下载页面

提示：如果没有用户名和密码，可以单击【创建账户】链接在网站上进行注册即可。

2. 安装 MySQL 8.0

MySQL 下载完成后，找到下载文件，双击运行进行安装，具体操作步骤如下：

步骤01 双击下载的 mysql-installer-community-8.0.28.0.msi 文件，如图 1.5 所示。

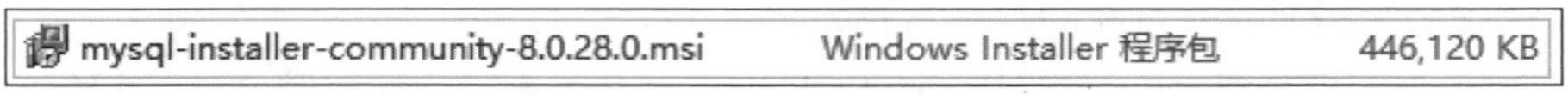

图 1.5　MySQL 安装文件名称

步骤02 打开【Choosing a Setup Type】（安装类型选择）窗口，在其中列出了 5 种安装类型，分别是 Developer Default（默认安装类型）、Server only（仅作为服务器）、Client only（仅作为客户端）、Full（完全安装）和 Custom（自定义安装类型）。这里选择【Custom】单选按钮，单击【Next】按钮，如图 1.6 所示。

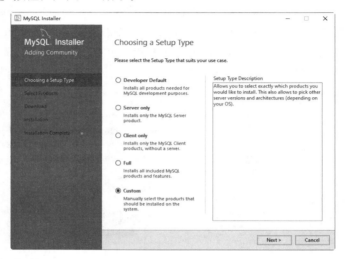

图 1.6　安装类型窗口

步骤03 打开【Select Products】（产品定制选择）窗口，选择【MySQL Server 8.0.28-x86】后，单击【添加】按钮，即可选择安装 MySQL 服务器。采用同样的方法，添加【Samples and Examples 8.0.28-x86】和【MySQL Documentation 8.0.28-x86】选项，如图 1.7 所示。

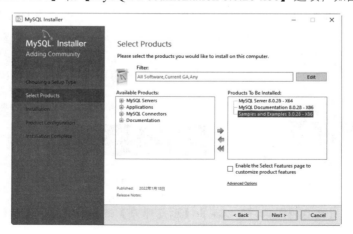

图 1.7　自定义安装组件窗口

步骤04 单击【Next】按钮，进入安装确认对话框，单击【Execute】按钮，如图 1.8 所示。

图 1.8　准备安装对话框

步骤 05　开始安装 MySQL 文件，安装完成后在【Status】（状态）列表下将显示 Complete（安装完成），如图 1.9 所示。

图 1.9　安装完成窗口

1.2.2　配置 MySQL 8.0

MySQL 安装完毕之后，需要对服务器进行配置。具体的配置步骤如下：

步骤 01　在 1.2.1 节的最后一步中，单击【Next】按钮，将进入产品信息窗口，如图 1.10 所示。

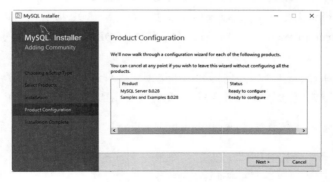

图 1.10　产品信息窗口

步骤 02　单击【Next】按钮，进入 MySQL 服务器类型配置窗口，采用默认设置，如图 1.11 所示。

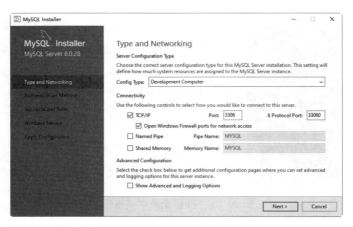

图 1.11　MySQL 服务器类型配置窗口

其中，【Config Type】选项用于设置服务器的类型。单击该选项右侧的下三角按钮，即可看到 3 个选项，如图 1.12 所示。

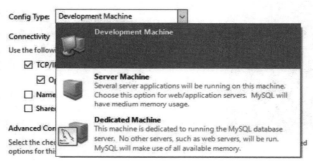

图 1.12　MySQL 服务器的类型

图 1.12 所示的界面中 3 个选项的具体含义如下：

（1）Development Machine（开发机器）：该选项代表典型的个人用桌面工作站。假定机器上运行着多个桌面应用程序，该选项会将 MySQL 服务器配置成使用最少的系统资源。

（2）Server Machine（服务器）：该选项代表服务器。MySQL 服务器可以同其他应用程序一起运行，例如 FTP、Email 和 Web 服务器，该选项会将 MySQL 服务器配置成使用适当比例的系统资源。

（3）Dedicated Machine（专用服务器）：该选项代表只运行 MySQL 服务的服务器。假定没有运行其他服务程序，该选项会将 MySQL 服务器配置成使用所有可用系统资源。

提示：作为初学者，建议选择【Development Machine】（开发机器）选项，这样占用系统的资源比较少。

步骤03 单击【Next】按钮，打开设置授权方式窗口。其中，第一个单选项的含义是 MySQL 8.0 提供的新的授权方式，采用 SHA256 基础的密码加密方法；第二个单选项的含义是传统授权方法（保留 5.x 版本兼容性）。这里选择第二个单选项，如图 1.13 所示。

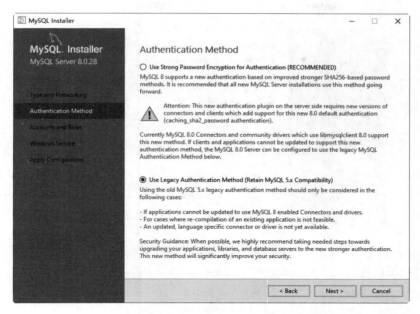

图 1.13　MySQL 服务器的授权方式

步骤 04 单击【Next】按钮，打开设置服务器的密码窗口，输入两次同样的登录密码，如图 1.14 所示。

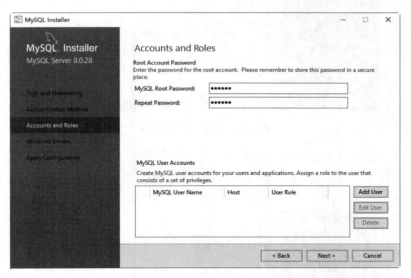

图 1.14　设置服务器的登录密码

提示：系统默认的用户名称为 root，如果想添加新用户，可以单击【Add User】（添加用户）按钮。

步骤 05 单击【Next】按钮，打开设置服务器名称窗口。本案例设置服务器名称为"MySQL"，如图 1.15 所示。

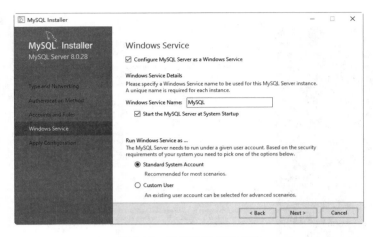

图 1.15　设置服务器的名称

步骤06 单击【Next】按钮，打开应用配置（Apply Configuration）服务器窗口，单击【Execute】按钮，如图 1.16 所示。

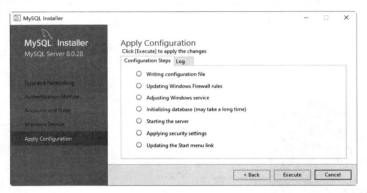

图 1.16　确认设置服务器

步骤07 系统自动配置 MySQL 服务器。配置完成后，单击【Finish】按钮，即可完成服务器的配置，如图 1.17 所示。其他的配置比较简单，采用默认值即可，这里不再重复讲解。

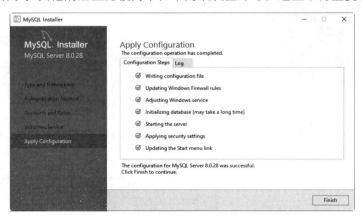

图 1.17　完成设置服务器

步骤 08 按键盘上的 Ctrl+Alt+Delete 组合键，打开【任务管理器】对话框，可以看到 MySQL 服务进程 mysqld.exe 已经启动了，如图 1.18 所示。

图 1.18　任务管理器窗口

至此，在 Windows 10 操作系统环境下安装 MySQL 的操作就完成了。

1.3　启动服务并登录 MySQL 数据库

MySQL 安装完毕之后，需要启动服务器进程，不然客户端无法连接数据库，客户端通过命令行工具登录数据库。本节将介绍启动 MySQL 服务器和登录 MySQL 的方法。

1.3.1　启动 MySQL 服务

在前面的配置过程中，已经将 MySQL 安装为 Windows 服务，当 Windows 启动、停止时，MySQL 也自动启动、停止。不过，用户还可以使用图形服务工具来控制 MySQL 服务器，或通过命令行使用 NET 命令。

可以通过 Windows 的服务管理器查看，具体的操作步骤如下：

步骤 01 右击 Windows 系统桌面（后文简称为桌面）上的【开始】按钮，选择【运行】菜单命令，在打开的【运行】对话框中输入"services.msc"命令，单击【确定】按钮，如图 1.19 所示。

步骤 02 打开 Windows 的服务管理器，在其中可以看到名为"MySQL"的服务项，其右边状态为"正在运行"，表明该服务已经启动，如图 1.20 所示。

图 1.19　【运行】对话框　　　　　　图 1.20　服务管理器窗口

1.3.2　登录 MySQL 数据库

当 MySQL 服务启动完成后，便可以通过客户端来登录 MySQL 数据库。在 Windows 操作系统下，可以通过两种方式登录 MySQL 数据库。

1. 以 Windows 命令行方式登录

具体的操作步骤如下：

步骤 01 右击桌面上的【开始】按钮，选择【运行】菜单命令，在打开的【运行】对话框中输入"cmd"命令，单击【确定】按钮，如图 1.21 所示。

步骤 02 打开 DOS 窗口，输入以下命令并按【Enter】键，如图 1.22 所示。

```
cd C:\Program Files\MySQL\MySQL Server 8.0\bin\
```

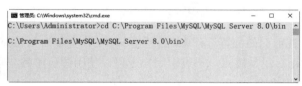

图 1.21　【运行】对话框　　　　　　图 1.22　DOS 窗口

步骤 03 在 DOS 窗口中可以通过登录命令连接到 MySQL 数据库。连接 MySQL 的命令格式为：

```
mysql -h hostname -u username -p
```

其中，mysql 为登录命令，–h 后面的参数是服务器的主机地址，在这里客户端和服务器在同一台机器上，所以输入"localhost"或者 IP 地址"127.0.0.1"；-u 后面跟登录数据库的用户名称，在这里为 root；-p 后面是用户登录密码。

接下来，输入如下命令：

```
mysql -h localhost -u root -p
```

按【Enter】键，系统会提示"Enter password"（输入密码），这里输入在前面配置向导中自己设置的密码，验证正确后，即可登录到 MySQL 数据库，如图 1.23 所示。

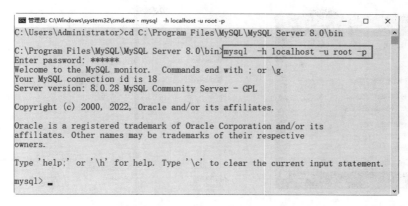

图 1.23　Windows 命令行登录窗口

提示：当窗口中出现如图 1.23 所示的说明信息，命令提示符变为"mysql>"时，表明已经成功登录 MySQL 服务器，可以开始对数据库进行操作了。读者后续对数据库的操作可以使用这个界面。

2. 使用 MySQL Command Line Client 登录

依次选择【开始】→【所有程序】→【MySQL】→【MySQL 8.0 Command Line Client】菜单命令，输入正确的密码之后，就可以登录到 MySQL 数据库了，如图 1.24 所示。

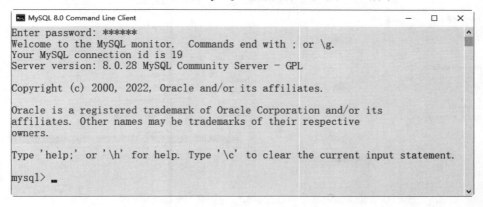

图 1.24　MySQL 命令行登录窗口

1.3.3　配置 Path 变量

在前面登录 MySQL 服务器的时候，因为没有把 MySQL 的 bin 目录添加到系统的环境变量里面，所以不能直接使用 MySQL 命令。如果每次登录都输入"cd C:\Program Files\MySQL\MySQL Server 8.0\bin"才能使用 MySQL 等其他命令工具，这样比较烦琐。

下面介绍手动配置 PATH 变量：

步骤01　在桌面上右击【此电脑】图标，在弹出的快捷菜单中选择【属性】菜单命令，如图 1.25 所示。

步骤02　打开【系统】窗口，单击【高级系统设置】链接，如图 1.26 所示。

图 1.25 "此电脑"右击快捷菜单　　　　　图 1.26 系统属性对话框

步骤 03 打开【系统属性】对话框，选择【高级】选项卡，然后单击【环境变量】按钮，如图 1.27 所示。

步骤 04 打开【环境变量】对话框，在系统变量列表中选择【Path】变量，如图 1.28 所示。

图 1.27 【系统属性】对话框　　　　　图 1.28 【环境变量】对话框

步骤 05 单击【编辑】按钮，在【编辑环境变量】对话框中，将 MySQL 应用程序的 bin 目录（C:\Program Files\MySQL\MySQL Server 8.0\bin）添加到变量值中，用分号将其与其他路径分隔开，如图 1.29 所示。

步骤 06 添加完成之后，单击【确定】按钮，这样就完成了配置 PATH 变量的操作，即可直接输入 MySQL 命令来登录数据库了。

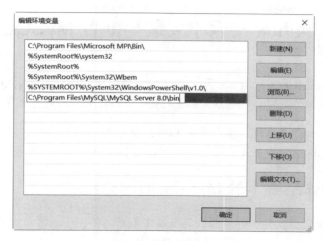

图 1.29 【编辑环境变量】对话框

1.4 MySQL 常用图形管理工具

MySQL 图形化管理工具极大地方便了数据库的操作与管理，常用的免费的图形化管理工具有 MySQL Workbench 和 phpMyAdmin。

1. MySQL Workbench

MySQL Workbench 是一款专门为用户提供创建、修改、执行和优化 MySQL 的可视化工具，并且提供给开发者一整套可视化的用于创建、编辑和管理 MySQL 查询和管理数据库连接的操作系统，可使开发人员轻松地管理数据。在可视化 MySQL 编辑工作模式下，用户创建表、删除表、修改表信息等只需要使用简单的可编辑列表就能完成。

MySQL Workbench 在数据库管理方面也提供了可视化操作的功能，为了方便管理，可以授予和收回用户权限，并且可以在数据库管理界面中查看数据库的状态，其中包括数据库中开启的客户端的数量、数据库缓存的大小以及管理数据库日志等信息。

目前 MySQL Workbench 提供了开源和商业化两个版本，同时支持 Windows 和 Linux 系统。

下载地址为：http://dev.MySQL.com/downloads/workbench/。

注意：读者也可以使用 Workbench 对数据库进行操作，由于是图形界面，操作结果比较清晰，有别于 MySQL 命令行方式的操作界面。

2. phpMyAdmin

phpMyAdmin 使用 PHP 编写，支持中文，必须安装在 Web 服务器中，通过 Web 方式控制和操作 MySQL 数据库。通过 phpMyAdmin 可以对数据库进行操作，例如建立、复制、删除数据等，管理数据库非常方便，局限在于对大数据库的备份和恢复不太方便。

下载地址：http://www.phpmyadmin.net/。

1.5 Linux 平台下安装与配置 MySQL 8.0

Linux 操作系统有众多的发行版本，不同的平台上需要安装不同的 MySQL 版本，MySQL 支持大多数主流的 Linux 版本。本节将以 Fedora 和 Red Hat 为例介绍 Linux 平台下 MySQL 的安装过程。

注意：对 MySQL 初学者来说，本书有关 Linux 操作的内容均可跳过，只关注 Windows 平台上的操作即可。对于 Linux 运维或开发人员来说，可以尝试直接在 Linux 平台学习 MySQL 数据库。

1.5.1 Linux 操作系统下的 MySQL 版本介绍

Linux 操作系统是自由软件和开放源代码发展中最著名的例子。经过世界各地计算机爱好者的共同努力，Linux 操作系统现已成为世界上使用最多的一种 UNIX 类操作系统，开发出超过 300 个发行版本，比较流行的版本有 Ubuntu、Debian GNU/Linux、Fedora、openSUSE 和 Red Hat。

目前 MySQL 主要支持的 Linux 版本为 Ubuntu、Debian、Fedora、Oracle、openSUSE 和 Red Hat 等。读者可以针对个人的喜好，选择使用不同的安装包，各平台的安装过程基本相同。

Linux 操作系统 MySQL 安装包分为以下 3 类。

- RPM：RPM 软件包是一种在 Linux 平台下的安装文件，通过安装命令可以很方便地安装与卸载。MySQL 的 RPM 安装文件包分为两个：服务器端和客户端，需要分别下载和安装。
- Generic Binaries：二进制软件包，经过编译生成的二进制文件软件包。
- 源码包：源码包是 MySQL 数据库的源代码，用户需要自己编译成二进制文件之后才能安装。

下面我们以 Fedora Linux 和 Red Hat Enterprise Linux 系统为例，简单介绍 MySQL 的安装。

1. Fedora Linux

Fedora Linux 是由 Fedora 项目社区开发、Red Hat 公司赞助的操作系统，其目标是创建一套新颖、多功能并且开放源代码的操作系统。Fedora 是商业化的 Red Hat Enterprise Linux 发行版的上游源码。Fedora 对于用户而言，是一套功能完备、更新快速的免费操作系统；而对赞助者 Red Hat 公司而言，它是许多新技术的测试平台，被认为可用的技术最终会加入到 Red Hat Enterprise Linux 中。Fedora 大约每六个月发布一次新版本，目前最新版本为 Fedora 36。MySQL 官方网站能够下载到支持 Fedora 34、35 的 MySQL 8.0.28 安装包。

读者可以在下载页面 http://dev.mysql.com/downloads/mysql/ 中选择【Fedora】操作系统，下载 MySQL 服务器和客户端的 RPM 包。

提示：下载页面中，MySQL Server 代表服务器端的 RPM 包，Client Utilities 代表客户端的 RPM 包。官方同时提供二进制和源码的 MySQL 安装包。

2. Red Hat Enterprise Linux

2004 年 4 月 30 日，Red Hat 公司正式停止对 Red Hat 9.0 版本的支持，标志着 Red Hat Linux 的

正式完结。Red Hat 公司不再开发桌面版的 Linux 发行包，转而集中力量开发服务器版，也就是 Red Hat Enterprise Linux 版。目前 Red Hat Enterprise Linux 9 为最新的版本。MySQL 官方网站能够下载到支持 Red Hat Enterprise Linux 6、7、8 的 MySQL 8.0.28 安装包。

读者可以在下载页面 http://dev.mysql.com/downloads/mysql/中选择【Red Hat Enterprise Linux / Oracle Linux】平台，根据 Red Hat 系统的版本以及不同的处理器架构，下载 MySQL 服务器端和客户端 RPM 包。

1.5.2　安装和配置 MySQL 的 RPM 包

MySQL 推荐使用 RPM 包进行 Linux 平台下的安装，从 MySQL 官方下载的 RPM 包能够在所有支持 RPM packages、glibc 2.3 的 Linux 系统下安装。本小节 Linux 系统使用 Red Hat Enterprise Linux 8，必要时读者可以使用虚拟机软件安装 Linux 进行试验。

通过 RPM 包安装成功之后，MySQL 服务器目录包括表 1.1 所示的子目录。

表 1.1　Linux 平台 MySQL 安装目录

文 件 夹	文件夹内容
/usr/bin	客户端和脚本
/usr/sbin	mysqld 服务器
/var/lib/mysql	日志文件和数据库
/usr/share/info	信息格式的手册
/usr/share/man	UNIX 帮助页
/usr/include/mysql	头文件
/usr/lib/mysql	库
/usr/share/mysql	错误消息、字符集、示例配置文件等

对于标准安装，只需要安装 MySQL-server 和 MySQL-client，下面演示通过 RPM 包进行安装。具体的操作步骤如下：

步骤01 进入下载页面 http://dev.mysql.com/downloads/mysql/，下载 RPM 包。在系统平台下拉列表中选择【Red Hat Enterprise Linux / Oracle Linux】选项。

步骤02 从 RPM 列表中选择要下载安装的包，单击【Download】按钮，开始下载安装文件。

步骤03 下载完成后，解压下载的 tar 包。

```
[root@localhost share]#tar -xvf MySQL-8.0.28-1.rhel5.i386.tar
MySQL-client-8.0.28-1.rhel5.i386.rpm
MySQL-devel-8.0.28-1.rhel5.i386.rpm
MySQL-embedded-8.0.28-1.rhel5.i386.rpm
MySQL-server-8.0.28-1.rhel5.i386.rpm
MySQL-shared-8.0.28-1.rhel5.i386.rpm
MySQL-test-8.0.28-1.rhel5.i386.rpm
```

tar 是 Linux/UNIX 系统上的一个打包工具，通过 tar -help 命令可以查看 tar 使用帮助。从中可以看到，解压出来的文件有 6 个。

（1）MySQL-client-8.0.28-1.rhel5.i386.rpm 是客户端的安装包。

（2）MySQL-server-8.0.28-1.rhel5.i386.rpm 是服务端的安装包。

（3）MySQL-devel-8.0.28-1.rhel5.i386.rpm 是包含开发用的库头文件安装的包。

（4）MySQL-shared-8.0.28-1.rhel5.i386.rpm 是包含 MySQL 的一些共享库文件的安装包。

（5）MySQL-test-8.0.28-1.rhel5.i386.rpm 是一些测试的安装包。

（6）MySQL-embedded-8.0.28-1.rhel5.i386.rpm 是嵌入式 MySQL 的安装包。

一般情况下，只需要安装 client 和 server 两个包。如果需要进行 C/C++ MySQL 相关开发，请安装 MySQL-devel-8.0.28-1.rhel5.i386.rpm。

步骤 04 切换到 root 用户。

```
[root@localhost share]$su - root
```

注意：此处也可以直接输入"su -"。符号"-"告诉系统在切换到 root 用户的时候，要初始化 root 的环境变量。然后按照提示输入 root 用户的密码，就可以完成切换 root 用户的操作。

步骤 05 安装 MySQL Server 8.0。

```
[root@localhost share]# rpm -ivh MySQL-server-8.0.28-1.rhel5.i386.rpm
Preparing...                ########################################### [100%]
1:MySQL-server              ########################################### [100%]
PLEASE REMEMBER TO SET A PASSWORD FOR THE MySQL root USER !
To do so, start the server, then issue the following commands:

/usr/bin/mysqladmin -u root password 'new-password'
/usr/bin/mysqladmin -u root -h localhost.localdomain password 'new-password'

Alternatively you can run:
/usr/bin/mysql_secure_installation
which will also give you the option of removing the test
databases and anonymous user created by default.  This is
strongly recommended for production servers.

See the manual for more instructions.

Please report any problems with the /usr/bin/mysqlbug script!
```

看到这些提示信息，说明 MySQL server 安装成功了。按照提示，执行/usr/bin/mysqladmin -u root password 'new-password'可以更改 root 用户密码；执行/usr/bin/mysql_secure_installation 会删除测试数据库和匿名用户；执行/usr/bin/mysqlbug script 会报告 bug。

注意：安装之前要查看计算机中是否已经安装旧版的 MySQL。如果有，最好先卸载，否则可能会产生冲突。查看旧版 MySQL 的命令如下：

```
[root@localhost share]# rpm -qa|grep -i mysql
mysql-5.0.77-4.el5_4.2
```

系统会显示机器上安装的旧版 MySQL 信息，如上面第 2 行所示。

然后，卸载 mysql-5.0.77-4.el5_4.2，命令如下：

```
[root@localhost share]# rpm -ev mysql-version-4.el5_4.2
```

步骤06 启动服务,输入命令如下:

```
[root@localhost share]# service mysql restart
MySQL server PID file could not be found!              [失败]
Starting MySQL...                                      [确定]
```

服务启动成功。

注意:从 MySQL 5.0 开始,MySQL 的服务名改为 mysql,而不是 4.* 的 mysqld。

MySQL 服务的操作命令是:

```
service mysql start|stop|restart|status。
```

start|stop|restart|status 这几个参数的意义如下:

- start: 启动服务。
- stop: 停止服务。
- restart: 重启服务。
- status: 查看服务状态。

步骤07 安装客户端,输入命令如下:

```
[root@localhost share]# rpm -ivh MySQL-client-8.0.28-1.rhel5.i386.rpm
Preparing...                ############################################# [100%]
1:MySQL-client               ############################################# [100%]
```

步骤08 安装成功之后,使用命令行登录。

```
[root@localhost share]# mysql -uroot -hlocalhost
Welcome to the MySQL monitor.  Commands end with ; or \g.
Your MySQL connection id is 1
Server version: 8.0.28 MySQL Community Server (GPL)

Copyright (c) 2000, 2022, Oracle and/or its affiliates. All rights reserved.

Oracle is a registered trademark of Oracle Corporation and/or its
affiliates. Other names may be trademarks of their respective
owners.

Type 'help;' or '\h' for help. Type '\c' to clear the current input statement.
```

读者看到上面的信息说明登录成功,接下来就可以对 MySQL 数据库进行操作了。

步骤09 更改 root 密码。

```
[root@localhost share]#/usr/bin/mysqladmin -u root password '123456'
```

执行完该命令,root 的密码被改为 123456。

步骤10 添加新的用户。

```
[root@localhost share]#mysql -u root -p123456 -hlocalhost
mysql> GRANT ALL PRIVILEGES ON *.* TO monty@localhost
       IDENTIFIED BY 'something' WITH GRANT OPTION;
```

1.5.3 安装和配置 MySQL 的源码包

进入下载页面 http://dev.mysql.com/downloads/mysql/#downloads，在安装平台下拉列表中选择【Source Code】选项。源码安装需要一些开发工具，分别如下：

（1）CMake（cross platform make），构建程序必需的一个跨平台构建工具。官方网址为 http://www.cmake.org/。

（2）一个好的 make 工具，MySQL 官方推荐使用 GNU make 3.75。GNU make 的下载地址为 http://www.gnu.org/software/make/。

（3）ANSI C++编译器，GCC 4.2.1 及以上版本。

（4）Perl，运行 test 版本所必需的工具包。

（5）rpm 包管理器，rpmbuild 工具。

编译安装，输入命令如下：

```
[root@localhost tmp]# rpmbuild --rebuild --clean MySQL-8.0.28-1.linux2.6.src.rpm
```

编译完成后会形成一个 rpm 包，然后按照 rpm 包的安装方法安装就可以了。如果是 MySQL 初学者，则不建议使用源码包进行安装。

第 2 章

数据库和数据表的基本操作

MySQL 安装好以后,首先需要创建数据库和数据表,这是使用 MySQL 各种功能的前提。在数据库中,数据表是最重要、最基本的操作对象,是数据存储的基本单位。数据表被定义为列的集合,数据在表中是按照行和列的格式来存储的。每一行代表一条唯一的记录,每一列代表记录中的一个域。本章将详细介绍数据库和数据表的基本操作。

2.1 创建数据库

MySQL 安装完成之后,将会在其 data 目录下自动创建几个必需的数据库,可以使用 "SHOW DATABASES;" 语句来查看当前所有存在的数据库,输入语句如下:

```
mysql> SHOW DATABASES;
+--------------------+
| Database           |
+--------------------+
| information_schema |
| mysql              |
| performance_schema |
| sakila             |
| sys                |
| world              |
+--------------------+
6 rows in set (0.04 sec)
```

可以看到,数据库列表中包含了 6 个数据库。其中数据库 mysql 是必需的,它描述用户访问权限。

创建数据库是在系统磁盘上划分一块区域用于数据的存储和管理,如果管理员在设置权限的时候为用户创建了数据库,则可以直接使用,否则需要自己创建数据库。MySQL 中创建数据库的基本 SQL 语法格式为:

```
CREATE DATABASE database_name;
```

"database_name"为要创建的数据库的名称，该名称不能与已经存在的数据库重名。

【例 2.1】创建测试数据库 test_db。输入如下语句：

```
CREATE DATABASE test_db;
```

数据库创建好之后，可以使用 SHOW CREATE DATABASE 语句查看数据库的定义。

【例 2.2】查看创建好的数据库 test_db 的定义。输入如下语句：

```
mysql> SHOW CREATE DATABASE test_db\G
*** 1. row ***
       Database: test_db
Create Database: CREATE DATABASE `test_db` /*!40100 DEFAULT CHARACTER SET utf8mb4 COLLATE utf8mb4_0900_ai_ci */ /*!80016 DEFAULT ENCRYPTION='N' */
```

可以看到，如果数据库创建成功，就将显示数据库的创建信息。

再次使用"SHOW DATABASES;"语句来查看当前所有存在的数据库，命令语句如下：

```
mysql> SHOW databases;
+--------------------+
| Database           |
+--------------------+
| information_schema |
| mysql              |
| performance_schema |
| sakila             |
| sys                |
| test_db            |
| world              |
+--------------------+
7 rows in set (0.05 sec)
```

可以看到，数据库列表中包含了刚刚创建的数据库 test_db 和其他已经存在的数据库的名称。

2.2　删除数据库

删除数据库是将已经存在的数据库从磁盘空间上清除，清除之后，数据库中的所有数据也将一同被删除。删除数据库语句和创建数据库的命令相似，MySQL 中删除数据库的基本语法格式为：

```
DROP DATABASE database_name;
```

其中，"database_name"为要删除的数据库的名称。若指定的数据库不存在，则会删除出错。

【例 2.3】删除测试数据库 test_db。输入如下语句：

```
DROP DATABASE test_db;
```

语句执行完毕之后，数据库 test_db 将被删除，再次使用 SHOW CREATE DATABASE 命令查看数据库的定义，结果如下：

```
mysql> SHOW CREATE DATABASE test_db\G
ERROR 1049 (42000): Unknown database 'test_db'
```

执行结果给出一条错误信息"ERROR 1049 (42000)：Unknown database 'test_db'"，即数据库 test_db 已不存在，说明删除成功了。

2.3 创建数据表

在创建完数据库之后，接下来的工作就是创建数据表。所谓创建数据表，指的是在已经创建好的数据库中建立新表。创建数据表的过程是规定数据列的属性的过程，同时也是实施数据完整性（包括实体完整性、引用完整性和域完整性等）约束的过程。本节将介绍创建数据表的语法形式以及如何添加主键约束、外键约束、非空约束等内容。

2.3.1 创建表的语法形式

数据表属于数据库，在创建数据表之前，应该使用语句"USE <数据库名>"指定操作在哪个数据库中进行。如果没有选择数据库，就会抛出"No database selected"的错误。

创建数据表的语句为 CREATE TABLE，语法规则如下：

```
CREATE TABLE <表名>
(
字段名1，数据类型 [列级别约束条件] [默认值],
字段名2，数据类型 [列级别约束条件] [默认值],
…
[表级别约束条件]
);
```

使用 CREATE TABLE 创建表时，必须指定以下信息：

（1）要创建的表的名称，不区分大小写，不能使用 SQL 语言中的关键字，如 DROP、ALTER、INSERT 等。

（2）数据表中每一列（字段）的名称和数据类型，如果创建多列，就要用逗号分隔开。

【例 2.4】创建员工表 tb_emp1，结构如表 2.1 所示。

表 2.1 tb_emp1 表结构

字段名称	数据类型	备 注
id	INT	员工编号
name	VARCHAR(25)	员工名称
deptId	INT	所在部门编号
salary	FLOAT	工资

① 创建数据库，SQL 语句如下：

```
CREATE DATABASE test_db;
```

② 选择创建表的数据库，SQL 语句如下：

```
USE test_db;
```

③ 创建 tb_emp1 表，SQL 语句为：

```
CREATE TABLE tb_emp1
(
id INT,
name VARCHAR(25),
deptId INT,
salary FLOAT
);
```

④ 语句执行成功后，便创建了一个名称为 tb_emp1 的数据表，使用"SHOW TABLES;"语句查看数据表是否创建成功，SQL 语句如下：

```
mysql> SHOW TABLES;
+--------------------+
| Tables_in_test_db  |
+--------------------+
| tb_emp1            |
+--------------------+
```

可以看到，test_db 数据库中已经有了数据表 tb_emp1，数据表创建成功。

2.3.2 使用主键约束

主键，又称主码，是表中一列或多列的组合。主键约束（Primary Key ConstraINT）要求主键列的数据唯一，并且不允许为空。主键能够唯一地标识表中的一条记录，可以结合外键来定义不同数据表之间的关系，并且可以加快数据库查询的速度。主键和记录之间的关系如同身份证和人之间的关系，它们之间是一一对应的。主键分为两种类型：单字段主键和多字段联合主键。

1. 单字段主键

主键由一个字段组成，SQL 语句格式分为以下两种情况。

（1）在定义列的同时指定主键，语法规则如下：

```
字段名 数据类型 PRIMARY KEY [默认值]
```

【例 2.5】定义数据表 tb_emp 2，其主键为 id，SQL 语句如下：

```
CREATE TABLE tb_emp2
(
id INT PRIMARY KEY,
name VARCHAR(25),
deptId INT,
salary FLOAT
);
```

（2）在定义完所有列之后指定主键。

```
[CONSTRAINT <约束名>] PRIMARY KEY [字段名]
```

【例 2.6】定义数据表 tb_emp 3，其主键为 id，SQL 语句如下：

```
CREATE TABLE tb_emp3
(
id INT,
name VARCHAR(25),
deptId INT,
salary FLOAT,
PRIMARY KEY(id)
);
```

上述两个例子执行后的结果是一样的，都会在 id 字段上设置主键约束。

2. 多字段联合主键

主键由多个字段联合组成，语法规则如下：

```
PRIMARY KEY [字段 1, 字段 2, …, 字段 n]
```

【例 2.7】定义数据表 tb_emp4，假设表中间没有主键 id，为了唯一确定一个员工，可以把 name、deptId 联合起来作为主键，SQL 语句如下：

```
CREATE TABLE tb_emp4
(
name VARCHAR(25),
deptId INT,
salary FLOAT,
PRIMARY KEY(name,deptId)
);
```

语句执行后，便创建了一个名称为 tb_emp4 的数据表，name 字段和 deptId 字段组合在一起成为 tb_emp4 的多字段联合主键。

2.3.3 使用外键约束

外键用来在两个表的数据之间建立连接，可以是一列或者多列。一个表可以有一个或多个外键。外键对应的是参照完整性，一个表的外键可以为空值，若不为空值，则每一个外键值必须等于另一个表中主键的某个值。

外键及其相关名词释义如下：

- 外键：它是表中的一个字段，虽然可以不是本表的主键，但要对应另外一个表的主键。外键的第一个作用是保证数据引用的完整性。定义外键后，不允许删除在另一个表中具有关联关系的行；第二个作用是保持数据的一致性、完整性。例如，部门表 tb_dept 的主键是 id，在员工表 tb_emp5 中有一个键 deptId 与这个 id 关联。
- 主表（父表）：对于两个具有关联关系的表而言，相关联字段中主键所在的那个表即是主表。
- 从表（子表）：对于两个具有关联关系的表而言，相关联字段中外键所在的那个表即是从表。

创建外键的语法规则如下：

```
[CONSTRAINT <外键名>] FOREIGN KEY 字段名 1 [ ,字段名 2,…]
REFERENCES <主表名> 主键列 1 [ ,主键列 2,…]
```

"外键名"为定义的外键约束的名称，一个表中不能有相同名称的外键；"字段名"为子表需要添加外键约束的字段列；"主表名"为被子表外键所依赖的表的名称；"主键列"为主表中定义的主键列，或者列组合。

【例 2.8】定义数据表 tb_emp5，并在 tb_emp5 表上创建外键约束。

创建一个部门表 tb_dept1，表结构如表 2.2 所示，SQL 语句如下：

```
CREATE TABLE tb_dept1
(
id INT PRIMARY KEY,
name VARCHAR(22) NOT NULL,
location VARCHAR(50)
);
```

表 2.2 tb_dept1 表结构

字段名称	数据类型	备注
id	INT	部门编号
name	VARCHAR(22)	部门名称
location	VARCHAR(50)	部门位置

定义数据表 tb_emp5，让它的字段 deptId 作为外键关联到 tb_dept1 的主键 id，SQL 语句为：

```
CREATE TABLE tb_emp5
(
id INT PRIMARY KEY,
name VARCHAR(25),
deptId INT,
salary FLOAT,
CONSTRAINT fk_emp_dept1 FOREIGN KEY(deptId) REFERENCES tb_dept1(id)
);
```

以上语句执行成功之后，在表 tb_emp5 上添加了名称为 fk_emp_dept1 的外键约束，外键名称为 deptId，其依赖于表 tb_dept1 的主键 id。

提示：关联指的是在关系型数据库中相关表之间的联系。它是通过相容或相同的属性或属性组来表示的。子表的外键必须关联父表的主键，且关联字段的数据类型必须匹配，如果类型不一样，则在创建子表时，就会出现错误 "ERROR 1005 (HY000): Can't create table' database.tablename' (errno: 150)"。

2.3.4 使用非空约束

非空约束（Not Null ConstraINT）指字段的值不能为空。对于使用了非空约束的字段，如果用户在添加数据时没有指定值，数据库系统会报错。

非空约束的语法规则如下：

```
字段名 数据类型 not null
```

【例 2.9】定义数据表 tb_emp6,指定员工的名称不能为空,SQL 语句如下:

```
CREATE TABLE tb_emp6
(
id INT PRIMARY KEY,
name VARCHAR(25) NOT NULL,
deptId INT,
salary FLOAT
);
```

执行后,在 tb_emp6 中创建了一个 name 字段,其插入值不能为空(NOT NULL)。

2.3.5 使用唯一性约束

唯一性约束(Unique ConstraINT)要求该列唯一,允许为空,但只能出现一个空值。唯一性约束可以确保一列或者几列不出现重复值。

唯一性约束的语法规则如下:

(1)在定义完列之后直接指定唯一约束,语法规则如下:

字段名 数据类型 UNIQUE

【例 2.10】定义数据表 tb_dept2,指定部门的名称唯一,SQL 语句如下:

```
CREATE TABLE tb_dept2
(
id INT PRIMARY KEY,
name VARCHAR(22) UNIQUE,
location VARCHAR(50)
);
```

(2)在定义完所有列之后指定唯一约束,语法规则如下:

[CONSTRAINT <约束名>] UNIQUE(<字段名>)

【例 2.11】定义数据表 tb_dept3,指定部门的名称唯一,SQL 语句如下:

```
CREATE TABLE tb_dept3
(
id INT PRIMARY KEY,
name VARCHAR(22),
location VARCHAR(50),
CONSTRAINT STH UNIQUE(name)
);
```

UNIQUE 和 PRIMARY KEY 的区别:一个表中可以有多个字段声明为 UNIQUE,但只能有一个 PRIMARY KEY;声明为 PRIMAY KEY 的列不允许有空值,但是声明为 UNIQUE 的字段允许空值(NULL)的存在。

2.3.6 使用默认约束

默认约束(Default ConstraINT)指定某列的默认值。如男性同学较多,性别就可以默认为"男"。如果插入一条新的记录时没有为这个字段赋值,那么系统会自动为这个字段赋值为"男"。

默认约束的语法规则如下:

字段名 数据类型 DEFAULT 默认值

【例2.12】定义数据表tb_emp7，指定员工的部门编号默认为1111，SQL语句如下:

```
CREATE TABLE tb_emp7
(
id INT PRIMARY KEY,
name VARCHAR(25) NOT NULL,
deptId INT DEFAULT 1111,
salary FLOAT
);
```

以上语句执行成功之后，表tb_emp7上的字段deptId拥有了一个默认的值1111，新插入的记录如果没有指定部门编号，则默认都为1111。

2.3.7 设置表的属性值自动增加

在数据库应用中，需要在每次插入新记录时，系统自动生成字段的主键值的话，可以通过为表主键添加AUTO_INCREMENT关键字来实现。默认情况下，在MySQL中AUTO_INCREMENT的初始值是1，每新增一条记录，字段值自动加1。一个表只能有一个字段使用AUTO_INCREMENT约束，且该字段必须为主键的一部分。AUTO_INCREMENT约束的字段可以是任何整数类型（TINYINT、SMALLIN、INT、BIGINT等）。

设置表的属性值自动增加的语法规则如下:

字段名 数据类型 AUTO_INCREMENT

【例2.13】定义数据表tb_emp8，指定员工的编号自动递增，SQL语句如下:

```
CREATE TABLE tb_emp8
(
id INT PRIMARY KEY AUTO_INCREMENT,
name VARCHAR(25) NOT NULL,
deptId INT,
salary FLOAT
);
```

上述例子执行后，会创建名称为tb_emp8的数据表。表tb_emp8中的id字段的值在添加记录的时候会自动增加，在插入记录的时候，默认的自增字段id的值从1开始，每次添加一条新记录，该值自动加1。

例如，执行如下插入语句:

```
mysql> INSERT INTO tb_emp8 (name,salary)
    -> VALUES('Lucy',1000), ('Lura',1200),('Kevin',1500);
```

提示：这里使用INSERT语句向表中插入记录的方法，并不是SQL的标准语法，这种语法不一定被其他的数据库支持，只能在MySQL中使用。

语句执行完后，tb_emp8表中增加3条记录，在这里并没有输入id的值，但系统已经自动添加该值。使用SELECT命令查看记录，如下所示:

```
mysql> SELECT * FROM tb_emp8;
+----+-------+--------+--------+
| id | name  | deptId | salary |
+----+-------+--------+--------+
|  1 | Lucy  | NULL   |   1000 |
|  2 | Lura  | NULL   |   1200 |
|  3 | Kevin | NULL   |   1500 |
+----+-------+--------+--------+
```

2.4 查看数据表结构

使用 SQL 语句创建好数据表之后，可以查看表结构的定义，以确认表的定义是否正确。在 MySQL 中，查看表结构可以使用 DESCRIBE 和 SHOW CREATE TABLE 语句。本节将针对这两个语句分别进行详细讲解。

2.4.1 查看表基本结构语句 DESCRIBE

DESCRIBE 语句可以查看表的字段信息，其中包括字段名、字段数据类型、是否为主键、是否有默认值等。语法规则如下：

DESCRIBE 表名;

或者简写为：

DESC 表名;

【例 2.14】分别使用 DESCRIBE 和 DESC 查看表 tb_dept1 和表 tb_emp1 的表结构。

查看 tb_dept1 表结构，SQL 语句如下：

```
mysql> DESCRIBE tb_dept1;
+----------+-------------+------+-----+---------+-------+
| Field    | Type        | Null | Key | Default | Extra |
+----------+-------------+------+-----+---------+-------+
| id       | int         | NO   | PRI | NULL    |       |
| name     | varchar(22) | NO   |     | NULL    |       |
| location | varchar(50) | YES  |     | NULL    |       |
+----------+-------------+------+-----+---------+-------+
```

查看 tb_emp1 表结构，SQL 语句如下：

```
mysql> DESC tb_emp1;
+--------+-------------+------+-----+---------+-------+
| Field  | Type        | Null | Key | Default | Extra |
+--------+-------------+------+-----+---------+-------+
| id     | int         | YES  |     | NULL    |       |
| name   | varchar(25) | YES  |     | NULL    |       |
| deptId | int         | YES  |     | NULL    |       |
| salary | float       | YES  |     | NULL    |       |
+--------+-------------+------+-----+---------+-------+
```

其中，各个字段的含义分别解释如下：

- NULL：表示该列是否可以存储 NULL 值。
- Key：表示该列是否已编制索引。PRI 表示该列是表主键的一部分；UNI 表示该列是 UNIQUE 索引的一部分；MUL 表示在列中某个给定值允许出现多次。
- Default：表示该列是否有默认值，若有的话指定值是多少。
- Extra：表示可以获取的与给定列有关的附加信息，例如 AUTO_INCREMENT 等。

2.4.2 查看表详细结构语句 SHOW CREATE TABLE

SHOW CREATE TABLE 语句可以用来显示创建表时的 CREATE TABLE 语句，语法格式如下：

```
SHOW CREATE TABLE <表名\G>;
```

提示：使用 SHOW CREATE TABLE 语句，不仅可以查看表创建时候的详细语句，还可以查看存储引擎和字符编码。

如果不加"\G"参数，显示的结果可能非常混乱，加上参数"\G"之后，可使显示结果更加直观，易于查看。

【例 2.15】使用 SHOW CREATE TABLE 查看表 tb_emp1 的详细信息，SQL 语句如下：

```
mysql> SHOW CREATE TABLE tb_emp1;
+---------+----------------------------------------------------------+
| Table   | Create Table                                             |
+---------+----------------------------------------------------------+
| tb_emp1 | CREATE TABLE `tb_emp1` (                                 |
|         |   `id` int DEFAULT NULL,                                 |
|         |   `name` varchar(25) DEFAULT NULL,                       |
|         |   `deptId` int DEFAULT NULL,                             |
|         |   `salary` float DEFAULT NULL                            |
|         | ) ENGINE=InnoDB DEFAULT CHARSET=utf8mb4 COLLATE=utf8mb4_0900_ai_ci |
+---------+----------------------------------------------------------+
```

使用参数"\G"之后的结果如下：

```
mysql> SHOW CREATE TABLE tb_emp1\G
*************************** 1. row ***************************
       Table: tb_emp1
Create Table: CREATE TABLE `tb_emp1` (
  `id` int DEFAULT NULL,
  `name` varchar(25) DEFAULT NULL,
  `deptId` int DEFAULT NULL,
  `salary` float DEFAULT NULL
) ENGINE=InnoDB DEFAULT CHARSET=utf8mb4 COLLATE=utf8mb4_0900_ai_ci
```

2.5 修改数据表

修改表指的是修改数据库中已经存在的数据表的结构。MySQL 使用 ALTER TABLE 语句修改表。常用的修改表的操作有：修改表名、修改字段数据类型或字段名、增加和删除字段、修改字段的排列位置、更改表的存储引擎、删除表的外键约束等。本节将对修改表的操作进行讲解。

2.5.1 修改表名

MySQL 通过 ALTER TABLE 语句来实现表名的修改，具体的语法规则如下：

```
ALTER TABLE <旧表名> RENAME [TO] <新表名>;
```

其中，TO 为可选参数，使用与否均不影响语句执行结果。

【例 2.16】将数据表 tb_dept3 改名为 tb_deptment3。

执行修改表名操作之前，使用 SHOW TABLES 查看数据库中所有的表。

```
mysql> SHOW TABLES;
+--------------------+
| Tables_in_test_db  |
+--------------------+
| tb_dept1           |
| tb_dept2           |
| tb_dept3           |
...省略部分内容
```

使用 ALTER TABLE 将表 tb_dept3 改名为 tb_deptment3，SQL 语句如下：

```
ALTER TABLE tb_dept3 RENAME tb_deptment3;
```

语句执行之后，检验表 tb_dept3 是否改名成功。使用 SHOW TABLES 查看数据库中的表，结果如下：

```
mysql> SHOW TABLES;
+--------------------+
| Tables_in_test_db  |
+--------------------+
| tb_dept            |
| tb_dept2           |
| tb_deptment3       |
...省略部分内容
```

经过比较可以看到，数据表列表中已经有了名称为 tb_deptment3 的表。

提示：读者可以在修改表名称时，使用 DESC 命令查看修改前后两个表的结构，修改表名并不修改表的结构，因此，修改名称后的表和修改名称前的表的结构必然是相同的。

2.5.2 修改字段的数据类型

修改字段的数据类型，就是把字段的数据类型转换成另一种数据类型。在 MySQL 中修改字段数据类型的语法规则如下：

```
ALTER TABLE <表名> MODIFY <字段名> <数据类型>
```

其中，"表名"指要修改数据类型的字段所在表的名称，"字段名"指需要修改的字段，"数据类型"指修改后字段的新数据类型。

【例2.17】将数据表tb_dept1中name字段的数据类型由VARCHAR(22)修改成VARCHAR(30)。

执行修改表名操作之前，使用DESC查看tb_dept1表结构，结果如下：

```
mysql> DESC tb_dept1;
+----------+-------------+------+-----+---------+-------+
| Field    | Type        | Null | Key | Default | Extra |
+----------+-------------+------+-----+---------+-------+
| id       | int         | NO   | PRI | NULL    |       |
| name     | varchar(22) | YES  |     | NULL    |       |
| location | varchar(50) | YES  |     | NULL    |       |
+----------+-------------+------+-----+---------+-------+
```

可以看到现在name字段的数据类型为VARCHAR(22)，下面修改其类型。输入如下SQL语句并执行：

```
ALTER TABLE tb_dept1 MODIFY name VARCHAR(30);
```

再次使用DESC查看表，结果如下：

```
mysql> DESC tb_dept1;
+----------+-------------+------+-----+---------+-------+
| Field    | Type        | Null | Key | Default | Extra |
+----------+-------------+------+-----+---------+-------+
| id       | int         | NO   | PRI | NULL    |       |
| name     | varchar(30) | YES  |     | NULL    |       |
| location | varchar(50) | YES  |     | NULL    |       |
+----------+-------------+------+-----+---------+-------+
```

语句执行之后，检验会发现表tb_dept1表中name字段的数据类型已经修改成了VARCHAR(30)，修改成功。

2.5.3 修改字段名

MySQL中修改表字段名的语法规则如下：

```
ALTER TABLE <表名> CHANGE <旧字段名> <新字段名> <新数据类型>;
```

其中，"旧字段名"指修改前的字段名；"新字段名"指修改后的字段名；"新数据类型"指修改后的数据类型，如果不需要修改字段的数据类型，将新数据类型设置成与原来一样即可，但数据类型不能为空。

【例2.18】将数据表tb_dept1中的location字段名称改为loc，数据类型保持不变，SQL语句如下：

```
ALTER TABLE tb_dept1 CHANGE location loc VARCHAR(50);
```

使用DESC查看表tb_dept1，会发现字段的名称已经修改成功，结果如下：

```
mysql> DESC tb_dept1;
+-------+------+------+-----+---------+-------+
| Field | Type | Null | Key | Default | Extra |
+-------+------+------+-----+---------+-------+
```

```
| id       | int         | NO  | PRI | NULL |  |
| name     | varchar(30) | YES |     | NULL |  |
| loc      | varchar(50) | YES |     | NULL |  |
```

【例 2.19】将数据表 tb_dept1 中的 loc 字段名称改为 location，同时将数据类型变为 VARCHAR(60)，SQL 语句如下：

```
ALTER TABLE tb_dept1 CHANGE loc location VARCHAR(60);
```

使用 DESC 查看表 tb_dept1，会发现字段的名称和数据类型均已经修改成功，结果如下：

```
mysql> DESC tb_dept1;
+----------+-------------+------+-----+---------+-------+
| Field    | Type        | Null | Key | Default | Extra |
+----------+-------------+------+-----+---------+-------+
| id       | int         | NO   | PRI | NULL    |       |
| name     | varchar(30) | YES  |     | NULL    |       |
| location | varchar(60) | YES  |     | NULL    |       |
+----------+-------------+------+-----+---------+-------+
```

提示：CHANGE 也可以只修改数据类型，实现和 MODIFY 同样的效果，方法是将 SQL 语句中的"新字段名"和"旧字段名"设置为相同的名称，只改变"数据类型"。

由于不同类型的数据在机器中存储的方式及长度并不相同，修改数据类型可能会影响到数据表中已有的数据记录，因此当数据库表中已经有数据时，不要轻易修改数据类型。

2.5.4 添加字段

随着业务需求的变化，可能需要在已经存在的表中添加新的字段。一个完整字段包括字段名、数据类型、完整性约束。添加字段的语法格式如下：

```
ALTER TABLE <表名> ADD <新字段名> <数据类型>
[约束条件] [FIRST | AFTER 已存在字段名];
```

新字段名为需要添加的字段的名称；"FIRST"为可选参数，其作用是将新添加的字段设置为表的第一个字段；"AFTER"为可选参数，其作用是将新添加的字段添加到指定的"已存在字段名"的后面。

提示："FIRST"或"AFTER 已存在字段名"用于指定新增字段在表中的位置，如果 SQL 语句中没有这两个参数，则默认将新添加的字段设置为数据表的最后一列。

1. 添加无完整性约束条件的字段

【例 2.20】在数据表 tb_dept1 中添加一个没有完整性约束的 INT 类型的字段 managerId（部门经理编号），SQL 语句如下：

```
ALTER TABLE tb_dept1 ADD managerId INT;
```

使用 DESC 查看表 tb_dept1，会发现在表的最后添加了一个名为 managerId 的 INT 类型的字段，结果如下：

```
mysql> DESC tb_dept1;
```

```
+-----------+-------------+------+-----+---------+-------+
| Field     | Type        | Null | Key | Default | Extra |
+-----------+-------------+------+-----+---------+-------+
| id        | int         | NO   | PRI | NULL    |       |
| name      | varchar(30) | YES  |     | NULL    |       |
| location  | varchar(60) | YES  |     | NULL    |       |
| managerId | int         | YES  |     | NULL    |       |
+-----------+-------------+------+-----+---------+-------+
```

2. 添加有完整性约束条件的字段

【例2.21】在数据表 tb_dept1 中添加一个不能为空的 VARCHAR(12)类型的字段 column1，SQL 语句如下：

```
ALTER TABLE tb_dept1 ADD column1 VARCHAR(12) not null;
```

使用 DESC 查看表 tb_dept1，会发现在表的最后添加了一个名为 column1 的 VARCHAR(12)类型且不为空的字段，结果如下：

```
mysql> DESC tb_dept1;
+-----------+-------------+------+-----+---------+-------+
| Field     | Type        | Null | Key | Default | Extra |
+-----------+-------------+------+-----+---------+-------+
| id        | int         | NO   | PRI | NULL    |       |
| name      | varchar(30) | YES  |     | NULL    |       |
| location  | varchar(60) | YES  |     | NULL    |       |
| managerId | int         | YES  |     | NULL    |       |
| column1   | varchar(12) | NO   |     | NULL    |       |
+-----------+-------------+------+-----+---------+-------+
```

3. 在表的第一列添加一个字段

【例2.22】在数据表 tb_dept1 中添加一个 INT 类型的字段 column2，SQL 语句如下：

```
ALTER TABLE tb_dept1 ADD column2 INT FIRST;
```

使用 DESC 查看表 tb_dept1，会发现在表第一列添加了一个名为 column2 的 INT 类型字段，结果如下：

```
mysql> DESC tb_dept1;
+-----------+-------------+------+-----+---------+-------+
| Field     | Type        | Null | Key | Default | Extra |
+-----------+-------------+------+-----+---------+-------+
| column2   | int         | YES  |     | NULL    |       |
| id        | int         | NO   | PRI | NULL    |       |
| name      | varchar(30) | YES  |     | NULL    |       |
| location  | varchar(60) | YES  |     | NULL    |       |
| managerId | int         | YES  |     | NULL    |       |
| column1   | varchar(12) | NO   |     | NULL    |       |
+-----------+-------------+------+-----+---------+-------+
```

4. 在表的指定列之后添加一个字段

【例2.23】在数据表 tb_dept1 中 name 列后添加一个 INT 类型的字段 column3，SQL 语句如下：

```
ALTER TABLE tb_dept1 ADD column3 INT AFTER name;
```

使用 DESC 查看表 tb_dept1，结果如下：

```
mysql> DESC tb_dept1;
+-----------+-------------+------+-----+---------+-------+
| Field     | Type        | Null | Key | Default | Extra |
+-----------+-------------+------+-----+---------+-------+
| column2   | int         | YES  |     | NULL    |       |
| id        | int         | NO   | PRI | NULL    |       |
| name      | varchar(30) | YES  |     | NULL    |       |
| column3   | int         | YES  |     | NULL    |       |
| location  | varchar(60) | YES  |     | NULL    |       |
| managerId | int         | YES  |     | NULL    |       |
| column1   | varchar(12) | NO   |     | NULL    |       |
+-----------+-------------+------+-----+---------+-------+
```

可以看到，tb_dept1 表中增加了一个名称为 column3 的字段，其位置在指定的 name 字段后面，添加字段成功。

2.5.5 删除字段

删除字段是将数据表中的某个字段从表中移除，语法格式如下：

ALTER TABLE <表名> DROP <字段名>；

"字段名"指需要从表中删除的字段的名称。

【例 2.24】删除数据表 tb_dept1 表中的 column2 字段。

首先，执行删除字段之前，使用 DESC 查看 tb_dept1 表结构，结果如下：

```
mysql> DESC tb_dept1;
+-----------+-------------+------+-----+---------+------+
| Field     | Type        | Null | Key | Default | Extr |
+-----------+-------------+------+-----+---------+------+
| column2   | int         | YES  |     | NULL    |      |
| id        | int         | NO   | PRI | NULL    |      |
| name      | varchar(30) | YES  |     | NULL    |      |
| column3   | int         | YES  |     | NULL    |      |
| location  | varchar(60) | YES  |     | NULL    |      |
| managerId | int         | YES  |     | NULL    |      |
| column1   | varchar(12) | NO   |     | NULL    |      |
+-----------+-------------+------+-----+---------+------+
```

删除 column2 字段，SQL 语句如下：

ALTER TABLE tb_dept1 DROP column2;

再次使用 DESC 查看表 tb_dept1，结果如下：

```
mysql> DESC tb_dept1;
+---------+-------------+------+-----+---------+------+
| Field   | Type        | Null | Key | Default | Extr |
+---------+-------------+------+-----+---------+------+
| id      | int         | NO   | PRI | NULL    |      |
| name    | varchar(30) | YES  |     | NULL    |      |
| column3 | int         | YES  |     | NULL    |      |
```

```
| location   | varchar(60) | YES |     | NULL |   |
| managerId  | int         | YES |     | NULL |   |
| column1    | varchar(12) | NO  |     | NULL |   |
```

可以看到，tb_dept1 表中已经不存在名称为 column2 的字段，说明删除字段成功。

2.5.6 修改字段的排列位置

对于一个数据表来说，在创建的时候，字段在表中的排列顺序就已经确定了，但表的结构也是可以改变的，可以通过 ALTER TABLE 来改变表中字段的相对位置。语法格式如下：

```
ALTER TABLE <表名> MODIFY <字段1> <数据类型> FIRST|AFTER <字段2>;
```

其中，"字段1"指要修改位置的字段；"数据类型"指"字段1"的数据类型；"FIRST"为可选参数，指将"字段1"修改为表的第一个字段；"AFTER 字段2"指将"字段1"插入到"字段2"后面。

1. 修改字段为表的第一个字段

【例 2.25】将数据表 tb_dept1 中的 column1 字段修改为表的第一个字段，SQL 语句如下：

```
ALTER TABLE tb_dept1 MODIFY column1 VARCHAR(12) FIRST;
```

使用 DESC 查看表 tb_dept1，发现字段 column1 已经被移至表的第一列，结果如下：

```
mysql> DESC tb_dept1;
+-----------+-------------+------+-----+---------+-------+
| Field     | Type        | Null | Key | Default | Extra |
+-----------+-------------+------+-----+---------+-------+
| column1   | varchar(12) | YES  |     | NULL    |       |
| id        | int         | NO   | PRI | NULL    |       |
| name      | varchar(30) | YES  |     | NULL    |       |
| column3   | int         | YES  |     | NULL    |       |
| location  | varchar(60) | YES  |     | NULL    |       |
| managerId | int         | YES  |     | NULL    |       |
+-----------+-------------+------+-----+---------+-------+
```

2. 修改字段到表的指定列之后

【例 2.26】将数据表 tb_dept1 中的 column1 字段插入到 location 字段后面，SQL 语句如下：

```
ALTER TABLE tb_dept1 MODIFY column1 VARCHAR(12) AFTER location;
```

使用 DESC 查看表 tb_dept1，结果如下：

```
mysql> DESC tb_dept1;
+-----------+-------------+------+-----+---------+-------+
| Field     | Type        | Null | Key | Default | Extra |
+-----------+-------------+------+-----+---------+-------+
| id        | int         | NO   | PRI | NULL    |       |
| name      | varchar(30) | YES  |     | NULL    |       |
| column3   | int         | YES  |     | NULL    |       |
| location  | varchar(60) | YES  |     | NULL    |       |
| column1   | varchar(12) | YES  |     | NULL    |       |
```

```
| managerId  | int          | YES  |     | NULL    |       |
+------------+--------------+------+-----+---------+-------+
```

可以看到，tb_dept1 表中的字段 column1 已经被移至 location 字段之后。

2.5.7 删除表的外键约束

对于数据库中定义的外键，如果不再需要，可以将其删除。一旦删除外键，就会解除主表和从表间的关联关系，MySQL 中删除外键的语法格式如下：

```
ALTER TABLE <表名> DROP FOREIGN KEY <外键约束名>
```

"外键约束名"指在定义表时 CONSTRAINT 关键字后面的参数，详细内容可参考 2.3.3 节。

【例 2.27】删除数据表 tb_emp9 中的外键约束。

首先创建表 tb_emp9，创建外键 deptId 关联 tb_dept1 表的主键 id，SQL 语句如下：

```
CREATE TABLE tb_emp9
(
id INT PRIMARY KEY,
name VARCHAR(25),
deptId INT,
salary FLOAT,
CONSTRAINT fk_emp_dept  FOREIGN KEY (deptId) REFERENCES tb_dept1(id)
);
```

使用 SHOW CREATE TABLE 查看表 tb_emp9 的结构，结果如下：

```
mysql> SHOW CREATE TABLE tb_emp9 \G
*** 1. row ***
       Table: tb_emp9
Create Table: CREATE TABLE `tb_emp9` (
  `id` INT NOT NULL,
  `name` varchar(25) DEFAULT NULL,
  `deptId` INT DEFAULT NULL,
  `salary` float DEFAULT NULL,
  PRIMARY KEY (`id`),
  KEY `fk_emp_dept` (`deptId`),
  CONSTRAINT `fk_emp_dept` FOREIGN KEY (`deptId`) REFERENCES `tb_dept1` (`id`)
) ENGINE=InnoDB DEFAULT CHARSET=utf8mb4 COLLATE=utf8mb4_0900_ai_ci
1 row in set (0.00 sec)
```

可以看到，已经成功添加了表的外键，下面删除外键约束，SQL 语句如下：

```
ALTER TABLE tb_emp9 DROP FOREIGN KEY fk_emp_dept;
```

执行完毕之后，将删除表 tb_emp9 的外键约束。使用 SHOW CREATE TABLE 再次查看表 tb_emp9 结构，结果如下：

```
mysql> SHOW CREATE TABLE tb_emp9 \G
*** 1. row ***
       Table: tb_emp9
Create Table: CREATE TABLE `tb_emp9` (
  `id` INT NOT NULL,
  `name` varchar(25) DEFAULT NULL,
```

```
  `deptId` INT DEFAULT NULL,
  `salary` float DEFAULT NULL,
  PRIMARY KEY (`id`),
  KEY `fk_emp_dept` (`deptId`)
) ENGINE=InnoDB DEFAULT CHARSET=utf8mb4 COLLATE=utf8mb4_0900_ai_ci
1 row in set (0.00 sec)
```

可以看到，tb_emp9 中已经不存在 FOREIGN KEY，原有的名称为 fk_emp_dept 的外键约束删除成功。

2.6 删除数据表

删除数据表就是将数据库中已经存在的表从数据库中删除。注意，在删除表的同时，表的定义和表中所有的数据均会被删除。因此，在进行删除操作前，最好对表中的数据做一个备份，以免造成无法挽回的后果。本节将详细讲解数据库表的删除方法。

2.6.1 删除没有被关联的表

在 MySQL 中，使用 DROP TABLE 可以一次删除一个或多个没有被其他表关联的数据表。语法格式如下：

```
DROP TABLE [IF EXISTS]表1, 表2, …, 表n;
```

其中，"表 n"指要删除的表的名称，后面可以同时删除多个表，只需将要删除的表名依次写在后面，相互之间用逗号隔开即可。如果要删除的数据表不存在，则 MySQL 会提示一条错误信息，"ERROR 1051 (42S02): Unknown table '表名'"。参数"IF EXISTS"用于在删除前判断删除的表是否存在，加上该参数后，再删除表的时候，如果表不存在，SQL 语句可以顺利执行，但是会发出警告（warning）。

在前面的例子中，已经创建了名为 tb_dept2 的数据表。如果没有，读者可输入语句，创建该表，SQL 语句如例 2.10 所示。下面使用删除语句将该表删除。

【例 2.28】删除数据表 tb_dept2，SQL 语句如下：

```
DROP TABLE IF EXISTS tb_dept2;
```

语句执行完毕之后，使用 SHOW TABLES 命令查看当前数据库中所有的表，SQL 语句如下：

```
mysql> SHOW TABLES;
+--------------------+
| Tables_in_test_db  |
+--------------------+
| tb_dept            |
| tb_deptment3       |
…省略部分内容
```

从执行结果可以看到，数据表列表中已经不存在名称为 tb_dept2 的表，删除操作成功。

2.6.2 删除被其他表关联的主表

在数据表之间存在外键关联的情况下，如果直接删除父表，结果会显示失败，原因是直接删除将破坏表的参照完整性。如果必须要删除，可以先删除与它关联的子表，再删除父表，只是这样就同时删除了两个表中的数据。有的情况下可能要保留子表，这时若要单独删除父表，只需将关联的表的外键约束条件取消，然后删除父表即可，下面讲解这种方法。

在数据库中创建两个关联表，首先创建表 tb_dept2，SQL 语句如下：

```
CREATE TABLE tb_dept2
(
id INT PRIMARY KEY,
name VARCHAR(22),
location VARCHAR(50)
);
```

接下来创建表 tb_emp，SQL 语句如下：

```
CREATE TABLE tb_emp
(
id INT PRIMARY KEY,
name VARCHAR(25),
deptId INT,
salary FLOAT,
CONSTRAINT fk_emp_dept FOREIGN KEY (deptId) REFERENCES tb_dept2(id)
);
```

使用 SHOW CREATE TABLE 命令查看表 tb_emp 的外键约束，结果如下：

```
mysql> SHOW CREATE TABLE tb_emp\G
*** 1. row ***
       Table: tb_emp
Create Table: CREATE TABLE `tb_emp` (
  `id` INT NOT NULL,
  `name` varchar(25) DEFAULT NULL,
  `deptId` INT DEFAULT NULL,
  `salary` float DEFAULT NULL,
  PRIMARY KEY (`id`),
  KEY `fk_emp_dept` (`deptId`),
  CONSTRAINT `fk_emp_dept` FOREIGN KEY (`deptId`) REFERENCES `tb_dept2` (`id`)
) ENGINE=InnoDB DEFAULT CHARSET=utf8mb4 COLLATE=utf8mb4_0900_ai_ci
```

可以看到，以上执行结果创建了两个关联表 tb_dept2 和表 tb_emp。其中，tb_emp 表为子表，具有名称为 fk_emp_dept 的外键约束；tb_dept2 为父表，其主键 id 被子表 tb_emp 所关联。

【例 2.29】删除被数据表 tb_emp 关联的数据表 tb_dept2。

首先试着直接删除父表 tb_dept2，输入如下删除语句：

```
mysql> DROP TABLE tb_dept2;
ERROR 3730 (HY000): Cannot drop table 'tb_dept2' referenced by a foreign key constraINT 'fk_emp_dept' on table 'tb_emp'.
```

如前所述，在存在外键约束的情况下，主表不能被直接删除。

接下来，解除关联子表 tb_emp 的外键约束，SQL 语句如下：

```
ALTER TABLE tb_emp DROP FOREIGN KEY fk_emp_dept;
```

语句成功执行后，将取消表 tb_emp 和表 tb_dept2 之间的关联关系。此时，可以输入删除语句，将原来的父表 tb_dept2 删除掉，SQL 语句如下：

```
DROP TABLE tb_dept2;
```

最后通过"SHOW TABLES;"查看数据表列表：

```
mysql> SHOW TABLES;
+--------------------+
| Tables_in_test_db  |
+--------------------+
| tb_dept            |
| tb_deptment3       |
...省略部分内容
```

可以看到，数据表列表中已经不存在名称为 tb_dept2 的表。

第 3 章

数据类型和运算符

数据库表由多列字段构成,每一个字段指定了不同的数据类型。指定字段的数据类型之后,也就决定了向字段插入的数据内容。例如,当要插入数值的时候,可以将它们存储为整数类型,也可以将它们存储为字符串类型。不同的数据类型也决定了 MySQL 在存储时使用的方式,以及在对其进行运算时使用的运算符号。本章将介绍 MySQL 中的数据类型和常见的运算符。

3.1 MySQL 数据类型介绍

MySQL 支持多种数据类型,主要有数值类型、日期/时间类型和字符串类型。

(1)数值类型:包括整数类型 TINYINT、SMALLINT、MEDIUMINT、INT 和 BIGINT,浮点小数数据类型 FLOAT 和 DOUBLE,定点小数类型 DECIMAL。

(2)日期/时间类型:包括 YEAR、TIME、DATE、DATETIME 和 TIMESTAMP。

(3)字符串类型:包括 CHAR、VARCHAR、BINARY、VARBINARY、BLOB、TEXT、ENUM 和 SET 等。字符串类型又分为文本字符串和二进制字符串。

3.1.1 整数类型

数值型数据类型主要用来存储数字,MySQL 提供了多种数值数据类型,不同的数据类型提供不同的取值范围,可以存储的值范围越大,其所需要的存储空间也会越大。MySQL 主要提供的整数类型有 TINYINT、SMALLINT、MEDIUMINT、INT(INTEGER)、BIGINT。整数类型的属性字段可以添加 AUTO_INCREMENT 自增约束条件。表 3.1 列出了 MySQL 中的数值类型。

表3.1 MySQL中的数值类型

类型名称	说 明	存储需求
TINYINT	很小的整数	1字节
SMALLINT	小的整数	2字节
MEDIUMINT	中等大小的整数	3字节
INT（INTEGER）	普通大小的整数	4字节
BIGINT	大整数	8字节

从表3.1中可以看到，不同类型整数存储所需的字节数是不同的，占用字节数最小的是TINYINT类型，占用字节最大的是BIGINT类型，相应的占用字节越多的类型所能表示的数值范围越大。根据占用字节数可以求出每一种数据类型的取值范围。例如，TINYINT需要1字节（8 bits）来存储，那么TINYINT无符号数的最大值为2^8-1（255）、TINYINT有符号数的最大值为2^7-1（127）。其他类型的整数的取值范围计算方法相同，如表3.2所示。

表3.2 不同整数类型的取值范围

数据类型	有 符 号	无 符 号
TINYINT	−128~127	0~255
SMALLINT	−32768~32767	0~65535
MEDIUMINT	−8388608~8388607	0~16777215
INT（INTEGER）	−2147483648~2147483647	0~4294967295
BIGINT	−9223372036854775808~9223372036854775807	0~18446744073709551615

【例3.1】创建表tmp1，其中字段x、y、z、m、n数据类型依次为TINYINT、SMALLINT、MEDIUMINT、INT、BIGINT，SQL语句如下：

```
CREATE TABLE tmp1 ( x TINYINT, y SMALLINT, z MEDIUMINT, m INT, n BIGINT );
```

执行成功之后，使用DESC查看表结构，结果如下：

```
mysql> DESC tmp1;
+-------+-----------+------+-----+---------+-------+
| Field | Type      | Null | Key | Default | Extra |
+-------+-----------+------+-----+---------+-------+
| x     | tinyint   | YES  |     | NULL    |       |
| y     | smallint  | YES  |     | NULL    |       |
| z     | mediumint | YES  |     | NULL    |       |
| m     | int       | YES  |     | NULL    |       |
| n     | bigint    | YES  |     | NULL    |       |
+-------+-----------+------+-----+---------+-------+
```

不同的整数类型有不同的取值范围，并且需要不同的存储空间，因此应该根据实际需要选择最合适的类型，这样有利于提高查询的效率和节省存储空间。

3.1.2 小数类型

MySQL 中使用浮点数和定点数来表示小数。浮点数类型有两种：单精度浮点类型（FLOAT）和双精度浮点类型（DOUBLE）。定点数类型只有 DECIMAL，定点数类型都可以用（M,N）来表示。其中，M 称为精度，表示总共的位数；N 称为标度，表示小数的位数。表 3.3 列出了 MySQL 中的小数类型和存储需求。

表 3.3　MySQL 中的小数类型和存储需求

类型名称	说明	存储需求
FLOAT	单精度浮点数	4 字节
DOUBLE	双精度浮点数	8 字节
DECIMAL(M,D)，DEC	压缩的"严格"定点数	M+2 字节

DECIMAL 类型不同于 FLOAT 和 DOUBLE，DECIMAL 实际是以串存储的，可能的最大取值范围与 DOUBLE 一样，但是其有效的取值范围由 M 和 D 的值决定。如果改变 M 而固定 D，则其取值范围将随 M 的变大而变大。从表 3.3 可以看到，DECIMAL 的存储空间并不是固定的，而由其精度值 M 决定的，占用 M+2 字节。

FLOAT 类型的取值范围如下：

- 有符号的取值范围：$-3.402823466E+38 \sim -1.175494351E-38$。
- 无符号的取值范围：0 和 $1.175494351E-38 \sim 3.402823466E+38$。

DOUBLE 类型的取值范围如下：

- 有符号的取值范围：$-1.7976931348623157E+308 \sim -2.2250738585072014E-308$。
- 无符号的取值范围：0 和 $2.2250738585072014E-308 \sim 1.7976931348623157E+308$。

【例 3.2】创建表 tmp2，其中字段 x、y、z 的数据类型依次为 FLOAT、DOUBLE 和 DECIMAL(5,1)，向表中插入数据 5.12、5.15 和 5.123，SQL 语句如下：

```
CREATE TABLE tmp2 (x FLOAT, y DOUBLE, z DECIMAL(5,1));
```

向表中插入数据：

```
mysql>INSERT INTO tmp2 VALUES(5.12, 5.15, 5.123);
```

可以看到，在插入数据时，MySQL 给出了一个警告信息，使用"SHOW WARNINGS;"语句查看警告信息：

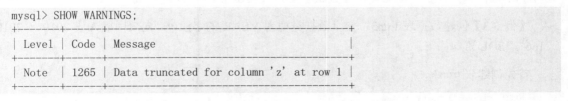

可以看到，给出了 z 字段数值被截断的警告。结果如下：

```
mysql> SELECT * FROM tmp2;
```

```
| x    | y    | z    |
+------+------+------+
| 5.1  | 5.2  | 5.1  |
+------+------+------+
```

3.1.3 日期与时间类型

MySQL 中有多种表示日期的数据类型，主要有 DATETIME、DATE、TIMESTAMP、TIME 和 YEAR。例如，当只记录年信息的时候，可以只使用 YEAR 类型，而没有必要使用 DATE。每一个类型都有合法的取值范围，当指定确实不合法的值时，系统将"零"值插入到数据库中。本节将介绍 MySQL 日期和时间类型的使用方法。表 3.4 列出了 MySQL 中的日期与时间类型。

表 3.4 日期与时间数据类型

类型名称	日期格式	日期范围	存储需求
YEAR	YYYY	1901~2155	1 字节
TIME	HH:MM:SS	−838:59:59~838:59:59	3 字节
DATE	YYYY-MM-DD	1000-01-01~9999-12-31	3 字节
DATETIME	YYYY-MM-DD HH:MM:SS	1000-01-01 00:00:00~9999-12-31 23:59:59	8 字节
TIMESTAMP	YYYY-MM-DD HH:MM:SS	1970-01-01 00:00:01 UTC~ 2038-01-19 03:14:07 UTC	4 字节

1. YEAR

YEAR 类型是一个单字节类型，用于表示年，在存储时只需要 1 字节。可以使用各种格式指定 YEAR 值，如下所示：

（1）以 4 位字符串或者 4 位数字格式表示的 YEAR，范围为 1901~2155。输入格式为"YYYY"或者 YYYY。例如，输入'2010'或 2010，插入到数据库的值均为 2010。

（2）以 2 位字符串格式表示的 YEAR，范围为 00~99。00~69 和 70~99 范围的值分别被转换为 2000~2069 和 1970~1999 范围的 YEAR 值。0 与 00 的作用相同。插入超过取值范围的值将被转换为 2000。

（3）以 2 位数字表示的 YEAR，范围为 1~99。1~69 和 70~99 范围的值分别被转换为 2001~2069 和 1970~1999 范围的 YEAR 值。注意：这里的 0 值将被转换为 0000，而不是 2000。

提示：两位整数范围与两位字符串范围稍有不同。例如：插入 2000 年，读者可能会使用数字格式的 0 表示 YEAR，实际上，插入数据库的值为 0000，而不是所希望的 2000。只有使用字符串格式的'0'或'00'，才可以被正确地解释为 2000。非法 YEAR 值将被转换为 0000。

【例 3.3】创建数据表 tmp3，定义数据类型为 YEAR 的字段 y，向表中插入值 2010、'2010'、'2166'，SQL 语句如下：

首先创建表 tmp3：

```
CREATE TABLE tmp3(y YEAR);
```

向表中插入数据：

```
mysql> INSERT INTO tmp3 values(2010),('2010');
```

再次向表中插入数据：

```
mysql> INSERT INTO tmp3 values ('2166');
ERROR 1264 (22003): Out of range value for column 'y' at row 1
```

语句执行之后，MySQL 给出了一条错误提示，使用 SHOW 查看错误信息：

```
mysql> SHOW WARNINGS;
+-------+------+-------------------------------------------------+
| Level | Code | Message                                         |
+-------+------+-------------------------------------------------+
| Error | 1264 | Out of range value for column 'y' at row 1;     |
+-------+------+-------------------------------------------------+
```

可以看到，插入的第 3 个值 '2166' 超过了 YEAR 类型的取值范围，此时不能正常执行插入操作，查看结果：

```
mysql> SELECT * FROM tmp3;
+------+
| y    |
+------+
| 2010 |
| 2010 |
+------+
```

由结果可以看到，当插入值为数值类型的 2010 或者字符串类型的 '2010' 时，都能正确地储存到数据库中；而当插入值 '2166' 时，由于超出了 YEAR 类型的取值范围，因此不能插入值。

【例 3.4】向 tmp3 表中 y 字段插入 2 位字符串表示的 YEAR 值，分别为 '0'、'00'、'77' 和 '10'，SQL 语句如下：

① 删除表中的数据：

```
DELETE FROM tmp3;
```

② 向表中插入数据：

```
INSERT INTO tmp3 values('0'),('00'),('77'),('10');
```

③ 查看结果：

```
mysql> SELECT * FROM tmp3;
+------+
| y    |
+------+
| 2000 |
| 2000 |
| 1977 |
| 2010 |
+------+
```

由结果可以看到，字符串 '0' 和 '00' 的作用相同，分别都转换成了 2000 年；'77' 转换为 1977；'10' 转换为 2010。

【例 3.5】向 tmp3 表中 y 字段插入 2 位数字表示的 YEAR 值，分别为 0、78 和 11，SQL 语句如下：

① 删除表中的数据：

```
DELETE FROM tmp3;
```

② 向表中插入数据：

```
INSERT INTO tmp3 values(0),(78),(11);
```

③ 查看结果：

```
mysql> SELECT * FROM tmp3;
+------+
| y    |
+------+
| 0000 |
| 1978 |
| 2011 |
+------+
```

由结果可以看到，0 被转换为 0000，78 被转换为 1978，11 被转换为 2011。

2. TIME

TIME 类型用在只需要时间信息的值，在存储时需要 3 字节，格式为"HH:MM:SS"。其中，HH 表示小时，MM 表示分钟，SS 表示秒。TIME 类型的取值范围为：-838:59:59~838:59:59，小时部分的取值范围会如此大的原因是 TIME 类型不仅可以用于表示一天的时间（必须小于 24 小时），还可以用于表示某个事件过去的时间或两个事件之间的时间间隔（可以大于 24 小时，或者为负）。可以使用各种格式指定 TIME 值。

（1）"D HH:MM:SS"格式的字符串。可以使用下面任何一种"非严格"的语法："HH:MM:SS"、"HH:MM"、"D HH:MM"、"D HH"或"SS"。这里的 D 表示日，可以取 0~24 之间的值。在插入数据库时，D 被转换为小时保存，格式为"D*24 + HH"。

（2）"HHMMSS"格式的、没有间隔符的字符串或者 HHMMSS 格式的数值，假定是有意义的时间。例如：'101112'被理解为 10:11:12，但'109712'是不合法的（它有一个没有意义的分钟部分），存储时将变为 00:00:00。

提示：为 TIME 列分配简写值时应该注意：如果没有冒号，MySQL 解释值时会假定最右边的两位表示秒（MySQL 解释 TIME 值为过去的时间而非当天的时间）。例如，读者可能认为'1112'和 1112 表示 11:12:00（11 点 12 分），但 MySQL 将它们解释为 00:11:12（11 分 12 秒）。同样地，'12' 和 12 被解释为 00:00:12。相反地，如果 TIME 值中使用冒号则肯定被看作当天的时间。也就是说，'11:12'表示 11:12:00，而不是 00:11:12。

【例 3.6】创建数据表 tmp4，定义数据类型为 TIME 的字段 t，向表中插入值'10:05:05'、'23:23'、'2 10:10'、'3 02'、'10'，SQL 语句如下：

① 创建表 tmp4：

```
CREATE TABLE tmp4(t TIME);
```

② 向表中插入数据：

```
mysql> INSERT INTO tmp4 values('10:05:05 '), ('23:23'), ('2 10:10'), ('3 02'),('10');
```

③ 查看结果：

```
mysql> SELECT * FROM tmp4;
+----------+
| t        |
+----------+
| 10:05:05 |
| 23:23:00 |
| 58:10:00 |
| 74:00:00 |
| 00:00:10 |
+----------+
```

由结果可以看到，'10:05:05'被转换为 10:05:05；'23:23'被转换为 23:23:00；'2 10:10'被转换为 58:10:00，'3 02'被转换为 74:00:00；'10'被转换成 00:00:10。

提示：在使用"D HH"格式时，小时一定要使用双位数值，如果是小于 10 的小时数，应在前面加 0。

【例 3.7】向表 tmp4 中插入值'101112'、111213、'0'、107010，SQL 语句如下：

① 删除表中的数据：

```
DELETE FROM tmp4;
```

② 向表中插入数据：

```
mysql>INSERT INTO tmp4 values('101112'),(111213),('0');
```

③ 再向表中插入数据：

```
mysql>INSERT INTO tmp4 values ( 107010);
ERROR 1292 (22007): Incorrect time value: '107010' for column 't' at row 1
```

可以看到，在插入数据时，MySQL 给出了一个错误提示信息，使用"SHOW WARNINGS;"查看错误信息，如下所示：

```
mysql> show warnings;
+-------+------+------------------------------------------------------------+
| Level | Code | Message                                                    |
+-------+------+------------------------------------------------------------+
| Error | 1292 | Incorrect time value: '107010' for column 't' at row 1     |
+-------+------+------------------------------------------------------------+
```

可以看到，第二次在插入记录的时候，数据超出了范围，原因是 107010 的分钟部分超过了 60（分钟部分是不会超过 60 的）。结果如下：

```
mysql> SELECT * FROM tmp4;
+----------+
| t        |
+----------+
| 10:11:12 |
| 11:12:13 |
```

```
| 00:00:00 |
+----------+
```

由结果可以看到，'101112' 被转换为 10:11:12；111213 被转换为 11:12:13；'0' 被转换为 00:00:00；因为 107010 是不合法的值，所以不能被插入。

也可以使用系统日期函数向 TIME 字段列插入值。

【例 3.8】向 tmp4 表中插入系统当前时间，SQL 语句如下：

① 删除表中的数据：

DELETE FROM tmp4;

② 向表中插入数据：

mysql> INSERT INTO tmp4 values (CURRENT_TIME) ,(NOW());

③ 查看结果：

```
mysql> SELECT * FROM tmp4;
+----------+
| t        |
+----------+
| 08:43:51 |
| 08:43:51 |
+----------+
```

由结果可以看到，获取系统当前的日期时间插入到 TIME 类型的列 t，因为读者输入语句的时间不确定，所以获取的值可能与这里的不同，但都是系统当前的日期时间值。

3. DATE 类型

DATE 类型用在仅需要日期值时，没有时间部分，在存储时需要 3 字节。日期格式为"YYYY-MM-DD"。其中，YYYY 表示年，MM 表示月，DD 表示日。在给 DATE 类型的字段赋值时，可以使用字符串类型或者数字类型的数据插入，只要符合 DATE 的日期格式即可。

（1）以"YYYY-MM-DD"或者"YYYYMMDD"字符串格式表示的日期，取值范围为 1000-01-01~9999-12-3。例如，输入"2012-12-31"或者"20121231"，插入数据库的日期都为 2012-12-31。

（2）以"YY-MM-DD"或者"YYMMDD"字符串格式表示的日期，在这里 YY 表示两位的年值。包含两位年值的日期会令人模糊，因为不知道世纪。MySQL 使用以下规则解释两位年值：00~69 范围的年值转换为 2000~2069；70~99 范围的年值转换为 1970~1999。例如，输入"12-12-31"，插入数据库的日期为 2012-12-31；输入"981231"，插入数据的日期为 1998-12-31。

（3）以 YY-MM-DD 或者 YYMMDD 数字格式表示的日期，与前面相似，00~69 范围的年值转换为 2000~2069，70~99 范围的年值转换为 1970~1999。例如，输入 12-12-31 插入数据库的日期为 2012-12-31；输入 981231，插入数据的日期为 1998-12-31。

（4）使用 CURRENT_DATE 或者 NOW()，插入当前系统日期。

【例 3.9】创建数据表 tmp5，定义数据类型为 DATE 的字段 d，向表中插入"YYYY-MM-DD"和"YYYYMMDD"字符串格式日期，SQL 语句如下：

① 创建表 tmp5：

```
MySQL> CREATE TABLE tmp5(d DATE);
```

② 向表中插入"YYYY-MM-DD"和"YYYYMMDD"字符串格式日期：

```
MySQL> INSERT INTO tmp5 values('1998-08-08'),('19980808'),('20101010');
```

③ 查看插入结果：

```
MySQL> SELECT * FROM tmp5;
+------------+
| d          |
+------------+
| 1998-08-08 |
| 1998-08-08 |
| 2010-10-10 |
+------------+
```

可以看到，各个不同类型的日期值都正确地插入到了数据表中。

【例 3.10】向 tmp5 表中插入"YY-MM-DD"和"YYMMDD"字符串格式日期，SQL 语句如下：

① 删除表中的数据：

```
DELETE FROM tmp5;
```

② 向表中插入"YY-MM-DD"和"YYMMDD"字符串格式日期：

```
mysql> INSERT INTO tmp5 values ('99-09-09'),('990909'),('000101'),('111111');
```

③ 查看插入结果：

```
mysql> SELECT * FROM tmp5;
+------------+
| d          |
+------------+
| 1999-09-09 |
| 1999-09-09 |
| 2000-01-01 |
| 2011-11-11 |
+------------+
```

【例 3.11】向 tmp5 表中插入 YYYYMMDD 和 YYMMDD 数字格式日期，SQL 语句如下：

① 删除表中的数据：

```
DELETE FROM tmp5;
```

② 向表中插入 YYYYMMDD 和 YYMMDD 数字格式日期：

```
mysql> INSERT INTO tmp5 values (19990909),(990909), (000101) ,(111111);
```

③ 查看插入结果：

```
mysql> SELECT * FROM tmp5;
+------------+
| d          |
+------------+
```

```
| 1999-09-09 |
| 1999-09-09 |
| 2000-01-01 |
| 2011-11-11 |
+------------+
```

【例 3.12】向 tmp5 表中插入系统当前日期，SQL 语句如下：

① 删除表中的数据：

DELETE FROM tmp5;

② 向表中插入系统当前日期：

mysql> INSERT INTO tmp5 values(CURRENT_DATE()),(NOW());

③ 查看插入结果：

```
mysql> SELECT * FROM tmp5;
+------------+
| d          |
+------------+
| 2022-03-09 |
| 2022-03-09 |
+------------+
```

CURRENT_DATE 只返回当前日期值，不包括时间部分；NOW()函数返回日期和时间值，在保存到数据库时，只保留了其日期部分。

提示：MySQL 允许"不严格"语法：任何标点符号都可以用作日期部分之间的间隔符。例如，"98-11-31" "98.11.31" "98/11/31" 和 "98@11@31" 是等价的，这些值都可以正确地插入到数据库中。

4. DATETIME

DATETIME 类型用于需要同时包含日期和时间信息的值，在存储时需要 8 字节。日期格式为 YYYY-MM-DD HH:MM:SS。其中，YYYY 表示年，MM 表示月，DD 表示日，HH 表示小时，MM 表示分钟，SS 表示秒。在给 DATETIME 类型的字段赋值时，可以使用字符串类型或者数字类型的数据插入，只要符合 DATETIME 的日期格式即可。

（1）以"YYYY-MM-DD HH:MM:SS"或者"YYYYMMDDHHMMSS"字符串格式表示的值，取值范围为 1000-01-01 00:00:00~9999-12-3 23:59:59。例如，输入"2012-12-31 05:05:05"或者"20121231050505"，插入数据库的 DATETIME 值都为 2012-12-31 05: 05: 05。

（2）以"YY-MM-DD HH:MM:SS"或者"YYMMDDHHMMSS"字符串格式表示的日期，在这里 YY 表示两位的年值。与前面相同，00~69 范围的年值转换为 2000~2069，70~99 范围的年值转换为 1970~1999。例如，输入"12-12-31 05:05:05"，插入数据库的 DATETIME 为 2012-12-31 05:05:05；输入"980505050505"，插入数据库的 DATETIME 为 1998-05-05 05: 05: 05。

（3）以 YYYYMMDDHHMMSS 或者 YYMMDDHHMMSS 数字格式表示的日期和时间。例如，输入"20121231050505"，插入数据库的 DATETIME 为 2012-12-31 05:05:05；输入"981231050505"，插入数据的 DATETIME 为 1998-12-31 05: 05: 05。

【例 3.13】创建数据表 tmp6，定义数据类型为 DATETIME 的字段 dt，向表中插入

"YYYY-MM-DD HH:MM:SS"和"YYYYMMDDHHMMSS"字符串格式日期和时间值,SQL 语句如下:

① 创建表 tmp6:

CREATE TABLE tmp6(dt DATETIME);

② 向表中插入"YYYY-MM-DD HH:MM:SS"和"YYYYMMDDHHMMSS"格式日期:

mysql> INSERT INTO tmp6 values('1998-08-08 08:08:08'),('19980808080808'),
('20101010101010');

③ 查看插入结果:

```
mysql> SELECT * FROM tmp6;
+---------------------+
| dt                  |
+---------------------+
| 1998-08-08 08:08:08 |
| 1998-08-08 08:08:08 |
| 2010-10-10 10:10:10 |
+---------------------+
```

可以看到,各个不同类型的日期值都正确地插入到了数据表中。

【例 3.14】向 tmp6 表中插入"YY-MM-DD HH:MM:SS"和"YYMMDDHHMMSS"字符串格式日期和时间值,SQL 语句如下:

① 删除表中的数据:

DELETE FROM tmp6;

② 向表中插入"YY-MM-DD HH:MM:SS"和"YYMMDDHHMMSS"格式日期:

mysql> INSERT INTO tmp6 values('99-09-09 09:09:09'),('990909090909'),
('101010101010');

③ 查看插入结果:

```
mysql> SELECT * FROM tmp6;
+---------------------+
| dt                  |
+---------------------+
| 1999-09-09 09:09:09 |
| 1999-09-09 09:09:09 |
| 2010-10-10 10:10:10 |
+---------------------+
```

【例 3.15】向 tmp6 表中插入"YYYYMMDDHHMMSS"和"YYMMDDHHMMSS"数字格式日期和时间值,SQL 语句如下:

① 删除表中的数据:

DELETE FROM tmp6;

② 向表中插入"YYYYMMDDHHMMSS"和"YYMMDDHHMMSS"数字格式日期和时间:

```
mysql> INSERT INTO tmp6 values(19990909090909), (101010101010);
```

③ 查看插入结果：

```
mysql> SELECT * FROM tmp6;
+---------------------+
| dt                  |
+---------------------+
| 1999-09-09 09:09:09 |
| 2010-10-10 10:10:10 |
+---------------------+
```

【例 3.16】向 tmp6 表中插入系统当前日期和时间值，SQL 语句如下：

① 删除表中的数据：

```
DELETE FROM tmp6;
```

② 向表中插入系统当前日期：

```
mysql> INSERT INTO tmp6 values( NOW() );
```

③ 查看插入结果：

```
mysql> SELECT * FROM tmp6;
+---------------------+
| dt                  |
+---------------------+
| 2022-03-15 17:07:30 |
+---------------------+
```

NOW()函数返回当前系统的日期和时间值，格式为"YYYY-MM-DD HH:MM:SS"。

提示：MySQL 允许"不严格"语法：任何标点符号都可以用作日期部分或时间部分之间的间隔符。例如，98-12-31 11:30:45、98.12.31 11+30+45、98/12/31 11*30*45 和 98@12@31 11^30^45 是等价的，这些值都可以正确地插入数据库。

5. TIMESTAMP

TIMESTAMP 的显示格式与 DATETIME 相同，显示宽度固定在 19 个字符，日期格式为"YYYY-MM-DD HH:MM:SS"，在存储时需要 4 字节。TIMESTAMP 列的取值范围小于 DATETIME 的取值范围，为 1970-01-01 00:00:01 UTC~2038-01-19 03:14:07 UTC。其中，UTC（Coordinated Universal Time）为世界标准时间，因此在插入数据时，要保证在合法的取值范围内。

【例 3.17】创建数据表 tmp7，定义数据类型为 TIMESTAMP 的字段 ts，向表中插入值 19950101010101、950505050505、1996-02-02 02:02:02、97@03@03 03@03@03、121212121212、NOW()，SQL 语句如下：

① 创建数据表 tmp7：

```
CREATE TABLE tmp7(ts TIMESTAMP);
```

② 向表中插入数据：

```
INSERT INTO tmp7 values ('19950101010101'),
```

```
('950505050505'),
('1996-02-02 02:02:02'),
('97@03@03 03@03@03'),
(121212121212),
( NOW() );
```

③ 查看插入结果：

```
mysql>SELECT * FROM tmp7;
+---------------------+
| ts                  |
+---------------------+
| 1995-01-01 01:01:01 |
| 1995-05-05 05:05:05 |
| 1996-02-02 02:02:02 |
| 1997-03-03 03:03:03 |
| 2012-12-12 12:12:12 |
| 2022-03-09 17:08:25 |
+---------------------+
```

由结果可以看到，"19950101010101"被转换为 1995-01-01 01:01:01；"950505050505"被转换为 1995-05-05 05:05:05；"1996-02-02 02:02:02"被转换为 1996-02-02 02:02:02；"97@03@03 03@03@03"被转换为 1997-03-03 03:03:03；121212121212 被转换为 2012-12-12 12:12:12；NOW()被转换为系统当前日期时间 2022-03-09 17:08:25。

提示：TIMESTAMP 与 DATETIME 除了存储字节和支持的范围不同外，还有一个最大的区别就是：DATETIME 在存储日期数据时，按实际输入的格式存储，即输入什么就存储什么，与时区无关；而 TIMESTAMP 值的存储是以 UTC（世界标准时间）格式保存的，存储时对当前时区进行转换，检索时再转换回当前时区。查询时，不同时区显示的时间值是不同的。

【例 3.18】向 tmp7 表中插入当前日期，查看插入值，更改时区为东 10 区，再次查看插入值，SQL 语句如下：

① 删除表中的数据：

```
DELETE FROM tmp7;
```

② 向表中插入系统当前日期：

```
mysql> INSERT INTO tmp7 values( NOW() );
```

③ 查看当前时区下的日期值：

```
mysql> SELECT * FROM tmp7;
+---------------------+
| ts                  |
+---------------------+
| 2022-03-09 17:12:20 |
+---------------------+
```

④ 查询结果为插入时的日期值。我国读者所在时区一般为东 8 区，下面修改当前时区为东 10 区，SQL 语句如下：

```
mysql> set time_zone='+10:00';
```

⑤ 再次查看插入时的日期值：

由结果可以看到，因为东 10 区时间比东 8 区快 2 个小时，所以查询的结果经过时区转换之后，显示的值增加了 2 小时。类似地，如果时区每减小一个值，则查询显示的日期中的小时数减 1。

提示：如果为一个 DATETIME 或 TIMESTAMP 对象分配一个 DATE 值，那么结果值的时间部分将被设置为"00:00:00"，因为 DATE 值未包含时间信息。如果为一个 DATE 对象分配一个 DATETIME 或 TIMESTAMP 值，那么结果值的时间部分将被删除，因为 DATE 值未包含时间信息。

3.1.4 文本字符串类型

字符串类型用来存储字符串数据，除了可以存储字符串数据之外，还可以存储其他数据，比如图片和声音的二进制数据。MySQL 支持两类字符型数据：文本字符串和二进制字符串。本小节主要讲解文本字符串类型。文本字符串可以进行区分或者不区分大小写的串比较，还可以进行模式匹配查找。在 MySQL 中，文本字符串类型是指 CHAR、VARCHAR、TEXT、ENUM 和 SET。表 3.5 列出了 MySQL 中的文本字符串数据类型。

表 3.5 MySQL 中文本字符串数据类型

类型名称	说 明	存储需求
CHAR(M)	固定长度非二进制字符串	M 字节，$1 \leq M \leq 255$
VARCHAR(M)	变长非二进制字符串	L+1 字节，在此 $L \leq M$ 和 $1 \leq M \leq 255$
TINYTEXT	非常小的非二进制字符串	L+1 字节，在此 $L < 2^8$
TEXT	小的非二进制字符串	L+2 字节，在此 $L < 2^{16}$
MEDIUMTEXT	中等大小的非二进制字符串	L+3 字节，在此 $L < 2^{24}$
LONGTEXT	大的非二进制字符串	L+4 字节，在此 $L < 2^{32}$
ENUM	枚举类型，只能有一个枚举字符串值	1 或 2 字节，取决于枚举值的数目（最大值为 65535）
SET	一个设置，字符串对象可以有零个或多个 SET 成员	1、2、3、4 或 8 字节，取决于集合成员的数量（最多为 64 个成员）

VARCHAR 和 TEXT 类型与下一小节讲到的 BLOB 都是变长类型，其存储需求取决于列值的实际长度（在前面的表格中用 L 表示）。例如，一个 VARCHAR(10)列能保存最大长度为 10 个字符的字符串，实际的存储需要是字符串的长度 L 加上 1 字节（记录字符串的长度）。对于字符"abcd"，L 是 4，而存储要求是 5 字节。本小节将介绍这些数据类型的作用以及在查询中使用这些类型的方法。

1. CHAR 和 VARCHAR 类型

CHAR(M)为固定长度字符串，在定义时指定字符串列长。当保存时在右侧填充空格，以达到指定的长度。M 表示列长度，M 的范围是 0~255 个字符。例如，CHAR(4)定义了一个固定长度的字符串列，其包含的字符个数最大为4。当检索到 CHAR 值时，尾部的空格将被删除。

VARCHAR(M)是长度可变的字符串，M 表示最大列长度。M 的范围是 0~65535。VARCHAR 的最大实际长度由最长的行的大小和使用的字符集确定，其实际占用的空间为字符串的实际长度加1。例如，VARCHAR(50)定义了一个最大长度为 50 的字符串，如果插入的字符串只有 10 个字符，则实际存储的字符串为 10 个字符和一个字符串结束字符。VARCHAR 在值保存和检索时尾部的空格仍保留。

【例 3.19】下面将不同字符串保存到 CHAR(4)和 VARCHAR(4)列，说明 CHAR 和 VARCHAR 之间的差别，如表 3.6 所示。

表 3.6　CHAR(4)与 VARCHAR(4)存储区别

插入值	CHAR(4)	存储需求	VARCHAR(4)	存储需求
' '	' '	4 字节	' '	1 字节
'ab'	'ab '	4 字节	'ab'	3 字节
'abc'	'abc'	4 字节	'abc'	4 字节
'abcd'	'abcd'	4 字节	'abcd'	5 字节
'abcdef'	'abcd'	4 字节	'abcd'	5 字节

对比结果可以看到，CHAR(4)定义了固定长度为 4 的列，不管存入的数据长度为多少，所占用的空间均为 4 个字节；VARCHAR(4)定义的列所占的字节数为实际长度加 1。

查询时，CHAR(4)和 VARCHAR(4)的值并不一定相同，如例 3.20 所示。

【例 3.20】创建 tmp8 表，定义字段 ch 和 vch 数据类型依次为 CHAR(4)、VARCHAR(4)，向表中插入数据'ab　'，SQL 语句如下：

① 创建表 tmp8：

CREATE TABLE tmp8(ch CHAR(4), vch VARCHAR(4));

② 输入数据：

INSERT INTO tmp8 VALUES('ab ', 'ab ');

③ 查询结果：

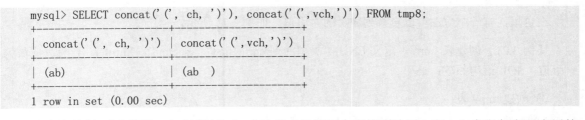

从查询结果可以看到，ch 在保存'ab　'时将末尾的两个空格删除了，而 vch 字段保留了末尾的两个空格。

提示：在表 3.6 中，最后一行的值只有在使用"不严格"模式时，字符串才会被截断插入；如果 MySQL 运行在"严格"模式，则超过列长度的值不会被保存，并且会出现错误信息"ERROR 1406(22001): Data too long for column"，即字符串长度超过指定长度，无法插入。

2. TEXT 类型

TEXT 列保存非二进制字符串，如文章内容、评论等。当保存或查询 TEXT 列的值时，不删除尾部空格。Text 类型分为 4 种：TINYTEXT、TEXT、MEDIUMTEXT 和 LONGTEXT。不同的 TEXT 类型的存储空间和数据长度不同。

（1）TINYTEXT 最大长度为 255（2^8-1）字符的 TEXT 列。

（2）TEXT 最大长度为 65535（$2^{16}-1$）字符的 TEXT 列。

（3）MEDIUMTEXT 最大长度为 16777215（$2^{24}-1$）字符的 TEXT 列。

（4）LONGTEXT 最大长度为 4294967295（$2^{32}-1$）或 4GB 字符的 TEXT 列。

3. ENUM 类型

ENUM 是一个字符串对象，其值为表创建时在列规定中枚举的一列值。语法格式如下：

字段名 ENUM('值1','值2',…,'值n')

其中，"字段名"指将要定义的字段，"值n"指枚举列表中的第 n 个值。ENUM 类型的字段在取值时，只能在指定的枚举列表中取，而且一次只能取一个。创建的成员中有空格时，其尾部的空格将自动被删除。ENUM 值在内部用整数表示，并且每个枚举值均有一个索引值：列表值所允许的成员值从 1 开始编号，MySQL 存储的就是这个索引编号。枚举最多可以有 65535 个元素。

例如，定义 ENUM 类型的列('first', 'second', 'third')，该列可以取的值和每个值的索引如表 3.7 所示。

表 3.7 ENUM 类型的取值范围

值	索引
NULL	NULL
''	0
first	1
second	2
third	3

ENUM 值依照列索引顺序排列，并且空字符串排在非空字符串前，NULL 值排在其他所有的枚举值前。这一点也可以从表 3.7 中看到。

在这里，有一个方法可以查看列成员的索引值，如例 3.21 所示。

【例 3.21】创建表 tmp9，定义 ENUM 类型的列 enm('first', 'second', 'third')，查看列成员的索引值，SQL 语句如下：

① 创建 tmp9 表：

```
CREATE TABLE tmp9(enm ENUM('first','second','third'));
```

② 插入各个列值：

```
INSERT INTO tmp9 values('first'),('second'),('third'),(NULL);
```

③ 查看索引值：

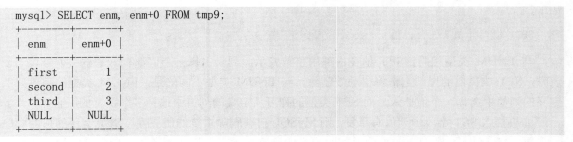

可以看到，这里的索引值和前面所述的相同。

提示：ENUM 列总有一个默认值：如果将 ENUM 列声明为 NULL，NULL 值则为该列的一个有效值，并且默认值为 NULL；如果 ENUM 列被声明为 NOT NULL，其默认值为允许的值列表的第 1 个元素。

【例 3.22】创建表 tmp10，定义 INT 类型的 soc 字段，ENUM 类型的字段 level，并且列表值为 ('excellent','good', 'bad')，向表 tmp10 中插入数据(70,'good')、(90,1)、(75,2)、(50,3)、(100,'best')，SQL 语句如下：

① 创建数据表：

```
CREATE TABLE tmp10 (soc INT, level enum('excellent', 'good','bad'));
```

② 插入数据：

```
INSERT INTO tmp10 values(70,'good'),(90,1),(75,2),(50,3);
```

③ 再次插入数据：

```
mysql>INSERT INTO tmp10 values (100,'best');
ERROR 1265 (01000): Data truncated for column 'level' at row 1
```

这里系统提示错误信息，可以看到，由于字符串值'best'不在 ENUM 列表中，所以对数据进行了阻止插入操作，查询结果如下：

```
mysql> SELECT * FROM tmp10;
+------+-----------+
| soc  | level     |
+------+-----------+
|   70 | good      |
|   90 | excellent |
|   75 | good      |
|   50 | bad       |
+------+-----------+
```

由结果可以看到，因为 ENUM 列表中的值在 MySQL 中都是以编号序列存储的，所以插入列表中的值'good'或者插入其对应序号 2 的结果是相同的。'best'不是列表中的值，因此不能插入数据。

4. SET 类型

SET 是一个字符串对象，可以有零个或多个值。SET 列最多可以有 64 个成员，其值为表创建时规定的一列值。指定包括多个 SET 成员的 SET 列值时，各成员之间用逗号（,）隔开。语法格式如下：

SET('值 1','值 2',…,'值 n')

与 ENUM 类型相同，SET 值在内部用整数表示，列表中每一个值都有一个索引编号。当创建表时，SET 成员值的尾部空格将自动被删除。与 ENUM 类型不同的是，ENUM 类型的字段只能从定义的列值中选择一个值插入，而 SET 类型的列可从定义的列值中选择多个字符的联合。

如果插入 SET 字段中列值有重复，则 MySQL 自动删除重复的值；插入 SET 字段的值的顺序并不重要，MySQL 会在存入数据库时按照定义的顺序显示；如果插入了不正确的值，默认情况下，MySQL 将忽视这些值，并给出警告。

【例 3.23】创建表 tmp11，定义 SET 类型的字段 s，取值列表为('a', 'b', 'c', 'd')，插入数据('a')、('a,b,a')、('c,a,d')、('a,x,b,y')，SQL 语句如下：

① 创建表 tmp11：

```
CREATE TABLE tmp11 (s SET('a', 'b', 'c', 'd'));
```

② 插入数据：

```
INSERT INTO tmp11 values('a'),('a,b,a'),('c,a,d');
```

③ 再次插入数据：

```
mysql>INSERT INTO tmp11 values ('a,x,b,y');
ERROR 1265 (01000): Data truncated for column 's' at row 1
```

由于插入了 SET 列不支持的值，所以 MySQL 给出错误提示。

④ 查看结果：

```
mysql> SELECT * FROM tmp11;
+-------+
| s     |
+-------+
| a     |
| a,b   |
| a,c,d |
+-------+
```

从结果可以看到，对于 SET 来说，如果插入的值是重复的，则只取一个，例如插入'a,b,a'，则结果为"a,b"；如果插入了不按顺序排列的值，则自动按顺序插入，例如插入'c,a,d'，结果为"a,c,d"；如果插入了不正确的值，那么该值将被阻止插入，例如插入值'a,x,b,y'。

3.1.5 二进制字符串类型

前面讲解了存储文本的字符串类型，这一小节将讲解 MySQL 中存储二进制数据的字符串类型特点及使用方法。MySQL 中的二进制数据类型有 BIT、BINARY、VARBINARY、TINYBLOB、BLOB、MEDIUMBLOB 和 LONGBLOB。表 3.8 列出了 MySQL 中的二进制数据类型。

表 3.8 MySQL 中的二进制字符串类型

类型名称	说 明	存储需求
BIT(M)	位字段类型	大约(M+7)/8 字节
BINARY(M)	固定长度二进制字符串	M 字节
VARBINARY(M)	可变长度二进制字符串	M+1 字节
TINYBLOB(M)	非常小的 BLOB	L+1 字节，在此 $L<2^8$
BLOB(M)	小 BLOB	L+2 字节，在此 $L<2^{16}$
MEDIUMBLOB(M)	中等大小的 BLOB	L+3 字节，在此 $L<2^{24}$
LONGBLOB(M)	非常大的 BLOB	L+4 字节，在此 $L<2^{32}$

1. BIT 类型

BIT 类型是位字段类型。M 表示每个值的位数，范围为 1~64。如果 M 被省略，默认为 1。如果为 BIT(M)列分配的值的长度小于 M 位，就在值的左边用 0 填充。例如，为 BIT(6)列分配一个值 b'101'，其效果与分配 b'000101'相同。BIT 数据类型用来保存位字段值。例如，以二进制的形式保存数据 13（13 的二进制形式为 1101），在这里需要位数至少为 4 位的 BIT 类型，即可以定义列类型为 BIT(4)，大于二进制 1111 的数据是不能插入 BIT(4)类型的字段中的。

【例 3.24】创建表 tmp12，定义 BIT(4)类型的字段 b，向表中插入数据 2、9、15。

① 创建表 tmp12：

CREATE TABLE tmp12(b BIT(4));

② 插入数据：

mysql> INSERT INTO tmp12 VALUES(2),(9),(15);

③ 查询插入结果：

mysql> SELECT BIN(b+0) FROM tmp12;

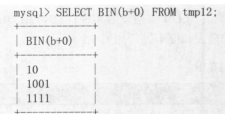

b+0 表示将二进制的结果转换为对应的数字的值，BIN()函数可以将数字转换为二进制。从结果可以看到，3 个数已被成功地插入表中。

提示：默认情况下，MySQL 不可以插入超出该列允许范围的值，因而插入的数据要确保插入的值在指定的范围内。

2. BINARY 和 VARBINARY 类型

BINARY 和 VARBINARY 类型类似于 CHAR 和 VARCHAR，不同的是它们包含二进制字节字符串。其使用的语法格式如下：

列名称 BINARY(M) 或者 VARBINARY(M)

BINARY 类型的长度是固定的，指定长度之后，不足最大长度的，将在它们右边填充"\0"补齐以达到指定长度。例如：指定列数据类型为 BINARY(3)，当插入"a"时，存储的内容实际为"a\0\0"；当插入"ab"时，实际存储的内容为"ab\0"；不管存储的内容是否达到指定的长度，其存储空间均为指定的值 M。

VARBINARY 类型的长度是可变的，指定好长度之后，其长度可以在 0 到最大值之间。例如：指定列数据类型为 VARBINARY(20)，如果插入的值的长度只有 10，则实际存储空间为 10 加 1，即实际占用的空间为字符串的实际长度加 1。

【例 3.25】创建表 tmp13，定义 BINARY(3)类型的字段 b 和 VARBINARY(3)类型的字段 vb，并向表中插入数据"5"，比较两个字段的存储空间。

① 创建表 tmp13：

```
CREATE TABLE tmp13(b binary(3), vb varbinary(3));
```

② 插入数据：

```
INSERT INTO tmp13 VALUES(5,5);
```

③ 查看两个字段存储数据的长度：

```
mysql> SELECT length(b), length(vb) FROM tmp13;
+-----------+------------+
| length(b) | length(vb) |
+-----------+------------+
|         3 |          1 |
+-----------+------------+
```

可以看到，b 字段的值数据长度为 3，而 vb 字段的数据长度仅为插入的一个字符的长度 1。

3. BLOB 类型

BLOB 是一个二进制大对象，用来存储可变数量的数据。BLOB 类型分为 4 种：TINYBLOB、BLOB、MEDIUMBLOB 和 LONGBLOB，它们可容纳值的最大长度不同，如表 3.9 所示。

表3.9 BLOB类型的存储范围

数据类型	存储范围
TINYBLOB	最大长度为 255（2^8-1）B
BLOB	最大长度为 65535（$2^{16}-1$）B
MEDIUMBLOB	最大长度为 16777215（$2^{24}-1$）B
LONGBLOB	最大长度为 4294967295（$2^{32}-1$）B 或 4GB

BLOB 列存储的是二进制字符串（字节字符串），TEXT 列存储的是非二进制字符串（字符字符串）。BLOB 列没有字符集，并且排序和比较基于列值字节的数值；TEXT 列有一个字符集，并且根据字符集对值进行排序和比较。

3.2 如何选择数据类型

MySQL 提供了大量的数据类型，为了优化存储、提高数据库性能，在任何情况下均应使用最精确的类型，即在所有可以表示该列值的类型中，该类型使用的存储最少。

1. 整数和浮点数

如果不需要小数部分，就使用整数来保存数据；如果需要表示小数部分，就使用浮点数类型。对于浮点数据列，存入的数值会对该列定义的小数位进行四舍五入。例如，假设列的值的范围为 1~99999，若使用整数，则使用 MEDIUMINT UNSIGNED 类型；若需要存储小数，则使用 FLOAT 类型。

浮点类型包括 FLOAT 和 DOUBLE 类型。DOUBLE 类型精度比 FLOAT 类型高，因此在对存储精度要求较高时应选择 DOUBLE 类型。

2. 浮点数和定点数

浮点数 FLOAT、DOUBLE 相对于定点数 DECIMAL 的优势是：在长度一定的情况下，浮点数能表示更大的数据范围。由于浮点数容易产生误差，所以对精确度要求比较高时，建议使用 DECIMAL 来存储。DECIMAL 在 MySQL 中是以字符串存储的，用于定义货币等对精确度要求较高的数据。在数据迁移时，float(M,D)是非标准 SQL 定义的，数据库迁移可能会出现问题，最好不要这样使用。另外，两个浮点数进行减法和比较运算时也容易出问题，因此在进行计算的时候，一定要小心。进行数值比较时，最好使用 DECIMAL 类型。

3. 日期与时间类型

MySQL 对于不同种类的日期和时间提供了很多数据类型，比如 YEAR 和 TIME。如果只需要记录年份，则使用 YEAR 类型；如果只记录时间，则使用 TIME 类型。

如果同时需要记录日期和时间，则可以使用 TIMESTAMP 或者 DATETIME 类型。由于 TIMESTAMP 列的取值范围小于 DATETIME 的取值范围，所以存储范围较大的日期最好使用 DATETIME。

TIMESTAMP 也有一个 DATETIME 不具备的属性。默认的情况下，当插入一条记录但并没有指定 TIMESTAMP 这个列值时，MySQL 会把 TIMESTAMP 列设为当前的时间。因此当需要插入记录的同时插入当前时间时，使用 TIMESTAMP 很方便。另外，TIMESTAMP 在存储空间上比 DATETIME 更有效。

4. CHAR 与 VARCHAR 的区别与选择

CHAR 和 VARCHAR 的区别如下：

- CHAR 是固定长度字符，VARCHAR 是可变长度字符。
- CHAR 会自动删除插入数据的尾部空格，VARCHAR 不会删除尾部空格。

CHAR 是固定长度，所以它的处理速度比 VARCHAR 的速度要快，但是它的缺点是浪费存储空间，所以对存储不大但在速度上有要求的数据可以使用 CHAR 类型，反之可以使用 VARCHAR 类型来实现。

存储引擎对于选择 CHAR 和 VARCHAR 的影响：

- 对于 MyISAM 存储引擎：最好使用固定长度的数据列来代替可变长度的数据列。这样可以使整个表静态化，从而使数据检索更快，用空间换时间。
- 对于 InnoDB 存储引擎：使用可变长度的数据列，因为 InnoDB 数据表的存储格式不分固定长度和可变长度，所以使用 CHAR 不一定比使用 VARCHAR 更好，但由于 VARCHAR 是按照实际的长度存储的，比较节省空间，所以对磁盘 I/O 和数据存储总量比较好。

5. ENUM 和 SET

ENUM 只能取单值，它的数据列表是一个枚举集合。它的合法取值列表最多允许有 65535 个成员。因此，在需要从多个值中选取一个时，可以使用 ENUM。比如：性别字段适合定义为 ENUM 类型，每次只能从"男"或"女"中取一个值。

SET 可取多值。它的合法取值列表最多允许有 64 个成员。空字符串也是一个合法的 SET 值。在需要取多个值的时候，适合使用 SET 类型，比如要存储一个人的兴趣爱好，最好使用 SET 类型。

ENUM 和 SET 的值是以字符串形式出现的，但在内部，MySQL 以数值的形式存储它们。

6. BLOB 和 TEXT

BLOB 是二进制字符串，TEXT 是非二进制字符串，两者均可存放大容量的信息。BLOB 主要存储图片、音频信息等，而 TEXT 只能存储纯文本文件。

3.3 常见运算符介绍

运算符连接表达式中的各个操作数，其作用是用来指明对操作数所进行的运算。运用运算符可以更加灵活地使用表中的数据，常见的运算符类型有算术运算符、比较运算符、逻辑运算符和位运算符。本节将介绍各种运算符的特点和使用方法。

3.3.1 运算符概述

运算符是一种让 MySQL 执行特定算术或逻辑操作的符号。MySQL 的内部运算符种类很丰富，主要有四大类，分别是算术运算符、比较运算符、逻辑运算符、位运算符。

1. 算术运算符

算术运算符用于各类数值运算，包括加（+）、减（-）、乘（*）、除（/）、求余（或称模运算，%）。

2. 比较运算符

比较运算符用于比较运算，包括大于（>）、小于（<）、等于（=）、大于或等于（>=）、小于或等于（<=）、不等于（!=），以及 IN、BETWEEN AND、IS NULL、GREATEST、LEAST、

LIKE、REGEXP 等。

3. 逻辑运算符

逻辑运算符的求值所得结果均为 1（True）、0（False），这类运算符包括逻辑非（NOT 或者!）、逻辑与（AND 或者&&）、逻辑或（OR 或者||）、逻辑异或（XOR）。

4. 位运算符

位运算符参与运算的操作数按二进制位进行运算，包括位与（&）、位或（|）、位非（~）、位异或（^）、左移（<<）、右移（>>）6 种。

接下来将详细地介绍 MySQL 中各种运算符的使用方法。

3.3.2 算术运算符

算术运算符是 MySQL 中最基本的运算符。MySQL 中的算术运算符如表 3.10 所示。

表 3.10 MySQL 中的算术运算符

运 算 符	作 用
+	加法运算
-	减法运算
*	乘法运算
/	除法运算，返回商
%	求余运算，返回余数

下面分别讨论不同算术运算符的使用方法。

【例 3.26】创建表 tmp14，定义数据类型为 INT 的字段 num，插入值 64，对 num 值进行算术运算。

① 创建表 tmp14：

CREATE TABLE tmp14(num INT);

② 向字段 num 插入数据 64：

INSERT INTO tmp14 value(64);

③ 对 num 值进行加法和减法运算：

```
mysql> SELECT num, num+10, num-3+5, num+5-3, num+36.5 FROM tmp14;
+-----+--------+---------+---------+----------+
| num | num+10 | num-3+5 | num+5-3 | num+36.5 |
+-----+--------+---------+---------+----------+
|  64 |     74 |      66 |      66 |    100.5 |
+-----+--------+---------+---------+----------+
```

由计算结果可以看到，对 num 字段的值进行加法和减法的运算时，"+"和"-"的优先级相同，先加后减或者先减后加的结果是相同的。

【例 3.27】对 tmp14 表中的 num 进行乘法、除法运算。

```
mysql> SELECT num, num *2, num /2, num/3, num%3 FROM tmp14;
+------+--------+---------+---------+-------+
| num  | num *2 | num /2  | num/3   | num%3 |
+------+--------+---------+---------+-------+
|  64  |  128   | 32.0000 | 21.3333 |   1   |
+------+--------+---------+---------+-------+
```

由计算结果可以看到，对 num 进行除法运算时，64 无法被 3 整除，MySQL 对 num/3 求商的结果保存到了小数点后面四位，结果为 21.3333；64 除以 3 的余数为 1，因此取余运算 num%3 的结果为 1。

在数学运算时，除数为 0 的除法是没有意义的，因此除法运算中的除数不能为 0，如果被 0 除，则返回结果为 NULL。

【例 3.28】用 0 除 num。

```
mysql> SELECT num, num / 0, num %0 FROM tmp14;
+------+---------+--------+
| num  | num / 0 | num %0 |
+------+---------+--------+
|  64  |  NULL   |  NULL  |
+------+---------+--------+
```

由计算结果可以看到，对 num 进行除法求商或者求余运算的结果均为 NULL。

3.3.3 比较运算符

一个比较运算符的结果总是 1、0 或者是 NULL。比较运算符经常在 SELECT 的查询条件子句中使用，用来查询满足指定条件的记录。MySQL 中的比较运算符如表 3.11 所示。

表 3.11　MySQL 中的比较运算符

运 算 符	作　　用
=	等于
<=>	安全等于
<>、!=	不等于
<=	小于或等于
>=	大于或等于
>	大于
IS NULL	判断一个值是否为 NULL
IS NOT NULL	判断一个值是否不为 NULL
LEAST	在有两个或多个参数时，返回最小值
GREATEST	当有两个或多个参数时，返回最大值
BETWEEN AND	判断一个值是否落在两个值之间
ISNULL	与 IS NULL 作用相同
IN	判断一个值是 IN 列表中的任意一个值
NOT IN	判断一个值不是 IN 列表中的任意一个值
LIKE	通配符匹配
REGEXP	正则表达式匹配

下面分别讨论不同比较运算符的使用方法。

1. 等于运算符（=）

等号（=）用来判断数字、字符串和表达式是否相等：如果相等，返回值为1，否则返回值为0。

【例3.29】使用"="进行相等判断，SQL语句如下：

```
mysql> SELECT 1=0, '2'=2, 2=2,'0.02'=0, 'b'='b', (1+3) = (2+2),NULL=NULL;
+-----+-------+-----+----------+---------+---------------+-----------+
| 1=0 | '2'=2 | 2=2 | '0.02'=0 | 'b'='b' | (1+3) = (2+2) | NULL=NULL |
+-----+-------+-----+----------+---------+---------------+-----------+
|  0  |   1   |  1  |    0     |    1    |       1       |   NULL    |
+-----+-------+-----+----------+---------+---------------+-----------+
```

由结果可以看到，在进行判断时，2=2和'2'=2的返回值相同，都为1。因为在进行判断时，MySQL自动进行了转换，把字符'2'转换成了数字2；'b'='b'为相同的字符比较，所以返回值为1；表达式1+3和表达式2+2的结果都为4，因此结果相等，返回值为1；由于"="不能用于空值NULL的判断，所以返回值为NULL。

数值比较时有如下规则：

（1）若有一个或两个参数为NULL，则比较运算的结果为NULL。
（2）若同一个比较运算中的两个参数都是字符串，则按照字符串进行比较。
（3）若两个参数均为整数，则按照整数进行比较。
（4）若用字符串和数字进行相等判断，则MySQL可以自动将字符串转换为数字。

2. 安全等于运算符（<=>）

这个操作符和=操作符执行相同的比较操作，不过<=>可以用来判断NULL值。在两个操作数均为NULL时，其返回值为1，而不为NULL；当一个操作数为NULL时，其返回值为0，而不为NULL。

【例3.30】使用"<=>"进行相等的判断，SQL语句如下：

```
mysql> SELECT 1<=>0, '2'<=>2, 2<=>2,'0.02'<=>0, 'b'<=>'b', (1+3)<=> (2+1),NULL<=>NULL;
+-------+---------+-------+------------+-----------+-----------------+-------------+
| 1<=>0 | '2'<=>2 | 2<=>2 | '0.02'<=>0 | 'b'<=>'b' | (1+3) <=> (2+1) | NULL<=>NULL |
+-------+---------+-------+------------+-----------+-----------------+-------------+
|   0   |    1    |   1   |     0      |     1     |        0        |      1      |
+-------+---------+-------+------------+-----------+-----------------+-------------+
```

由结果可以看到，"<=>"在执行比较操作时和"="的作用相似，唯一的区别是"<=>"可以用来对NULL进行判断，两者都为NULL时返回值为1。

3. 不等于运算符（<>或者 !=）

"<>"或者"!="用于判断数字、字符串、表达式是否相等：如果不相等，返回值为1，否则返回值为0。这两个运算符不能用于判断空值NULL。

【例3.31】使用">"和"!="进行不相等的判断，SQL语句如下：

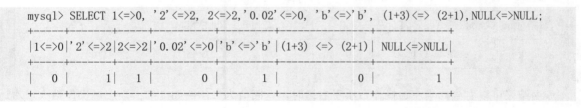

```
mysql> SELECT 'good'<>'god', 1<>2, 4!=4, 5.5!=5, (1+3)!=(2+1),NULL<>NULL;
```

```
+---------------+------+------+--------+-------------+-----------+
| 'good'<>'god' | 1<>2 | 4!=4 | 5.5!=5 | (1+3)!=(2+1)| NULL<>NULL|
+---------------+------+------+--------+-------------+-----------+
|       1       |   1  |   0  |    1   |      1      |   NULL    |
+---------------+------+------+--------+-------------+-----------+
```

由结果可以看到，上面两个不等于运算符的作用相同，都可以进行数字、字符串、表达式的比较判断。

4. 小于或等于运算符（<=）

"<="用来判断左边的操作数是否小于或等于右边的操作数：如果小于或等于，返回值为1，否则返回值为0。"<="不能用于判断空值NULL。

【例3.32】使用"<="进行比较判断，SQL语句如下：

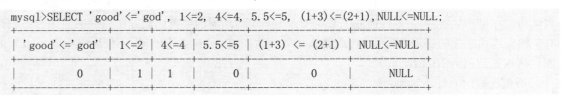

由结果可以看到，左边操作数小于或等于右边时，返回值为1，例如4<=4；当左边操作数大于右边时，返回值为0，例如'good'<='god'（'good'第3个位置的"o"字符在字母表中的顺序大于'god'中第3个位置的"d"字符，因此返回值为0）；比较两个NULL值时将返回NULL。

5. 小于运算符（<）

"<"运算符用来判断左边的操作数是否小于右边的操作数：如果是，则返回值为1，否则返回值为0。"<"不能用于判断空值NULL。

【例3.33】使用"<"进行比较判断，SQL语句如下：

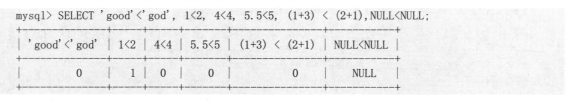

由结果可以看到，当左边操作数小于右边时，返回值为1，例如1<2；当左边操作数大于右边时，返回值为0，例如'good'<'god'（'good'第3个位置的"o"字符在字母表中的顺序大于'god'中第3个位置的"d"字符，因此返回值为0）；比较两个NULL值时将返回NULL。

6. 大于或等于运算符（>=）

">="运算符用来判断左边的操作数是否大于或等于右边的操作数：如果是，则返回值为1；否则返回值为0。">="不能用于判断空值NULL。

【例3.34】使用">="进行比较判断，SQL语句如下：

```
MySQL> SELECT 'good'>='god', 1>=2, 4>=4, 5.5>=5, (1+3) >= (2+1),NULL>=NULL;
```

```
| 'good'>='god' | 1>=2 | 4>=4 | 5.5>=5 | (1+3) >= (2+1) | NULL>=NULL |
+---------------+------+------+--------+----------------+------------+
|       1       |   0  |   1  |    1   |        1       |    NULL    |
+---------------+------+------+--------+----------------+------------+
```

由结果可以看到，左边操作数大于或等于右边时，返回值为 1，例如 4>=4；当左边操作数小于右边时，返回值为 0，例如 1>=2；比较两个 NULL 值时将返回 NULL。

7. 大于运算符（>）

">" 运算符用来判断左边的操作数是否大于右边的操作数：如果是，则返回值为 1；否则返回值为 0。">" 不能用于判断空值 NULL。

【例 3.35】使用 ">" 进行比较判断，SQL 语句如下：

```
mysql> SELECT 'good'>'god', 1>2, 4>4, 5.5>5, (1+3) > (2+1), NULL>NULL;
+--------------+-----+-----+-------+---------------+-----------+
| 'good'>'god' | 1>2 | 4>4 | 5.5>5 | (1+3) > (2+1) | NULL>NULL |
+--------------+-----+-----+-------+---------------+-----------+
|      1       |  0  |  0  |   1   |       1       |    NULL   |
+--------------+-----+-----+-------+---------------+-----------+
```

由结果可以看到，左边操作数大于右边时，返回值为 1，例如 5.5>5；左边操作数小于右边时，返回 0，例如 1>2；比较两个 NULL 值时将返回 NULL。

8. IS NULL（ISNULL）和 IS NOT NULL 运算符

IS NULL 和 ISNULL 检验一个值是否为 NULL：如果为 NULL，返回值为 1，否则返回值为 0。IS NOT NULL 检验一个值是否为非 NULL：如果是非 NULL，返回值为 1，否则返回值为 0。

【例 3.36】使用 IS NULL、ISNULL 和 IS NOT NULL 判断 NULL 值和非 NULL 值，SQL 语句如下：

```
mysql> SELECT NULL IS NULL, ISNULL(NULL),ISNULL(10), 10 IS NOT NULL;
+--------------+--------------+------------+----------------+
| NULL IS NULL | ISNULL(NULL) | ISNULL(10) | 10 IS NOT NULL |
+--------------+--------------+------------+----------------+
|       1      |       1      |      0     |        1       |
+--------------+--------------+------------+----------------+
```

由结果可以看到，IS NULL 和 ISNULL 的作用相同，只是格式不同。ISNULL 和 IS NOT NULL 的返回值正好相反。

9. BETWEEN...AND...运算符

语法格式为：expr BETWEEN min AND max。假如 expr 大于等于 min 且小于等于 max，则 BETWEEN 的返回值为 1，否则返回值为 0。

【例 3.37】使用 BETWEEN...AND...进行值区间判断，输入 SQL 语句如下：

```
mysql> SELECT 4 BETWEEN 2 AND 5, 4 BETWEEN 4 AND 6,12 BETWEEN 9 AND 10;
+-------------------+-------------------+---------------------+
| 4 BETWEEN 2 AND 5 | 4 BETWEEN 4 AND 6 | 12 BETWEEN 9 AND 10 |
+-------------------+-------------------+---------------------+
```

```
|         1         |         1         |         0         |
+-------------------+-------------------+-------------------+

mysql> SELECT 'x' BETWEEN 'f' AND 'g', 'b' BETWEEN 'a' AND 'c';
+--------------------------+--------------------------+
| 'x' BETWEEN 'f' AND 'g'  | 'b' BETWEEN 'a' AND 'c'  |
+--------------------------+--------------------------+
|            0             |            1             |
+--------------------------+--------------------------+
```

由结果可以看到，4 在端点值区间内或者等于其中一个端点值时，BETWEEN…AND…表达式返回值为 1；12 并不在指定区间内，因此返回值为 0；对于字符串类型的比较，按字母表中字母顺序进行比较，"x"不在指定的字母区间内，因此返回值为 0，而"b"位于指定字母区间内，因此返回值为 1。

10. LEAST 运算符

语法格式为：LEAST(值 1,值 2,…,值 n)。其中，"值 n"表示参数列表中有 n 个值。在有两个或多个参数的情况下，返回最小值。假如任意一个自变量为 NULL，则 LEAST()的返回值为 NULL。

【例 3.38】使用 LEAST 运算符进行大小判断，SQL 语句如下：

```
mysql> SELECT least(2,0), least(20.0,3.0,100.5), least('a','c','b'), least(10,NULL);
+------------+-----------------------+--------------------+----------------+
| least(2,0) | least(20.0,3.0,100.5) | least('a','c','b') | least(10,NULL) |
+------------+-----------------------+--------------------+----------------+
|     0      |         3.0           |         a          |      NULL      |
+------------+-----------------------+--------------------+----------------+
```

由结果可以看到，当参数是整数或者浮点数时，LEAST 将返回其中最小的值；当参数为字符串时，返回字母表中顺序最靠前的字符；当比较值列表中有 NULL 时，不能判断大小，返回值为 NULL。

11. GREATEST (value1,value2,…)

语法格式为：GREATEST(值 1, 值 2,…,值 n)。其中，n 表示参数列表中有 n 个值。当有两个或多个参数时，返回值为最大值。假如任意一个自变量为 NULL，则 GREATEST()的返回值为 NULL。

【例 3.39】使用 GREATEST 运算符进行大小判断，SQL 语句如下：

```
mysql> SELECT greatest(2,0), greatest(20.0,3.0,100.5),
greatest('a','c','b'), greatest(10,NULL);
+---------------+--------------------------+-----------------------+-------------------+
| greatest(2,0) | greatest(20.0,3.0,100.5) | greatest('a','c','b') | greatest(10,NULL) |
+---------------+--------------------------+-----------------------+-------------------+
|       2       |          100.5           |           c           |       NULL        |
+---------------+--------------------------+-----------------------+-------------------+
```

由结果可以看到，当参数中是整数或者浮点数时，GREATEST 将返回其中最大的值；当参数为字符串时，返回字母表中顺序最靠后的字符；当比较值列表中有 NULL 时，不能判断大小，返回值为 NULL。

12. IN、NOT IN 运算符

IN 运算符用来判断操作数是否为 IN 列表中的其中一个值：如果是，返回值为 1，否则返回值为 0。

NOT IN 运算符用来判断表达式是否为 IN 列表中的其中一个值：如果不是，返回值为 1，否则返回值为 0。

【例 3.40】使用 IN、NOT IN 运算符进行判断，SQL 语句如下：

```
mysql> SELECT 2 IN (1,3,5,'thks'), 'thks' IN (1,3,5,'thks');
+---------------------+--------------------------+
| 2 IN (1,3,5,'thks') | 'thks' IN (1,3,5,'thks') |
+---------------------+--------------------------+
|                   0 |                        1 |
+---------------------+--------------------------+

mysql> SELECT 2 NOT IN (1,3,5,'thks'), 'thks' NOT IN (1,3,5,'thks');
+-------------------------+------------------------------+
| 2 NOT IN (1,3,5,'thks') | 'thks' NOT IN (1,3,5,'thks') |
+-------------------------+------------------------------+
|                       1 |                            0 |
+-------------------------+------------------------------+
```

由结果可以看到，IN 和 NOT IN 的返回值正好相反。

在左侧表达式为 NULL 的情况下，或是表中找不到匹配项并且表中一个表达式为 NULL 的情况下，IN 的返回值均为 NULL。

【例 3.41】存在 NULL 值时的 IN 查询，SQL 语句如下：

```
mysql> SELECT NULL IN (1,3,5,'thks'), 10 IN (1,3,NULL,'thks');
+------------------------+-------------------------+
| NULL IN (1,3,5,'thks') | 10 IN (1,3,NULL,'thks') |
+------------------------+-------------------------+
|                   NULL |                    NULL |
+------------------------+-------------------------+
```

IN()语法也可用于在 SELECT 语句中进行嵌套子查询，在后面的章节中将会讲到。

13. LIKE

LIKE 运算符用来匹配字符串，语法格式为：expr LIKE 匹配条件。如果 expr 满足匹配条件，则返回值为 1（True）；如果不匹配，则返回值为 0（False）。expr 或匹配条件中任何一个为 NULL，则结果为 NULL。

LIKE 运算符在进行匹配时，可以使用下面的两种通配符：

（1）"%"：匹配任何数目的字符，甚至包括零字符。
（2）"_"：只能匹配一个字符。

【例 3.42】使用运算符 LIKE 进行字符串匹配运算，SQL 语句如下：

```
mysql> SELECT 'stud' LIKE 'stud','stud' LIKE 'stu_','stud' LIKE '%d','stud' LIKE 't___','s' LIKE NULL;
```

```
+------------------+------------------+------------------+--------------------+---------------+
|'stud' LIKE 'stud'|'stud' LIKE 'stu_'|'stud' LIKE '%d' |'stud' LIKE 't _ _ _'|'s' LIKE NULL |
+------------------+------------------+------------------+--------------------+---------------+
|        1         |        1         |        1         |         0          |     NULL      |
+------------------+------------------+------------------+--------------------+---------------+
```

由结果可以看到，指定匹配字符串为"stud"。"stud"表示直接匹配"stud"字符串，满足匹配条件，返回 1；"stu_"表示匹配以 stu 开头的长度为 4 个字符的字符串，"stud"正好是 4 个字符，满足匹配条件，因此匹配成功，返回 1；"%d"表示匹配以字母"d"结尾的字符串，"stud"满足匹配条件，匹配成功，返回 1；"t _ _ _"表示匹配以"t"开头的长度为 4 个字符的字符串，"stud"不满足匹配条件，因此返回 0；当字符"s"与 NULL 匹配时，结果为 NULL。

14. REGEXP

REGEXP 运算符用来匹配字符串，语法格式为：expr REGEXP 匹配条件。如果 expr 满足匹配条件，则返回 1；如果不满足，则返回 0。若 expr 或匹配条件任意一个为 NULL，则结果为 NULL。

REGEXP 运算符在进行匹配时，常用的有下面几种通配符：

（1）"^"：匹配以该字符后面的字符开头的字符串。
（2）"$"：匹配以该字符后面的字符结尾的字符串。
（3）"."：匹配任何一个单字符。
（4）"[...]"：匹配在方括号内的任何字符。例如："[abc]"匹配"a""b"或"c"。为了命名字符的范围，使用一个"-"。"[a-z]"匹配任何字母，而"[0-9]"匹配任何数字。
（5）"*"：匹配零个或多个在它前面的字符。例如："x*"匹配任何数量的"x"字符，"[0-9]*"匹配任何数量的数字，而"*"匹配任何数量的任何字符。

【例 3.43】使用运算符 REGEXP 进行字符串匹配运算，SQL 语句如下：

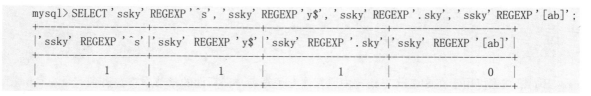

由结果可以看到，指定匹配字符串为"ssky"。"^s"表示匹配任何以字母"s"开头的字符串，因此满足匹配条件，返回 1；"y$"表示任何以字母"y"结尾的字符串，因此满足匹配条件，返回 1；".sky"匹配任何以"sky"结尾、字符长度为 4 的字符串，满足匹配条件，返回 1；"[ab]"匹配任何包含字母"a"或者"b"的字符串，指定字符串中既没有字母"a"也没有字母"b"，因此不满足匹配条件，返回 0。

提示：正则表达式是一个可以进行复杂查询的强大工具。相对于 LIKE 字符串匹配，它可以使用更多的通配符类型，查询结果更加灵活。读者可以参考相关的书籍或资料，深入学习正则表达式的写法，在这里就不详细介绍了。在后面的章节中，将会介绍如何使用正则表达式查询表中的记录。

3.3.4 逻辑运算符

在 SQL 中，所有逻辑运算符的求值所得结果均为 True、False 或 NULL。在 MySQL 中，它们体现为 1（True）、0（False）和 NULL。逻辑运算符大多数都与不同的数据库 SQL 通用。MySQL 中的逻辑运算符如表 3.12 所示。

表 3.12 MySQL 中的逻辑运算符

运算符	作用
NOT 或者 !	逻辑非
AND 或者 &&	逻辑与
OR 或者 ||	逻辑或
XOR	逻辑异或

接下来，分别讨论不同的逻辑运算符的使用方法。

1. NOT 或者!

逻辑非运算符 NOT 或者！表示当操作数为 0 时，所得值为 1；当操作数为非零值时，所得值为 0；当操作数为 NULL 时，所得的返回值为 NULL。

【例 3.44】分别使用非运算符"NOT"和"！"进行逻辑判断，SQL 语句如下：

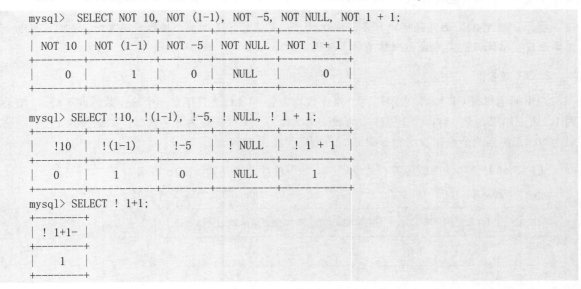

由结果可以看到，前 4 列"NOT"和"！"的返回值都相同。为什么最后 1 列会出现不同的值呢？这是因为"NOT"与"！"的优先级不同。"NOT"的优先级低于"+"，因此"NOT 1+1"相当于"NOT(1+1)"，先计算"1+1"，然后再进行 NOT 运算，因为操作数不为 0，因此 NOT 1 + 1 的结果是 0；相反，"！"的优先级要高于"+"运算，因此"！1+1"相当于"(!1)+1"，先计算"!1"，结果为 0，再加 1，最后结果为 1。

提示：读者在使用运算符运算时，一定要注意不同运算符的优先级不同。如果不能确定计算顺序，最好使用括号，以保证运算结果的正确性。

2. AND 或者&&

逻辑与运算符 AND 或者&&表示当所有操作数均为非零值并且不为 NULL 时，计算所得结果为 1；当一个或多个操作数为 0 时，所得结果为 0；其余情况返回值为 NULL。

【例 3.45】 分别使用与运算符"AND"和"&&"进行逻辑判断，SQL 语句如下：

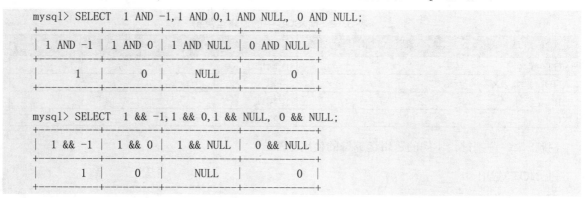

由结果可以看到，"AND"和"&&"的作用相同。"1 AND -1"中没有 0 或者 NULL，因此结果为 1；"1 AND 0"中有操作数 0，因此结果为 0；"1 AND NULL"中虽然有 NULL，但是没有操作数 0，返回结果为 NULL。

提示："AND"运算符可以有多个操作数，需要注意的是：多个操作数运算时，AND 两边一定要使用空格隔开，不然会影响结果的正确性。

3. OR 或者||

逻辑或运算符 OR 或者||表示当两个操作数均为非 NULL 值且任意一个操作数为非零值时，结果为 1，否则结果为 0；当有一个操作数为 NULL，且另一个操作数为非零值时，则结果为 1，否则结果为 NULL；当两个操作数均为 NULL 时，则所得结果为 NULL。

【例 3.46】 分别使用或运算符"OR"和"||"进行逻辑判断，SQL 语句如下：

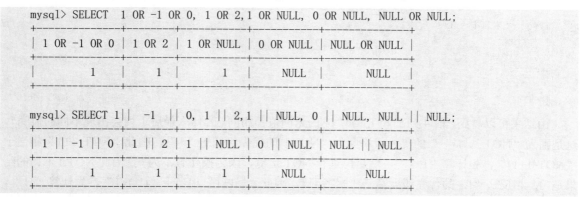

由结果可以看到，"OR"和"||"的作用相同。"1 OR -1 OR 0"中有 0，但同时包含有非 0 的值 1 和-1，返回结果为 1；"1 OR 2"中没有操作数 0，返回结果为 1；"1 OR NULL"中虽然有 NULL，但是有操作数 1，返回结果为 1；"0 OR NULL"中没有非 0 值，并且有 NULL，返回结果为 NULL；

"NULL OR NULL"中只有 NULL，返回结果为 NULL。

4. XOR

逻辑异或运算符 XOR 表示当任意一个操作数为 NULL 时，返回值为 NULL；对于非 NULL 的操作数，如果两个操作数都是非 0 值或者都是 0 值，则返回结果为 0；如果一个为 0 值、另一个为非 0 值，返回结果为 1。

【例 3.47】使用异或运算符"XOR"进行逻辑判断，SQL 语句如下：

```
mysql> SELECT 1 XOR 1, 0 XOR 0, 1 XOR 0, 1 XOR NULL, 1 XOR 1 XOR 1;
+---------+---------+---------+------------+---------------+
| 1 XOR 1 | 0 XOR 0 | 1 XOR 0 | 1 XOR NULL | 1 XOR 1 XOR 1 |
+---------+---------+---------+------------+---------------+
|       0 |       0 |       1 |       NULL |             1 |
+---------+---------+---------+------------+---------------+
```

由结果可以看到，在"1 XOR 1"和"0 XOR 0"中，运算符两边的操作数都为非零值或者都是零值，因此返回 0；在"1 XOR 0"中，两边的操作数一个为 0 值、一个为非 0 值，返回结果为 1；在"1 XOR NULL"中，有一个操作数为 NULL，返回结果为 NULL；在"1 XOR 1 XOR 1"中，有多个操作数，运算符相同，因此运算顺序从左到右依次计算，"1 XOR 1"的结果为 0，再与 1 进行异或运算，最终结果为 1。

提示：a XOR b 的计算等同于(a AND (NOT b))或者((NOT a)AND b)。

3.3.5 位运算符

位运算符是在二进制数上进行计算的运算符。位运算符先将操作数变成二进制数，然后进行位运算，最后将计算结果从二进制变回十进制数。MySQL 中提供的位运算符有按位或（|）、按位与（&）、按位异或（^）、按位左移（<<）、按位右移（>>）和按位取反（~），如表 3.13 所示。

表 3.13 MySQL 中的位运算符

运 算 符	作 用
|	位或
&	位与
^	位异或
<<	位左移
>>	位右移
~	位取反，反转所有比特位

接下来，分别讨论不同的位运算符的使用方法。

1. 位或运算符（|）

位或运算的实质是将参与运算的几个数据按照对应的二进制数逐位进行逻辑或运算。对应的二进制位有一个或两个为 1，则该位的运算结果为 1，否则为 0。

【例 3.48】使用位或运算符进行运算，SQL 语句如下：

```
mysql> SELECT 10 | 15, 9 | 4 | 2;
+---------+-----------+
| 10 | 15 | 9 | 4 | 2 |
+---------+-----------+
|     15  |       15  |
+---------+-----------+
```

10 的二进制数值为 1010，15 的二进制数值为 1111，按位或运算之后，结果为 1111，即整数 15；9 的二进制数值为 1001，4 的二进制数值为 0100，2 的二进制数值为 0010，按位或运算之后，结果为 1111，即整数 15。其结果为一个 64 位无符号整数。

2. 位与运算符（&）

位与运算的实质是将参与运算的几个操作数按照对应的二进制数逐位进行逻辑与运算。对应的二进制位都为 1，则该位的运算结果为 1，否则为 0。

【例 3.49】 使用位与运算符进行运算，SQL 语句如下：

```
mysql> SELECT 10 & 15, 9 &4& 2;
+---------+-----------+
| 10 & 15 | 9 &4& 2   |
+---------+-----------+
|      10 |         0 |
+---------+-----------+
```

10 的二进制数值为 1010，15 的二进制数值为 1111，按位与运算之后，结果为 1010，即整数 10；9 的二进制数值为 1001，4 的二进制数值为 0100，2 的二进制数值为 0010，按位与运算之后，结果为 0000，即整数 0。其结果为一个 64 位无符号整数。

3. 位异或运算符（^）

位异或运算的实质是将参与运算的两个数据按照对应的二进制数逐位进行逻辑异或运算。对应位的二进制数不同时，对应位的结果为 1。如果两个对应位数都为 0 或者都为 1，则对应位的结果为 0。

【例 3.50】 使用位异或运算符进行运算，SQL 语句如下：

```
mysql> SELECT 10 ^ 15, 1 ^0, 1 ^ 1;
+---------+------+-------+
| 10 ^ 15 | 1 ^0 | 1 ^ 1 |
+---------+------+-------+
|       5 |    1 |     0 |
+---------+------+-------+
```

10 的二进制数值为 1010，15 的二进制数值为 1111，按位异或运算之后，结果为 0101，即整数 5；1 的二进制数值为 0001，0 的二进制数值为 0000，按位异或运算之后，结果为 0001；1 和 1 本身二进制位完全相同，因此结果为 0。

4. 位左移运算符（<<）

位左移运算符<<使指定的二进制值的所有位都左移指定的位数。左移指定位数之后，左边高位的数值将被移出并丢弃，右边低位空出的位置用 0 补齐。语法格式为：expr<<n。其中，n 指定值 expr

要移位的位数。

【例 3.51】使用位左移运算符进行运算，SQL 语句如下：

```
mysql> SELECT 1<<2, 4<<2;
+------+------+
| 1<<2 | 4<<2 |
+------+------+
|    4 |   16 |
+------+------+
```

1 的二进制值为 0000 0001，左移两位之后变成 0000 0100，即十进制整数 4；十进制 4 左移两位之后变成 0001 0000，即变成十进制的 16。

5. 位右移运算符（>>）

位右移运算符>>使指定的二进制值的所有位都右移指定的位数。右移指定位数之后，右边低位的数值将被移出并丢弃，左边高位空出的位置用 0 补齐。语法格式为：expr>>n。其中，n 指定值 expr 要移位的位数。

【例 3.52】使用位右移运算符进行运算，SQL 语句如下：

```
mysql> SELECT 1>>1, 16>>2;
+------+-------+
| 1>>1 | 16>>2 |
+------+-------+
|    0 |     4 |
+------+-------+
```

1 的二进制值为 0000 0001，右移 1 位之后变成 0000 0000，即十进制整数 0；16 的二进制值为 0001 0000 右移两位之后变成 0000 0100，即变成十进制的 4。

6. 位取反运算符（~）

位取反运算的实质是将参与运算的数据按照对应的二进制数逐位反转，即 1 取反后变为 0、0 取反后变为 1。

【例 3.53】使用位取反运算符进行运算，SQL 语句如下：

```
mysql> SELECT 5 & ~1;
+--------+
| 5 & ~1 |
+--------+
|      4 |
+--------+
```

在逻辑运算 5&~1 中，由于位取反运算符"~"的级别高于位与运算符"&"，因此先对 1 进行取反操作，取反之后，除了最低位为 0 外其他位都为 1，即 1110，然后与十进制数值 5 进行与运算，结果为 0100，即整数 4。

提示：MySQL 经过位运算之后的数值是一个 64 位的无符号整数，1 的二进制数值表示为最右边位为 1、其他位均为 0，取反操作之后，除了最低位为 0 外，其他位均变为 1。

可以使用 BIN()函数查看 1 取反之后的结果，SQL 语句如下：

```
mysql> SELECT BIN(~1);
+------------------------------------------------------------------+
| BIN(~1)                                                          |
+------------------------------------------------------------------+
| 1111111111111111111111111111111111111111111111111111111111111110 |
+------------------------------------------------------------------+
```

至此，读者就明白例 3.53 是如何计算结果的了。

3.3.6 运算符的优先级

运算符的优先级决定了不同的运算符在表达式中计算的先后顺序。表 3.14 列出了 MySQL 中的各类运算符及其优先级。

表3.14 运算符按优先级由低到高排列

优 先 级	运 算 符
最低	=（赋值运算），:=
	‖，OR
	XOR
	&&，AND
	NOT
	BETWEEN，CASE，WHEN，THEN，ELSE
	=（比较运算），<=>, >=, >, <=, <, <>, != , IS, LIKE, REGEXP, IN
	|
	&
	<<, >>
	-, +
	*, /（DIV），%（MOD）
	^
	-（负号），~（位反转）
最高	!

可以看到，不同运算符的优先级是不同的。一般情况下，级别高的运算符先进行计算，如果级别相同，则按表达式从左到右的顺序依次计算。在无法确定优先级的情况下，可以使用英文圆括号()来改变优先级，这样会使计算过程更加清晰。

第 4 章

MySQL 函数

MySQL 提供了众多功能强大、方便易用的函数。使用这些函数，可以极大地提高用户对数据库的管理效率。MySQL 中的函数包括数学函数、字符串函数、日期和时间函数、条件判断函数、系统信息函数和加密函数等函数。本章将介绍 MySQL 中这些函数的功能和用法。

4.1 MySQL 函数简介

函数表示对输入参数值返回一个具有特定关系的值。MySQL 提供了丰富的函数，在进行数据库管理以及数据的查询和操作时，会经常用到各种函数。通过对数据的处理，数据库功能可以变得更加强大，可以更加灵活地满足不同用户的需求。各类函数从功能方面主要分为数学函数、字符串函数、日期和时间函数、条件判断函数、系统信息函数和加密函数等函数。本章将分类介绍不同函数的使用方法。

4.2 数学函数

数学函数主要用来处理数值数据，数学函数主要有绝对值函数、三角函数（包括正弦函数、余弦函数、正切函数、余切函数等）、对数函数、随机数函数等。在产生错误时，数学函数将会返回空值 NULL。本节将介绍各种数学函数的作用和用法。

4.2.1 绝对值函数 ABS(x)和返回圆周率的函数 PI()

ABS(X)返回 X 的绝对值。

【例 4.1】求 2、-3.3 和-33 的绝对值，输入语句如下：

```
mysql>SELECT ABS(2), ABS(-3.3), ABS(-33);
+--------+-----------+----------+
| ABS(2) | ABS(-3.3) | ABS(-33) |
+--------+-----------+----------+
|      2 |       3.3 |       33 |
+--------+-----------+----------+
```

正数的绝对值为其本身，2 的绝对值为 2；负数的绝对值为其相反数，-3.3 的绝对值为 3.3；-33 的绝对值为 33。

PI()返回圆周率 π 的值。默认显示小数位数是 6 位。

【例 4.2】返回圆周率值，输入语句如下：

```
mysql> SELECT pi();
+----------+
| pi()     |
+----------+
| 3.141593 |
+----------+
```

返回结果保留了 7 位有效数字。

4.2.2 平方根函数 SQRT(x)和求余函数 MOD(x,y)

SQRT(x)返回非负数 x 的二次平方根。

【例 4.3】求 9、40 和-49 的二次平方根，输入语句如下：

```
mysql> SELECT SQRT(9), SQRT(40), SQRT(-49);
+---------+-------------------+-----------+
| SQRT(9) | SQRT(40)          | SQRT(-49) |
+---------+-------------------+-----------+
|       3 | 6.324555320336759 |      NULL |
+---------+-------------------+-----------+
```

3 的平方等于 9，因此 9 的二次平方根为 3；40 的平方根为 6.324555320336759；负数没有平方根，因此-49 的平方根返回的结果为 NULL。

MOD(x,y)返回 x 被 y 除后的余数，MOD()对于带有小数部分的数值也起作用，它返回除法运算后的精确余数。

【例 4.4】对(31,8)、(234, 10)、(45.5,6)进行求余运算，输入语句如下：

```
mysql> SELECT MOD(31,8),MOD(234, 10),MOD(45.5,6);
+-----------+--------------+-------------+
| MOD(31,8) | MOD(234, 10) | MOD(45.5,6) |
+-----------+--------------+-------------+
|         7 |            4 |         3.5 |
+-----------+--------------+-------------+
```

4.2.3 获取整数的函数 CEIL(x)、CEILING(x)和 FLOOR(x)

CEIL(x)和 CEILING(x)的意义相同，返回不小于 x 的最小整数值，返回值转化为一个 BIGINT。

【例 4.5】使用 CEILING 函数返回最小整数，输入语句如下：

```
mysql> SELECT CEIL(-3.35),CEILING(3.35);
+-------------+---------------+
| CEIL(-3.35) | CEILING(3.35) |
+-------------+---------------+
|          -3 |             4 |
+-------------+---------------+
```

-3.35 为负数，不小于-3.35 的最小整数为-3，因此返回值为-3；不小于 3.35 的最小整数为 4，因此返回值为 4。

FLOOR(x)返回不大于 x 的最大整数值，返回值转化为一个 BIGINT。

【例 4.6】使用 FLOOR 函数返回最大整数，输入语句如下：

```
mysql> SELECT FLOOR(-3.35), FLOOR(3.35);
+--------------+-------------+
| FLOOR(-3.35) | FLOOR(3.35) |
+--------------+-------------+
|           -4 |           3 |
+--------------+-------------+
```

-3.35 为负数，不大于-3.35 的最大整数为-4，因此返回值为-4；不大于 3.35 的最大整数为 3，因此返回值为 3。

4.2.4 获取随机数的函数 RAND()和 RAND(x)

RAND(x)返回一个随机浮点值 v，范围在 0 到 1 之间（0 ≤ v ≤ 1.0）。若已指定一个整数参数 x，则它被用作种子值，用来产生重复序列。

【例 4.7】使用 RAND()函数产生随机数，输入语句如下：

```
mysql> SELECT RAND(),RAND(),RAND();
+---------------------+---------------------+---------------------+
| RAND()              | RAND()              | RAND()              |
+---------------------+---------------------+---------------------+
| 0.1380485446546679  | 0.45428662510056667 | 0.8572875222724746  |
+---------------------+---------------------+---------------------+
```

可以看到，不带参数的 RAND()每次产生的随机数值是不同的。

【例 4.8】使用 RAND(x)函数产生随机数，输入语句如下：

```
mysql> SELECT RAND(10),RAND(10),RAND(11);
+--------------------+--------------------+-------------------+
| RAND(10)           | RAND(10)           | RAND(11)          |
+--------------------+--------------------+-------------------+
| 0.6570515219653505 | 0.6570515219653505 | 0.907234631392392 |
+--------------------+--------------------+-------------------+
```

可以看到，当 RAND(x)的参数相同时，将产生相同的随机数，不同的 x 产生的随机数值不同。

4.2.5 函数 ROUND(x)、ROUND(x,y)和 TRUNCATE(x,y)

ROUND(x)返回最接近于参数 x 的整数，对 x 值进行四舍五入。

【例 4.9】使用 ROUND(x)函数对操作数进行四舍五入，输入语句如下：

```
mysql> SELECT ROUND(-1.14),ROUND(-1.67), ROUND(1.14),ROUND(1.66);
+--------------+--------------+-------------+-------------+
| ROUND(-1.14) | ROUND(-1.67) | ROUND(1.14) | ROUND(1.66) |
+--------------+--------------+-------------+-------------+
|           -1 |           -2 |           1 |           2 |
+--------------+--------------+-------------+-------------+
```

可以看到，四舍五入处理之后，只保留了各个值的整数部分。

ROUND(x,y)返回最接近于参数 x 的数，其值保留到小数点后面 y 位，若 y 为负值，则将保留 x 值到小数点左边 y 位。

【例 4.10】使用 ROUND(x,y)函数对操作数 x 进行四舍五入操作，结果保留小数点后面 y 位，输入语句如下：

```
mysql> SELECT ROUND(1.38, 1), ROUND(1.38, 0), ROUND(232.38, -1), ROUND(232.38,-2);
+----------------+----------------+-------------------+------------------+
| ROUND(1.38, 1) | ROUND(1.38, 0) | ROUND(232.38, -1) | ROUND(232.38,-2) |
+----------------+----------------+-------------------+------------------+
|            1.4 |              1 |               230 |              200 |
+----------------+----------------+-------------------+------------------+
```

ROUND(1.38,1)保留小数点后面 1 位，四舍五入的结果为 1.4；ROUND(1.38,0)保留小数点后面 0 位，即返回四舍五入后的整数值；ROUND(23.38,-1)和 ROUND (232.38,-2)分别保留小数点左边 1 位和 2 位。

提示：y 值为负数时，保留的小数点左边的相应位数直接保存为 0，不进行四舍五入。

TRUNCATE(x,y)返回被舍去至小数点后 y 位的数字 x。若 y 的值为 0，则结果不带有小数点或不带有小数部分。若 y 设为负数，则截去（归零）x 小数点左起第 y 位开始后面所有低位的值。

【例 4.11】使用 TRUNCATE(x,y)函数对操作数进行截取操作，结果保留小数点后面指定 y 位，输入语句如下：

```
mysql> SELECT TRUNCATE(1.31,1), TRUNCATE(1.99,1), TRUNCATE(1.99,0), TRUNCATE(19.99,-1);
+------------------+------------------+------------------+--------------------+
| TRUNCATE(1.31,1) | TRUNCATE(1.99,1) | TRUNCATE(1.99,0) | TRUNCATE(19.99,-1) |
+------------------+------------------+------------------+--------------------+
|              1.3 |              1.9 |                1 |                 10 |
+------------------+------------------+------------------+--------------------+
```

TRUNCATE(1.31,1)和 TRUNCATE(1.99,1)都保留小数点后 1 位数字，返回值分别为 1.3 和 1.9；TRUNCATE(1.99,0)返回整数部分值 1；TRUNCATE(19.99,-1)截去小数点左边第 1 位后面的值，并将整数部分的 1 位数字设置 0，结果为 10。

提示：ROUND(x,y)函数在截取值的时候会四舍五入，而 TRUNCATE (x,y)直接截取值，并不进行四舍五入。

4.2.6 符号函数 SIGN(x)

SIGN(x)返回参数的符号，x 的值分别为负数、零或正数时，返回结果依次为-1、0 或 1。

【例 4.12】使用 SIGN 函数返回参数的符号，输入语句如下：

```
mysql> SELECT SIGN(-21),SIGN(0), SIGN(21);
+-----------+---------+----------+
| SIGN(-21) | SIGN(0) | SIGN(21) |
+-----------+---------+----------+
|        -1 |       0 |        1 |
+-----------+---------+----------+
```

SIGN(-21)返回-1；SIGN(0)返回 0；SIGN(21)返回 1。

4.2.7 幂运算函数 POW(x,y)、POWER(x,y)和 EXP(x)

POW(x,y)或者 POWER(x,y)函数返回 x 的 y 次方（乘方）的结果值。

【例 4.13】使用 POW 和 POWER 函数进行乘方运算，输入语句如下：

```
mysql> SELECT POW(2,2), POWER(2,2),POW(2,-2), POWER(2,-2);
+----------+------------+-----------+-------------+
| POW(2,2) | POWER(2,2) | POW(2,-2) | POWER(2,-2) |
+----------+------------+-----------+-------------+
|        4 |          4 |      0.25 |        0.25 |
+----------+------------+-----------+-------------+
```

可以看到，POW 和 POWER 的结果是相同的，POW(2,2)和 POWER(2,2)返回 2 的 2 次方，结果都是 4；POW(2,-2)和 POWER(2,-2)都返回 2 的-2 次方，结果为 4 的倒数，即 0.25。

EXP(x)返回 e 的 x 乘方后的值。

【例 4.14】使用 EXP 函数计算 e 的乘方，输入语句如下：

```
mysql> SELECT EXP(3),EXP(-3),EXP(0);
+--------------------+----------------------+--------+
| EXP(3)             | EXP(-3)              | EXP(0) |
+--------------------+----------------------+--------+
| 20.085536923187668 | 0.049787068367863944 |      1 |
+--------------------+----------------------+--------+
```

EXP(3)返回以 e 为底的 3 次方，结果为 20.085536923187668；EXP(-3)返回以 e 为底的-3 次方，结果为 0.049787068367863944；EXP(0)返回以 e 为底的 0 次方，结果为 1。

4.2.8 对数运算函数 LOG(x)和 LOG10(x)

LOG(x)返回 x 的自然对数，x 相对于基数 e 的对数。

【例 4.15】使用 LOG(x)函数计算自然对数，输入语句如下：

```
mysql> SELECT LOG(3), LOG(-3);
+--------------------+---------+
| LOG(3)             | LOG(-3) |
```

```
+--------------------+--------+
| 1.0986122886681098 |  NULL  |
+--------------------+--------+
```

对数定义域不能为负数，因此 LOG(-3)返回结果为 NULL。

LOG10(x)返回 x 的基数为 10 的对数。

【例 4.16】 使用 LOG10 计算以 10 为基数的对数，输入语句如下：

```
mysql> SELECT LOG10(2), LOG10(100), LOG10(-100);
+--------------------+-----------+-------------+
| LOG10(2)           | LOG10(100)| LOG10(-100) |
+--------------------+-----------+-------------+
| 0.3010299956639812 |     2     |    NULL     |
+--------------------+-----------+-------------+
```

10 的 2 次乘方等于 100，因此 LOG10(100)返回结果为 2；LOG10(-100)定义域非负，因此返回 NULL。

4.2.9 角度与弧度相互转换的函数 RADIANS(x)和 DEGREES(x)

RADIANS(x)将参数 x 由角度转化为弧度。

【例 4.17】 使用 RADIANS 将角度转换为弧度，输入语句如下：

```
mysql> SELECT RADIANS(90),RADIANS(180);
+--------------------+-------------------+
| RADIANS(90)        | RADIANS(180)      |
+--------------------+-------------------+
| 1.5707963267948966 | 3.141592653589793 |
+--------------------+-------------------+
```

DEGREES(x)将参数 x 由弧度转化为角度。

【例 4.18】 使用 DEGREES 将弧度转换为角度，输入语句如下：

```
mysql> SELECT DEGREES(PI()), DEGREES(PI() / 2);
+---------------+-------------------+
| DEGREES(PI()) | DEGREES(PI() / 2) |
+---------------+-------------------+
|      180      |        90         |
+---------------+-------------------+
```

4.2.10 正弦函数 SIN(x)和反正弦函数 ASIN(x)

SIN(x)返回 x 正弦，其中 x 为弧度值。

【例 4.19】 使用 SIN 函数计算正弦值，输入语句如下：

```
mysql> SELECT SIN(1), ROUND(SIN(PI()));
+--------------------+------------------+
| SIN(1)             | ROUND(SIN(PI())) |
+--------------------+------------------+
| 0.8414709848078965 |        0         |
+--------------------+------------------+
```

ASIN(x)返回 x 的反正弦,即正弦为 x 的值。若 x 不在-1 到 1 的范围之内,则返回 NULL。

【例 4.20】使用 ASIN 函数计算反正弦值,输入语句如下:

```
mysql> SELECT ASIN(0.8414709848078965), ASIN(3);
+--------------------------+---------+
| ASIN(0.8414709848078965) | ASIN(3) |
+--------------------------+---------+
|                        1 |    NULL |
+--------------------------+---------+
```

由结果可以看到,函数 ASIN 和 SIN 互为反函数;ASIN(3)中的参数 3 超出了正弦值的范围,因此返回 NULL。

4.2.11 余弦函数 COS(x)和反余弦函数 ACOS(x)

COS(x)返回 x 的余弦,其中 x 为弧度值。

【例 4.21】使用 COS 函数计算余弦值,输入语句如下:

```
mysql> SELECT COS(0),COS(PI()),COS(1);
+--------+-----------+--------------------+
| COS(0) | COS(PI()) | COS(1)             |
+--------+-----------+--------------------+
|      1 |        -1 | 0.5403023058681398 |
+--------+-----------+--------------------+
```

由结果可以看到,COS(0)值为 1;COS(PI())值为-1;COS(1)值为 0.5403023058681398。

ACOS(x)返回 x 的反余弦,即余弦是 x 的值。若 x 不在-1~1 的范围之内,则返回 NULL。

【例 4.22】使用 ACOS 函数计算反余弦值,输入语句如下:

```
mysql> SELECT ACOS(1),ACOS(0), ROUND(ACOS(0.5403023058681398));
+---------+--------------------+---------------------------------+
| ACOS(1) | ACOS(0)            | ROUND(ACOS(0.5403023058681398)) |
+---------+--------------------+---------------------------------+
|       0 | 1.5707963267948966 |                               1 |
+---------+--------------------+---------------------------------+
```

由结果可以看到,函数 ACOS 和 COS 互为反函数。

4.2.12 正切函数、反正切函数和余切函数

TAN(x)返回 x 的正切,其中 x 为给定的弧度值。

【例 4.23】使用 TAN 函数计算正切值,输入语句如下:

```
mysql> SELECT TAN(0.3), ROUND(TAN(PI()/4));
+--------------------+--------------------+
| TAN(0.3)           | ROUND(TAN(PI()/4)) |
+--------------------+--------------------+
| 0.309336249960962325 |                1 |
+--------------------+--------------------+
```

ATAN(x)返回 x 的反正切，即正切为 x 的值。

【例 4.24】使用 ATAN 函数计算反正切值，输入语句如下：

```
mysql> SELECT ATAN(0.30933624960962325), ATAN(1);
+---------------------------+--------------------+
| ATAN(0.30933624960962325) | ATAN(1)            |
+---------------------------+--------------------+
|                       0.3 | 0.7853981633974483 |
+---------------------------+--------------------+
```

由结果可以看到，函数 ATAN 和 TAN 互为反函数。

COT(x)返回 x 的余切。

【例 4.25】使用 COT()函数计算余切值，输入语句如下：

```
mysql> SELECT COT(0.3), 1/TAN(0.3),COT(PI() / 4);
+--------------------+--------------------+--------------------+
| COT(0.3)           | 1/TAN(0.3)         | COT(PI() / 4)      |
+--------------------+--------------------+--------------------+
| 3.2327281437658275 | 3.2327281437658275 | 1.0000000000000002 |
+--------------------+--------------------+--------------------+
```

由结果可以看到，函数 COT 和 TAN 互为倒函数。

4.3 字符串函数

字符串函数主要用来处理数据库中的字符串数据。MySQL 中的字符串函数有字符串长度计算函数、字符串合并函数、字符串替换函数、字符串比较函数、查找指定字符串位置函数等。本节将介绍各种字符串函数的作用和用法。

4.3.1 计算字符串字符数的函数和字符串长度的函数

CHAR_LENGTH(str)返回值为字符串 str 所包含的字符个数。一个多字节字符算作一个单字符。

【例 4.26】使用 CHAR_LENGTH 函数计算字符串字符个数，输入语句如下：

```
mysql> SELECT CHAR_LENGTH('date'), CHAR_LENGTH('egg');
+---------------------+--------------------+
| CHAR_LENGTH('date') | CHAR_LENGTH('egg') |
+---------------------+--------------------+
|                   4 |                  3 |
+---------------------+--------------------+
```

LENGTH(str)返回值为字符串的字节长度，使用 utf8（UNICODE 的一种变长字符编码，又称万国码）编码字符集时，一个汉字是 3 字节，一个数字或字母为 1 字节。

【例 4.27】使用 LENGTH 函数计算字符串长度，输入语句如下：

```
mysql> SELECT LENGTH('date'), LENGTH('egg');
+----------------+---------------+
| LENGTH('date') | LENGTH('egg') |
+----------------+---------------+
|              4 |             3 |
+----------------+---------------+
```

可以看到，计算的结果与 CHAR_LENGTH 相同，因为英文字符的个数和所占的字节相同，一个字符占 1 字节。

4.3.2 合并字符串函数 CONCAT(s1,s2,…)、CONCAT_WS(x,s1,s2,…)

CONCAT(s1,s2,…)返回结果为连接参数产生的字符串，或许有一个或多个参数。如有任何一个参数为 NULL，则返回值为 NULL；如果所有参数均为非二进制字符串，则结果为非二进制字符串；如果自变量中含有任一二进制字符串，则结果为一个二进制字符串。

【例 4.28】使用 CONCAT 函数连接字符串，输入语句如下：

```
mysql> SELECT CONCAT('My SQL', '8.0'),CONCAT('My',NULL, 'SQL');
+-------------------------+--------------------------+
| CONCAT('My SQL', '8.0') | CONCAT('My',NULL, 'SQL') |
+-------------------------+--------------------------+
| My SQL8.0               | NULL                     |
+-------------------------+--------------------------+
```

CONCAT('My SQL', '8.0')返回两个字符串连接后的字符串；CONCAT('My',NULL, 'SQL')中有一个参数为 NULL，因此返回结果为 NULL。

在 CONCAT_WS(x,s1,s2,…)中，CONCAT_WS 代表 CONCAT With Separator，是 CONCAT() 的特殊形式。第一个参数 x 是其他参数的分隔符，分隔符的位置放在要连接的两个字符串之间。分隔符可以是一个字符串，也可以是其他参数。如果分隔符为 NULL，则结果为 NULL。函数会忽略任何分隔符参数后的 NULL 值。

【例 4.29】使用 CONCAT_WS 函数连接带分隔符的字符串，输入语句如下：

```
mysql> SELECT CONCAT_WS('-', '1st','2nd', '3rd'), CONCAT_WS('*', '1st', NULL, '3rd');
+------------------------------------+------------------------------------+
| CONCAT_WS('-', '1st','2nd', '3rd') | CONCAT_WS('*', '1st', NULL, '3rd') |
+------------------------------------+------------------------------------+
| 1st-2nd-3rd                        | 1st*3rd                            |
+------------------------------------+------------------------------------+
```

CONCAT_WS('-', '1st','2nd', '3rd')使用分隔符 "-" 将 3 个字符串连接成一个字符串，结果为 "1st-2nd-3rd"；CONCAT_WS('*', '1st', NULL, '3rd')使用分隔符 "*" 将两个字符串连接成一个字符串，同时忽略 NULL 值。

4.3.3 替换字符串的函数 INSERT(s1,x,len,s2)

INSERT(s1,x,len,s2)返回字符串 s1，其子字符串起始于 x 位置和被字符串 s2 取代的 len 字符。如果 x 超过字符串长度，则返回值为原始字符串。如果 len 的长度大于其他字符串的长度，则从位置 x 开始替换。若任何一个参数为 NULL，则返回值为 NULL。

【例 4.30】使用 INSERT 函数进行字符串替代操作，输入语句如下：

```
MySQL> SELECT INSERT('Quest', 2, 4, 'What') AS col1, INSERT('Quest', -1, 4, 'What') AS col2,
INSERT('Quest', 3, 100, 'Wh') AS col3;
+-------+-------+--------+
| col1  | col2  | col3   |
+-------+-------+--------+
| QWhat | Quest | QuWhat |
+-------+-------+--------+
```

第一个函数 INSERT('Quest', 2, 4, 'What')将"Quest"第 2 个字符开始、长度为 4 的字符串替换为 What，结果为"QWhat"；第二个函数 INSERT('Quest', -1, 4, 'What')中起始位置-1 超出了字符串长度，直接返回原字符；第三个函数 INSERT('Quest', 3, 100, 'What')替换长度超出了原字符串长度，则从第 3 个字符开始，截取后面所有的字符，并替换为指定字符 What，结果为"QuWhat"。

4.3.4 字母大小写转换函数

LOWER (str)或者 LCASE (str)可以将字符串 str 中的字母字符全部转换成小写字母。

【例 4.31】使用 LOWER 函数或者 LCASE 函数，将字符串中所有字母字符转换为小写，输入语句如下：

```
mysql> SELECT LOWER('BEAUTIFUL'), LCASE('Well');
+--------------------+---------------+
| LOWER('BEAUTIFUL') | LCASE('Well') |
+--------------------+---------------+
| beautiful          | well          |
+--------------------+---------------+
```

由结果可以看到，原来所有字母为大写的，全部转换为小写，如"BEAUTIFUL"，转换之后为"beautiful"；大小写字母混合的字符串，小写不变，大写字母转换为小写字母，如"WelL"，转换之后为"well"。

UPPER(str)或者 UCASE(str)可以将字符串 str 中的字母字符全部转换成大写字母。

【例 4.32】使用 UPPER 函数或者 UCASE 函数，将字符串中所有字母字符转换为大写，输入语句如下：

```
mysql> SELECT UPPER('black'), UCASE('BLacK');
+----------------+----------------+
| UPPER('black') | UCASE('BLacK') |
+----------------+----------------+
| BLACK          | BLACK          |
+----------------+----------------+
```

由结果可以看到，原来所有字母字符为小写的，全部转换为大写，如"black"，转换之后为"BLACK"；大小写字母混合的字符串，大写不变，小写字母转换为大写字母，如"BLacK"，转换之后为"BLACK"。

4.3.5 获取指定长度的字符串的函数 LEFT(s,n)和 RIGHT(s,n)

LEFT(s,n)返回字符串 s 开始的最左边 n 个字符。

【例 4.33】使用 LEFT 函数返回字符串中左边的字符，输入语句如下：

```
mysql> SELECT LEFT('football', 5);
+---------------------+
| LEFT('football', 5) |
+---------------------+
| footb               |
+---------------------+
```

函数返回字符串"football"左边开始的、长度为 5 的子字符串，结果为"footb"。
RIGHT(s,n)返回字符串 str 最右边的 n 个字符。

【例 4.34】使用 RIGHT 函数返回字符串中右边的字符，输入语句如下：

```
MySQL> SELECT RIGHT('football', 4);
+----------------------+
| RIGHT('football', 4) |
+----------------------+
| ball                 |
+----------------------+
```

函数返回字符串"football"右边开始的、长度为 4 的子字符串，结果为"ball"。

4.3.6 填充字符串的函数 LPAD(s1,len,s2)和 RPAD(s1,len,s2)

LPAD(s1,len,s2)返回字符串 s1，其左边由字符串 s2 填补到 len 字符长度。假如 s1 的长度大于 len，则返回值被缩短至 len 字符。

【例 4.35】使用 LPAD 函数对字符串进行填充操作，输入语句如下：

```
MySQL> SELECT LPAD('hello',4,'??'), LPAD('hello',10,'??');
+----------------------+-----------------------+
| LPAD('hello',4,'??') | LPAD('hello',10,'??') |
+----------------------+-----------------------+
| hell                 | ?????hello            |
+----------------------+-----------------------+
```

字符串"hello"长度大于 4，不需要填充，因此 LPAD('hello',4,'??')只返回被缩短的、长度为 4 的子串"hell"；字符串"hello"长度小于 10，LPAD('hello',10,'??')返回结果为"?????hello"，左侧填充"?"，长度为 10。

RPAD(s1,len,s2)返回字符串 s1，其右边被字符串 s2 填补至 len 字符长度。假如字符串 s1 的长度大于 len，则返回值被缩短到 len 字符长度。

【例 4.36】使用 RPAD 函数对字符串进行填充操作，输入语句如下：

```
mysql> SELECT RPAD('hello',4,'?'), RPAD('hello',10,'?');
+---------------------+----------------------+
| RPAD('hello',4,'?') | RPAD('hello',10,'?') |
+---------------------+----------------------+
```

```
| hell              | hello?????             |
```

字符串"hello"长度大于 4，不需要填充，因此 RPAD('hello',4,'?')只返回被缩短的、长度为 4 的子串"hell"；字符串"hello"长度小于 10，RPAD('hello',10,'?')返回结果为"hello?????"，右侧填充"？"，长度为 10。

4.3.7 删除空格的函数 LTRIM(s)、RTRIM(s)和 TRIM(s)

LTRIM(s)返回字符串 s，字符串左侧空格字符被删除。

【例 4.37】使用 LTRIM 函数删除字符串左边的空格，输入语句如下：

```
mysql> SELECT '( book )',CONCAT('(',LTRIM('  book  '),')');
+------------+------------------------------------+
| ( book )   | CONCAT('(',LTRIM('  book  '),')')  |
+------------+------------------------------------+
| ( book )   | (book  )                           |
+------------+------------------------------------+
```

LTRIM 只删除字符串左边的空格，而右边的空格不会被删除，" book "删除左边空格之后的结果为"book "。

RTRIM(s)返回字符串 s，字符串右侧空格字符被删除。

【例 4.38】使用 RTRIM 函数删除字符串右边的空格，输入语句如下：

```
mysql> SELECT '( book )',CONCAT('(', RTRIM ('  book  '),')');
+------------+-------------------------------------+
| ( book )   | CONCAT('(', RTRIM ('  book  '),')') |
+------------+-------------------------------------+
| ( book )   | (  book)                            |
+------------+-------------------------------------+
```

RTRIM 只删除字符串右边的空格，左边的空格不会被删除，" book "删除右边空格之后的结果为" book"。

TRIM(s)删除字符串 s 左右两侧的空格。

【例 4.39】使用 TRIM 函数删除字符串两侧的空格，使用语句如下：

```
mysql> SELECT '( book )',CONCAT('(', TRIM('  book  '),')');
+------------+-----------------------------------+
| ( book )   | CONCAT('(', TRIM('  book  '),')') |
+------------+-----------------------------------+
| ( book )   | (book)                            |
+------------+-----------------------------------+
```

可以看到，函数执行之后字符串" book "两边的空格都被删除，结果为"book"。

4.3.8 删除指定字符串的函数 TRIM(s1 FROM s)

TRIM(s1 FROM s)删除字符串 s 中两端所有的子字符串 s1。s1 为可选项，在未指定的情况下，会删除空格。

【例 4.40】 使用 TRIM(s1 FROM s)函数删除字符串两端指定的字符，输入语句如下：

```
mysql> SELECT TRIM('xy' FROM 'xyxboxyokxxyxy') ;
+--------------------------------+
| TRIM('xy' FROM 'xyxboxyokxxyxy') |
+--------------------------------+
| xboxyokx                       |
+--------------------------------+
```

删除字符串"xyxboxyokxxyxy"两端的重复字符串"xy"，而中间的"xy"并不删除，结果为"xboxyokx"。

4.3.9 重复生成字符串的函数 REPEAT(s,n)

REPEAT(s,n)返回一个由重复的字符串 s 组成的字符串，字符串 s 的数目等于 n。若 n<=0，则返回一个空字符串。若 s 或 n 为 NULL，则返回 NULL。

【例 4.41】 使用 REPEAT 函数重复生成相同的字符串，输入语句如下：

```
mysql> SELECT REPEAT('mysql', 3);
+--------------------+
| REPEAT('mysql', 3) |
+--------------------+
| mysqlmysqlmysql    |
+--------------------+
```

REPEAT('MySQL', 3)函数返回的字符串由 3 个重复的"mysql"字符串组成。

4.3.10 空格函数 SPACE(n)和替换函数 REPLACE(s,s1,s2)

SPACE(n)返回一个由 n 个空格组成的字符串。

【例 4.42】 使用 SPACE 函数生成由空格组成的字符串，输入语句如下：

```
mysql> SELECT CONCAT('(', SPACE(6), ')');
+----------------------------+
| CONCAT('(', SPACE(6), ')') |
+----------------------------+
| (      )                   |
+----------------------------+
```

SPACE(6)返回的字符串由 6 个空格组成。

REPLACE(s,s1,s2)使用字符串 s2 替代字符串 s 中所有的字符串 s1。

【例 4.43】 使用 REPLACE 函数进行字符串替代操作，输入语句如下：

```
mysql> SELECT REPLACE('xxx.mysql.com', 'x', 'w');
+------------------------------------+
| REPLACE('xxx.mysql.com', 'x', 'w') |
+------------------------------------+
| www.mysql.com                      |
+------------------------------------+
```

REPLACE('xxx.MySQL.com', 'x', 'w')将"xxx.mysql.com"字符串中的"x"字符替换为"w"字

符,结果为"www.mysql.com"。

4.3.11 比较字符串大小的函数 STRCMP(s1,s2)

STRCMP(s1,s2)比较字符串 s1 与 s2 的大小。若 s1 与 s2 所有的字符均相同,则返回 0;若根据当前分类次序,第一个参数小于第二个,则返回-1;其他情况返回 1。

【例 4.44】使用 STRCMP 函数比较字符串大小,输入语句如下:

```
mysql> SELECT STRCMP('txt', 'txt2'),STRCMP('txt2', 'txt'), STRCMP('txt', 'txt');
+-----------------------+-----------------------+----------------------+
| STRCMP('txt', 'txt2') | STRCMP('txt2', 'txt') | STRCMP('txt', 'txt') |
+-----------------------+-----------------------+----------------------+
|                    -1 |                     1 |                    0 |
+-----------------------+-----------------------+----------------------+
```

"txt"小于"txt2",因此 STRCMP('txt', 'txt2')返回结果为-1,STRCMP('txt2', 'txt')返回结果为1;"txt"与"txt"相等,因此 STRCMP('txt', 'txt')返回结果为 0。

4.3.12 获取子串的函数 SUBSTRING(s,n,len)和 MID(s,n,len)

SUBSTRING(s,n,len)带有 len 参数的格式,从字符串 s 返回一个长度与 len 字符相同的子字符串,起始于位置 n。也可能对 n 使用一个负值,如果是这种情况,则子字符串的位置起始于字符串结尾的 n 字符,即倒数第 n 个字符,而不是字符串的开头位置。

【例 4.45】使用 SUBSTRING 函数获取指定位置处的子字符串,输入语句如下:

```
MySQL> SELECT SUBSTRING('breakfast',5) AS col1, SUBSTRING('breakfast',5,3) AS col2,SUBSTRING('lunch', -3) AS col3,SUBSTRING('lunch', -5, 3) AS col4;
+-------+------+------+------+
| col1  | col2 | col3 | col4 |
+-------+------+------+------+
| kfast | kfa  | nch  | lun  |
+-------+------+------+------+
```

SUBSTRING('breakfast',5)返回从第 5 个位置开始到字符串结尾的子字符串,结果为"kfast";SUBSTRING('breakfast',5,3)返回从第 5 个位置开始长度为 3 的子字符串,结果为"kfa";SUBSTRING('lunch', -3)返回从结尾开始第 3 个位置到字符串结尾的子字符串,结果为"nch";SUBSTRING('lunch', -5, 3)返回从结尾开始第 5 个位置,即字符串开头起、长度为 3 的子字符串,结果为"lun"。

MID(s,n,len)函数与 SUBSTRING(s,n,len)的作用相同。

【例 4.46】使用 MID 函数获取指定位置处的子字符串,输入语句如下:

```
MySQL> SELECT MID('breakfast',5) as col1, MID('breakfast',5,3) as col2,MID('lunch', -3) as col3, MID('lunch', -5, 3) as col4;
+-------+------+------+------+
| col1  | col2 | col3 | col4 |
+-------+------+------+------+
| kfast | kfa  | nch  | lun  |
```

```
+--------+--------+--------+--------+
```

可以看到 MID 和 SUBSTRING 的处理结果是一样的。

提示：如果对 len 使用的是一个小于 1 的值，则结果始终为空字符串。

4.3.13 匹配子串开始位置的函数

LOCATE(str1,str)、POSITION(str1 IN str)和 INSTR(str, str1)3 个函数的作用相同，返回子字符串 str1 在字符串 str 中的开始位置。

【例 4.47】使用 LOCATE、POSITION、INSTR 函数，查找字符串中指定子字符串的开始位置，输入语句如下：

```
mysql> SELECT LOCATE('ball','football'),POSITION('ball' IN 'football'), INSTR ('football', 'ball');
+---------------------------+--------------------------------+----------------------------+
| LOCATE('ball','football') | POSITION('ball' IN 'football') | INSTR ('football', 'ball') |
+---------------------------+--------------------------------+----------------------------+
|                         5 |                              5 |                          5 |
+---------------------------+--------------------------------+----------------------------+
```

子字符串"ball"在字符串"football"中从第 5 个字母位置开始，因此 3 个函数返回结果都为 5。

4.3.14 字符串逆序的函数 REVERSE(s)

REVERSE(s)函数将字符串 s 反转，返回的字符串的顺序和 s 字符串顺序相反。

【例 4.48】使用 REVERSE 函数反转字符串，输入语句如下：

```
mysql> SELECT REVERSE('abc');
+----------------+
| REVERSE('abc') |
+----------------+
| cba            |
+----------------+
```

可以看到，字符串"abc"经过 REVERSE 函数处理之后，所有字符顺序被反转，结果为"cba"。

4.3.15 返回指定位置的字符串的函数

对于 ELT(N,字符串 1,字符串 2,字符串 3,...,字符串 N)函数，若 N=1，则返回值为字符串 1；若 N=2，则返回值为字符串 2；以此类推；若 N 小于 1 或大于参数的个数，则返回值为 NULL。

【例 4.49】使用 ELT 函数返回指定位置字符串，输入语句如下：

```
mysql> SELECT ELT(3,'1st','2nd','3rd'), ELT(3,'net','os');
+--------------------------+-------------------+
| ELT(3,'1st','2nd','3rd') | ELT(3,'net','os') |
+--------------------------+-------------------+
| 3rd                      | NULL              |
+--------------------------+-------------------+
```

由结果可以看到，ELT(3,'1st','2nd','3rd')返回第 3 个位置的字符串"3rd"；ELT(3,'net','os')指定

返回字符串位置超出参数个数，返回 NULL。

4.3.16 返回指定字符串位置的函数 FIELD(s,s1,s2,…,sn)

FIELD(s,s1,s2,…,sn)函数返回字符串 s 在列表 s1,s2,…,sn 中第一次出现的位置，在找不到 s 的情况下，返回值为 0。如果 s 为 NULL，则返回值为 0，原因是 NULL 不能与任何值进行同等比较。

【例 4.50】使用 FIELD 函数返回指定字符串第一次出现的位置，输入语句如下：

```
mysql> SELECT FIELD('Hi', 'hihi', 'Hey', 'Hi', 'bas') as col1, FIELD('Hi', 'Hey', 'Lo', 'Hilo', 'foo') as col2;
+------+------+
| col1 | col2 |
+------+------+
|    3 |    0 |
+------+------+
```

在 FIELD('Hi', 'hihi', 'Hey', 'Hi', 'bas')函数中，字符串"Hi"出现在列表的第 3 个字符串位置，因此返回结果为 3；FIELD('Hi', 'Hey', 'Lo', 'Hilo', 'foo')列表中没有字符串"Hi"，因此返回结果为 0。

4.3.17 返回子串位置的函数 FIND_IN_SET(s1,s2)

FIND_IN_SET(s1,s2)函数返回字符串 s1 在字符串列表 s2 中出现的位置，字符串列表是一个由多个逗号","分开的字符串组成的列表。如果 s1 不在 s2 中或 s2 为空字符串，则返回值为 0。如果任意一个参数为 NULL，则返回值为 NULL。这个函数在第一个参数包含一个逗号","的情况下，将无法正常运行。

【例 4.51】使用 FIND_IN_SET()函数返回子字符串在字符串列表中的位置，输入语句如下：

```
mysql> SELECT FIND_IN_SET('Hi','hihi,Hey,Hi,bas');
+-------------------------------------+
| FIND_IN_SET('Hi','hihi,Hey,Hi,bas') |
+-------------------------------------+
|                                   3 |
+-------------------------------------+
```

虽然 FIND_IN_SET()和 FIELD()两个函数格式不同，但作用类似，都可以返回指定字符串在字符串列表中的位置。

4.3.18 选取字符串的函数 MAKE_SET(x,s1,s2,…,sn)

MAKE_SET(x,s1,s2,…,sn)函数按 x 的二进制数从 s1,s2,…,sn 中选取字符串。例如，5 的二进制是 0101，这个二进制从右往左的第 1 位和第 3 位是 1，所以选取 s1 和 s3。s1,s2,…,sn 中的 NULL 值不会被添加到结果中。

【例 4.52】使用 MAKE_SET 函数根据二进制位选取指定字符串，输入语句如下：

```
mysql> SELECT MAKE_SET(1,'a','b','c') as col1, MAKE_SET(1 | 4,'hello','nice','world') as col2,
MAKE_SET(1 | 4,'hello','nice',NULL,'world') as col3, MAKE_SET(0,'a','b','c') as col4;
+------+------+------+------+
| col1 | col2 | col3 | col4 |
+------+------+------+------+
```

```
| a     | hello,world | hello |       |
```

1 的二进制值为 0001，4 的二进制值为 0100，1 与 4 进行或操作之后的二进制值为 0101，从右到左第 1 位和第 3 位为 1。MAKE_SET(1,'a','b','c')返回第 1 个字符串；SET(1 | 4,'hello','nice','world')返回从左端开始第 1 个和第 3 个字符串组成的字符串；NULL 不会添加到结果中，因此 SET(1 | 4,'hello','nice',NULL,'world')只返回第 1 个字符串"hello"；SET(0,'a','b','c')返回空字符串。

4.4 日期和时间函数

日期和时间函数主要用来处理日期和时间值，一般的日期函数除了使用 DATE 类型的参数外，也可以使用 DATETIME 或者 TIMESTAMP 类型的参数，但会忽略这些值的时间部分。相同地，以 TIME 类型值为参数的函数，可以接受 TIMESTAMP 类型的参数，但会忽略日期部分，许多日期函数可以同时接受数字和字符串类型的两种参数。本节将介绍各种日期和时间函数的作用和用法。

4.4.1 获取当前日期的函数和获取当前时间的函数

CURDATE()和 CURRENT_DATE()函数的作用相同，将当前日期按照 YYYY-MM-DD 或 YYYYMMDD 格式的值返回，具体格式根据函数用于字符串或是数字的语境而定。

【例 4.53】使用日期函数获取系统当前日期，输入语句如下：

```
mysql> SELECT CURDATE(),CURRENT_DATE(), CURDATE() + 0;
+------------+----------------+---------------+
| CURDATE()  | CURRENT_DATE() | CURDATE() + 0 |
+------------+----------------+---------------+
| 2022-03-16 | 2022-03-16     |      20220316 |
+------------+----------------+---------------+
```

可以看到，两个函数作用相同，都返回了相同的系统当前日期，"CURDATE() + 0"将当前日期值转换为数值型。

CURTIME()和 CURRENT_TIME()函数的作用相同，将当前时间以 HH:MM:SS 或 HHMMSS 的格式返回，具体格式根据函数用于字符串或是数字的语境而定。

【例 4.54】使用时间函数获取系统当前时间，输入语句如下：

```
mysql> SELECT CURTIME(),CURRENT_TIME(),CURTIME() + 0;
+-----------+----------------+---------------+
| CURTIME() | CURRENT_TIME() | CURTIME() + 0 |
+-----------+----------------+---------------+
| 18:03:22  | 18:03:22       |        180322 |
+-----------+----------------+---------------+
```

可以看到，两个函数的作用相同，都返回了相同的系统当前时间，"CURTIME () + 0"将当前时间值转换为数值型。

4.4.2 获取当前日期和时间的函数

CURRENT_TIMESTAMP()、LOCALTIME()、NOW()和 SYSDATE() 4 个函数的作用相同,均返回当前日期和时间值,格式为 YYYY-MM-DD HH:MM:SS 或 YYYYMMDDHHMMSS,具体格式根据函数用于字符串或数字的语境而定。

【例 4.55】使用日期时间函数获取当前系统日期和时间,输入语句如下:

```
mysql>SELECT CURRENT_TIMESTAMP(),LOCALTIME(),NOW(),SYSDATE();
+---------------------+---------------------+---------------------+---------------------+
| CURRENT_TIMESTAMP() | LOCALTIME()         | NOW()               | SYSDATE()           |
+---------------------+---------------------+---------------------+---------------------+
| 2022-03-16 12:06:09 | 2022-03-16 12:06:09 | 2022-03-16 12:06:09 | 2022-03-16 12:06:09 |
+---------------------+---------------------+---------------------+---------------------+
```

可以看到,4 个函数返回的结果是相同的。

4.4.3 UNIX 时间戳函数

UNIX_TIMESTAMP(date)函数若无参数调用,则返回一个 UNIX 时间戳(即"1970-01-01 00:00:00"GMT 之后的秒数)作为无符号整数。其中,GMT(Greenwich Mean Time)为格林尼治标准时间。若用 date 来调用 UNIX_TIMESTAMP(),它会将参数值以"1970-01-01 00:00:00"(GMT)后的秒数的形式返回。date 可以是一个 DATE 字符串、DATETIME 字符串、TIMESTAMP 或一个当地时间的 YYMMDD 或 YYYYMMDD 格式的数字。

【例 4.56】使用 UNIX_TIMESTAMP 函数返回 UNIX 格式的时间戳,输入语句如下:

```
mysql> SELECT UNIX_TIMESTAMP(), UNIX_TIMESTAMP(NOW()), NOW();
+------------------+-----------------------+---------------------+
| UNIX_TIMESTAMP() | UNIX_TIMESTAMP(NOW()) | NOW()               |
+------------------+-----------------------+---------------------+
| 1647403611       | 1647403611            | 2022-03-16 12:06:51 |
+------------------+-----------------------+---------------------+
```

FROM_UNIXTIME(date)函数把 UNIX 时间戳转换为普通格式的时间,与 UNIX_TIMESTAMP(date)函数互为反函数。

【例 4.57】使用 FROM_UNIXTIME 函数将 UNIX 时间戳转换为普通格式时间,输入语句如下:

```
mysql> SELECT FROM_UNIXTIME('1541844424');
+----------------------------+
| FROM_UNIXTIME('1541844424')|
+----------------------------+
| 2018-11-10 18:07:04.000000 |
+----------------------------+
```

可以看到,FROM_UNIXTIME('1541844424')与例 4.56 中 UNIX_TIMESTAMP(NOW())的结果正好相反,即两个函数互为反函数。

4.4.4 返回 UTC 日期的函数和返回 UTC 时间的函数

UTC_DATE()函数返回当前 UTC(世界标准时间)日期值,其格式为 YYYY-MM-DD 或

YYYYMMDD，具体格式取决于函数是否用在字符串或数字语境中。

【例 4.58】使用 UTC_DATE()函数返回当前 UTC 日期值，输入语句如下：

```
mysql> SELECT UTC_DATE(), UTC_DATE() + 0;
+------------+----------------+
| UTC_DATE() | UTC_DATE() + 0 |
+------------+----------------+
| 2022-03-16 |       20220316 |
+------------+----------------+
```

UTC_DATE()函数返回值为当前时区的日期值。

UTC_TIME()返回当前 UTC 时间值，其格式为 HH:MM:SS 或 HHMMSS，具体格式取决于函数是否用在字符串或数字语境中。

【例 4.59】使用 UTC_TIME()函数返回当前 UTC 时间值，输入语句如下：

```
mysql> SELECT UTC_TIME(), UTC_TIME() + 0;
+------------+----------------+
| UTC_TIME() | UTC_TIME() + 0 |
+------------+----------------+
| 10:08:22   |         100822 |
+------------+----------------+
```

UTC_TIME()返回当前时区的时间值。

4.4.5 获取月份的函数 MONTH(date)和 MONTHNAME(date)

MONTH(date)函数返回 date 对应的月份，范围值为 1~12。

【例 4.60】使用 MONTH()函数返回指定日期中的月份，输入语句如下：

```
mysql> SELECT MONTH('2020-02-13');
+---------------------+
| MONTH('2020-02-13') |
+---------------------+
|                   2 |
+---------------------+
```

MONTHNAME(date)函数返回日期 date 对应月份的英文全名。

【例 4.61】使用 MONTHNAME()函数返回指定日期中月份的名称，输入语句如下：

```
mysql> SELECT MONTHNAME('2018-02-13');
+-------------------------+
| MONTHNAME('2018-02-13') |
+-------------------------+
| February                |
+-------------------------+
```

4.4.6 获取星期的函数 DAYNAME(d)、DAYOFWEEK(d)和 WEEKDAY(d)

DAYNAME(d)函数返回参数 d 对应的工作日的英文名称，例如 Sunday、Monday 等。

【例 4.62】使用 DAYNAME()函数返回指定日期的工作日名称，输入语句如下：

```
mysql> SELECT DAYNAME('2018-10-10');
```

```
+------------------------+
| DAYNAME('2018-10-10')  |
+------------------------+
| Wednesday              |
+------------------------+
```

可以看到，2018 年 10 月 10 日是星期三，因此返回结果为 Wednesday。

DAYOFWEEK(d)函数返回参数 d 对应的一周中的索引（位置，1 表示周日，2 表示周一，…，7 表示周六）。

【例 4.63】使用 DAYOFWEEK()函数返回日期对应的周索引，输入语句如下：

```
mysql> SELECT DAYOFWEEK('2018-10-10');
+-------------------------+
| DAYOFWEEK('2018-10-10') |
+-------------------------+
|                       4 |
+-------------------------+
```

由【例 4.63】可知，2018 年 10 月 10 日为周三，因此返回其对应的索引值，结果为 4。

WEEKDAY(d)返回参数 d 对应的工作日索引：0 表示周一，1 表示周二，…，6 表示周日。

【例 4.64】使用 WEEKDAY()函数返回日期对应的工作日索引，输入语句如下：

```
mysql>SELECT WEEKDAY('2018-10-10 22:23:00'), WEEKDAY('2018-11-11');
+--------------------------------+-----------------------+
| WEEKDAY('2018-10-10 22:23:00') | WEEKDAY('2018-11-11') |
+--------------------------------+-----------------------+
|                              2 |                     6 |
+--------------------------------+-----------------------+
```

可以看到，WEEKDAY()和 DAYOFWEEK()函数都是返回指定日期在某一周内的位置，只是索引编号不同。

4.4.7 获取星期数的函数 WEEK(d)和 WEEKOFYEAR(d)

WEEK(d)计算日期 d 是一年中的第几周。WEEK()的双参数形式允许指定该星期是否起始于周日或周一，以及返回值的范围是否为 0~53 或 1~53。若 Mode 参数被省略，则使用 default_week_format 系统自变量的值，可参考表 4.1。

表 4.1 WEEK(d)函数中 Mode 参数取值

Mode	一周的第一天	范围	Week 1 为第一周...
0	周日	0~53	本年度中有一个周日
1	周一	0~53	本年度中有 3 天以上
2	周日	1~53	本年度中有一个周日
3	周一	1~53	本年度中有 3 天以上
4	周日	0~53	本年度中有 3 天以上
5	周一	0~53	本年度中有一个周一
6	周日	1~53	本年度中有 3 天以上
7	周一	1~53	本年度中有一个周一

【例 4.65】使用 WEEK()函数查询指定日期是一年中的第几周，输入语句如下：

```
mysql>SELECT WEEK('2018-02-20'),WEEK('2018-02-20',0), WEEK('2018-02-20',1);
+--------------------+----------------------+----------------------+
| WEEK('2018-02-20') | WEEK('2018-02-20',0) | WEEK('2018-02-20',1) |
+--------------------+----------------------+----------------------+
|                  7 |                    7 |                    8 |
+--------------------+----------------------+----------------------+
```

可以看到，WEEK('2018-02-20')使用一个参数，其第二个参数为 default_week_format 默认值，MySQL 中该值默认为 0，指定一周的第一天为周日，因此和 WEEK('2018-02-20',0)返回结果相同；WEEK('2018-02-20',1)中第二个参数为 1，指定一周的第一天为周一，返回值为第 8 周。可以看到，第二个参数不同，返回的结果也不同，使用不同的参数的原因是不同地区和国家的习惯不同，每周的第一天并不相同。

WEEKOFYEAR(d)计算某天位于一年中的第几周，范围是 1~53，相当于 WEEK(d,3)。

【例 4.66】使用 WEEKOFYEAR()查询指定日期是一年中的第几周，输入语句如下：

```
mysql> SELECT WEEK('2018-01-20',3), WEEKOFYEAR('2018-01-20');
+----------------------+--------------------------+
| WEEK('2018-01-20',3) | WEEKOFYEAR('2018-01-20') |
+----------------------+--------------------------+
|                    3 |                        3 |
+----------------------+--------------------------+
```

可以看到，两个函数返回结果相同。

4.4.8　获取天数的函数 DAYOFYEAR(d)和 DAYOFMONTH(d)

DAYOFYEAR(d)函数返回日期 d 是一年中的第几天，范围是 1~366。

【例 4.67】使用 DAYOFYEAR()函数返回指定日期在一年中的位置，输入语句如下：

```
mysql> SELECT DAYOFYEAR('2018-02-20');
+-------------------------+
| DAYOFYEAR('2018-02-20') |
+-------------------------+
|                      51 |
+-------------------------+
```

1 月份 31 天，再加上 2 月份的 20 天，因此返回结果为 51。

DAYOFMONTH(d)函数返回日期 d 是一个月中的第几天，范围是 1~31。

【例 4.68】使用 DAYOFMONTH()函数返回指定日期在一个月中的位置，输入语句如下：

```
mysql> SELECT DAYOFMONTH('2018-08-20');
+--------------------------+
| DAYOFMONTH('2018-08-20') |
+--------------------------+
|                       20 |
+--------------------------+
```

结果显而易见。

4.4.9 获取年份、季度、小时、分钟和秒钟的函数

YEAR(date) 返回日期 date 对应的年份，范围是 1970~2069。

【例 4.69】 使用 YEAR()函数返回指定日期对应的年份，输入语句如下：

```
mysql>SELECT YEAR('18-02-03'),YEAR('96-02-03');
+------------------+------------------+
| YEAR('18-02-03') | YEAR('96-02-03') |
+------------------+------------------+
|       2018       |       1996       |
+------------------+------------------+
```

提示："00~69"转换为"2000~2069"，"70~99"转换为"1970~1999"。

QUARTER(date)返回日期 date 对应的一年中的季度值，范围是 1~4。

【例 4.70】 使用 QUARTER()函数返回指定日期对应的季度，输入语句如下：

```
mysql> SELECT QUARTER('18-04-01');
+---------------------+
| QUARTER('18-04-01') |
+---------------------+
|          2          |
+---------------------+
```

MINUTE(time)返回 time 对应的分钟数，范围是 0~59。

【例 4.71】 使用 MINUTE()函数返回指定时间的分钟值，输入语句如下：

```
mysql> SELECT MINUTE('18-02-03 10:10:03');
+-----------------------------+
| MINUTE('18-02-03 10:10:03') |
+-----------------------------+
|             10              |
+-----------------------------+
```

SECOND(time)返回 time 对应的秒数，范围是 0~59。

【例 4.72】 使用 SECOND()函数返回指定时间的秒值，输入语句如下：

```
mysql> SELECT SECOND('10:05:03');
+--------------------+
| SECOND('10:05:03') |
+--------------------+
|         3          |
+--------------------+
```

4.4.10 获取日期的指定值的函数 EXTRACT(type FROM date)

EXTRACT(type FROM date)函数所使用的时间间隔类型说明符与 DATE_ADD()或 DATE_SUB()使用的相同，但它从日期中提取一部分，而不是执行日期运算。

【例 4.73】使用 EXTRACT 函数提取日期或者时间值，输入语句如下：

```
mysql> SELECT EXTRACT(YEAR FROM '2018-07-02') AS col1, EXTRACT(YEAR_MONTH FROM '2018-07-12
01:02:03') AS col2, EXTRACT(DAY_MINUTE FROM '2018-07-12 01:02:03') AS col3;
+------+--------+--------+
| col1 | col2   | col3   |
+------+--------+--------+
| 2018 | 201807 | 120102 |
+------+--------+--------+
```

type 值为 YEAR 时，只返回年值，结果为 2018；type 值为 YEAR_MONTH 时返回年与月份，结果为 201807；type 值为 DAY_MINUTE 时，返回日、小时和分钟值，结果为 120102。

4.4.11 时间和秒钟转换的函数

TIME_TO_SEC(time)返回已转化为秒的 time 参数。转换公式为：小时×3600+分钟×60+秒。

【例 4.74】使用 TIME_TO_SEC 函数将时间值转换为秒值，输入语句如下：

```
mysql> SELECT TIME_TO_SEC('23:23:00');
+-------------------------+
| TIME_TO_SEC('23:23:00') |
+-------------------------+
|                   84180 |
+-------------------------+
```

SEC_TO_TIME(seconds)函数返回被转化为小时、分钟和秒数的 seconds 参数值，其格式为 HH:MM:SS 或 HHMMSS，具体格式根据该函数用在字符串或数字的语境而定。

【例 4.75】使用 SEC_TO_TIME()函数将秒值转换为时间格式，输入语句如下：

```
mysql> SELECT SEC_TO_TIME(2345),SEC_TO_TIME(2345)+0, TIME_TO_SEC('23:23:00'),
SEC_TO_TIME(84180);
+-------------------+---------------------+-------------------------+--------------------+
| SEC_TO_TIME(2345) | SEC_TO_TIME(2345)+0 | TIME_TO_SEC('23:23:00') | SEC_TO_TIME(84180) |
+-------------------+---------------------+-------------------------+--------------------+
| 00:39:05          |                3905 |                   84180 | 23:23:00           |
+-------------------+---------------------+-------------------------+--------------------+
```

可以看到，SEC_TO_TIME 函数返回值加上 0 值之后变成了小数值；TIME_TO_SEC 正好和 SEC_TO_TIME 互为反函数。

4.4.12 计算日期和时间的函数

计算日期和时间的函数有 DATE_ADD()、ADDDATE()、DATE_SUB()、SUBDATE()、ADDTIME()、SUBTIME()和 DATE_DIFF()。

在 DATE_ADD(date,INTERVAL expr type)和 DATE_SUB(date,INTERVAL expr type)中，参数 date 是一个 DATETIME 或 DATE 值，用来指定起始时间。expr 是一个表达式，用来指定从起始日期添加或减去的时间间隔值。对于负值的时间间隔，expr 以一个负号"－"开头。type 为关键词，指示了表达式被解释的方式。表 4.2 显示了 type 和 expr 参数的关系。

表 4.2　MySQL 中计算日期和时间的格式

type 值	预期的 expr 格式
MICROSECOND	MICROSECONDS
SECOND	SECONDS
MINUTE	MINUTES
HOUR	HOURS
DAY	DAYS
WEEK	WEEKS
MONTH	MONTHS
QUARTER	QUARTERS
YEAR	YEARS
SECOND_MICROSECOND	'SECONDS.MICROSECONDS'
MINUTE_MICROSECOND	'MINUTES.MICROSECONDS'
MINUTE_SECOND	'MINUTES:SECONDS'
HOUR_MICROSECOND	'HOURS.MICROSECONDS'
HOUR_SECOND	'HOURS:MINUTES:SECONDS'
HOUR_MINUTE	'HOURS:MINUTES'
DAY_MICROSECOND	'DAYS.MICROSECONDS'
DAY_SECOND	'DAYS HOURS:MINUTES:SECONDS'
DAY_MINUTE	'DAYS HOURS:MINUTES'
DAY_HOUR	'DAYS HOURS'
YEAR_MONTH	'YEARS-MONTHS'

若 date 参数是一个 DATE 值，计算只会包括 YEAR、MONTH 和 DAY 部分（没有时间部分），其结果是一个 DATE 值；否则，结果将是一个 DATETIME 值。

DATE_ADD(date,INTERVAL expr type)和 ADDDATE(date,INTERVAL expr type)两个函数的作用相同，执行日期的加运算。

【例 4.76】使用 DATE_ADD()和 ADDDATE()函数执行日期的加操作，输入语句如下：

```
mysql> SELECT DATE_ADD('2010-12-31 23:59:59', INTERVAL 1 SECOND) AS col1, ADDDATE('2010-12-31 23:59:59', INTERVAL 1 SECOND) AS col2, DATE_ADD('2010-12-31 23:59:59', INTERVAL '1:1' MINUTE_SECOND) AS col3;
+---------------------+---------------------+---------------------+
| col1                | col2                | col3                |
+---------------------+---------------------+---------------------+
| 2011-01-01 00:00:00 | 2011-01-01 00:00:00 | 2011-01-01 00:01:00 |
+---------------------+---------------------+---------------------+
```

由结果可以看到，DATE_ADD('2010-12-31 23:59:59', INTERVAL 1 SECOND)和 ADDDATE('2010-12-31 23:59:59', INTERVAL 1 SECOND)两个函数执行的结果是相同的，将时间增加 1 秒后返回，结果都为"2011-01-01 00:00:00"；DATE_ADD('2010-12-31 23:59:59', INTERVAL '1:1' MINUTE_SECOND)日期运算类型是 MINUTE_SECOND，将指定时间增加 1 分 1 秒后返回，结果为"2011-01-01 00:01:00"。

DATE_SUB(date,INTERVAL expr type)或者 SUBDATE(date,INTERVAL expr type)两个函数的作用相同，执行日期的减运算。

【例4.77】使用 DATE_SUB 和 SUBDATE 函数执行日期的减操作，输入语句如下：

```
mysql> SELECT DATE_SUB('2011-01-02', INTERVAL 31 DAY) AS col1, SUBDATE('2011-01-02', INTERVAL
31 DAY) AS col2, DATE_SUB('2011-01-01 00:01:00',INTERVAL '0 0:1:1' DAY_SECOND) AS col3;
+------------+------------+---------------------+
| col1       | col2       | col3                |
+------------+------------+---------------------+
| 2010-12-02 | 2010-12-02 | 2010-12-31 23:59:59 |
+------------+------------+---------------------+
```

由结果可以看到，DATE_SUB('2011-01-02', INTERVAL 31 DAY)和 SUBDATE('2011-01-02', INTERVAL 31 DAY)两个函数执行的结果是相同的，将日期值减少 31 天后返回，结果都为"2010-12-02"；DATE_SUB('2011-01-01 00:01:00',INTERVAL '0 0:1:1' DAY_SECOND)函数将指定日期减少 1 天，时间减少 1 分 1 秒后返回，结果为"2010-12-31 23:59:59"。

提示：DATE_ADD 和 DATE_SUB 在指定修改的时间段时，也可以指定负值，负值代表相减，即返回以前的日期和时间。

ADDTIME(date,expr)函数将 expr 值添加到 date，并返回修改后的值，date 是一个日期或者日期时间表达式，而 expr 是一个时间表达式。

【例4.78】使用 ADDTIME 进行时间的加操作，输入语句如下：

```
mysql>SELECT ADDTIME('2000-12-31 23:59:59','1:1:1'), ADDTIME('02:02:02','02:00:00');
+----------------------------------------+--------------------------------+
| ADDTIME('2000-12-31 23:59:59','1:1:1') | ADDTIME('02:02:02','02:00:00') |
+----------------------------------------+--------------------------------+
| 2001-01-01 01:01:00                    | 04:02:02                       |
+----------------------------------------+--------------------------------+
```

可以看到，将"2000-12-31 23:59:59"的时间部分值增加 1 小时 1 分钟 1 秒后的日期变为"2001-01-01 01:01:00"；"02:02:02"增加两小时后的时间为"04:02:02"。

SUBTIME(date,expr)函数将 date 减去 expr 值，并返回修改后的值。其中，date 是一个日期或者日期时间表达式，而 expr 是一个时间表达式。

【例4.79】使用 SUBTIME()函数执行时间的减操作，输入语句如下：

```
mysql> SELECT SUBTIME('2000-12-31 23:59:59','1:1:1'), SUBTIME('02:02:02','02:00:00');
+----------------------------------------+--------------------------------+
| SUBTIME('2000-12-31 23:59:59','1:1:1') | SUBTIME('02:02:02','02:00:00') |
+----------------------------------------+--------------------------------+
| 2000-12-31 22:58:58                    | 00:02:02                       |
+----------------------------------------+--------------------------------+
```

可以看到，将"2000-12-31 23:59:59"的时间部分值减少 1 小时 1 分钟 1 秒后的日期变为"2000-12-31 22:58:58"；"02:02:02"减少两小时的时间为"00:02:02"。

DATEDIFF(date1,date2)函数返回起始时间 date1 和结束时间 date2 之间的天数。date1 和 date2 为日期或日期时间表达式，计算中只用到这些值的日期部分。

【例4.80】使用 DATEDIFF()函数计算两个日期之间的间隔天数，输入语句如下：

```
mysql> SELECT DATEDIFF('2010-12-31 23:59:59','2010-12-30') AS col1, DATEDIFF('2010-11-30
```

```
23:59:59','2010-12-31') AS col2;
+------+------+
| col1 | col2 |
+------+------+
|    1 |  -31 |
+------+------+
```

DATEDIFF()函数返回 date1-date2 后的值,因此 DATEDIFF('2010-12-31 23:59:59','2010-12-30') 返回值为 1;DATEDIFF('2010-11-30 23:59:59','2010-12-31')返回值为-31。

4.4.13 将日期和时间格式化的函数

DATE_FORMAT(date,format)根据 format 指定的格式显示 date 值。主要 format 格式如表 4.3 所示。

表 4.3 DATE_FORMAT 时间日期格式

说明符	说明
%a	工作日的缩写名称(Sun...Sat)
%b	月份的缩写名称(Jan...Dec)
%c	月份,数字形式(0...12)
%D	以英文后缀表示月中的几号(1st, 2nd...)
%d	该月日期,数字形式(00...31)
%e	该月日期,数字形式(0...31)
%f	微秒(000000...999999)
%H	以 2 位数表示 24 小时(00...23)
%h, %I	以 2 位数表示 12 小时(01...12)
%i	分钟,数字形式(00...59)
%j	一年中的天数(001...366)
%k	以 24(0...23)小时表示时间
%l	以 12(1...12)小时表示时间
%M	月份名称(January...December)
%m	月份,数字形式(00...12)
%p	上午(AM)或下午(PM)
%r	时间,12 小时制(小时 hh:分钟 mm:秒数 ss 后加 AM 或 PM)
%S,%s	以 2 位数形式表示秒(00...59)
%T	时间,24 小时制(小时 hh:分钟 mm:秒数 ss)
%U	周(00...53),其中周日为每周的第一天
%u	周(00...53),其中周一为每周的第一天
%V	周(01...53),其中周日为每周的第一天;和 %X 同时使用
%v	周(01...53),其中周一为每周的第一天;和 %x 同时使用
%W	工作日名称(周日...周六)
%w	一周中的每日(0=周日...6=周六)
%X	该周的年份,其中周日为每周的第一天;数字形式,4 位数;和%V 同时使用

(续表)

说明符	说 明
%x	该周的年份,其中周一为每周的第一天;数字形式,4位数;和%v同时使用
%Y	4位数形式表示年份
%y	2位数形式表示年份
%%	标识符%

【例4.81】使用 DATE_FORMAT()函数格式化输出日期和时间值,输入语句如下:

```
mysql> SELECT DATE_FORMAT('1997-10-04 22:23:00', '%W %M %Y') AS col1, DATE_FORMAT('1997-10-04 22:23:00','%D %y %a %d %m %b %j') AS col2;
+----------------------+-------------------------------+
| col1                 | col2                          |
+----------------------+-------------------------------+
| Saturday October 1997| 4th 97 Sat 04 10 Oct 277      |
+----------------------+-------------------------------+

mysql> SELECT DATE_FORMAT('1997-10-04 22:23:00', '%H:%i:%s') AS col3, DATE_FORMAT('1999-01-01', '%X %V') AS col4;
+----------+---------+
| col3     | col4    |
+----------+---------+
| 22:23:00 | 1998 52 |
+----------+---------+
```

可以看到"1997-10-04 22:23:00"分别按照不同参数转换成了不同格式的日期值和时间值。

TIME_FORMAT(time,format)根据表达式 format 的要求显示时间 time。表达式 format 指定了显示的格式。因为 TIME_FORMAT(time,format)只处理时间,所以 format 只使用时间格式。

【例4.82】使用 TIME_FORMAT()函数格式化输入的时间值,输入语句如下:

```
mysql>SELECT TIME_FORMAT('16:00:00', '%H %k %h %I %l');
+-------------------------------------------+
| TIME_FORMAT('16:00:00', '%H %k %h %I %l') |
+-------------------------------------------+
| 16 16 04 04 4                             |
+-------------------------------------------+
```

TIME_FORMAT 只处理时间值,可以看到,"16:00:00"按照不同的格式参数转换为不同格式的时间值。

GET_FORMAT(val_type, format_type)返回日期时间字符串的显示格式,val_type 表示日期数据类型,包括 DATE、DATETIME 和 TIME;format_type 表示格式化显示类型,包括 EUR、INTERVAL、ISO、JIS、USA。GET_FORMAT 根据两个值类型组合返回的字符串显示格式如表4.4所示。

表4.4 GET_FORMAT 返回的格式字符串

值 类 型	格式化类型	显示格式字符串
DATE	EUR	%d.%m.%Y
DATE	INTERVAL	%Y%m%d
DATE	ISO	%Y-%m-%d

(续表)

值 类 型	格式化类型	显示格式字符串
DATE	JIS	%Y-%m-%d
DATE	USA	%m.%d.%Y
TIME	EUR	%H.%i.%s
TIME	INTERVAL	%H%i%s
TIME	ISO	%H:%i:%s
TIME	JIS	%H:%i:%s
TIME	USA	%h:%i:%s %p
DATETIME	EUR	%Y-%m-%d %H.%i.%s
DATETIME	INTERVAL	%Y%m%d%H%i%s
DATETIME	ISO	%Y-%m-%d %H:%i:%s
DATETIME	JIS	%Y-%m-%d %H:%i:%s
DATETIME	USA	%Y-%m-%d %H.%i.%s

【例 4.83】使用 GET_FORMAT()函数显示不同格式化类型下的格式字符串,输入语句如下:

```
mysql> SELECT GET_FORMAT(DATE,'EUR'), GET_FORMAT(DATE,'USA');
+------------------------+------------------------+
| GET_FORMAT(DATE,'EUR') | GET_FORMAT(DATE,'USA') |
+------------------------+------------------------+
| %d.%m.%Y               | %m.%d.%Y               |
+------------------------+------------------------+
```

可以看到,不同类型的格式化字符串并不相同。

【例 4.84】在 DATE_FORMAT()函数中,使用 GET_FORMAT 函数返回的格式字符串来显示指定的日期值,输入语句如下:

```
mysql> SELECT DATE_FORMAT('2000-10-05 22:23:00', GET_FORMAT(DATE,'USA') );
+-------------------------------------------------------------+
| DATE_FORMAT('2000-10-05 22:23:00', GET_FORMAT(DATE,'USA') ) |
+-------------------------------------------------------------+
| 10.05.2000                                                  |
+-------------------------------------------------------------+
```

GET_FORMAT(DATE,'USA')返回的格式字符串为%m.%d.%Y,对照表 4.3 中的显示格式,%m 以数字形式显示月份,%d 以数字形式显示日,%Y 以 4 位数字形式显示年,因此结果为 10.05.2000。

4.5 条件判断函数

条件判断函数也称为控制流程函数,根据满足的不同条件,执行相应的流程。MySQL 中进行条件判断的函数有 IF、IFNULL 和 CASE。本节将分别介绍各个条件判断函数的用法。

4.5.1 IF(expr,v1,v2)函数

对于 IF(expr, v1, v2)函数,如果表达式 expr 是 TRUE(expr <> 0 and expr <> NULL),则返回

值为 v1，否则返回值为 v2。IF()的返回值为数字值或字符串值，具体情况视其所在的语境而定。

【例 4.85】使用 IF()函数进行条件判断，输入语句如下：

```
mysql> SELECT IF(1>2,2,3), IF(1<2,'yes ','no'), IF(STRCMP('test','test1'),'no','yes');
+-------------+----------------------+----------------------------------------+
| IF(1>2,2,3) | IF(1<2,'yes ','no')  | IF(STRCMP('test','test1'),'no','yes')  |
+-------------+----------------------+----------------------------------------+
|           3 | yes                  | no                                     |
+-------------+----------------------+----------------------------------------+
```

1>2 的结果为 FALSE，IF(1>2,2,3)返回第 2 个表达式的值；1<2 的结果为 TRUE，IF(1<2,'yes ','no')返回第一个表达式的值；"test"小于"test1"，结果为 true，故 IF(STRCMP('test','test1'),'no','yes')返回第一个表达式的值。

提示：如果 v1 或 v2 中只有一个明确是 NULL，则 IF()函数的结果类型为非 NULL 表达式的结果类型。

4.5.2　IFNULL(v1,v2)函数

对于 IFNULL(v1,v2)函数，假如 v1 不为 NULL，则 IFNULL()的返回值为 v1，否则其返回值为 v2。IFNULL()的返回值是数字或者字符串，具体情况取决于其所在的语境。

【例 4.86】使用 IFNULL()函数进行条件判断，输入语句如下：

```
mysql> SELECT IFNULL(1,2), IFNULL(NULL,10), IFNULL(1/0,'wrong');
+-------------+-----------------+---------------------+
| IFNULL(1,2) | IFNULL(NULL,10) | IFNULL(1/0,'wrong') |
+-------------+-----------------+---------------------+
|           1 |              10 | wrong               |
+-------------+-----------------+---------------------+
```

IFNULL(1,2)虽然第二个值也不为空，但返回结果依然是第一个值；IFNULL(NULL,10)第一个值为空，因此返回 10；"1/0"的结果为空，因此 IFNULL(1/0,'wrong')返回字符串"wrong"。

4.5.3　CASE 函数

CASE 函数的格式是 CASE expr WHEN v1 THEN r1 [WHEN v2 THEN r2]…[ELSE rn+1] END，如果 expr 值等于某个 vn，则返回对应位置 THEN 后面的结果；如果与所有值都不相等，则返回 ELSE 后面的 rn+1。

【例 4.87】使用 CASE value WHEN 语句执行分支操作，输入语句如下：

```
mysql> SELECT CASE 2 WHEN 1 THEN 'one' WHEN 2 THEN 'two' ELSE 'more' END;
+------------------------------------------------------------+
| CASE 2 WHEN 1 THEN 'one' WHEN 2 THEN 'two' ELSE 'more' END |
+------------------------------------------------------------+
| two                                                        |
+------------------------------------------------------------+
```

CASE 后面的值为 2，与第二条分支语句 WHEN 后面的值相等，因此返回结果为"two"。

对于 CASE WHEN v1 THEN r1 [WHEN v2 THEN r2]… ELSE rn+1] END，某个 vn 值为 TRUE 时，返回对应位置 THEN 后面的结果；如果所有值都不为 TRUE，则返回 ELSE 后的 rn+1。

【例 4.88】使用 CASE WHEN 语句执行分支操作，输入语句如下：

```
mysql> SELECT CASE WHEN 1<0 THEN 'true' ELSE 'false' END;
+--------------------------------------------+
| CASE WHEN 1<0 THEN 'true' ELSE 'false' END |
+--------------------------------------------+
| false                                      |
+--------------------------------------------+
```

1<0 的结果为 FALSE，因此函数返回值为 ELSE 后面的"false"。

提示：一个 CASE 表达式的默认返回值类型是任何返回值的相容集合类型，但具体情况视其所在的语境而定。如果用在字符串语境中，则返回结果为字符串。如果用在数字语境中，则返回结果为十进制值、实数值或整数值。

4.6 系统信息函数

本节将介绍常用的系统信息函数。MySQL 中的系统信息有数据库的版本号、当前用户名和连接数、系统字符集、最后一个自动生成的 ID 值等。

4.6.1 获取 MySQL 版本号、连接数和数据库名的函数

VERSION()返回表示 MySQL 服务器版本的字符串，这个字符串使用 utf8 字符集。

【例 4.89】查看当前 MySQL 版本号，输入语句如下：

```
mysql> SELECT VERSION();
+-----------+
| VERSION() |
+-----------+
| 8.0.2     |
+-----------+
```

CONNECTION_ID()返回 MySQL 服务器当前连接的次数，每个连接都有各自唯一的 ID。

【例 4.90】查看当前用户的连接数，输入语句如下：

```
mysql> SELECT CONNECTION_ID();
+-----------------+
| CONNECTION_ID() |
+-----------------+
|              23 |
+-----------------+
```

这里返回 23，返回值根据登录的次数会有所不同。

```
SHOW PROCESSLIST;
SHOW FULL PROCESSLIST;
```

processlist 命令的输出结果显示了有哪些线程在运行，不仅可以查看当前所有的连接数，还可以查看当前的连接状态、帮助识别出有问题的查询语句等。

如果是 root 账号，能看到所有用户的当前连接；如果是其他普通账号，则只能看到自己占用的连接。show processlist 只列出前 100 条，如果想全部列出，可使用 show full processlist 命令。

【例 4.91】使用 SHOW PROCESSLIST 命令输出当前用户的连接信息，输入语句如下：

```
MySQL> SHOW PROCESSLIST;
+----+-----------------+-----------------+------+---------+-------+-------------------------+-----------------+
|Id  |User             |Host             |db    |Command  |Time   |State                    |Info             |
+----+-----------------+-----------------+------+---------+-------+-------------------------+-----------------+
|4   |event_scheduler  |localhost        |NULL  |Daemon   |15274  |Waiting on empty queue   |NULL             |
|23  |root             |localhost:58788  |NULL  |Query    |0      |starting                 |SHOW PROCESSLIST |
+----+-----------------+-----------------+------+---------+-------+-------------------------+-----------------+
```

各个列的含义和用途：

（1）Id 列，用户登录 MySQL 时，系统分配的是"connection id"。

（2）User 列，显示当前用户。如果不是 root，这个命令就只显示用户权限范围内的 SQL 语句。

（3）Host 列，显示这个语句是从哪个 IP 的哪个端口上发出的，可以用来追踪出现问题语句的用户。

（4）db 列，显示这个进程目前连接的是哪个数据库。

（5）Command 列，显示当前连接执行的命令，一般取值为休眠（Sleep）、查询（Query）、连接（Connect）。

（6）Time 列，显示这个状态持续的时间，单位是秒。

（7）State 列，显示使用当前连接的 SQL 语句的状态，很重要的列。后续会有所有状态的描述，State 只是语句执行中的某一个状态。一个 SQL 语句，以查询为例，可能需要经过 Copying to tmp table、Sorting result、Sending data 等状态才可以完成。

（8）Info 列，显示这个 SQL 语句，是判断问题语句的一个重要依据。

打开另一个命令行窗口登录 MySQL，此时将会有 2 个连接，在第 2 个登录界面的命令行下再次输入 SHOW PROCESSLIST，结果如下：

```
mysql> SHOW PROCESSLIST;
+----+------+----------------+------+---------+------+-------+------------------+
| Id | User | Host           | db   | Command | Time | State | Info             |
+----+------+----------------+------+---------+------+-------+------------------+
| 1  | root | localhost:3602 | NULL | Sleep   | 38   |       | NULL             |
| 2  | root | localhost:3272 | NULL | Query   | 0    | NULL  | show processlist |
+----+------+----------------+------+---------+------+-------+------------------+
```

可以看到，当前活动用户登录的连接 Id 为 2，正在执行的 Command（操作命令）是 Query（查询），使用的查询命令为 SHOW PROCESSLIST；而连接 Id 为 1 的用户目前没有对数据进行操作，即处于 Sleep 操作，而且已经过了 38s。

DATABASE()和 SCHEMA()函数返回使用 utf8 字符集的默认（当前）数据库名。

【例 4.92】查看当前使用的数据库，输入语句如下：

```
mysql> SELECT DATABASE(),SCHEMA();
```

```
+------------+------------+
| DATABASE() | SCHEMA()   |
+------------+------------+
| test_db    | test_db    |
+------------+------------+
```

可以看到，两个函数的作用相同。

4.6.2 获取用户名的函数

USER()、CURRENT_USER()、CURRENT_USER()、SYSTEM_USER()和 SESSION_USER()这几个函数返回当前被 MySQL 服务器验证的用户名和主机名组合。这个值符合确定当前登录用户存取权限的 MySQL 账户。一般情况下，这几个函数的返回值是相同的。

【例 4.93】获取当前登录用户名称，输入语句如下：

```
mysql> SELECT USER(), CURRENT_USER(), SYSTEM_USER();
+----------------+----------------+----------------+
| USER()         | CURRENT_USER() | SYSTEM_USER()  |
+----------------+----------------+----------------+
| root@localhost | root@localhost | root@localhost |
+----------------+----------------+----------------+
```

返回结果值表明了当前账户连接服务器时的用户名及所连接的客户主机，root 为当前登录的用户名，localhost 为登录的主机名。

4.6.3 获取字符串的字符集和排序方式的函数

CHARSET(str)返回字符串 str 自变量的字符集。

【例 4.94】使用 CHARSET()函数返回字符串使用的字符集，输入语句如下：

```
mysql> SELECT CHARSET('abc'), CHARSET(CONVERT('abc' USING latin1)), CHARSET(VERSION());
+----------------+--------------------------------------+--------------------+
| CHARSET('abc') | CHARSET(CONVERT('abc' USING latin1)) | CHARSET(VERSION()) |
+----------------+--------------------------------------+--------------------+
| utf8mb4        | latin1                               | utf8mb3            |
+----------------+--------------------------------------+--------------------+
```

CHARSET('abc')返回系统默认的字符集 utf8mb4；CHARSET(CONVERT('abc' USING latin1))返回的字符集为 latin1；前面介绍过，VERSION()返回的字符串使用 utf8 字符集，因此 CHARSET 返回结果为 utf8mb3。

COLLATION(str)返回字符串 str 的字符排列方式。

【例 4.95】使用 COLLATION()函数返回字符串的排列方式，输入语句如下：

```
mysql> SELECT COLLATION('abc'),COLLATION(CONVERT('abc' USING utf8));
+-------------------+--------------------------------------+
| COLLATION('abc')  | COLLATION(CONVERT('abc' USING utf8)) |
+-------------------+--------------------------------------+
| utf8mb4_0900_ai_ci| utf8_general_ci                      |
+-------------------+--------------------------------------+
```

可以看到，使用不同字符集时字符串的排列方式不同。

4.6.4 获取最后一个自动生成的 ID 值的函数

LAST_INSERT_ID()函数返回最后生成的 AUTO_INCREMENT 值。

【例 4.96】使用 SELECT LAST_INSERT_ID 查看最后一个自动生成的列值。

（1）一次插入一条记录

首先创建表 worker，其 Id 字段带有 AUTO_INCREMENT 约束，输入语句如下：

```
mysql> CREATE TABLE worker(Id INT AUTO_INCREMENT NOT NULL PRIMARY KEY, Name VARCHAR(30));
```

分别单独向表 worker 中插入两条记录：

```
mysql> INSERT INTO worker VALUES(NULL, 'jimy');
mysql> INSERT INTO worker VALUES(NULL, 'Tom');
mysql> SELECT * FROM worker;
+----+------+
| Id | Name |
+----+------+
|  1 | jimy |
|  2 | Tom  |
+----+------+
```

查看已经插入的数据可以发现，最后一条插入的记录的 Id 字段值为 2，使用 LAST_INSERT_ID() 查看最后自动生成的 Id 值：

```
mysql> SELECT LAST_INSERT_ID();
+------------------+
| LAST_INSERT_ID() |
+------------------+
|                2 |
+------------------+
```

可以看到，一次插入一条记录时，返回值为最后一条插入记录的 Id 值。

（2）一次同时插入多条记录

接下来，向表中插入多条记录，输入语句如下：

```
mysql> INSERT INTO worker VALUES (NULL,'Kevin'),(NULL,'Michal'),(NULL,'Nick');
```

查询已经插入的记录：

```
mysql> SELECT * FROM worker;
+----+--------+
| Id | Name   |
+----+--------+
|  1 | jimy   |
|  2 | Tom    |
|  3 | Kevin  |
|  4 | Michal |
|  5 | Nick   |
+----+--------+
```

可以看到最后一条记录的 Id 字段值为 5，使用 LAST_INSERT_ID()查看最后自动生成的 Id 值：

```
mysql> SELECT LAST_INSERT_ID();
+------------------+
| LAST_INSERT_ID() |
+------------------+
|                3 |
+------------------+
```

结果显示，LAST_INSERT_ID 值不是 5 而是 3，这是为什么呢？在向数据表中插入一条新记录时，LAST_INSERT_ID()返回带有 AUTO_INCREMENT 约束的字段最新生成的值 2；继续向表中同时添加 3 条记录，读者可能以为这时 LAST_INSERT_ID 值为 5，可显示结果却为 3，这是因为当使用一条 INSERT 语句插入多行时，LAST_INSERT_ID()只返回插入的第一行数据时产生的值，在这里为第 3 条记录。之所以这样，是因为这使依靠其他服务器复制同样的 INSERT 语句变得简单。

提示：LAST_INSERT_ID 是与数据表无关的，如果向表 a 插入数据后再向表 b 插入数据，那么 LAST_INSERT_ID 返回表 b 中的 Id 值。

4.7 加密函数

加密函数主要用来对数据进行加密处理，以保证某些重要数据不被别人获取。这些函数在保证数据库安全时非常有用。本节将介绍各种加密函数的作用和使用方法。

4.7.1 加密函数 MD5(str)

MD5(str)为字符串算出一个 MD5 128 比特校验和。该值以 32 位十六进制数字的二进制字符串形式返回，若参数为 NULL，则会返回 NULL。

【例 4.97】使用 MD5 函数加密字符串，输入语句如下：

```
mysql> SELECT MD5 ('mypwd');
+----------------------------------+
| MD5 ('mypwd')                    |
+----------------------------------+
| 318bcb4be908d0da6448a0db76908d78 |
+----------------------------------+
```

可以看到，"mypwd"经 MD5 加密后的结果为 318bcb4be908d0da6448a0db76908d78。

4.7.2 加密函数 SHA(str)

SHA(str)使用原明文密码 str 计算并返回加密后的密码字符串，当参数为 NULL 时，返回 NULL。SHA 加密算法比 MD5 更加安全。

【例 4.98】使用 SHA 函数加密密码，输入语句如下：

```
mysql> SELECT SHA('tom123456');
```

```
| SHA('tom123456')                         |
+------------------------------------------+
| 8218b487f490cb484f45c31403eb1f597a2b531a |
```

4.7.3 加密函数 SHA2(str, hash_length)

SHA2(str, hash_length)使用 hash_length 作为长度，加密 str。hash_length 支持的值为 224、256、384、512 和 0。其中，0 等同于 256。

【例 4.99】使用 SHA2 加密字符串，输入语句如下：

```
mysql> SELECT SHA2('tom123456',0) A,sha2('tom123456',256) B\G
*************************** 1. row ***************************
A: 9242a986a9edbd14a60450e9284a372efeff7e9f6209f675fdc4457f55de5e27
B: 9242a986a9edbd14a60450e9284a372efeff7e9f6209f675fdc4457f55de5e27
```

可以看到，hash_length 的值为 256 和 0 时，结果都是一样的。

4.8 其他函数

本节将要介绍的函数不能笼统地归为一类，但是这些函数也非常有用，例如重复指定操作函数、改变字符集函数、IP 地址与数字转换函数等。本节将介绍这些函数的作用和用法。

4.8.1 格式化函数 FORMAT(x,n)

FORMAT(x,n)将数字 x 格式化，并以四舍五入的方式保留小数点后 n 位，结果以字符串的形式返回。若 n 为 0，则返回结果函数不含小数部分。

【例 4.100】使用 FORMAT 函数格式化数字，保留小数点位数为指定值，输入语句如下：

```
MySQL> SELECT FORMAT(12332.123456, 4), FORMAT(12332.1,4), FORMAT(12332.2,0);
+-------------------------+-------------------+-------------------+
| FORMAT(12332.123456, 4) | FORMAT(12332.1,4) | FORMAT(12332.2,0) |
+-------------------------+-------------------+-------------------+
| 12,332.1235             | 12,332.1000       | 12,332            |
+-------------------------+-------------------+-------------------+
```

由结果可以看到，FORMAT(12332.123456, 4)保留 4 位小数点后的数字，并进行四舍五入，结果为 12332.1235；FORMAT(12332.1,4)保留 4 位小数值，位数不够的用 0 补齐；FORMAT(12332.2,0)不保留小数位值，返回结果为整数 12332。

4.8.2 不同进制的数字进行转换的函数

CONV(N, from_base, to_base)函数进行不同进制数之间的转换，返回值为数值 N 的字符串表示，由 from_base 进制转化为 to_base 进制。如有任意一个参数为 NULL，则返回值为 NULL。自变量 N 被理解为一个整数，但是可以被指定为一个整数或字符串。进制数最小基数为 2，最大基数为 36。

【例 4.101】 使用 CONV 函数在不同进制数值之间转换，输入语句如下：

```
mysql> SELECT CONV('a',16,2), CONV(15,10,2), CONV(15,10,8), CONV(15,10,16);
```

CONV('a',16,2)	CONV(15,10,2)	CONV(15,10,8)	CONV(15,10,16)
1010	1111	17	F

CONV('a',16,2) 将十六进制的 a 转换为二进制表示的数值，十六进制的 a 表示十进制的数值 10，二进制的数值 1010 正好等于十进制的数值 10；CONV(15,10,2)将十进制的数值 15 转换为二进制值，结果为 1111；CONV(15,10,8)将十进制的数值 15 转换为八进制值，结果为 17；CONV(15,10,16)将十进制的数值 15 转换为十六进制值，结果为 F。

进制说明：

- 二进制，采用 0 和 1 两个数字来表示的数。它以 2 为基数，逢二进一。
- 八进制，采用 0、1、2、3、4、5、6、7 八个数字，以数字 0 开头，逢八进一。
- 十进制，采用 0~9，共 10 个数字表示，逢十进一。
- 十六进制，由 0~9、A~F 组成，以数字 0x 开头。它与十进制的对应关系是：0~9 对应 0~9，A~F 对应 10~15。

4.8.3 IP 地址与数字相互转换的函数

INET_ATON(expr)给出一个作为字符串的网络地址的点地址表示，返回一个代表该地址数值的整数。地址可以是 4 或 8bit 地址。

【例 4.102】 使用 INET_ATON 函数将字符串网络点地址转换为数值网络地址，输入语句如下：

```
mysql> SELECT INET_ATON('209.207.224.40');
```

INET_ATON('209.207.224.40')
3520061480

INET_NTOA(expr)函数给定一个数字网络地址（4 或 8bit），返回作为字符串的该地址的点地址表示。

【例 4.103】 使用 INET_NTOA 函数将数值网络地址转换为字符串网络点地址，输入语句如下：

```
mysql> SELECT INET_NTOA(3520061480);
```

INET_NTOA(3520061480)
209.207.224.40

可以看到，INET_NTOA 和 INET_ATON 互为反函数。

4.8.4 加锁函数和解锁函数

GET_LOCK(str,timeout)函数设法使用字符串 str 给定的名字得到一个锁,超时为 timeout 秒。若成功得到锁,则返回 1;若操作超时,则返回 0;若发生错误,则返回 NULL。假如有一个用 GET_LOCK() 得到的锁,当执行 RELEASE_LOCK()函数或连接断开(正常或非正常)时,这个锁就会解除。

RELEASE_LOCK(str)函数解开被 GET_LOCK()获取的、用字符串 str 所命名的锁。若锁被解开,则返回 1;若该线程尚未创建锁,则返回 0(此时锁没有被解开);若命名的锁不存在,则返回 NULL。若该锁从未被 GET_LOCK()的调用获取,或锁已经被提前解开,则该锁不存在。

IS_FREE_LOCK(str)函数检查名为 str 的锁是否可以使用(换言之,没有被封锁)。若锁可以使用,则返回 1(没有人在用这个锁);若这个锁正在被使用,则返回 0;出现错误,则返回 NULL(诸如不正确的参数)。

IS_USED_LOCK(str)函数检查名为 str 的锁是否正在被其他人使用(换言之,被封锁)。若被封锁,则返回使用该锁的客户端的连接标识符(connection ID),否则返回 NULL。

【例 4.104】使用加锁、解锁函数,输入语句如下:

```
mysql> SELECT GET_LOCK('lock1',10) AS GetLock, IS_USED_LOCK('lock1') AS ISUsedLock,
IS_FREE_LOCK('lock1') AS ISFreeLock, RELEASE_LOCK('lock1') AS ReleaseLock;
+---------+------------+------------+-------------+
| GetLock | ISUsedLock | ISFreeLock | ReleaseLock |
+---------+------------+------------+-------------+
|    1    |     23     |     0      |      1      |
+---------+------------+------------+-------------+
```

GET_LOCK('lock1',10)返回结果为 1,说明成功得到了一个名称为"lock1"的锁,持续时间为 10 秒。

IS_USED_LOCK('lock1')返回结果为当前连接 ID,表示名称为"lock1"的锁正在被使用。

IS_FREE_LOCK('lock1')返回结果为 0,说明名称为"lock1"的锁正在被使用。

RELEASE_LOCK('lock1')返回值为 1,说明解锁成功。

4.8.5 重复执行指定操作的函数

BENCHMARK(count,expr)函数重复 count 次执行表达式 expr。它可以用于计算 MySQL 处理表达式的速度。结果值通常为 0(0 只是表示处理过程很快,并不是没有花费时间)。另一个作用是它可以在 MySQL 客户端内部报告语句执行的时间。

【例 4.105】使用 BENCHMARK 重复执行指定函数。

首先,使用 PASSWORD 函数加密密码,输入语句如下:

```
mysql> SELECT SHA('newpwd');
+------------------------------------------+
| SHA('newpwd')                            |
+------------------------------------------+
| c7f005b657906521157aa3fc261afe886d51f792 |
+------------------------------------------+
1 row in set (0.00 sec)
```

可以看到，PASSWORD 执行花费时间为 0.00sec。下面使用 BENCHMARK 函数重复执行 PASSWORD 操作 500000 次：

```
mysql> SELECT BENCHMARK(500000, SHA('newpwd'));
+----------------------------------+
| BENCHMARK(500000, SHA('newpwd')) |
+----------------------------------+
|                                0 |
+----------------------------------+
1 row in set (0.29 sec)
```

由此可以看出，使用 BENCHMARK 执行 500000 次的时间为 0.29 sec，明显比执行一次的时间延长了。

提示：BENCHMARK 报告的时间是客户端经过的时间，而不是在服务器端的 CPU 时间，每次执行后报告的时间并不一定相同。读者可以多次执行该语句进行验证。

4.8.6 改变字符集的函数

带有 USING 参数的 CONVERT(...USING...)函数用来在不同的字符集之间转化数据。

【例 4.106】使用 CONVERT()函数改变字符串的默认字符集，输入语句如下：

```
MySQL> SELECT CHARSET('string'),CHARSET(CONVERT('string' USING latin1));
+-------------------+----------------------------------------+
| CHARSET('string') | CHARSET(CONVERT('string' USING latin1)) |
+-------------------+----------------------------------------+
| utf8mb4           | latin1                                 |
+-------------------+----------------------------------------+
```

字符串"string"的字符集默认为 utf8mb4，通过 CONVERT 将其默认字符集改为 latin1。

4.8.7 改变数据类型的函数

CAST(x , AS type)和 CONVERT(x, type)函数将一个类型的值转换为另一个类型的值，可转换的 type 有 BINARY、CHAR(n)、DATE、TIME、DATETIME、DECIMAL、SIGNED、UNSIGNED。

【例 4.107】使用 CAST 和 CONVERT 函数进行数据类型的转换，SQL 语句如下：

```
mysql> SELECT CAST(100 AS CHAR(2)), CONVERT('2018-10-01 12:12:12',TIME);
+----------------------+-------------------------------------+
| CAST(100 AS CHAR(2)) | CONVERT('2018-10-01 12:12:12',TIME) |
+----------------------+-------------------------------------+
| 10                   | 12:12:12                            |
+----------------------+-------------------------------------+
```

可以看到，CAST(100 AS CHAR(2))将整数数据 100 转换为带有两个显示宽度的字符串类型，结果为"10"；CONVERT('2018-10-01 12:12:12',TIME)将 DATETIME 类型的时间值转换为 TIME 类型，结果为"12:12:12"。

4.9 窗口函数

在 MySQL 8.0 版本之前，没有提供排名函数，所以当需要在查询当中实现排名时，必须手写@变量，使用起来比较麻烦。

在 MySQL 8.0 版本中，新增了一个窗口函数，用它可以实现很多新的查询方式。窗口函数类似于 SUM()、COUNT()这样的集合函数，但它并不会将多行查询结果合并为一行，而是将结果放回多行当中。也就是说，窗口函数是不需要 GROUP BY 的。

下面通过案例来讲解通过窗口函数实现排名效果的方法。

创建公司部门表 branch，包含部门的名称和部门人数两个字段，创建语句如下：

```
mysql> CREATE TABLE branch
       (
       name char(255) NOT NULL,
       brcount INT(11) NOT NULL
       );
mysql> INSERT INTO branch(name,brcount)
       VALUES('branch1',5),
       ('branch2',10),
       ('branch3',8),
       ('branch4',20),
       ('branch5',9);
```

查询数据表 branch 中的数据：

```
mysql> SELECT * FROM branch;
+---------+---------+
| name    | brcount |
+---------+---------+
| branch1 |       5 |
| branch2 |      10 |
| branch3 |       8 |
| branch4 |      20 |
| branch5 |       9 |
+---------+---------+
```

对公司部门人数按从小到大进行排名，可以利用窗口函数来实现：

```
mysql> SELECT *, rank() OVER w1 AS 'rank' FROM branch  window w1 AS (ORDER BY brcount);
+---------+---------+------+
| name    | brcount | rank |
+---------+---------+------+
| branch1 |       5 |    1 |
| branch3 |       8 |    2 |
| branch5 |       9 |    3 |
| branch2 |      10 |    4 |
| branch4 |      20 |    5 |
+---------+---------+------+
```

这里创建了名称为 w1 的窗口函数，规定对 brcount 字段进行排序，然后在 SELECT 子句中对窗口函数 w1 执行 rank()方法，将结果输出为 rank 字段。

需要注意，这里的 window w1 是可选的。例如，在每一行中加入员工的总数，可以这样操作：

```
mysql> SELECT *, SUM(brcount) over() as total_count FROM branch;
+---------+---------+-------------+
| name    | brcount | total_count |
+---------+---------+-------------+
| branch1 |    5    |     52      |
| branch2 |   10    |     52      |
| branch3 |    8    |     52      |
| branch4 |   20    |     52      |
| branch5 |    9    |     52      |
+---------+---------+-------------+
```

可以一次性查询出每个部门的员工人数占总人数的百分比，查询结果如下：

```
mysql> SELECT *,(brcount)/(sum(brcount) over()) AS rate FROM branch;
+---------+---------+--------+
| name    | brcount | rate   |
+---------+---------+--------+
| branch1 |    5    | 0.0962 |
| branch2 |   10    | 0.1923 |
| branch3 |    8    | 0.1538 |
| branch4 |   20    | 0.3846 |
| branch5 |    9    | 0.1731 |
+---------+---------+--------+
```

第 5 章

查询数据

数据库管理系统的一个重要功能就是数据查询,数据查询不应只是简单返回数据库中存储的数据,还应该根据需要对数据进行筛选,以及确定数据以什么样的格式显示。MySQL 提供了功能强大、灵活的语句来实现这些操作。本章将介绍如何使用 SELECT 语句查询数据表中的一列或多列数据、使用集合函数显示查询结果、连接查询、子查询以及使用正则表达式进行查询等。

5.1 基本查询语句

MySQL 从数据表中查询数据的基本语句为 SELECT 语句。SELECT 语句的基本格式是:

```
SELECT
        {* | <字段列表>}
        [
            FROM <表 1>,<表 2>…
            [WHERE <表达式>
            [GROUP BY <group by definition>]
            [HAVING <expression> [{<operator> <expression>}…]]
            [ORDER BY <order by definition>]
            [LIMIT [<offset>,] <row count>]
        ]
SELECT [字段 1,字段 2,…, 字段 n]
FROM [表或视图]
WHERE [查询条件];
```

其中,各条子句的含义如下:

- {*|<字段列表>}:包含星号通配符和字段列表,表示查询的字段。其中,字段列表至少包含一个字段名称,如果要查询多个字段,多个字段之间用逗号隔开,最后一个字段后不加逗号。
- FROM <表 1>,<表 2>…: 表 1 和表 2 表示查询数据的来源,可以是单个或者多个。

- WHERE 子句：是可选项，如果选择该项，将限定查询行必须满足的查询条件。
- GROUP BY <字段>：该子句告诉 MySQL 如何显示查询出来的数据，并按照指定的字段分组。
- [ORDER BY <字段 >]：该子句告诉 MySQL 按什么样的顺序显示查询出来的数据，可以进行的排序有升序（ASC）、降序（DESC）。
- [LIMIT [<offset>,] <row count>]：该子句告诉 MySQL 每次显示查询出来的数据条数。

SELECT 的可选参数比较多，读者可能无法一下完全理解。不要紧，接下来将从最简单的参数开始，一步一步深入学习之后，读者会对各个参数的作用有清晰的认识。

下面以一个例子说明如何使用 SELECT 从单个表中获取数据。

首先定义数据表，输入语句如下：

```
CREATE TABLE fruits
(
f_id char(10) NOT NULL,
s_id INT NOT NULL,
f_name char(255) NOT NULL,
f_price decimal(8,2) NOT NULL,
PRIMARY KEY(f_id)
);
```

为了演示如何使用 SELECT 语句，需要插入如下数据：

```
mysql> INSERT INTO fruits (f_id, s_id, f_name, f_price)
       VALUES('a1', 101,'apple',5.2),
       ('b1',101,'blackberry', 10.2),
       ('bs1',102,'orange', 11.2),
       ('bs2',105,'melon',8.2),
       ('t1',102,'banana', 10.3),
       ('t2',102,'grape', 5.3),
       ('o2',103,'coconut', 9.2),
       ('c0',101,'cherry', 3.2),
       ('a2',103, 'apricot',2.2),
       ('l2',104,'lemon', 6.4),
       ('b2',104,'berry', 7.6),
       ('m1',106,'mango', 15.7),
       ('m2',105,'xbabay', 2.6),
       ('t4',107,'xbababa', 3.6),
       ('m3',105,'xxtt', 11.6),
       ('b5',107,'xxxx', 3.6);
```

使用 SELECT 语句查询 f_id 字段的数据：

```
mysql> SELECT f_id, f_name FROM fruits;
+------+------------+
| f_id | f_name     |
+------+------------+
| a1   | apple      |
| a2   | apricot    |
| b1   | blackberry |
| b2   | berry      |
| b5   | xxxx       |
| bs1  | orange     |
```

```
| bs2    | melon    |
| c0     | cherry   |
| l2     | lemon    |
| m1     | mango    |
| m2     | xbabay   |
| m3     | xxtt     |
| o2     | coconut  |
| t1     | banana   |
| t2     | grape    |
| t4     | xbababa  |
+--------+----------+
```

该语句的执行过程是，SELECT 语句决定了要查询的列值，在这里查询 f_id 和 f_name 两个字段的值。FROM 子句指定了数据的来源，这里指定数据表 fruits，因此返回结果为 fruits 表中 f_id 和 f_name 两个字段下所有的数据。其显示顺序为添加到表中的顺序。

5.2　单表查询

单表查询是指从一张数据表中查询所需的数据。本节将介绍单表查询中各种基本的查询方式，主要有查询所有字段、查询指定字段、查询指定记录、查询空值、多条件的查询、对查询结果进行排序等。

5.2.1　查询所有字段

1. 在 SELECT 语句中使用星号（*）通配符查询所有字段

SELECT 查询记录最简单的形式是从一个表中检索所有记录，实现的方法是使用星号（*）通配符指定查找所有列的名称。语法格式如下：

SELECT * FROM 表名；

【例 5.1】从 fruits 表中检索所有字段的数据，SQL 语句如下：

```
mysql> SELECT * FROM fruits;
+------+------+------------+---------+
| f_id | s_id | f_name     | f_price |
+------+------+------------+---------+
| a1   | 101  | apple      |   5.20  |
| a2   | 103  | apricot    |   2.20  |
| b1   | 101  | blackberry |  10.20  |
| b2   | 104  | berry      |   7.60  |
| b5   | 107  | xxxx       |   3.60  |
| bs1  | 102  | orange     |  11.20  |
| bs2  | 105  | melon      |   8.20  |
| c0   | 101  | cherry     |   3.20  |
| l2   | 104  | lemon      |   6.40  |
| m1   | 106  | mango      |  15.70  |
| m2   | 105  | xbabay     |   2.60  |
| m3   | 105  | xxtt       |  11.60  |
| o2   | 103  | coconut    |   9.20  |
```

```
| t1    | 102   | banana    | 10.30 |
| t2    | 102   | grape     |  5.30 |
| t4    | 107   | xbababa   |  3.60 |
```

可以看到，使用星号（*）通配符时，将返回所有列，列按照定义表时候的顺序显示。

2. 在 SELECT 语句中指定所有字段

下面介绍另外一种查询所有字段值的方法。根据前面 SELECT 语句的格式，SELECT 关键字后面的字段名为将要查找的数据，因此可以将表中所有字段的名称跟在 SELECT 子句后面，如果忘记了字段名称，可以使用 DESC 命令查看表的结构。有时候，表中的字段可能比较多，不一定能记得所有字段的名称，因此该方法会很不方便，不建议使用。例如，查询 fruits 表中的所有数据，SQL 语句也可以书写如下：

```
SELECT f_id, s_id ,f_name, f_price FROM fruits;
```

查询结果与例 5.1 相同。

提示：一般情况下，除非需要使用表中所有的字段数据，否则最好不要使用通配符"*"。使用通配符虽然可以节省输入查询语句的时间，但是获取不需要的列数据通常会降低查询和所使用的应用程序的效率。通配符的优势是，当不知道所需要的列的名称时，可以通过它获取名称。

5.2.2 查询指定字段

1. 查询单个字段

查询表中的某一个字段，语法格式为：

```
SELECT 列名 FROM 表名;
```

【例 5.2】查询 fruits 表中 f_name 列所有的水果名称，SQL 语句如下：

```
SELECT f_name FROM fruits;
```

该语句使用 SELECT 语句从 fruits 表中获取名称为 f_name 字段下的所有水果名称，指定字段的名称紧跟在 SELECT 关键字之后，查询结果如下：

```
mysql> SELECT f_name FROM fruits;
+------------+
| f_name     |
+------------+
| apple      |
| apricot    |
| blackberry |
| berry      |
| xxxx       |
| orange     |
| melon      |
| cherry     |
| lemon      |
| mango      |
| xbabay     |
```

```
| xxtt        |
| coconut     |
| banana      |
| grape       |
| xbababa     |
+-------------+
```

输出结果显示了 fruits 表中 f_name 字段下的所有数据。

2. 查询多个字段

使用 SELECT 语句，可以获取多个字段下的数据，只需要在关键字 SELECT 后面指定要查找的字段的名称，不同字段名称之间用英文逗号（,）分隔开，最后一个字段后面不需要加逗号，语法格式如下：

SELECT 字段名1, 字段名2, ..., 字段名n FROM 表名;

【例 5.3】从 fruits 表中获取 f_name 和 f_price 两列，SQL 语句如下：

SELECT f_name, f_price FROM fruits;

该语句使用 SELECT 语句从 fruits 表中获取名称为 f_name 和 f_price 两个字段下的所有水果名称和价格，两个字段之间用逗号分隔开，查询结果如下：

```
mysql> SELECT f_name, f_price FROM fruits;
+------------+---------+
| f_name     | f_price |
+------------+---------+
| apple      |    5.20 |
| apricot    |    2.20 |
| blackberry |   10.20 |
| berry      |    7.60 |
| xxxx       |    3.60 |
| orange     |   11.20 |
| melon      |    8.20 |
| cherry     |    3.20 |
| lemon      |    6.40 |
| mango      |   15.70 |
| xbabay     |    2.60 |
| xxtt       |   11.60 |
| coconut    |    9.20 |
| banana     |   10.30 |
| grape      |    5.30 |
| xbababa    |    3.60 |
+------------+---------+
```

输出结果显示了 fruits 表中 f_name 和 f_price 两个字段下的所有数据。

提示：MySQL 中的 SQL 语句是不区分大小写的，因此 SELECT 和 select 的作用是相同的，但是，许多开发人员习惯将关键字大写、数据列和表名小写，读者也应该养成一个良好的编程习惯，这样写出来的代码更容易阅读和维护。

5.2.3 查询指定记录

数据库中包含大量的数据，根据特殊要求，可能只需要查询表中的指定数据，即对数据进行过滤。在 SELECT 语句中，通过 WHERE 子句可以对数据进行过滤，语法格式为：

```
SELECT 字段名1,字段名2,…,字段名n
FROM 表名
WHERE 查询条件
```

在 WHERE 子句中，MySQL 提供了一系列的条件判断符，查询结果如表 5.1 所示。

表 5.1 WHERE 条件判断符

操 作 符	说　明
=	相等
<>, !=	不相等
<	小于
<=	小于或等于
>	大于
>=	大于或等于
BETWEEN	位于两值之间

【例 5.4】查询价格为 10.2 元的水果的名称，SQL 语句如下：

```
SELECT f_name, f_price FROM fruits  WHERE f_price = 10.2;
```

该语句使用 SELECT 语句从 fruits 表中获取价格等于 10.2 的水果的数据，从查询结果可以看到，价格是 10.2 的水果的名称是 blackberry，其他的均不满足查询条件，查询结果如下：

```
+------------+---------+
| f_name     | f_price |
+------------+---------+
| blackberry | 10.20   |
+------------+---------+
```

本例采用了简单的相等过滤，查询一个指定列 f_price 具有值 10.20。相等还可以用来比较字符串，下面给出一个例子：

【例 5.5】查找名称为 "apple" 的水果的价格，SQL 语句如下：

```
SELECT f_name, f_price FROM fruits WHERE f_name = 'apple';
```

该语句使用 SELECT 命令从 fruits 表中获取名称为 "apple" 的水果的价格，从查询结果可以看到只有名称为 "apple" 行被返回，其他的均不满足查询条件。

```
+--------+---------+
| f_name | f_price |
+--------+---------+
| apple  | 5.20    |
+--------+---------+
```

【例 5.6】查询价格小于 10 的水果的名称，SQL 语句如下：

```
SELECT f_name, f_price FROM fruits WHERE f_price < 10;
```

该语句使用 SELECT 命令从 fruits 表中获取价格低于 10 的水果名称，即 f_price 小于 10 的水果信息被返回，查询结果如下：

```
+---------+---------+
| f_name  | f_price |
+---------+---------+
| apple   |    5.20 |
| apricot |    2.20 |
| berry   |    7.60 |
| xxxx    |    3.60 |
| melon   |    8.20 |
| cherry  |    3.20 |
| lemon   |    6.40 |
| xbabay  |    2.60 |
| coconut |    9.20 |
| grape   |    5.30 |
| xbababa |    3.60 |
+---------+---------+
```

可以看到查询结果中所有记录的 f_price 字段的值均小于 10.00 元，而大于等于 10.00 元的记录没有被返回。

5.2.4　带 IN 关键字的查询

IN 操作符用来查询满足指定范围内的条件的记录，使用 IN 操作符，将所有检索条件用括号括起来，检索条件之间用逗号分隔开，只要满足条件范围内的一个值即为匹配项。

【例 5.7】查询 s_id 为 101 和 102 的记录，SQL 语句如下：

SELECT s_id, f_name, f_price FROM fruits WHERE s_id IN (101,102) ORDER BY f_name;

查询结果如下：

```
+------+------------+---------+
| s_id | f_name     | f_price |
+------+------------+---------+
|  101 | apple      |    5.20 |
|  102 | banana     |   10.30 |
|  101 | blackberry |   10.20 |
|  101 | cherry     |    3.20 |
|  102 | grape      |    5.30 |
|  102 | orange     |   11.20 |
+------+------------+---------+
```

相反，可以使用关键字 NOT 来检索不在条件范围内的记录。

【例 5.8】查询所有 s_id 不等于 101 也不等于 102 的记录，SQL 语句如下：

SELECT s_id, f_name, f_price FROM fruits WHERE s_id NOT IN (101,102) ORDER BY f_name;

查询结果如下：

```
+------+--------+---------+
| s_id | f_name | f_price |
+------+--------+---------+
```

```
| 103 | apricot  |  2.20 |
| 104 | berry    |  7.60 |
| 103 | coconut  |  9.20 |
| 104 | lemon    |  6.40 |
| 106 | mango    | 15.70 |
| 105 | melon    |  8.20 |
| 107 | xbababa  |  3.60 |
| 105 | xbabay   |  2.60 |
| 105 | xxtt     | 11.60 |
| 107 | xxxx     |  3.60 |
```

可以看到，该语句在 IN 关键字前面加上 NOT 关键字，这使得查询的结果与前面例 5.7 的结果正好相反，前面例子检索了 s_id 等于 101 和 102 的记录，而这里所要求查询的记录中 s_id 字段值不等于这两个值中的任何一个。

5.2.5　带 BETWEEN…AND…的范围查询

BETWEEN…AND…用来查询某个范围内的值，该操作符需要两个参数，即范围的开始值和结束值，如果字段值满足指定范围的查询条件，则这些记录被返回。

【例 5.9】查询价格在 2.00 元到 10.20 元之间的水果名称和价格，SQL 语句如下：

```
SELECT f_name, f_price FROM fruits WHERE f_price BETWEEN 2.00 AND 10.20;
```

查询结果如下：

```
+------------+---------+
| f_name     | f_price |
+------------+---------+
| apple      |    5.20 |
| apricot    |    2.20 |
| blackberry |   10.20 |
| berry      |    7.60 |
| xxxx       |    3.60 |
| melon      |    8.20 |
| cherry     |    3.20 |
| lemon      |    6.40 |
| xbabay     |    2.60 |
| coconut    |    9.20 |
| grape      |    5.30 |
| xbababa    |    3.60 |
+------------+---------+
```

可以看到，返回结果包含了价格从 2.00 元到 10.20 元之间的字段值，并且端点值 10.20 也包括在返回结果中，即 BETWEEN 匹配范围中的所有值，包括开始值和结束值。

BETWEEN…AND…操作符前可以加关键字 NOT，表示指定范围之外的值，如果字段值不满足指定的范围内的值，则这些记录被返回。

【例 5.10】查询价格在 2.00 元到 10.20 元之外的水果名称和价格，SQL 语句如下：

```
SELECT f_name, f_price FROM fruits WHERE f_price NOT BETWEEN 2.00 AND 10.20;
```

查询结果如下：

```
+--------+---------+
| f_name | f_price |
+--------+---------+
| orange |  11.20  |
| mango  |  15.70  |
| xxtt   |  11.60  |
| banana |  10.30  |
+--------+---------+
```

由结果可以看到，返回的记录只有 f_price 字段大于 10.20 的，其实，f_price 字段小于 2.00 的记录也满足查询条件。因此，如果表中有 f_price 字段小于 2.00 的记录，也应当作为查询结果。

5.2.6　带 LIKE 的字符匹配查询

在前面的检索操作中讲解了如何查询多个字段的记录，如何进行比较查询或查询一个条件范围内的记录。如果要查找所有包含字符"ge"的水果名称，该如何查找呢？简单的比较操作在这里已经行不通了，这时需要使用通配符进行匹配查找，通过创建查找模式对表中的数据进行比较。执行这个任务的关键字是 LIKE。

通配符是一种在 SQL 的 WHERE 条件子句中拥有特殊意思的字符。SQL 语句中支持多种通配符，可以和 LIKE 一起使用的通配符有"%"和"_"。

1. 百分号通配符"%"，匹配任意长度的字符，甚至包括零字符

【例 5.11】查找所有以"b"字母开头的水果，SQL 语句如下：

SELECT f_id, f_name FROM fruits WHERE f_name LIKE 'b%';

查询结果如下：

```
+------+------------+
| f_id | f_name     |
+------+------------+
| b1   | blackberry |
| b2   | berry      |
| t1   | banana     |
+------+------------+
```

该语句查询的结果返回所有以"b"开头的水果的 id 和 name，"%"告诉 MySQL，返回所有以字母"b"开头的记录，不管"b"后面有多少个字符。

在搜索匹配时通配符"%"可以放在不同位置，如例 5.12 所示。

【例 5.12】在 fruits 表中，查询 f_name 中包含字母"g"的记录，SQL 语句如下：

SELECT f_id, f_name FROM fruits WHERE f_name LIKE '%g%';

查询结果如下：

```
+------+--------+
| f_id | f_name |
+------+--------+
```

```
| bs1   | orange |
| m1    | mango  |
| t2    | grape  |
+-------+--------+
```

该语句查询字符串中包含字母"g"的水果名称,只要名字中有字符"g",不管前面或后面有多少个字符,都满足查询条件。

【例 5.13】查询以"b"开头并以"y"结尾的水果的名称,SQL 语句如下:

SELECT f_name FROM fruits WHERE f_name LIKE 'b%y';

查询结果如下:

```
+------------+
| f_name     |
+------------+
| blackberry |
| berry      |
+------------+
```

通过以上查询结果可以看到,"%"用于匹配在指定位置的任意数目的字符。

2. 下划线通配符"_",一次只能匹配任意一个字符

另一个非常有用的通配符是下划线通配符"_"。该通配符的用法和"%"相同,区别是"%"可以匹配多个字符,而"_"只能匹配任意单个字符。如果要匹配多个字符,则需要使用相同个数的"_",并且各个"_"之间没有空格。

【例 5.14】在 fruits 表中,查询以字母"y"结尾,且"y"前面只有 4 个字母的记录,SQL 语句如下:

SELECT f_id, f_name FROM fruits WHERE f_name LIKE '_ _ _ _y';

查询结果如下:

```
+------+--------+
| f_id | f_name |
+------+--------+
| b2   | berry  |
+------+--------+
```

从结果可以看到,以"y"结尾且前面只有 4 个字母的记录只有一条。其他记录的 f_name 字段也有以"y"结尾的,但其总的字符串长度不为 5,因此不在返回结果中。

5.2.7 查询空值

数据表创建的时候,设计者可以指定某列中是否包含空值(NULL)。空值不同于 0,也不同于空字符串。空值一般表示数据未知、不适用或将在以后添加数据。在 SELECT 语句中使用 IS NULL 子句,可以查询某字段内容为空记录。

下面在数据库中创建数据表 customers,该表中包含了本章例子需要用到的数据。

CREATE TABLE customers

```
(
c_id int NOT NULL AUTO_INCREMENT,
c_name char(50) NOT NULL,
c_address char(50) NULL,
c_city char(50) NULL,
c_zip char(10) NULL,
c_contact char(50) NULL,
c_email char(255) NULL,
PRIMARY KEY (c_id)
);
```

为了演示，需要插入数据，执行以下语句：

```
INSERT INTO customers(c_id, c_name, c_address, c_city,
c_zip,  c_contact, c_email)
VALUES(10001, 'RedHook', '200 Street ', 'Tianjin',
 '300000',   'LiMing', 'LMing@163.com'),
(10002, 'Stars', '333 Fromage Lane',
 'Dalian', '116000',  'Zhangbo','Jerry@hotmail.com'),
(10003, 'Netbhood', '1 Sunny Place', 'Qingdao', '266000',
 'LuoCong',  NULL),
(10004, 'JOTO', '829 Riverside Drive', 'Haikou',
 '570000',  'YangShan', 'sam@hotmail.com');
```

【例 5.15】查询 customers 表中 c_email 为空的记录的 c_id、c_name 和 c_email 字段值，SQL 语句如下：

```
SELECT c_id, c_name,c_email FROM customers WHERE c_email IS NULL;
```

查询结果如下：

```
+-------+----------+---------+
| c_id  | c_name   | c_email |
+-------+----------+---------+
| 10003 | Netbhood | NULL    |
+-------+----------+---------+
```

可以看到，显示 customers 表中字段 c_email 的值为 NULL 的记录，满足查询条件。

与 IS NULL 相反的是 NOT NULL，该关键字查找字段不为空的记录。

【例 5.16】查询 customers 表中 c_email 不为空的记录的 c_id、c_name 和 c_email 字段值，SQL 语句如下：

```
SELECT c_id, c_name,c_email FROM customers WHERE c_email IS NOT NULL;
```

查询结果如下：

```
+-------+---------+-------------------+
| c_id  | c_name  | c_email           |
+-------+---------+-------------------+
| 10001 | RedHook | LMing@163.com     |
| 10002 | Stars   | Jerry@hotmail.com |
| 10004 | JOTO    | sam@hotmail.com   |
+-------+---------+-------------------+
```

可以看到，查询出来的记录的 c_email 字段都不为空值。

5.2.8 带 AND 的多条件查询

使用 SELECT 查询时，可以增加查询的限制条件，这样可以使查询的结果更加精确。MySQL 在 WHERE 子句中使用 AND 操作符限定只有满足所有查询条件的记录才会被返回。可以使用 AND 连接两个甚至多个查询条件，多个条件表达式之间用 AND 分开。

【例 5.17】在 fruits 表中查询 s_id = 101 并且 f_price 大于等于 5 的水果 id、价格和名称，SQL 语句如下：

```
SELECT f_id, f_price, f_name FROM fruits WHERE s_id = '101' AND f_price >=5;
```

查询结果如下：

```
+------+---------+-----------+
| f_id | f_price | f_name    |
+------+---------+-----------+
| a1   |    5.20 | apple     |
| b1   |   10.20 | blackberry|
+------+---------+-----------+
```

前面的语句检索了 s_id=101 的水果供应商所有价格大于等于 5 元的水果名称和价格。WHERE 子句中的条件分为两部分，AND 关键字指示 MySQL 返回所有同时满足两个条件的行。id=101 的水果供应商提供的水果如果价格小于 5，或者是 id 不等于 101 的水果供应商提供的水果（不管其价格为多少），均不是要查询的结果。

提示：上述例子的 WHERE 子句中只包含了一个 AND 语句，把两个过滤条件组合在一起。实际上可以添加多个 AND 过滤条件，增加条件的同时增加一个 AND 关键字。

【例 5.18】在 fruits 表中查询 s_id = 101 或者 102，并且 f_price 大于等于 5、f_name='apple' 的水果价格和名称，SQL 语句如下：

```
SELECT f_id, f_price, f_name FROM fruits WHERE s_id IN('101','102') AND f_price >= 5 AND f_name = 'apple';
```

查询结果如下：

```
+------+---------+--------+
| f_id | f_price | f_name |
+------+---------+--------+
| a1   |    5.20 | apple  |
+------+---------+--------+
```

可以看到，符合查询条件的返回记录只有一条。

5.2.9 带 OR 的多条件查询

与 AND 相反，在 WHERE 子句中使用 OR 操作符，表示只需要满足其中一个条件的记录即可返回。OR 也可以连接两个甚至多个查询条件，多个条件表达式之间用 OR 分开。

【例 5.19】查询 s_id=101 或者 s_id=102 的水果供应商的 f_price 和 f_name，SQL 语句如下：

```
SELECT s_id, f_name, f_price FROM fruits WHERE s_id = 101 OR s_id = 102;
```

查询结果如下：

```
+------+------------+---------+
| s_id | f_name     | f_price |
+------+------------+---------+
|  101 | apple      |    5.20 |
|  101 | blackberry |   10.20 |
|  102 | orange     |   11.20 |
|  101 | cherry     |    3.20 |
|  102 | banana     |   10.30 |
|  102 | grape      |    5.30 |
+------+------------+---------+
```

结果显示了 s_id=101 和 s_id=102 的水果供应商的水果名称和价格，OR 操作符告诉 MySQL 检索的时候只需要满足其中的一个条件，不需要全部都满足。如果这里使用 AND 的话，将检索不到符合条件的数据。

在这里，也可以使用 IN 操作符实现与 OR 相同的功能，看下面的例子。

【例 5.20】查询 s_id=101 或者 s_id=102 的水果供应商的 f_price 和 f_name，SQL 语句如下：

SELECT s_id, f_name, f_price FROM fruits WHERE s_id IN(101, 102);

查询结果如下：

```
+------+------------+---------+
| s_id | f_name     | f_price |
+------+------------+---------+
|  101 | apple      |    5.20 |
|  101 | blackberry |   10.20 |
|  102 | orange     |   11.20 |
|  101 | cherry     |    3.20 |
|  102 | banana     |   10.30 |
|  102 | grape      |    5.30 |
+------+------------+---------+
```

在这里可以看到，OR 操作符和 IN 操作符使用后的结果是一样的，它们可以实现相同的功能，但是使用 IN 操作符使得检索语句更加简洁明了，并且 IN 执行的速度要快于 OR。更重要的是，使用 IN 操作符可以执行更加复杂的嵌套查询（后面章节将会讲解）。

提示：OR 可以和 AND 一起使用，但是在使用时要注意两者的优先级，由于 AND 的优先级高于 OR，因此先对 AND 两边的操作数进行操作，再与 OR 中的操作数结合。

5.2.10 查询结果不重复

从前面的例子可以看到，SELECT 查询返回所有匹配的行。例如，查询 fruits 表中所有的 s_id，其结果为：

```
+------+
| s_id |
+------+
|  101 |
|  103 |
|  101 |
```

```
|  104  |
|  107  |
|  102  |
|  105  |
|  101  |
|  104  |
|  106  |
|  105  |
|  105  |
|  103  |
|  102  |
|  102  |
|  107  |
+-------+
```

可以看到查询结果返回了 16 条记录，其中有一些重复的 s_id 值。有时出于对数据分析的要求，需要消除重复的记录值，该如何操作呢？在 SELECT 语句中，可以使用 DISTINCT 关键字指示 MySQL 消除重复的记录值。语法格式为：

```
SELECT DISTINCT 字段名 FROM 表名;
```

【例 5.21】查询 fruits 表中 s_id 字段的值，返回 s_id 字段值且不得重复，SQL 语句如下：

```
SELECT DISTINCT s_id FROM fruits;
```

查询结果如下：

```
+-------+
| s_id  |
+-------+
|  101  |
|  103  |
|  104  |
|  107  |
|  102  |
|  105  |
|  106  |
+-------+
```

可以看到，这次查询结果只返回了 7 条记录的 s_id 值，且不再有重复的值，SELECT DISTINCT s_id 告诉 MySQL 只返回不同的 s_id 行。

5.2.11　对查询结果排序

从前面的查询结果，读者会发现有些字段的值是没有任何顺序的，MySQL 可以通过在 SELECT 语句中使用 ORDER BY 子句对查询的结果进行排序。

1. 单列排序

例如，查询 f_name 字段，查询结果如下：

```
mysql> SELECT f_name FROM fruits;
+-------------+
| f_name      |
```

```
+-------------+
| apple       |
| apricot     |
| blackberry  |
| berry       |
| xxxx        |
| orange      |
| melon       |
| cherry      |
| lemon       |
| mango       |
| xbabay      |
| xxtt        |
| coconut     |
| banana      |
| grape       |
| xbababa     |
+-------------+
```

可以看到，查询的数据并没有以一种特定的顺序显示，如果没有对它们进行排序，就将根据它们插入到数据表中的顺序来显示。

下面使用 ORDER BY 子句对指定的列数据进行排序。

【例 5.22】查询 fruits 表的 f_name 字段值，并对其进行排序，SQL 语句如下：

```
mysql> SELECT f_name FROM fruits ORDER BY f_name;
+-------------+
| f_name      |
+-------------+
| apple       |
| apricot     |
| banana      |
| berry       |
| blackberry  |
| cherry      |
| coconut     |
| grape       |
| lemon       |
| mango       |
| melon       |
| orange      |
| xbababa     |
| xbabay      |
| xxtt        |
| xxxx        |
+-------------+
```

该语句查询的结果和前面的语句相同，不同的是，通过指定 ORDER BY 子句，MySQL 对查询的 f_name 列的数据按字母表的顺序进行了升序排列。

2. 多列排序

有时，需要根据多列值进行排序。比如，如果要显示一个学生列表，可能会有多个学生的姓氏是相同的，因此还需要根据学生的名进行排序。对多列数据进行排序，要将需要排序的列之间用逗号隔开。

【例 5.23】查询 fruits 表中的 f_name 和 f_price 字段，先按 f_name 排序，再按 f_price 排序，SQL 语句如下：

```
SELECT f_name, f_price FROM fruits ORDER BY f_name, f_price;
```

查询结果如下：

```
+------------+---------+
| f_name     | f_price |
+------------+---------+
| apple      |    5.20 |
| apricot    |    2.20 |
| banana     |   10.30 |
| berry      |    7.60 |
| blackberry |   10.20 |
| cherry     |    3.20 |
| coconut    |    9.20 |
| grape      |    5.30 |
| lemon      |    6.40 |
| mango      |   15.70 |
| melon      |    8.20 |
| orange     |   11.20 |
| xbababa    |    3.60 |
| xbabay     |    2.60 |
| xxtt       |   11.60 |
| xxxx       |    3.60 |
+------------+---------+
```

提示：在对多列进行排序的时候，首先排序的第一列必须有相同的列值，才会对第二列进行排序。如果第一列数据中所有值都是唯一的，将不再对第二列进行排序。

3. 指定排序方向

默认情况下，查询数据按字母升序进行排序（A~Z），但还可以使用 ORDER BY 子句对查询结果进行降序排序（Z~A）。这可以通过关键字 DESC 实现，下面的例子表明了如何进行降序排序。

【例 5.24】查询 fruits 表中的 f_name 和 f_price 字段，对结果按 f_price 降序方式排序，SQL 语句如下：

```
SELECT f_name, f_price FROM fruits ORDER BY f_price DESC;
```

查询结果如下：

```
+------------+---------+
| f_name     | f_price |
+------------+---------+
| mango      |   15.70 |
| xxtt       |   11.60 |
| orange     |   11.20 |
| banana     |   10.30 |
| blackberry |   10.20 |
| coconut    |    9.20 |
| melon      |    8.20 |
```

```
| berry    |  7.60 |
| lemon    |  6.40 |
| grape    |  5.30 |
| apple    |  5.20 |
| xxxx     |  3.60 |
| xbababa  |  3.60 |
| cherry   |  3.20 |
| xbabay   |  2.60 |
| apricot  |  2.20 |
```

提示：与 DESC 相反的是 ASC（升序），将字段列中的数据按字母表顺序升序排列。实际上，在排序的时候 ASC 是默认的排序方式，所以加不加都可以。

也可以对多列进行不同的顺序排序，如例 5.25 所示。

【例 5.25】查询 fruits 表，先按 f_price 降序排列，再按 f_name 字段升序排列，SQL 语句如下：

SELECT f_price, f_name FROM fruits ORDER BY f_price DESC, f_name;

查询结果如下：

```
| f_price | f_name     |
|   15.70 | mango      |
|   11.60 | xxtt       |
|   11.20 | orange     |
|   10.30 | banana     |
|   10.20 | blackberry |
|    9.20 | coconut    |
|    8.20 | melon      |
|    7.60 | berry      |
|    6.40 | lemon      |
|    5.30 | grape      |
|    5.20 | apple      |
|    3.60 | xbababa    |
|    3.60 | xxxx       |
|    3.20 | cherry     |
|    2.60 | xbabay     |
|    2.20 | apricot    |
```

DESC 排序方式只应用到直接位于其前面的字段上，由结果可以看出。

提示：DESC 关键字只对其前面的列进行降序排列，这里只对 f_price 排序，而并没有对 f_name 进行排序，因此，f_price 按降序排序，而 f_name 列仍按升序排序。如果要对多列都进行降序排序，必须在每一列的列名后面加 DESC 关键字。

5.2.12 分组查询

分组查询是对数据按照某个或多个字段进行分组。MySQL 中使用 GROUP BY 关键字对数据进行分组，基本语法形式为：

```
[GROUP BY  字段] [HAVING <条件表达式>]
```

字段值为进行分组时所依据的列名称;"HAVING <条件表达式>"指定满足表达式限定条件的结果将被显示。

1. **创建分组**

GROUP BY 关键字通常和集合函数一起使用,比如 MAX()、MIN()、COUNT()、SUM()、AVG()。例如,要返回每个水果供应商提供的水果种类,这时就要在分组过程中使用 COUNT()函数,把数据分为多个逻辑组,并对每个组进行集合计算。

【例 5.26】根据 s_id 对 fruits 表中的数据进行分组,SQL 语句如下:
```
SELECT s_id, COUNT(*) AS Total FROM fruits GROUP BY s_id;
```

查询结果如下:

```
+------+-------+
| s_id | Total |
+------+-------+
|  101 |     3 |
|  103 |     2 |
|  104 |     2 |
|  107 |     2 |
|  102 |     3 |
|  105 |     3 |
|  106 |     1 |
+------+-------+
```

查询结果显示,s_id 表示供应商的 ID,Total 字段使用 COUNT()函数计算得出,GROUP BY 子句按照 s_id 排序并对数据分组,可以看到 ID 为 101、102、105 的供应商分别提供 3 种水果,ID 为 103、104、107 的供应商分别提供 2 种水果,ID 为 106 的供应商只提供 1 种水果。

如果要查看每个供应商提供的水果的种类名称,该怎么办呢?在 MySQL 中,可以在 GROUP BY 子句中使用 GROUP_CONCAT()函数,将每个分组中各个字段的值显示出来。

【例 5.27】根据 s_id 对 fruits 表中的数据进行分组,将每个供应商的水果名称显示出来,SQL 语句如下:
```
SELECT s_id, GROUP_CONCAT(f_name) AS Names FROM fruits GROUP BY s_id;
```

查询结果如下:

```
+------+--------------------------+
| s_id | Names                    |
+------+--------------------------+
|  101 | apple, blackberry, cherry|
|  102 | orange, banana, grape    |
|  103 | apricot, coconut         |
|  104 | berry, lemon             |
|  105 | melon, xbabay, xxtt      |
|  106 | mango                    |
|  107 | xxxx, xbababa            |
+------+--------------------------+
```

由结果可以看到,GROUP_CONCAT()函数将每个分组中的名称显示出来了,其名称的个数与 COUNT()函数计算出来的相同。

2. 使用 HAVING 过滤分组

GROUP BY 可以和 HAVING 一起限定所查询的记录需要满足的条件,只有满足条件的分组才会被显示。

【例 5.28】根据 s_id 对 fruits 表中的数据进行分组,并显示水果种类大于 1 的分组信息,SQL 语句如下:

SELECT s_id, GROUP_CONCAT(f_name) AS Names FROM fruits GROUP BY s_id HAVING COUNT(f_name) > 1;

查询结果如下:

```
+------+--------------------------+
| s_id | Names                    |
+------+--------------------------+
| 101  | apple, blackberry, cherry|
| 102  | orange, banana, grape    |
| 103  | apricot, coconut         |
| 104  | berry, lemon             |
| 105  | melon, xbabay, xxtt      |
| 107  | xxxx, xbababa            |
+------+--------------------------+
```

由结果可以看到,ID 为 101、102、103、104、105、107 的供应商提供的水果种类大于 1,满足 HAVING 子句条件,因此出现在返回结果中;而 ID 为 106 的供应商的水果种类等于 1,不满足限定条件,因此不在返回结果中。

提示:HAVING 关键字与 WHERE 关键字都是用来过滤数据的,两者有什么区别呢?其中重要的一点是,HAVING 在数据分组之后进行过滤来选择分组,而 WHERE 在分组之前来选择记录。另外,WHERE 排除的记录不再包括在分组中。

3. 在 GROUP BY 子句中使用 WITH ROLLUP

使用 WITH ROLLUP 关键字之后,在所有查询出的分组记录之后增加一条记录,该记录计算查询出的所有记录的总和,即统计记录数量。

【例 5.29】根据 s_id 对 fruits 表中的数据进行分组,并显示记录数量,SQL 语句如下:

SELECT s_id, COUNT(*) AS Total FROM fruits GROUP BY s_id WITH ROLLUP;

查询结果如下:

```
+------+-------+
| s_id | Total |
+------+-------+
| 101  |   3   |
| 102  |   3   |
| 103  |   2   |
| 104  |   2   |
| 105  |   3   |
| 106  |   1   |
```

```
| 107   |   2 |
| NULL  |  16 |
+-------+-----+
```

由结果可以看到，通过 GROUP BY 分组之后，在显示结果的最后面新添加了一行，该行 Total 列的值正好是上面所有数值之和。

4. 多字段分组

使用 GROUP BY 可以对多个字段进行分组，GROUP BY 关键字后面跟需要分组的字段，MySQL 根据多字段的值来进行层次分组，分组层次从左到右，即先按第 1 个字段分组，然后在第 1 个字段值相同的记录中再根据第 2 个字段的值进行分组，以此类推。

【例 5.30】根据 s_id 和 f_name 字段对 fruits 表中的数据进行分组，SQL 语句如下：

```
mysql> SELECT * FROM fruits group by s_id,f_name;
```

查询结果如下：

```
+------+------+------------+---------+
| f_id | s_id | f_name     | f_price |
+------+------+------------+---------+
| a1   | 101  | apple      |   5.20  |
| a2   | 103  | apricot    |   2.20  |
| b1   | 101  | blackberry |  10.20  |
| b2   | 104  | berry      |   7.60  |
| b5   | 107  | xxxx       |   3.60  |
| bs1  | 102  | orange     |  11.20  |
| bs2  | 105  | melon      |   8.20  |
| c0   | 101  | cherry     |   3.20  |
| l2   | 104  | lemon      |   6.40  |
| m1   | 106  | mango      |  15.70  |
| m2   | 105  | xbabay     |   2.60  |
| m3   | 105  | xxtt       |  11.60  |
| o2   | 103  | coconut    |   9.20  |
| t1   | 102  | banana     |  10.30  |
| t2   | 102  | grape      |   5.30  |
| t4   | 107  | xbababa    |   3.60  |
+------+------+------------+---------+
```

由结果可以看到，查询记录先按照 s_id 进行分组，再对 f_name 字段按不同的取值进行分组。

5. GROUP BY 和 ORDER BY 一起使用

某些情况下需要对分组进行排序，在前面的介绍中，ORDER BY 用来对查询的记录排序，如果和 GROUP BY 一起使用可以完成对分组的排序。

为了演示效果，首先创建数据表，SQL 语句如下：

```
CREATE TABLE orderitems
(
o_num        int            NOT NULL,
o_item       int            NOT NULL,
f_id         char(10)       NOT NULL,
quantity     int            NOT NULL,
item_price   decimal(8,2)   NOT NULL,
```

```
PRIMARY KEY (o_num,o_item)
);
```

然后插入演示数据。SQL 语句如下：

```
INSERT INTO orderitems(o_num, o_item, f_id, quantity, item_price)
VALUES(30001, 1, 'a1', 10, 5.2),
(30001, 2, 'b2', 3, 7.6),
(30001, 3, 'bs1', 5, 11.2),
(30001, 4, 'bs2', 15, 9.2),
(30002, 1, 'b3', 2, 20.0),
(30003, 1, 'c0', 100, 10),
(30004, 1, 'o2', 50, 2.50),
(30005, 1, 'c0', 5, 10),
(30005, 2, 'b1', 10, 8.99),
(30005, 3, 'a2', 10, 2.2),
(30005, 4, 'm1', 5, 14.99);
```

【例 5.31】查询订单价格大于 100 的订单号和总订单价格，SQL 语句如下：

```
SELECT o_num,  SUM(quantity * item_price) AS orderTotal FROM orderitems
GROUP BY o_num HAVING SUM(quantity*item_price) >= 100;
```

查询结果如下：

```
+-------+-----------+
| o_num | orderTotal |
+-------+-----------+
| 30001 |    268.80 |
| 30003 |   1000.00 |
| 30004 |    125.00 |
| 30005 |    236.85 |
+-------+-----------+
```

可以看到，返回的结果中 orderTotal 列的总订单价格并没有按照一定顺序显示，接下来使用 ORDER BY 关键字按总订单价格排序显示结果，SQL 语句如下：

```
SELECT o_num,  SUM(quantity * item_price) AS orderTotal FROM orderitems
GROUP BY o_num HAVING SUM(quantity*item_price) >= 100 ORDER BY orderTotal;
```

查询结果如下：

```
+-------+-----------+
| o_num | orderTotal |
+-------+-----------+
| 30004 |    125.00 |
| 30005 |    236.85 |
| 30001 |    268.80 |
| 30003 |   1000.00 |
+-------+-----------+
```

由结果可以看到，GROUP BY 子句按订单号对数据进行分组，SUM()函数便可以返回总的订单价格，HAVING 子句对分组数据进行过滤，使得只返回总价格大于 100 的订单，最后使用 ORDER BY 子句排序输出。

提示：当使用 ROLLUP 时，不能同时使用 ORDER BY 子句进行结果排序，即 ROLLUP 和 ORDER BY 是互相排斥的。

5.2.13 使用 LIMIT 限制查询结果的数量

SELECT 返回所有匹配的行，有可能是表中所有的行，若仅仅需要返回第一行或者前几行，可使用 LIMIT 关键字，基本语法格式如下：

LIMIT [位置偏移量,] 行数

第一个"位置偏移量"参数指示 MySQL 从哪一行开始显示，是一个可选参数，如果不指定"位置偏移量"，将会从表中的第一条记录开始（第一条记录的位置偏移量是 0，第二条记录的位置偏移量是 1，以此类推）；第二个参数"行数"指示返回的记录条数。

【例 5.32】显示 fruits 表查询结果的前 4 行，SQL 语句如下：

SELECT * From fruits LIMIT 4;

查询结果如下：

```
+------+------+------------+---------+
| f_id | s_id | f_name     | f_price |
+------+------+------------+---------+
| a1   | 101  | apple      |    5.20 |
| a2   | 103  | apricot    |    2.20 |
| b1   | 101  | blackberry |   10.20 |
| b2   | 104  | berry      |    7.60 |
+------+------+------------+---------+
```

由结果可以看到，该语句没有指定返回记录的"位置偏移量"参数，显示结果从第一行开始，"行数"参数为 4，因此返回的结果为表中的前 4 行记录。

如果指定返回记录的开始位置，那么返回结果为从"位置偏移量"参数开始的指定行数，"行数"参数指定返回的记录条数。

【例 5.33】在 fruits 表中，使用 LIMIT 子句，返回从第 5 个记录开始的、行数长度为 3 的记录，SQL 语句如下：

SELECT * From fruits LIMIT 4, 3;

查询结果如下：

```
+------+------+--------+---------+
| f_id | s_id | f_name | f_price |
+------+------+--------+---------+
| b5   | 107  | xxxx   |    3.60 |
| bs1  | 102  | orange |   11.20 |
| bs2  | 105  | melon  |    8.20 |
+------+------+--------+---------+
```

由结果可以看到，该语句指示 MySQL 返回从第 5 条记录行开始之后的 3 条记录。第一个数字 4 表示从第 5 行开始（位置偏移量从 0 开始，第 5 行的位置偏移量为 4），第二个数字 3 表示返回的行数。

所以，带一个参数的 LIMIT 指定从查询结果的首行开始，唯一的参数表示返回的行数，即"LIMIT n"与"LIMIT 0,n"等价。带两个参数的 LIMIT 可以返回从任何一个位置开始的指定的行数。

返回第一行时，位置偏移量是 0。因此，"LIMIT 1, 1"将返回第二行，而不是第一行。

提示：MySQL 8.0 中可以使用"LIMIT 4 OFFSET 3"，意思是获取从第 5 条记录开始后面的 3 条记录，和"LIMIT 4,3"返回的结果相同。

5.3 使用集合函数查询

有时候并不需要返回实际表中的数据，而只是对数据进行总结。MySQL 提供一些查询功能，可以对获取的数据进行分析和报告。这些函数的功能有：计算数据表中记录行数的总数、计算某个字段列下数据的总和，以及计算表中某个字段下的最大值、最小值或者平均值。本节将介绍这些函数及其用法，这些聚合函数的名称和作用如表 5.2 所示。

表 5.2　MySQL 聚合函数

函　　数	作　　用
AVG()	返回某列的平均值
COUNT()	返回某列的行数
MAX()	返回某列的最大值
MIN()	返回某列的最小值
SUM()	返回某列值的和

接下来，将详细介绍各个函数的使用方法。

5.3.1　COUNT()函数

COUNT()函数统计数据表中包含的记录行的总数，或者根据查询结果返回列中包含的数据行数。其使用方法有两种：

- COUNT(*)：计算表中总的行数，不管某列是否有数值或者为空值。
- COUNT(字段名)：计算指定列下总的行数，计算时将忽略空值的行。

【例 5.34】查询 customers 表中总的行数，SQL 语句如下：

```
mysql> SELECT COUNT(*) AS cust_num FROM customers;
+----------+
| cust_num |
+----------+
|        4 |
+----------+
```

由查询结果可以看到，COUNT(*)返回 customers 表中记录的总行数，不管其值是什么，返回的总数的名称都为 cust_num。

【例 5.35】查询 customers 表中有电子邮箱的顾客的总数，SQL 语句如下：

```
mysql> SELECT COUNT(c_email) AS email_num  FROM customers;
+-----------+
| email_num |
+-----------+
|         3 |
+-----------+
```

由查询结果可以看到，表中 5 个 customer 只有 3 个有 email，customer 的 email 为空值 NULL 的记录没有被 COUNT()函数计算。

提示：两个例子中不同的数值说明了两种方式在计算总数的时候对待 NULL 值的方式不同：指定列的值为空的行被 COUNT()函数忽略；如果不指定列，而在 COUNT()函数中使用星号"*"，则所有记录都不忽略。

前面介绍分组查询的时候，介绍了如何使用 COUNT()函数与 GROUP BY 关键字一起来计算不同分组中的记录总数。

【例 5.36】在 orderitems 表中，使用 COUNT()函数统计不同订单号中订购的水果种类，SQL 语句如下：

```
mysql> SELECT o_num, COUNT(f_id) FROM orderitems GROUP BY o_num;
+-------+-------------+
| o_num | COUNT(f_id) |
+-------+-------------+
| 30001 |           4 |
| 30002 |           1 |
| 30003 |           1 |
| 30004 |           1 |
| 30005 |           4 |
+-------+-------------+
```

从查询结果可以看到，GROUP BY 关键字先按照订单号进行分组，然后计算每个分组中的总记录数。

5.3.2　SUM()函数

SUM()是一个求总和的函数，返回指定列值的总和。

【例 5.37】在 orderitems 表中，查询 30005 号订单一共购买的水果总量，SQL 语句如下：

```
mysql>SELECT SUM(quantity) AS items_total FROM orderitems WHERE o_num = 30005;
+-------------+
| items_total |
+-------------+
|          30 |
+-------------+
```

由查询结果可以看到，SUM(quantity)函数返回订单中所有水果数量之和，WHERE 子句指定查询的订单号为 30005。

SUM()函数可以与 GROUP BY 子句一起使用，用来计算每个分组的总和。

【例 5.38】 在 orderitems 表中，使用 SUM()函数统计不同订单号中订购的水果总量，SQL 语句如下：

```
mysql> SELECT o_num, SUM(quantity) AS items_total FROM orderitems GROUP BY o_num;
+-------+-------------+
| o_num | items_total |
+-------+-------------+
| 30001 |          33 |
| 30002 |           2 |
| 30003 |         100 |
| 30004 |          50 |
| 30005 |          30 |
+-------+-------------+
```

由查询结果可以看到，GROUP BY 按照订单号 o_num 进行分组，SUM()函数计算每个分组中订购的水果的总量。

提示：SUM()函数在计算时，忽略列值为 NULL 的行。

5.3.3 AVG()函数

AVG()函数通过计算返回的行数和每一行数据的和，求得指定列数据的平均值。

【例 5.39】 在 fruits 表中，查询 s_id=103 的供应商的水果价格的平均值，SQL 语句如下：

```
mysql> SELECT AVG(f_price) AS avg_price FROM fruits WHERE s_id = 103;
+-----------+
| avg_price |
+-----------+
|  5.700000 |
+-----------+
```

该例中，查询语句增加了一个 WHERE 子句，并且添加了查询过滤条件，只查询 s_id=103 的记录中的 f_price。因此，通过 AVG()函数计算的结果只是指定的供应商水果的价格平均值，而不是市场上所有水果价格的平均值。

AVG()函数可以与 GROUP BY 一起使用，用来计算每个分组的平均值。

【例 5.40】 在 fruits 表中，查询每一个供应商的水果价格的平均值，SQL 语句如下：

```
mysql> SELECT s_id,AVG(f_price) AS avg_price FROM fruits GROUP BY s_id;
+------+-----------+
| s_id | avg_price |
+------+-----------+
|  101 |  6.200000 |
|  103 |  5.700000 |
|  104 |  7.000000 |
|  107 |  3.600000 |
|  102 |  8.933333 |
|  105 |  7.466667 |
|  106 | 15.700000 |
+------+-----------+
```

GROUP BY 关键字根据 s_id 字段对记录进行分组，然后计算出每个分组的平均值，这种分组求平均值的方法非常有用，例如求不同班级学生成绩的平均值、求不同部门工人的平均工资、求各地的年平均气温等。

提示：AVG()函数使用时，其参数为需要计算的列名称。如果想要得到多个列的多个平均值，则需要在每一列上使用 AVG()函数。

5.3.4　MAX()函数

MAX()函数返回指定列中的最大值。

【例 5.41】在 fruits 表中，查找市场上价格最高的水果值，SQL 语句如下：

```
mysql>SELECT MAX(f_price) AS max_price FROM fruits;
+-----------+
| max_price |
+-----------+
|     15.70 |
+-----------+
```

由结果可以看到，MAX()函数查询出了 f_price 字段的最大值 15.70。

MAX()函数也可以和 GROUP BY 关键字一起使用，求每个分组中的最大值。

【例 5.42】在 fruits 表中，查找不同供应商提供的价格最高的水果值，SQL 语句如下：

```
mysql> SELECT s_id, MAX(f_price) AS max_price FROM fruits GROUP BY s_id;
+------+-----------+
| s_id | max_price |
+------+-----------+
|  101 |     10.20 |
|  103 |      9.20 |
|  104 |      7.60 |
|  107 |      3.60 |
|  102 |     11.20 |
|  105 |     11.60 |
|  106 |     15.70 |
+------+-----------+
```

由结果可以看到，GROUP BY 关键字根据 s_id 字段对记录进行分组，然后计算出每个分组中的最大值。

MAX()函数不仅适用于查找数值类型，也可应用于字符类型。

【例 5.43】在 fruits 表中，查找 f_name 的最大值，SQL 语句如下：

```
mysql> SELECT MAX(f_name) FROM fruits;
+-------------+
| MAX(f_name) |
+-------------+
| xxxx        |
+-------------+
```

由结果可以看到，MAX()函数可以对字母进行大小判断，并返回最大的字符或者字符串值。

提示：MAX()函数除了用来找出最大的列值或日期值之外，还可以返回任意列中的最大值，包括返回字符类型的最大值。在对字符类型数据进行比较时，按照字符的 ASCII 码值大小进行比较，从 a~z，a 的 ASCII 码最小，z 的最大。在比较时，先比较第一个字母，如果相等，继续比较下一个字符，一直到两个字符不相等或者字符结束为止。例如，"b"与"t"比较时，"t"为最大值；"bcd"与"bca"比较时，"bcd"为最大值。

5.3.5 MIN()函数

MIN()函数返回查询列中的最小值。

【例 5.44】在 fruits 表中，查找市场上价格最低的水果值，SQL 语句如下：

```
mysql>SELECT MIN(f_price) AS min_price FROM fruits;
+-----------+
| min_price |
+-----------+
|    2.20   |
+-----------+
```

由结果可以看到，MIN ()函数查询出了 f_price 字段的最小值 2.20。

MIN()函数也可以和 GROUP BY 关键字一起使用，求出每个分组中的最小值。

【例 5.45】在 fruits 表中，查找不同供应商提供的价格最低的水果值，SQL 语句如下：

```
mysql> SELECT s_id, MIN(f_price) AS min_price  FROM fruits GROUP BY s_id;
+------+-----------+
| s_id | min_price |
+------+-----------+
|  101 |    3.20   |
|  103 |    2.20   |
|  104 |    6.40   |
|  107 |    3.60   |
|  102 |    5.30   |
|  105 |    2.60   |
|  106 |   15.70   |
+------+-----------+
```

由结果可以看到，GROUP BY 关键字根据 s_id 字段对记录进行分组，然后计算出每个分组中的最小值。MIN()函数与 MAX()函数类似，不仅适用于查找数值类型，也可应用于字符类型。

5.4 连接查询

连接是关系数据库模型的主要特点。连接查询是关系数据库中最主要的查询，主要包括内连接、外连接等。通过连接运算符可以实现多个表查询。在关系数据库管理系统中，表建立时各数据之间的关系不必确定，常把一个实体的所有信息存放在一个表中。在查询数据时，通过连接操作查询出存放在多个表中的不同实体的信息。当两个或多个表中存在相同意义的字段时，便可以通过这些字段对不同的表进行连接查询。本节将介绍多表之间的内连接查询、外连接查询以及复合条件连接查询。

5.4.1 内连接查询

内连接(INNER JOIN)使用比较运算符进行表间某(些)列数据的比较操作,并列出这些表中与连接条件相匹配的数据行,组合成新的记录,也就是说,在内连接查询中,只有满足条件的记录才能出现在结果关系中。

为了演示的需要,首先创建数据表 suppliers,SQL 语句如下:

```
CREATE TABLE suppliers
(
s_id int NOT NULL AUTO_INCREMENT,
s_name char(50) NOT NULL,
s_city char(50) NULL,
s_zip char(10) NULL,
s_call CHAR(50) NOT NULL,
PRIMARY KEY (s_id)
);
```

插入需要演示的数据,SQL 语句如下:

```
INSERT INTO suppliers(s_id, s_name, s_city, s_zip, s_call)
VALUES(101,'FastFruit Inc.','Tianjin','300000','48075'),
(102,'LT Supplies','Chongqing','400000','44333'),
(103,'ACME','Shanghai','200000','90046'),
(104,'FNK Inc.','Zhongshan','528437','11111'),
(105,'Good Set','Taiyuan','030000','22222'),
(106,'Just Eat Ours','Beijing','010','45678'),
(107,'DK Inc.','Zhengzhou','450000','33332');
```

【例 5.46】在 fruits 表和 suppliers 表之间使用内连接查询。

查询之前,查看两个表的结构:

```
mysql> DESC fruits;
+---------+--------------+------+-----+---------+-------+
| Field   | Type         | Null | Key | Default | Extra |
+---------+--------------+------+-----+---------+-------+
| f_id    | char(10)     | NO   | PRI | NULL    |       |
| s_id    | int          | NO   |     | NULL    |       |
| f_name  | char(255)    | NO   |     | NULL    |       |
| f_price | decimal(8,2) | NO   |     | NULL    |       |
+---------+--------------+------+-----+---------+-------+

mysql> DESC suppliers;
+--------+----------+------+-----+---------+----------------+
| Field  | Type     | Null | Key | Default | Extra          |
+--------+----------+------+-----+---------+----------------+
| s_id   | int      | NO   | PRI | NULL    | auto_increment |
| s_name | char(50) | NO   |     | NULL    |                |
| s_city | char(50) | YES  |     | NULL    |                |
| s_zip  | char(10) | YES  |     | NULL    |                |
| s_call | char(50) | NO   |     | NULL    |                |
+--------+----------+------+-----+---------+----------------+
```

由结果可以看到,fruits 表和 suppliers 表中都有相同数据类型的字段 s_id,两个表通过 s_id 字段建立联系。接下来从 fruits 表中查询 f_name、f_price 字段，从 suppliers 表中查询 s_id、s_name，SQL 语句如下：

```
mysql> SELECT suppliers.s_id, s_name,f_name, f_price  FROM fruits ,suppliers WHERE fruits.s_id
= suppliers.s_id;
+------+----------------+------------+---------+
| s_id | s_name         | f_name     | f_price |
+------+----------------+------------+---------+
| 101  | FastFruit Inc. | apple      |  5.20   |
| 103  | ACME           | apricot    |  2.20   |
| 101  | FastFruit Inc. | blackberry | 10.20   |
| 104  | FNK Inc.       | berry      |  7.60   |
| 107  | DK Inc.        | xxxx       |  3.60   |
| 102  | LT Supplies    | orange     | 11.20   |
| 105  | Good Set       | melon      |  8.20   |
| 101  | FastFruit Inc. | cherry     |  3.20   |
| 104  | FNK Inc.       | lemon      |  6.40   |
| 106  | Just Eat Ours  | mango      | 15.70   |
| 105  | Good Set       | xbabay     |  2.60   |
| 105  | Good Set       | xxtt       | 11.60   |
| 103  | ACME           | coconut    |  9.20   |
| 102  | LT Supplies    | banana     | 10.30   |
| 102  | LT Supplies    | grape      |  5.30   |
| 107  | DK Inc.        | xbababa    |  3.60   |
+------+----------------+------------+---------+
```

在这里，SELECT 语句与前面所介绍的一个最大的差别是：SELECT 后面指定的列分别属于两个不同的表，f_name、f_price 在表 fruits 中，而另外两个字段在表 suppliers 中；同时 FROM 子句列出了两个表 fruits 和 suppliers。WHERE 子句在这里作为过滤条件，指明只有两个表中的 s_id 字段值相等的时候才符合连接查询的条件。从返回的结果可以看到，显示的记录是由两个表中的不同列值组成的新记录。

提示：因为 fruits 表和 suppliers 表中有相同的字段 s_id，因此在比较的时候需要完全限定表名（格式为"表名.列名"），如果只给出 s_id，MySQL 将不知道指的是哪一个，并返回错误信息。

下面的内连接查询语句返回与前面例子完全相同的结果。

【例 5.47】在 fruits 表和 suppliers 表之间，使用 INNER JOIN 语法进行内连接查询，SQL 语句如下：

```
mysql> SELECT suppliers.s_id, s_name,f_name, f_price  FROM fruits INNER JOIN suppliers ON
fruits.s_id = suppliers.s_id;
+------+----------------+------------+---------+
| s_id | s_name         | f_name     | f_price |
+------+----------------+------------+---------+
| 101  | FastFruit Inc. | apple      |  5.20   |
| 103  | ACME           | apricot    |  2.20   |
| 101  | FastFruit Inc. | blackberry | 10.20   |
| 104  | FNK Inc.       | berry      |  7.60   |
| 107  | DK Inc.        | xxxx       |  3.60   |
| 102  | LT Supplies    | orange     | 11.20   |
| 105  | Good Set       | melon      |  8.20   |
| 101  | FastFruit Inc. | cherry     |  3.20   |
```

```
| 104 | FNK Inc.       | lemon   |  6.40 |
| 106 | Just Eat Ours  | mango   | 15.70 |
| 105 | Good Set       | xbabay  |  2.60 |
| 105 | Good Set       | xxtt    | 11.60 |
| 103 | ACME           | coconut |  9.20 |
| 102 | LT Supplies    | banana  | 10.30 |
| 102 | LT Supplies    | grape   |  5.30 |
| 107 | DK Inc.        | xbababa |  3.60 |
+-----+----------------+---------+-------+
```

在这里的查询语句中，两个表之间的关系通过 INNER JOIN 指定。使用这种语法的时候，连接的条件使用 ON 子句而不是 WHERE，ON 和 WHERE 后面指定的条件相同。

提示：使用 WHERE 子句定义连接条件比较简单明了，而 INNER JOIN 语法是 ANSI SQL 的标准规范，使用 INNER JOIN 连接语法能够确保不会忘记连接条件，而且 WHERE 子句在某些时候会影响查询的性能。

如果在一个连接查询中，涉及的两个表都是同一个表，这种查询称为自连接查询。自连接是一种特殊的内连接，它是指相互连接的表在物理上为同一张表，但可以在逻辑上分为两张表。

【例 5.48】 查询 f_id= 'a1'的水果供应商提供的水果种类，SQL 语句如下：

```
mysql> SELECT f1.f_id, f1.f_name  FROM fruits AS f1, fruits AS f2  WHERE f1.s_id = f2.s_id AND
f2.f_id = 'a1';
+------+------------+
| f_id | f_name     |
+------+------------+
| a1   | apple      |
| b1   | blackberry |
| c0   | cherry     |
+------+------------+
```

此处查询的两个表是相同的表，为了防止产生二义性，对表使用了别名，fruits 表第 1 次出现的别名为 f1，第 2 次出现的别名为 f2；使用 SELECT 语句返回列时，明确指出返回以 f1 为前缀的列的全名；WHERE 连接两个表，并按照第 2 个表的 f_id 对数据进行过滤，返回所需数据。

5.4.2　外连接查询

外连接查询将查询多个表中相关联的行；内连接时，返回查询结果集合中仅是符合查询条件和连接条件的行。有时候需要包含没有关联的行中数据，即返回查询结果集合中不仅包含符合连接条件的行，还包括左表（左外连接或左连接）、右表（右外连接或右连接）或两个边接表（全外连接）中的所有数据行。外连接分为左外连接（左连接）和右外连接（右连接）：

- LEFT JOIN（左连接）：返回包括左表中的所有记录和右表中连接字段相等的记录。
- RIGHT JOIN（右连接）：返回包括右表中的所有记录和左表中连接字段相等的记录。

1. LEFT JOIN 左连接

左连接的结果包括 LEFT OUTER 子句中指定的左表的所有行,而不仅仅是连接列所匹配的行。如果左表的某行在右表中没有匹配行，则在相关联的结果行中，右表的所有选择列表列均为空值。

首先创建表 orders，SQL 语句如下：

```
CREATE TABLE orders
(
o_num int NOT NULL AUTO_INCREMENT,
o_date datetime NOT NULL,
c_id int NOT NULL,
PRIMARY KEY (o_num)
);
```

插入需要演示的数据，SQL 语句如下：

```
INSERT INTO orders(o_num, o_date, c_id)
VALUES(30001, '2008-09-01', 10001),
(30002, '2008-09-12', 10003),
(30003, '2008-09-30', 10004),
(30004, '2008-10-03', 10005),
(30005, '2008-10-08', 10001);
```

【例 5.49】在 customers 表和 orders 表中，查询所有客户，包括没有订单的客户，SQL 语句如下：

```
mysql> SELECT customers.c_id, orders.o_num  FROM customers LEFT OUTER JOIN orders ON
customers.c_id = orders.c_id;
+-------+-------+
| c_id  | o_num |
+-------+-------+
| 10001 | 30001 |
| 10003 | 30002 |
| 10004 | 30003 |
| 10001 | 30005 |
| 10002 |  NULL |
+-------+-------+
```

结果显示了 5 条记录，ID 等于 10002 的客户目前并没有下订单，所以对应的 orders 表中并没有该客户的订单信息，所以该条记录只取出了 customers 表中相应的值，而从 orders 表中取出的值为空值 NULL。

2. RIGHT JOIN 右连接

右连接是左连接的反向连接，将返回右表的所有行。如果右表的某行在左表中没有匹配行，左表将返回空值。

【例 5.50】在 customers 表和 orders 表中，查询所有订单，包括没有客户的订单，SQL 语句如下：

```
mysql> SELECT customers.c_id, orders.o_num  FROM customers RIGHT OUTER JOIN orders ON
customers.c_id = orders.c_id;
+-------+-------+
| c_id  | o_num |
+-------+-------+
| 10001 | 30001 |
| 10003 | 30002 |
| 10004 | 30003 |
|  NULL | 30004 |
```

```
| 10001  | 30005  |
+--------+--------+
```

结果显示了 5 条记录，订单号等于 30004 的订单的客户可能由于某种原因取消了该订单，对应的 customers 表中并没有该客户的信息，所以该条记录只取出了 orders 表中相应的值，而从 customers 表中取出的值为空值 NULL。

5.4.3 复合条件连接查询

复合条件连接查询是在连接查询的过程中，通过添加过滤条件限制查询的结果，使查询的结果更加准确。

【例 5.51】在 customers 表和 orders 表中，使用 INNER JOIN 语法查询 customers 表中 ID 为 10001 的客户的订单信息，SQL 语句如下：

```
mysql> SELECT customers.c_id, orders.o_num  FROM customers INNER JOIN orders
  ON customers.c_id = orders.c_id AND customers.c_id = 10001;
+--------+--------+
| c_id   | o_num  |
+--------+--------+
| 10001  | 30001  |
| 10001  | 30005  |
+--------+--------+
```

结果显示，在连接查询时指定查询客户 ID 为 10001 的订单信息，添加了过滤条件之后返回的结果将会变少，因此返回结果只有两条记录。

使用连接查询，并对查询的结果进行排序。

【例 5.52】在 fruits 表和 suppliers 表之间，使用 INNER JOIN 语法进行内连接查询，并对查询结果排序，SQL 语句如下：

```
mysql> SELECT suppliers.s_id, s_name,f_name, f_price FROM fruits INNER JOIN suppliers ON
fruits.s_id = suppliers.s_id ORDER BY fruits.s_id;
+------+---------------+------------+---------+
| s_id | s_name        | f_name     | f_price |
+------+---------------+------------+---------+
| 101  | FastFruit Inc.| apple      |    5.20 |
| 101  | FastFruit Inc.| blackberry |   10.20 |
| 101  | FastFruit Inc.| cherry     |    3.20 |
| 102  | LT Supplies   | orange     |   11.20 |
| 102  | LT Supplies   | banana     |   10.30 |
| 102  | LT Supplies   | grape      |    5.30 |
| 103  | ACME          | apricot    |    2.20 |
| 103  | ACME          | coconut    |    9.20 |
| 104  | FNK Inc.      | berry      |    7.60 |
| 104  | FNK Inc.      | lemon      |    6.40 |
| 105  | Good Set      | melon      |    8.20 |
| 105  | Good Set      | xbabay     |    2.60 |
| 105  | Good Set      | xxtt       |   11.60 |
| 106  | Just Eat Ours | mango      |   15.70 |
| 107  | DK Inc.       | xxxx       |    3.60 |
| 107  | DK Inc.       | xbababa    |    3.60 |
```

由结果可以看到,内连接查询的结果按照 suppliers.s_id 字段进行了升序排序。

5.5 子查询

子查询指一个查询语句嵌套在另一个查询语句内部的查询,这个特性从 MySQL 4.1 开始引入。在 SELECT 子句中先计算子查询,子查询结果作为外层另一个查询的过滤条件,查询可以基于一个表或者多个表。子查询中常用的操作符有 ANY(SOME)、ALL、IN、EXISTS。子查询可以添加到 SELECT、UPDATE 和 DELETE 语句中,而且可以进行多层嵌套。子查询中也可以使用比较运算符,如"<""<="">"">="和"!="等。本节将介绍如何在 SELECT 语句中嵌套子查询。

5.5.1 带 ANY、SOME 关键字的子查询

ANY 和 SOME 关键字是同义词,表示满足其中任一条件,它们允许创建一个表达式对子查询的返回值列表进行比较,只要满足内层子查询中的任何一个比较条件,就返回一个结果作为外层查询的条件。

下面定义两个表 tbl1 和 tbl2:

```
CREATE table tbl1 ( num1 INT NOT NULL);
CREATE table tbl2 ( num2 INT NOT NULL);
```

分别向两个表中插入数据:

```
INSERT INTO tbl1 values(1), (5), (13), (27);
INSERT INTO tbl2 values(6), (14), (11), (20);
```

ANY 关键字接在一个比较操作符的后面,表示若与子查询返回的任何值比较为 TRUE,则返回 TRUE。

【例 5.53】返回 tbl2 表的所有 num2 列,然后将 tbl1 中的 num1 的值与之进行比较,只要大于 num2 的任何 1 个值,即为符合查询条件的结果。

```
mysql> SELECT num1 FROM tbl1 WHERE num1 > ANY (SELECT num2 FROM tbl2);
+------+
| num1 |
+------+
|   13 |
|   27 |
+------+
```

在子查询中,返回的是 tbl2 表的所有 num2 列结果(6,14,11,20),然后将 tbl1 中的 num1 列的值与之进行比较,只要大于 num2 列的任意一个数即为符合条件的结果。

5.5.2 带 ALL 关键字的子查询

ALL 关键字与 ANY 和 SOME 不同,使用 ALL 时需要同时满足所有内层查询的条件。例如,

修改前面的例子，用 ALL 关键字替换 ANY。

ALL 关键字接在一个比较操作符的后面，表示与子查询返回的所有值比较为 TRUE，则返回 TRUE。

【例 5.54】返回 tbl1 表中比 tbl2 表的 num2 列所有值都大的值，SQL 语句如下：

```
mysql> SELECT num1 FROM tbl1 WHERE num1 > ALL (SELECT num2 FROM tbl2);
+------+
| num1 |
+------+
|   27 |
+------+
```

在子查询中，返回的是 tbl2 的所有 num2 列结果（6,14,11,20），然后将 tbl1 中的 num1 列的值与之进行比较，大于所有 num2 列值的 num1 值只有 27，因此返回结果为 27。

5.5.3　带 EXISTS 关键字的子查询

EXISTS 关键字后面的参数是一个任意的子查询，系统对子查询进行运算以判断它是否返回行，如果至少返回一行，那么 EXISTS 的结果为 true，此时外层查询语句将进行查询；如果子查询没有返回任何行，那么 EXISTS 返回的结果是 false，此时外层语句将不进行查询。

【例 5.55】查询 suppliers 表中是否存在 s_id=107 的供应商，如果存在，则查询 fruits 表中的记录，SQL 语句如下：

```
mysql> SELECT * FROM fruits WHERE EXISTS (SELECT s_name FROM suppliers WHERE s_id = 107);
+------+------+-----------+---------+
| f_id | s_id | f_name    | f_price |
+------+------+-----------+---------+
| a1   | 101  | apple     |    5.20 |
| a2   | 103  | apricot   |    2.20 |
| b1   | 101  | blackberry|   10.20 |
| b2   | 104  | berry     |    7.60 |
| b5   | 107  | xxxx      |    3.60 |
| bs1  | 102  | orange    |   11.20 |
| bs2  | 105  | melon     |    8.20 |
| c0   | 101  | cherry    |    3.20 |
| l2   | 104  | lemon     |    6.40 |
| m1   | 106  | mango     |   15.70 |
| m2   | 105  | xbabay    |    2.60 |
| m3   | 105  | xxtt      |   11.60 |
| o2   | 103  | coconut   |    9.20 |
| t1   | 102  | banana    |   10.30 |
| t2   | 102  | grape     |    5.30 |
| t4   | 107  | xbababa   |    3.60 |
+------+------+-----------+---------+
```

由结果可以看到，内层查询结果表明 suppliers 表中存在 s_id=107 的记录，因此 EXISTS 表达式返回 true；外层查询语句接收 true 之后对表 fruits 进行查询，返回所有的记录。

EXISTS 关键字可以和条件表达式一起使用。

【例 5.56】查询 suppliers 表中是否存在 s_id=107 的供应商，如果存在，则查询 fruits 表中的 f_price 大于 10.20 的记录，SQL 语句如下：

```
mysql> SELECT * FROM fruits WHERE f_price>10.20 AND EXISTS  (SELECT s_name FROM suppliers WHERE s_id = 107);
+------+------+--------+---------+
| f_id | s_id | f_name | f_price |
+------+------+--------+---------+
| bs1  | 102  | orange | 11.20   |
| m1   | 106  | mango  | 15.70   |
| m3   | 105  | xxtt   | 11.60   |
| t1   | 102  | banana | 10.30   |
+------+------+--------+---------+
```

由结果可以看到，内层查询结果表明 suppliers 表中存在 s_id=107 的记录，因此 EXISTS 表达式返回 true；外层查询语句接收 true 之后，根据查询条件 f_price > 10.20 对 fruits 表进行查询，返回结果为 4 条 f_price 大于 10.20 的记录。

NOT EXISTS 与 EXISTS 使用方法相同，返回的结果相反。子查询如果至少返回一行，那么 NOT EXISTS 的结果为 false，此时外层查询语句将不进行查询；如果子查询没有返回任何行，那么 NOT EXISTS 返回的结果是 true，此时外层语句将进行查询。

【例 5.57】查询 suppliers 表中是否存在 s_id=107 的供应商，如果不存在，则查询 fruits 表中的记录，SQL 语句如下：

```
mysql> SELECT * FROM fruits WHERE NOT EXISTS  (SELECT s_name FROM suppliers WHERE s_id = 107);
Empty set (0.00 sec)
```

查询语句 SELECT s_name FROM suppliers WHERE s_id = 107，对 suppliers 表进行查询返回了一条记录，NOT EXISTS 表达式返回 false，外层表达式接收 false，将不再查询 fruits 表中的记录。

提示：EXISTS 和 NOT EXISTS 的结果只取决于是否会返回行，而不取决于这些行的内容，所以这个子查询输入列表通常是无关紧要的。

5.5.4 带 IN 关键字的子查询

IN 关键字进行子查询时，内层查询语句仅仅返回一个数据列，这个数据列里的值将提供给外层查询语句进行比较操作。

【例 5.58】在 orderitems 表中，查询 f_id 为 c0 的订单号，并根据订单号查询具有订单号的客户 c_id，SQL 语句如下：

```
mysql> SELECT c_id FROM orders WHERE o_num IN (SELECT o_num  FROM orderitems WHERE f_id = 'c0');
+-------+
| c_id  |
+-------+
| 10004 |
| 10001 |
+-------+
```

查询结果的 c_id 有两个值，分别为 10001 和 10004。上述查询过程可以分步执行，首先内层子查询查出 orderitems 表中符合条件的订单号，单独执行内查询，查询结果如下：

```
mysql> SELECT o_num  FROM orderitems WHERE f_id = 'c0';
+-------+
| o_num |
+-------+
| 30003 |
| 30005 |
+-------+
```

可以看到，符合条件的 o_num 列的值有两个：30003 和 30005，然后执行外层查询，在 orders 表中查询订单号等于 30003 或 30005 的客户 c_id。嵌套子查询语句还可以写为如下形式，实现相同的效果：

```
mysql> SELECT c_id FROM orders WHERE o_num IN (30003, 30005);
+-------+
| c_id  |
+-------+
| 10004 |
| 10001 |
+-------+
```

这个例子说明在处理 SELECT 语句的时候，MySQL 实际上执行了两个操作过程，即先执行内层子查询，再执行外层查询，内层子查询的结果作为外部查询的比较条件。

SELECT 语句中可以使用 NOT IN 关键字，其作用与 IN 正好相反。

【例 5.59】与前一个例子类似，但是在 SELECT 语句中使用 NOT IN 关键字，SQL 语句如下：

```
mysql> SELECT c_id FROM orders WHERE o_num NOT IN (SELECT o_num  FROM orderitems WHERE f_id = 'c0');
+-------+
| c_id  |
+-------+
| 10001 |
| 10003 |
| 10005 |
+-------+
```

这里返回的结果有 3 条记录，由前面例子可以看到，子查询返回的订单值有两个，即 30003 和 30005，但为什么这里还有值为 10001 的 c_id 呢？这是因为 c_id 等于 10001 的客户的订单不止一个，可以查看订单表 orders 中的记录。

```
mysql> SELECT * FROM orders;
+-------+---------------------+-------+
| o_num | o_date              | c_id  |
+-------+---------------------+-------+
| 30001 | 2008-09-01 00:00:00 | 10001 |
| 30002 | 2008-09-12 00:00:00 | 10003 |
| 30003 | 2008-09-30 00:00:00 | 10004 |
| 30004 | 2008-10-03 00:00:00 | 10005 |
| 30005 | 2008-10-08 00:00:00 | 10001 |
+-------+---------------------+-------+
```

可以看到，虽然排除了订单号为 30003 和 30005 的客户 c_id，但是 o_num 为 30001 的订单与

30005 都是 10001 号客户的订单。所以结果中只是排除了订单号，但是仍然有可能选择同一个客户。

提示：子查询的功能也可以通过连接查询完成，但是子查询使得 MySQL 代码更容易阅读和编写。

5.5.5 带比较运算符的子查询

在前面介绍的带 ANY、ALL 关键字的子查询时使用了 ">" 比较运算符，子查询时还可以使用其他的比较运算符，如 "<" "<=" "=" ">=" 和 "!=" 等。

【例 5.60】在 suppliers 表中查询 s_city 等于 "Tianjin" 的供应商 s_id，然后在 fruits 表中查询所有该供应商提供的水果的种类，SQL 语句如下：

```
SELECT s_id, f_name FROM fruits WHERE s_id = (SELECT s1.s_id FROM suppliers AS s1 WHERE s1.s_city = 'Tianjin');
```

该嵌套查询首先在 suppliers 表中查找 s_city 等于 "Tianjin" 的供应商的 s_id，单独执行子查询查看 s_id 的值，执行下面的操作过程：

```
mysql> SELECT s1.s_id FROM suppliers AS s1 WHERE s1.s_city = 'Tianjin';
+------+
| s_id |
+------+
|  101 |
+------+
```

然后在外层查询时，在 fruits 表中查找 s_id 等于 101 的供应商提供的水果的种类，查询结果如下：

```
mysql> SELECT s_id, f_name FROM fruits WHERE s_id = (SELECT s1.s_id FROM suppliers AS s1 WHERE s1.s_city = 'Tianjin');
+------+------------+
| s_id | f_name     |
+------+------------+
|  101 | apple      |
|  101 | blackberry |
|  101 | cherry     |
+------+------------+
```

结果表明，"Tianjin" 地区的供应商提供的水果种类有 3 种，分别为 "apple" "blackberry" "cherry"。

【例 5.61】在 suppliers 表中查询 s_city 等于 "Tianjin" 的供应商 s_id，然后在 fruits 表中查询所有非该供应商提供的水果的种类，SQL 语句如下：

```
mysql> SELECT s_id, f_name FROM fruits WHERE s_id <> (SELECT s1.s_id FROM suppliers AS s1 WHERE s1.s_city = 'Tianjin');
+------+---------+
| s_id | f_name  |
+------+---------+
|  103 | apricot |
|  104 | berry   |
|  107 | xxxx    |
|  102 | orange  |
|  105 | melon   |
|  104 | lemon   |
```

```
| 106 | mango   |
| 105 | xbabay  |
| 105 | xxtt    |
| 103 | coconut |
| 102 | banana  |
| 102 | grape   |
| 107 | xbababa |
```

该嵌套查询执行过程与前面相同，在这里使用了不等于"<>"运算符，因此返回的结果和前面正好相反。

5.6 合并查询结果

利用 UNION 关键字，可以给出多条 SELECT 语句，并将它们的结果组合成单个结果集。合并时，两个表对应的列数和数据类型必须相同。各个 SELECT 语句之间使用 UNION 或 UNION ALL 关键字分隔。UNION 不使用关键字 ALL，执行的时候删除重复的记录，所有返回的行都是唯一的；使用关键字 ALL 的作用是不删除重复行也不对结果进行自动排序。基本语法格式如下：

```
SELECT column,... FROM table1
UNION [ALL]
SELECT column,... FROM table2
```

【例 5.62】查询所有价格小于 9 的水果的信息，查询 s_id 等于 101 和 103 所有的水果的信息，使用 UNION 连接查询结果，SQL 语句如下：

```
SELECT s_id, f_name, f_price FROM fruits WHERE f_price < 9.0 UNION SELECT s_id, f_name, f_price
FROM fruits WHERE s_id IN(101,103);
```

合并查询结果如下：

```
| s_id | f_name     | f_price |
|------|------------|---------|
| 101  | apple      | 5.20    |
| 103  | apricot    | 2.20    |
| 104  | berry      | 7.60    |
| 107  | xxxx       | 3.60    |
| 105  | melon      | 8.20    |
| 101  | cherry     | 3.20    |
| 104  | lemon      | 6.40    |
| 105  | xbabay     | 2.60    |
| 102  | grape      | 5.30    |
| 107  | xbababa    | 3.60    |
| 101  | blackberry | 10.20   |
| 103  | coconut    | 9.20    |
```

如前所述，UNION 将多个 SELECT 语句的结果组合成一个结果集合。可以分开查看每个 SELECT 语句的结果：

```
mysql> SELECT s_id, f_name, f_price FROM fruits WHERE f_price < 9.0;
+------+---------+---------+
| s_id | f_name  | f_price |
+------+---------+---------+
|  101 | apple   |    5.20 |
|  103 | apricot |    2.20 |
|  104 | berry   |    7.60 |
|  107 | xxxx    |    3.60 |
|  105 | melon   |    8.20 |
|  101 | cherry  |    3.20 |
|  104 | lemon   |    6.40 |
|  105 | xbabay  |    2.60 |
|  102 | grape   |    5.30 |
|  107 | xbababa |    3.60 |
+------+---------+---------+

mysql> SELECT s_id, f_name, f_price FROM fruits WHERE s_id IN(101,103);
+------+------------+---------+
| s_id | f_name     | f_price |
+------+------------+---------+
|  101 | apple      |    5.20 |
|  103 | apricot    |    2.20 |
|  101 | blackberry |   10.20 |
|  101 | cherry     |    3.20 |
|  103 | coconut    |    9.20 |
+------+------------+---------+
```

由分开查询的结果可以看到，第 1 条 SELECT 语句查询价格小于 9 的水果，第 2 条 SELECT 语句查询供应商 101 和 103 提供的水果。使用 UNION 将两条 SELECT 语句分隔开，执行完毕之后把输出结果组合成单个的结果集，并删除重复的记录。

使用 UNION ALL 包含重复的行，在前面的例子中，分开查询时，两个返回结果中有相同的记录。UNION 从查询结果集中自动去除了重复的行，如果要返回所有匹配行，而不进行删除，可以使用 UNION ALL。

【例 5.63】查询所有价格小于 9 的水果的信息，查询 s_id 等于 101 和 103 的所有水果的信息，使用 UNION ALL 连接查询结果，SQL 语句如下：

```
SELECT s_id, f_name, f_price FROM fruits WHERE f_price < 9.0 UNION ALL
SELECT s_id, f_name, f_price FROM fruits WHERE s_id IN(101,103);
```

查询结果如下：

```
+------+---------+---------+
| s_id | f_name  | f_price |
+------+---------+---------+
|  101 | apple   |    5.20 |
|  103 | apricot |    2.20 |
|  104 | berry   |    7.60 |
|  107 | xxxx    |    3.60 |
|  105 | melon   |    8.20 |
|  101 | cherry  |    3.20 |
|  104 | lemon   |    6.40 |
|  105 | xbabay  |    2.60 |
```

```
| 102 | grape      |  5.30 |
| 107 | xbababa    |  3.60 |
| 101 | apple      |  5.20 |
| 103 | apricot    |  2.20 |
| 101 | blackberry | 10.20 |
| 101 | cherry     |  3.20 |
| 103 | coconut    |  9.20 |
```

由结果可以看到，这里总的记录数等于两条 SELECT 语句返回的记录数之和，连接查询结果并没有去除重复的行。

提示：UNION 和 UNION ALL 的区别：使用 UNION ALL 的功能是不删除重复行，加上 ALL 关键字语句执行时所需要的资源少，所以尽可能地使用它，因此知道有重复行但是想保留这些行，确定查询结果中不会有重复数据或者不需要删除重复数据的时候，应当使用 UNION ALL 以提高查询效率。

5.7 为表和字段取别名

在前面介绍分组查询、集合函数查询和嵌套子查询章节中，读者可能注意到有的地方使用 AS 关键字，为查询结果中的某一列指定一个特定的名字。在内连接查询时，则对相同的表 fruits 分别指定两个不同的名字，这里可以为字段或者表取一个别名，在查询时，使用别名替代其指定的内容。本节将介绍如何为字段和表创建别名以及如何使用别名。

5.7.1 为表取别名

当表名字很长或者执行一些特殊查询时，为了方便操作或者需要多次使用相同的表时，可以为表指定别名，用这个别名替代表原来的名称。为表取别名的基本语法格式为：

表名 [AS] 表别名

"表名"为数据库中存储的数据表的名称，"表别名"为查询时指定的表的新名称，AS 关键字为可选参数。

【例 5.64】为 orders 表取别名 o，查询 30001 订单的下单日期，SQL 语句如下：

SELECT * FROM orders AS o WHERE o.o_num = 30001;

在这里 orders AS o 代码表示为 orders 表取别名为 o，指定过滤条件时直接使用 o 代替 orders，查询结果如下：

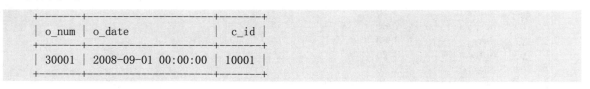

【例 5.65】为 customers 和 orders 表分别取别名，并进行连接查询，SQL 语句如下：

```
mysql> SELECT c.c_id, o.o_num FROM customers AS c LEFT OUTER JOIN orders AS o ON c.c_id = o.c_id;
+-------+-------+
| c_id  | o_num |
+-------+-------+
| 10001 | 30001 |
| 10001 | 30005 |
| 10002 | NULL  |
| 10003 | 30002 |
| 10004 | 30003 |
+-------+-------+
```

由结果看到，MySQL 可以同时为多个表取别名，而且表别名可以放在不同的位置，如 WHERE 子句、SELECT 列表、ON 子句以及 ORDER BY 子句等。

在前面介绍内连接查询时，我们知道自连接是一种特殊的内连接，在连接查询中的两个表都是同一个表，其查询语句如下：

```
mysql> SELECT f1.f_id, f1.f_name FROM fruits AS f1, fruits AS f2 WHERE f1.s_id = f2.s_id AND
f2.f_id = 'a1';
+------+------------+
| f_id | f_name     |
+------+------------+
| a1   | apple      |
| b1   | blackberry |
| c0   | cherry     |
+------+------------+
```

在这里，如果不使用表别名，MySQL 将不知道引用的是哪个 fruits 表实例，这是表别名一个非常有用的地方。

提示：在为表取别名时，要保证不能与数据库中其他表的名称冲突。

5.7.2　为字段取别名

从本章和前面各章节的例子中可以看到，在使用 SELECT 语句显示查询结果时，MySQL 会显示每个 SELECT 后面指定的输出列，在有些情况下，显示的列的名称会很长或者名称不够直观，MySQL 可以指定列别名，替换字段或表达式。为字段取别名的基本语法格式为：

列名 [AS] 列别名

"列名"为表中字段定义的名称，"列别名"为字段新的名称，AS 关键字为可选参数。

【例 5.66】查询 fruits 表，为 f_name 取别名 fruit_name，f_price 取别名 fruit_price，为 fruits 表取别名 f1，查询表中 f_price < 8 的水果的名称，SQL 语句如下：

```
mysql> SELECT f1.f_name AS fruit_name, f1.f_price AS fruit_price FROM fruits AS f1 WHERE
f1.f_price < 8;
+------------+-------------+
| fruit_name | fruit_price |
+------------+-------------+
| apple      |        5.20 |
| apricot    |        2.20 |
```

```
| berry   |    7.60 |
| xxxx    |    3.60 |
| cherry  |    3.20 |
| lemon   |    6.40 |
| xbabay  |    2.60 |
| grape   |    5.30 |
| xbababa |    3.60 |
+---------+---------+
```

也可以为 SELECT 子句中的计算字段取别名。例如，对使用 COUNT 聚合函数或者 CONCAT 等系统函数执行的结果字段取别名。

【例 5.67】查询 suppliers 表中字段 s_name 和 s_city，使用 CONCAT 函数连接这两个字段值，并取列别名为 suppliers_title。

如果没有对连接后的值取别名，其显示列名称将会不够直观，SQL 语句如下：

```
mysql> SELECT CONCAT(TRIM(s_name) , ' (', TRIM(s_city), ')') FROM suppliers
 ORDER BY s_name;
+------------------------------------------------+
| CONCAT(TRIM(s_name) , ' (', TRIM(s_city), ')') |
+------------------------------------------------+
| ACME (Shanghai)                                |
| DK Inc. (Zhengzhou)                            |
| FastFruit Inc. (Tianjin)                       |
| FNK Inc. (Zhongshan)                           |
| Good Set (Taiyuang)                            |
| Just Eat Ours (Beijing)                        |
| LT Supplies (Chongqing)                        |
+------------------------------------------------+
```

由结果可以看到，显示结果的列名称为 SELECT 子句后面的计算字段，实际上计算之后的列是没有名字的，这样的结果让人很不容易理解，如果为字段取一个别名，将会使结果清晰，SQL 语句如下：

```
mysql> SELECT CONCAT(TRIM(s_name) , ' (', TRIM(s_city), ')') AS suppliers_title FROM suppliers
ORDER BY s_name;
+--------------------------+
| suppliers_title          |
+--------------------------+
| ACME (Shanghai)          |
| DK Inc. (Zhengzhou)      |
| FastFruit Inc. (Tianjin) |
| FNK Inc. (Zhongshan)     |
| Good Set (Taiyuang)      |
| Just Eat Ours (Beijing)  |
| LT Supplies (Chongqing)  |
+--------------------------+
```

由结果可以看到，SELECT 子句计算字段值之后增加了 AS suppliers_title，它指示 MySQL 为计算字段创建一个别名 suppliers_title，显示结果为指定的列别名，这样就增强了查询结果的可读性。

提示：表别名只在执行查询的时候使用，并不在返回结果中显示，而列别名定义之后，将返回给客户端显示，显示的结果字段为字段列的别名。

5.8　使用正则表达式查询

正则表达式通常被用来检索或替换那些符合某个模式的文本内容，根据指定的匹配模式匹配文本中符合要求的特殊字符串。例如，从一个文本文件中提取电话号码，查找一篇文章中重复的单词或者替换用户输入的某些敏感词语等，这些情形都可以使用正则表达式。正则表达式强大而且灵活，可以应用于非常复杂的查询。

MySQL 中使用 REGEXP 关键字指定正则表达式的字符匹配模式。表 5.3 列出了 REGEXP 操作符中常用字符匹配列表。

表 5.3　正则表达式常用字符匹配列表

选项	说明	例子	匹配值示例
^	匹配文本的开始字符	'^b'匹配以字母 b 开头的字符串	book, big, banana, bike
$	匹配文本的结束字符	'st$'匹配以 st 结尾的字符串	test, resist, persist
.	匹配任何单个字符	'b.t'匹配任何 b 和 t 之间有一个字符的字符串	bit, bat, but, bite
*	匹配零个或多个在它前面的字符	'f*n'匹配字符 n 前面有任意个字符 f 的字符串	fn, fan, faan, fabcn
+	匹配前面的字符 1 次或多次	'ba+'匹配以 b 开头后面紧跟至少有一个 a 的字符串	ba, bay, bare, battle
<字符串>	匹配包含指定的字符串的文本	'fa'匹配包含 fa 的字符串	fan, afa, faad
[字符集合]	匹配字符集合中的任何一个字符	'[xz]' 匹配包含 x 或者 z 的字符串	dizzy, zebra, x-ray, extra
[^]	匹配不在括号中的任何字符	'[^abc]'匹配任何不包含 a、b 或 c 的字符串	desk, fox, f8ke
字符串{n,}	匹配前面的字符串至少 n 次	b{2}匹配 2 个或更多的 b	bbb, bbbb, bbbbbbb
字符串{n,m}	匹配前面的字符串至少 n 次，至多 m 次。如果 n 为 0，此参数为可选参数	b{2,4}匹配含最少 2 个、最多 4 个 b 的字符串	bb, bbb, bbbb

下面将详细介绍在 MySQL 中如何使用正则表达式。

5.8.1　查询以特定字符或字符串开头的记录

字符"^"匹配以特定字符或者字符串开头的文本。

【例 5.68】在 fruits 表中，查询 f_name 字段以字母"b"开头的记录，SQL 语句如下：

```
mysql> SELECT * FROM fruits WHERE f_name REGEXP '^b';
+-------+--------+---------+
```

```
+------+------+------------+---------+
| f_id | s_id | f_name     | f_price |
+------+------+------------+---------+
| b1   | 101  | blackberry |  10.20  |
| b2   | 104  | berry      |   7.60  |
| t1   | 102  | banana     |  10.30  |
+------+------+------------+---------+
```

fruits 表中有 3 条记录的 f_name 字段值是以字母 b 开头的,返回结果有 3 条记录。

【例 5.69】在 fruits 表中,查询 f_name 字段以 "be" 开头的记录,SQL 语句如下:

```
mysql> SELECT * FROM fruits WHERE f_name REGEXP '^be';
+------+------+--------+---------+
| f_id | s_id | f_name | f_price |
+------+------+--------+---------+
| b2   | 104  | berry  |   7.60  |
+------+------+--------+---------+
```

只有 berry 是以 "be" 开头的,所以查询结果中只有 1 条记录。

5.8.2 查询以特定字符或字符串结尾的记录

字符 "$" 匹配以特定字符或者字符串结尾的文本。

【例 5.70】在 fruits 表中,查询 f_name 字段以字母 "y" 结尾的记录,SQL 语句如下:

```
mysql> SELECT * FROM fruits WHERE f_name REGEXP 'y$';
+------+------+------------+---------+
| f_id | s_id | f_name     | f_price |
+------+------+------------+---------+
| b1   | 101  | blackberry |  10.20  |
| b2   | 104  | berry      |   7.60  |
| c0   | 101  | cherry     |   3.20  |
| m2   | 105  | xbabay     |   2.60  |
+------+------+------------+---------+
```

fruits 表中有 4 条记录的 f_name 字段值是以字母 "y" 结尾的,返回结果有 4 条记录。

【例 5.71】在 fruits 表中,查询 f_name 字段以字符串 "rry" 结尾的记录,SQL 语句如下:

```
mysql> SELECT * FROM fruits WHERE f_name REGEXP 'rry$';
+------+------+------------+---------+
| f_id | s_id | f_name     | f_price |
+------+------+------------+---------+
| b1   | 101  | blackberry |  10.20  |
| b2   | 104  | berry      |   7.60  |
| c0   | 101  | cherry     |   3.20  |
+------+------+------------+---------+
```

fruits 表中有 3 条记录的 f_name 字段值是以字符串 "rry" 结尾的,返回结果有 3 条记录。

5.8.3 用符号 "." 来替代字符串中的任意一个字符

字符 "." 匹配任意一个字符。

【例 5.72】在 fruits 表中，查询 f_name 字段值包含字母 "a" 与 "g" 且两个字母之间只有一个字母的记录，SQL 语句如下：

```
mysql> SELECT * FROM fruits WHERE f_name REGEXP 'a.g';
+------+------+--------+---------+
| f_id | s_id | f_name | f_price |
+------+------+--------+---------+
| bs1  | 102  | orange |  11.20  |
| m1   | 106  | mango  |  15.70  |
+------+------+--------+---------+
```

查询语句中 "a.g" 指定匹配字符中要有字母 a 和 g，且两个字母之间包含单个字符，并不限定匹配的字符的位置和所在查询字符串的总长度，因此 orange 和 mango 都符合匹配条件。

5.8.4 使用 "*" 和 "+" 来匹配多个字符

星号 "*" 匹配前面的字符任意多次，包括 0 次。加号 "+" 匹配前面的字符至少一次。

【例 5.73】在 fruits 表中，查询 f_name 字段值以字母 "b" 开头且 "b" 后面出现字母 "a" 的记录，SQL 语句如下：

```
mysql> SELECT * FROM fruits WHERE f_name REGEXP '^ba*';
+------+------+------------+---------+
| f_id | s_id | f_name     | f_price |
+------+------+------------+---------+
| b1   | 101  | blackberry |  10.20  |
| b2   | 104  | berry      |   7.60  |
| t1   | 102  | banana     |  10.30  |
+------+------+------------+---------+
```

星号 "*" 可以匹配任意多个字符，blackberry 和 berry 中字母 b 后面并没有出现字母 a，但是也满足匹配条件。

【例 5.74】在 fruits 表中，查询 f_name 字段值以字母 "b" 开头且 "b" 后面出现字母 "a" 至少一次的记录，SQL 语句如下：

```
mysql> SELECT * FROM fruits WHERE f_name REGEXP '^ba+';
+------+------+--------+---------+
| f_id | s_id | f_name | f_price |
+------+------+--------+---------+
| t1   | 102  | banana |  10.30  |
+------+------+--------+---------+
```

"a+" 匹配字母 a 至少一次，只有 banana 满足匹配条件。

5.8.5 匹配指定字符串

正则表达式可以匹配指定字符串，只要这个字符串在查询文本中即可，如要匹配多个字符串，多个字符串之间使用分隔符 "|" 隔开。

【例 5.75】在 fruits 表中，查询 f_name 字段值包含字符串"on"的记录，SQL 语句如下：

```
mysql> SELECT * FROM fruits WHERE f_name REGEXP 'on';
+------+------+---------+---------+
| f_id | s_id | f_name  | f_price |
+------+------+---------+---------+
| bs2  | 105  | melon   | 8.20    |
| l2   | 104  | lemon   | 6.40    |
| o2   | 103  | coconut | 9.20    |
+------+------+---------+---------+
```

可以看到，f_name 字段的 melon、lemon 和 coconut 3 个值中都包含有字符串"on"，满足匹配条件。

【例 5.76】在 fruits 表中，查询 f_name 字段值包含字符串"on"或者"ap"的记录，SQL 语句如下：

```
mysql> SELECT * FROM fruits WHERE f_name REGEXP 'on|ap';
+------+------+---------+---------+
| f_id | s_id | f_name  | f_price |
+------+------+---------+---------+
| a1   | 101  | apple   | 5.20    |
| a2   | 103  | apricot | 2.20    |
| bs2  | 105  | melon   | 8.20    |
| l2   | 104  | lemon   | 6.40    |
| o2   | 103  | coconut | 9.20    |
| t2   | 102  | grape   | 5.30    |
+------+------+---------+---------+
```

可以看到，f_name 字段的 melon、lemon 和 coconut 3 个值中都包含有字符串"on"，apple 和 apricot 值中包含字符串"ap"，满足匹配条件。

提示：之前介绍过，LIKE 运算符也可以匹配指定的字符串，但与 REGEXP（正则表达式）不同，LIKE 匹配的字符串如果在文本中间出现，则找不到它，相应的行也不会返回。REGEXP 在文本内进行匹配，如果被匹配的字符串在文本中出现，REGEXP 将会找到它，相应的行也会被返回。对比结果如例 5.77 所示。

【例 5.77】在 fruits 表中，使用 LIKE 运算符查询 f_name 字段值为"on"的记录，SQL 语句如下：

```
mysql> SELECT * FROM fruits WHERE f_name LIKE 'on';
Empty set (0.00 sec)
```

f_name 字段没有值为"on"的记录，返回结果为空。读者可以体会一下两者的区别。

5.8.6 匹配指定字符中的任意一个

方括号"[]"指定一个字符集合，只匹配其中任何一个字符，即为所查找的文本。

【例 5.78】在 fruits 表中，查找 f_name 字段中包含字母"o"或者"t"的记录，SQL 语句如下：

```
mysql> SELECT * FROM fruits WHERE f_name REGEXP '[ot]';
```

```
+------+------+---------+---------+
| f_id | s_id | f_name  | f_price |
+------+------+---------+---------+
| a2   | 103  | apricot |    2.20 |
| bs1  | 102  | orange  |   11.20 |
| bs2  | 105  | melon   |    8.20 |
| l2   | 104  | lemon   |    6.40 |
| m1   | 106  | mango   |   15.70 |
| m3   | 105  | xxtt    |   11.60 |
| o2   | 103  | coconut |    9.20 |
+------+------+---------+---------+
```

由查询结果可以看到，所有返回的记录的 f_name 字段的值中都包含有字母 o 或者 t，或者两个都有。

方括号"[]"还可以指定数值集合。

【例 5.79】在 fruits 表中，查询 s_id 字段中包含 4、5 或者 6 的记录，SQL 语句如下：

```
mysql> SELECT * FROM fruits WHERE s_id REGEXP '[456]';
+------+------+--------+---------+
| f_id | s_id | f_name | f_price |
+------+------+--------+---------+
| b2   | 104  | berry  |    7.60 |
| bs2  | 105  | melon  |    8.20 |
| l2   | 104  | lemon  |    6.40 |
| m1   | 106  | mango  |   15.70 |
| m2   | 105  | xbabay |    2.60 |
| m3   | 105  | xxtt   |   11.60 |
+------+------+--------+---------+
```

在查询结果中，s_id 字段值中只要有 3 个数字中的 1 个，即为匹配记录字段。

匹配集合"[456]"也可以写成"[4-6]"，即指定集合区间。例如，"[a-z]"表示集合区间为从 a~z 的字母，"[0-9]"表示集合区间为所有数字，读者可以自行修改本例测试一下。

5.8.7 匹配指定字符以外的字符

"[^字符集合]"匹配不在指定集合中的任何字符。

【例 5.80】在 fruits 表中，查询 f_id 字段中包含字母 a~e 和数字 1~2 以外字符的记录，SQL 语句如下：

```
mysql> SELECT * FROM fruits WHERE f_id REGEXP '[^a-e1-2]';
+------+------+---------+---------+
| f_id | s_id | f_name  | f_price |
+------+------+---------+---------+
| b5   | 107  | xxxx    |    3.60 |
| bs1  | 102  | orange  |   11.20 |
| bs2  | 105  | melon   |    8.20 |
| c0   | 101  | cherry  |    3.20 |
| l2   | 104  | lemon   |    6.40 |
| m1   | 106  | mango   |   15.70 |
| m2   | 105  | xbabay  |    2.60 |
| m3   | 105  | xxtt    |   11.60 |
| o2   | 103  | coconut |    9.20 |
```

```
| t1  | 102 | banana  | 10.30 |
| t2  | 102 | grape   |  5.30 |
| t4  | 107 | xbababa |  3.60 |
```

返回记录中的 f_id 字段值中包含指定字母和数字以外的值，如 s、m、o、t 等，这些字母均不在 a~e 与 1~2 之间，满足匹配条件。

5.8.8 使用{n,}或者{n,m}来指定字符串连续出现的次数

"字符串{n,}"表示至少匹配 n 次前面的字符；"字符串{n,m}"表示匹配前面的字符串不少于 n 次，不多于 m 次。例如，a{2,}表示字母 a 连续出现至少 2 次，也可以大于 2 次；a{2,4}表示字母 a 连续出现最少 2 次，最多不能超过 4 次。

【例 5.81】在 fruits 表中，查询 f_name 字段值出现字母"x"至少 2 次的记录，SQL 语句如下：

```
mysql> SELECT * FROM fruits WHERE f_name REGEXP 'x{2,}';
+------+------+--------+---------+
| f_id | s_id | f_name | f_price |
+------+------+--------+---------+
| b5   | 107  | xxxx   |  3.60   |
| m3   | 105  | xxtt   | 11.60   |
+------+------+--------+---------+
```

可以看到，f_name 字段的"xxxx"值包含了 4 个字母"x"，"xxtt"值包含两个字母"x"，均为满足匹配条件的记录。

【例 5.82】在 fruits 表中，查询 f_name 字段值出现字符串"ba"最少 1 次、最多 3 次的记录，SQL 语句如下：

```
mysql> SELECT * FROM fruits WHERE f_name REGEXP 'ba{1,3}';
+------+------+---------+---------+
| f_id | s_id | f_name  | f_price |
+------+------+---------+---------+
| m2   | 105  | xbabay  |  2.60   |
| t1   | 102  | banana  | 10.30   |
| t4   | 107  | xbababa |  3.60   |
+------+------+---------+---------+
```

可以看到，f_name 字段的"xbabay"值中"ba"出现了 2 次，"banana"值中出现了 1 次，"xbababa"值中出现了 3 次，都是满足匹配条件的记录。

5.9 通用表表达式

通用表表达式简称为 CTE（Common Table Expressions）。CTE 是命名的临时结果集，作用范围是当前语句。CTE 可以理解成一个可以复用的子查询，当然跟子查询还是有点区别的，CTE 可以引用其他 CTE，但子查询不能引用其他子查询。

CTE 的语法格式如下：

```
with_clause:
    WITH [RECURSIVE]
        cte_name [(col_name [, col_name] ...)] AS (subquery)
        [, cte_name [(col_name [, col_name] ...)] AS (subquery)] ...
```

使用 WITH 语句创建 CTE 的情况如下：

（1）SELECT、UPDATE、DELETE 语句的开头：

```
WITH ... SELECT ...
WITH ... UPDATE ...
WITH ... DELETE ...
```

（2）在子查询的开头：

```
SELECT ... WHERE id IN (WITH ... SELECT ...) ...
SELECT * FROM (WITH ... SELECT ...) AS dt ...
```

（3）紧接 SELECT，在包含 SELECT 语句的语句之前：

```
INSERT ... WITH ... SELECT ...
REPLACE ... WITH ... SELECT ...
CREATE TABLE ... WITH ... SELECT ...
CREATE VIEW ... WITH ... SELECT ...
DECLARE CURSOR ... WITH ... SELECT ...
EXPLAIN ... WITH ... SELECT ...
```

下面通过示例来讲解通用表表达式的使用方法。

创建商品表 goods，该数据表包含上下级关系的数据，具体字段包含商品编号（id）、商品名称（name）、上级商品的编号（gid）。创建语句如下：

```
CREATE TABLE goods(
id int(11),
name varchar(30),
gid int(11),
PRIMARY KEY (`id`));
```

插入演示数据：

```
INSERT INTO goods (id, name, gid) VALUES (1, '商品', 0);
INSERT INTO goods (id, name, gid) VALUES (2, '水果', 1);
INSERT INTO goods (id, name, gid) VALUES (3, '蔬菜', 1);
INSERT INTO goods (id, name, gid) VALUES (4, '苹果', 2);
INSERT INTO goods (id, name, gid) VALUES (5, '香蕉', 2);
INSERT INTO goods (id, name, gid) VALUES (6, '菠菜', 3);
INSERT INTO goods (id, name, gid) VALUES (7, '萝卜', 3);
```

下面开始查询每个商品对应的上级商品名称。

这里使用子查询的方式：

```
mysql> SELECT g.*, (SELECT name FROM goods where id = g.gid) as pname FROM goods AS g;
+----+------+-----+-------+
| id | name | gid | pname |
+----+------+-----+-------+
|  1 | 商品 |   0 | NULL  |
|  2 | 水果 |   1 | 商品  |
|  3 | 蔬菜 |   1 | 商品  |
```

```
|  4 | 苹果 |   2 | 水果   |
|  5 | 香蕉 |   2 | 水果   |
|  6 | 菠菜 |   3 | 蔬菜   |
|  7 | 萝卜 |   3 | 蔬菜   |
+----+------+-----+--------+
```

接着使用 CTE 的方式，完成上述功能：

```
mysql> WITH cte as (SELECT * FROM goods) SELECT g.*, (SELECT cte.name FROM cte WHERE cte.id = g.gid) AS gname FROM goods AS g;
+----+------+-----+-------+
| id | name | gid | gname |
+----+------+-----+-------+
|  1 | 商品 |   0 | NULL  |
|  2 | 水果 |   1 | 商品  |
|  3 | 蔬菜 |   1 | 商品  |
|  4 | 苹果 |   2 | 水果  |
|  5 | 香蕉 |   2 | 水果  |
|  6 | 菠菜 |   3 | 蔬菜  |
|  7 | 萝卜 |   3 | 蔬菜  |
+----+------+-----+-------+
```

从结果可以看出，CTE 是一个可以重复使用的结果集。相比于子查询，CTE 的效率会更高，因为非递归的 CTE 只会查询一次并可以重复使用。

CTE 可以引用其他 CTE 的结果。例如，下面的语句中，cte2 就引用了 cte1 中的结果。

```
mysql> with cte1 as (select * from goods), cte2 as (select g.*, cte1.name as gname from goods as g left join cte1 on g.gid = cte1.id) select * from cte2;
+----+------+-----+-------+
| id | name | gid | gname |
+----+------+-----+-------+
|  1 | 商品 |   0 | NULL  |
|  2 | 水果 |   1 | 商品  |
|  3 | 蔬菜 |   1 | 商品  |
|  4 | 苹果 |   2 | 水果  |
|  5 | 香蕉 |   2 | 水果  |
|  6 | 菠菜 |   3 | 蔬菜  |
|  7 | 萝卜 |   3 | 蔬菜  |
+----+------+-----+-------+
```

还有一种特殊的 CTE，就是递归 CTE，其子查询会引用自身。WITH 子句必须以 WITH RECURSIVE 开头。

CTE 递归子查询包括两部分：seed 查询和 recursive 查询，中间由 union [all] 或 union distinct 分隔。seed 查询会被执行一次，以创建初始数据子集。recursive 查询会被重复执行以返回数据子集，直到获得完整结果集。当迭代不会生成任何新行时，递归会停止。可以参看下面的示例：

```
mysql> WITH RECURSIVE cte(n) AS (SELECT 1 UNION ALL SELECT n + 1 FROM cte WHERE n < 8) SELECT * FROM cte;
+------+
| n    |
+------+
|    1 |
|    2 |
|    3 |
```

```
|   4    |
|   5    |
|   6    |
|   7    |
|   8    |
+--------+
```

上面的语句会递归显示 8 行，每行分别显示 1~8 数字。

递归的过程如下：

（1）首先执行 SELECT 1 得到结果 1，即当前 n 的值为 1。

（2）接着执行 SELECT N + 1 FROM cte WHERE n < 8，因为当前 n 为 1，所以 WHERE 条件成立，生成新行，SELECT n + 1 得到结果 2，即当前 n 的值为 2。

（3）继续执行 SELECT n + 1 FROM cte WHERE n < 8，因为当前 n 为 2，所以 WHERE 条件成立，生成新行，SELECT n + 1 得到结果 3，即当前 n 的值为 3。

（4）一直递归下去。

（5）直到当 n 为 8 时，where 条件不成立，无法生成新行，递归停止。

下面使用递归 CTE 来查询每个商品到顶级商品的层次。

```
mysql>with recursive cte as (select id, name, cast('0' as char(255)) as path from goods where
gid = 0 union all select goods.id, goods.name, concat(cte.path, ',', cte.id) as path from goods inner
join cte on goods.gid = cte.id) select * from cte;
+------+------+-------+
| id   | name | path  |
+------+------+-------+
|   1  | 商品 | 0     |
|   2  | 水果 | 0,1   |
|   3  | 蔬菜 | 0,1   |
|   4  | 苹果 | 0,1,2 |
|   5  | 香蕉 | 0,1,2 |
|   6  | 菠菜 | 0,1,3 |
|   7  | 萝卜 | 0,1,3 |
+------+------+-------+
```

查询一个指定商品的所有父级商品。

```
mysql> with recursive cte as (select id, name, gid from goods where id = 7
union all select goods.id, goods.name, goods.gid from goods inner join cte on cte.gid =
goods.id)select * from cte;
+-----+------+-----+
| id  | name | gid |
+-----+------+-----+
|  7  | 萝卜 |  3  |
|  3  | 蔬菜 |  1  |
|  1  | 商品 |  0  |
+-----+------+-----+
```

第 6 章

插入、更新与删除数据

存储在系统中的数据是数据库管理系统（DBMS）的核心，数据库被设计用来管理数据的存储、访问和维护数据的完整性。MySQL 中提供了功能丰富的数据库管理语句，包括有效地向数据库中插入数据的 INSERT 语句、更新数据的 UPDATE 语句，以及当数据不再使用时删除数据的 DELETE 语句。本章将详细介绍在 MySQL 中如何使用这些语句操作数据。

6.1 插入数据

在使用数据库之前，数据库中必须有数据，MySQL 中使用 INSERT 语句向数据库表中插入新的数据记录。可以插入的方式有插入完整的记录、插入记录的一部分、插入多条记录、插入另一个查询的结果，下面将分别介绍这些内容。

6.1.1 为表的所有字段插入数据

使用基本的 INSERT 语句插入数据需要指定表名称和插入新记录中的值。基本语法格式为：

INSERT INTO table_name (column_list) VALUES (value_list);

table_name 指定要插入数据的表名，column_list 指定要插入数据的那些列，value_list 指定每个列需要对应插入的数据。注意，使用该语句时字段列和数据值的数量必须相同。

本章将使用示例表 person，创建语句如下：

```
CREATE TABLE person
(
id INT UNSIGNED NOT NULL AUTO_INCREMENT,
name CHAR(40) NOT NULL DEFAULT '',
age INT NOT NULL DEFAULT 0,
info CHAR(50) NULL,
PRIMARY KEY (id)
```

);
```

向表中所有字段插入值的方法有两种：一种是指定所有字段名，另一种是完全不指定字段名。

【例 6.1】在 person 表中，插入一条新记录，id 值为 1，name 值为 Green，age 值为 21，info 值为 Lawyer，SQL 语句如下：

执行插入操作之前，使用 SELECT 语句查看表中的数据：

```
mysql> SELECT * FROM person;
Empty set (0.00 sec)
```

结果显示当前表为空，没有数据，接下来执行插入操作：

```
mysql> INSERT INTO person (id ,name, age , info) VALUES (1,'Green', 21, 'Lawyer');
```

语句执行完毕，查看执行结果：

```
mysql> SELECT * FROM person;
+----+-------+-----+--------+
| id | name  | age | info   |
+----+-------+-----+--------+
|  1 | Green |  21 | Lawyer |
+----+-------+-----+--------+
```

可以看到插入记录成功了。在插入数据时，指定了 person 表的所有字段，因此将为每一个字段插入新的值。

INSERT 语句后面的列名称顺序可以不是 person 表定义时的顺序。即插入数据时，不需要按照表定义的顺序插入，只要保证值的顺序与列字段的顺序相同就可以，如例 6.2 所示。

【例 6.2】在 person 表中，插入一条新记录，id 值为 2，name 值为 Suse，age 值为 22，info 值为 dancer，SQL 语句如下：

```
mysql> INSERT INTO person (age ,name, id , info) VALUES (22, 'Suse', 2, 'dancer');
```

语句执行完毕，查看执行结果：

```
mysql> SELECT * FROM person;
+----+-------+-----+--------+
| id | name  | age | info   |
+----+-------+-----+--------+
|  1 | Green |  21 | Lawyer |
|  2 | Suse  |  22 | dancer |
+----+-------+-----+--------+
```

由结果可以看到，INSERT 语句成功插入了这条记录。

使用 INSERT 插入数据时，允许列名称列表 column_list 为空，此时，值列表中需要为表的每一个字段指定值，并且值的顺序必须和数据表中字段定义时的顺序相同，如例 6.3 所示。

【例 6.3】在 person 表中，插入一条新记录，id 值为 3，name 值为 Mary，age 值为 24，info 值为 Musician，SQL 语句如下：

```
mysql> INSERT INTO person  VALUES (3,'Mary', 24, 'Musician');
```
```

语句执行完毕，查看执行结果：

```
mysql> SELECT * FROM person;
+----+-------+------+---------+
| id | name  | age  | info    |
+----+-------+------+---------+
|  1 | Green |   21 | Lawyer  |
|  2 | Suse  |   22 | dancer  |
|  3 | Mary  |   24 | Musician|
+----+-------+------+---------+
```

可以看到插入记录成功。数据库中增加了一条 id 为 3 的记录，其他字段值为指定的插入值。本例的 INSERT 语句中没有指定字段列表，只有一个值列表。在这种情况下，值列表为每一个字段列指定插入值，并且这些值的顺序必须和 person 表中字段定义的顺序相同。

提示：虽然使用 INSERT 插入数据时可以忽略插入数据的列名称，但是值如果不包含列名称，那么 VALUES 关键字后面的值不仅要求完整，而且顺序必须和表定义时列的顺序相同。如果表的结构被修改，对列进行增加、删除或者位置改变操作，这些操作将使得用这种方式插入数据时的顺序也同时改变。如果指定列名称，则不会受到表结构改变的影响。

6.1.2　为表的指定字段插入数据

为表的指定字段插入数据，就是在 INSERT 语句中只向部分字段插入值，而其他字段的值为表定义时的默认值。

【例 6.4】 在 person 表中，插入一条新记录，name 值为 Willam，age 值为 20，info 值为 sports man，SQL 语句如下：

```
mysql> INSERT INTO person (name, age,info) VALUES('Willam', 20, 'sports man');
```

提示信息表示插入这条记录成功了。使用 SELECT 查询表中的记录，查询结果如下：

```
mysql> SELECT * FROM person;
+----+--------+------+------------+
| id | name   | age  | info       |
+----+--------+------+------------+
|  1 | Green  |   21 | Lawyer     |
|  2 | Suse   |   22 | dancer     |
|  3 | Mary   |   24 | Musician   |
|  4 | Willam |   20 | sports man |
+----+--------+------+------------+
```

可以看到插入记录成功。查询结果显示，该 id 字段自动添加了一个整数值 4。这里，id 字段为表的主键，不能为空，系统会自动为该字段插入自增的序列值。在插入记录时，如果某些字段没有指定插入值，MySQL 将插入该字段定义时的默认值。下面的例子说明在没有指定列字段时，插入默认值。

【例 6.5】在 person 表中，插入一条新记录，name 值为 laura，age 值为 25，SQL 语句如下：

```
mysql> INSERT INTO person (name, age ) VALUES ('Laura', 25);
```

语句执行完毕，查看执行结果：

```
mysql> SELECT * FROM person;
+----+--------+------+------------+
| id | name   | age  | info       |
+----+--------+------+------------+
|  1 | Green  |  21  | Lawyer     |
|  2 | Suse   |  22  | dancer     |
|  3 | Mary   |  24  | Musician   |
|  4 | Willam |  20  | sports man |
|  5 | Laura  |  25  | NULL       |
+----+--------+------+------------+
```

可以看到，在本例插入语句中，没有指定 info 字段值，查询结果显示，info 字段在定义时默认为 NULL，因此系统自动为该字段插入空值。

提示：要保证每个插入值的类型和对应列的数据类型匹配，如果类型不同，将无法插入，并且 MySQL 会产生错误。

6.1.3 同时插入多条记录

INSERT 语句可以同时向数据表中插入多条记录，插入时指定多个值列表，每个值列表之间用逗号分隔开，基本语法格式如下：

```
INSERT INTO table_name (column_list)
VALUES (value_list1), (value_list2),...,(value_listn);
```

"value_list1,value_list2,…,value_listn" 表示第 1,2,…,n 个插入记录的字段的值列表。

【例 6.6】在 person 表中，在 name、age 和 info 字段指定插入值，同时插入 3 条新记录，SQL 语句如下：

```
INSERT INTO person(name, age, info) VALUES ('Evans', 27, 'secretary'),
('Dale', 22, 'cook'), ('Edison', 28, 'singer');
```

语句执行完毕，查看执行结果：

```
mysql> SELECT * FROM person;
+----+--------+------+------------+
| id | name   | age  | info       |
+----+--------+------+------------+
|  1 | Green  |  21  | Lawyer     |
|  2 | Suse   |  22  | dancer     |
|  3 | Mary   |  24  | Musician   |
|  4 | Willam |  20  | sports man |
|  5 | Laura  |  25  | NULL       |
|  6 | Evans  |  27  | secretary  |
|  7 | Dale   |  22  | cook       |
|  8 | Edison |  28  | singer     |
+----+--------+------+------------+
```

由结果可以看到，INSERT 语句执行后，person 表中添加了 3 条记录，其 name 和 age 字段分别为指定的值，id 字段为 MySQL 添加的默认的自增值。

使用 INSERT 同时插入多条记录时，MySQL 会返回一些在执行单行插入时没有的额外信息，这些信息的含义如下：

- Records：表明插入的记录条数。
- Duplicates：表明插入时被忽略的记录，原因可能是这些记录包含了重复的主键值。
- Warnings：表明有问题的数据值，例如发生数据类型转换。

【例 6.7】在 person 表中，不指定插入列表，同时插入 2 条新记录，SQL 语句如下：

```
INSERT INTO person VALUES (9,'Harry',21, 'magician'), (NULL,'Harriet',19, 'pianist');
```

语句执行完毕，查看执行结果：

```
mysql> SELECT * FROM person;
+----+---------+-----+------------+
| id | name    | age | info       |
+----+---------+-----+------------+
|  1 | Green   |  21 | Lawyer     |
|  2 | Suse    |  22 | dancer     |
|  3 | Mary    |  24 | Musician   |
|  4 | Willam  |  20 | sports man |
|  5 | Laura   |  25 | NULL       |
|  6 | Evans   |  27 | secretary  |
|  7 | Dale    |  22 | cook       |
|  8 | Edison  |  28 | singer     |
|  9 | Harry   |  21 | magician   |
| 10 | Harriet |  19 | pianist    |
+----+---------+-----+------------+
```

由结果可以看到，INSERT 语句执行后，person 表中添加了 2 条记录，与前面介绍单个 INSERT 语法不同，person 表名后面没有指定插入字段列表，因此 VALUES 关键字后面的多个值列表都要为每一条记录的每一个字段列指定插入值，并且这些值的顺序必须和 person 表中字段定义的顺序相同，带有 AUTO_INCREMENT 属性的 id 字段插入 NULL 值，系统会自动为该字段插入唯一的自增编号。

提示：一个同时插入多行记录的 INSERT 语句等同于多个单行插入的 INSERT 语句，但是多行的 INSERT 语句在处理过程中效率更高。因为 MySQL 执行单条 INSERT 语句插入多行数据比使用多条 INSERT 语句快，所以在插入多条记录时最好选择使用单条 INSERT 语句的方式插入。

6.1.4 将查询结果插入到表中

INSERT 语句用来给数据表插入记录时指定插入记录的列值。INSERT 还可以将 SELECT 语句查询的结果插入到表中，如果想要从另外一个表中合并个人信息到 person 表，不需要把每一条记录的值一个一个输入，只需要使用一条 INSERT 语句和一条 SELECT 语句组成的组合语句，即可快速地从一个或多个表中向一个表中插入多行。基本语法格式如下：

```
INSERT INTO table_name1 (column_list1)
SELECT (column_list2) FROM table_name2 WHERE (condition)
```

table_name1 指定待插入数据的表；column_list1 指定待插入表中要插入数据的哪些列；table_name2 指定插入数据是从哪个表中查询出来的；column_list2 指定数据来源表的查询列，该列表必须和 column_list1 列表中的字段个数相同、数据类型相同；condition 指定 SELECT 语句的查询条件。

【例 6.8】从 person_old 表中查询所有的记录，并将其插入到 person 表中。

首先，创建一个名为 person_old 的数据表，其表结构与 person 结构相同，SQL 语句如下：

```
CREATE TABLE person_old
(
id INT UNSIGNED NOT NULL AUTO_INCREMENT,
name CHAR(40) NOT NULL DEFAULT '',
age INT NOT NULL DEFAULT 0,
info CHAR(50) NULL,
PRIMARY KEY (id)
);
```

向 person_old 表中添加两条记录：

```
mysql> INSERT INTO person_old VALUES (11,'Harry',20, 'student'), (12,'Beckham',31, 'police');

mysql> SELECT * FROM person_old;
+----+---------+-----+---------+
| id | name    | age | info    |
+----+---------+-----+---------+
| 11 | Harry   |  20 | student |
| 12 | Beckham |  31 | police  |
+----+---------+-----+---------+
```

可以看到，插入记录成功，peson_old 表中现在有两条记录。接下来将 person_old 表中所有的记录插入 person 表中，SQL 语句如下：

```
INSERT INTO person(id, name, age, info) SELECT id, name, age, info FROM person_old;
```

语句执行完毕，查看执行结果：

```
mysql> SELECT * FROM person;
+----+---------+-----+------------+
| id | name    | age | info       |
+----+---------+-----+------------+
|  1 | Green   |  21 | Lawyer     |
|  2 | Suse    |  22 | dancer     |
|  3 | Mary    |  24 | Musician   |
|  4 | Willam  |  20 | sports man |
|  5 | Laura   |  25 | NULL       |
|  6 | Evans   |  27 | secretary  |
|  7 | Dale    |  22 | cook       |
|  8 | Edison  |  28 | singer     |
|  9 | Harry   |  21 | magician   |
| 10 | Harriet |  19 | pianist    |
| 11 | Harry   |  20 | student    |
| 12 | Beckham |  31 | police     |
+----+---------+-----+------------+
```

由结果可以看到，INSERT 语句执行后，person 表中多了两条记录，这两条记录和 person_old 表中的记录完全相同，数据转移成功。这里的 id 字段为自增的主键，在插入的时候要保证该字段值的唯一性，如果不能确定，可以在插入的时候忽略该字段，只插入其他字段的值。

提示：这个例子中使用的 person_old 表和 person 表的定义相同，事实上，MySQL 不关心 SELECT 返回的列名，它根据列的位置进行插入，SELECT 的第 1 列对应待插入表的第 1 列，第 2 列对应待插入表的第 2 列……即使不同列名的表之间也可以方便地转移数据。

6.2 更新数据

表中有数据之后，接下来可以对数据进行更新操作，MySQL 中使用 UPDATE 语句更新表中的记录，可以更新特定的行或者同时更新所有的行。基本语法结构如下：

```
UPDATE table_name
SET column_name1 = value1,column_name2=value2,…,column_namen=valuen
WHERE (condition);
```

column_name1,column_name2,…,column_namen 为指定更新的字段的名称；value1,value2,…,valuen 为相对应的指定字段的更新值；condition 指定更新的记录需要满足的条件。更新多列时，每个"列-值"对之间用逗号隔开，最后一列之后不需要逗号。

【例 6.9】在 person 表中，更新 id 值为 11 的记录，将 age 字段值改为 15，将 name 字段值改为 LiMing，SQL 语句如下：

```
UPDATE person SET age = 15, name='LiMing' WHERE id = 11;
```

更新操作执行前可以使用 SELECT 语句查看当前的数据：

```
mysql> SELECT * FROM person WHERE id=11;
+----+-------+-----+---------+
| id | name  | age | info    |
+----+-------+-----+---------+
| 11 | Harry | 20  | student |
+----+-------+-----+---------+
```

由结果可以看到，更新之前 id 等于 11 的记录的 name 字段值为 harry、age 字段值为 20。下面使用 UPDATE 语句更新数据：

```
mysql> UPDATE person SET age = 15, name='LiMing' WHERE id = 11;
```

语句执行完毕，查看执行结果：

```
mysql> SELECT * FROM person WHERE id=11;
+----+--------+-----+---------+
| id | name   | age | info    |
+----+--------+-----+---------+
| 11 | LiMing | 15  | student |
+----+--------+-----+---------+
```

由结果可以看到，id 等于 11 的记录中的 name 和 age 字段的值已经成功地被修改为指定值。

提示：保证 UPDATE 以 WHERE 子句结束，通过 WHERE 子句指定被更新的记录所需要满足的条件，如果忽略 WHERE 子句，MySQL 将更新表中所有的行。

【例 6.10】在 person 表中，更新 age 值为 19~22 的记录，将 info 字段值都改为 student，SQL 语句如下：

```
UPDATE person SET info='student' WHERE id BETWEEN 19 AND 22;
```

更新操作执行之前，可以使用 SELECT 语句查看当前的数据：

```
mysql> SELECT * FROM person WHERE age BETWEEN 19 AND 22;
+----+---------+-----+------------+
| id | name    | age | info       |
+----+---------+-----+------------+
|  1 | Green   |  21 | Lawyer     |
|  2 | Suse    |  22 | dancer     |
|  4 | Willam  |  20 | sports man |
|  7 | Dale    |  22 | cook       |
|  9 | Harry   |  21 | magician   |
| 10 | Harriet |  19 | pianist    |
+----+---------+-----+------------+
```

可以看到，这些 age 字段值在 19~22 之间的记录的 info 字段值各不相同。下面使用 UPDATE 语句更新数据，语句执行结果如下：

```
mysql> UPDATE person SET info='student' WHERE age BETWEEN 19 AND 22;
```

语句执行完毕，查看执行结果：

```
mysql> SELECT * FROM person WHERE age BETWEEN 19 AND 22;
+----+---------+-----+---------+
| id | name    | age | info    |
+----+---------+-----+---------+
|  1 | Green   |  21 | student |
|  2 | Suse    |  22 | student |
|  4 | Willam  |  20 | student |
|  7 | Dale    |  22 | student |
|  9 | Harry   |  21 | student |
| 10 | Harriet |  19 | student |
+----+---------+-----+---------+
```

由结果可以看到，UPDATE 执行后，成功将表中符合条件的 6 条记录的 info 字段值都改为 student。

6.3 删除数据

从数据表中删除数据使用 DELETE 语句，DELETE 语句允许 WHERE 子句指定删除条件。DELETE 语句基本语法格式如下：

```
DELETE FROM table_name [WHERE <condition>];
```

table_name 指定要执行删除操作的表；"[WHERE <condition>]"为可选参数，指定删除条件，如果没有 WHERE 子句，DELETE 语句将删除表中的所有记录。

【例 6.11】在 person 表中，删除 id 等于 11 的记录，SQL 语句如下：

执行删除操作前，使用 SELECT 语句查看当前 id=11 的记录：

```
mysql> SELECT * FROM person WHERE id=11;
+----+-------+-----+---------+
| id | name  | age | info    |
+----+-------+-----+---------+
| 11 | LiMing|  15 | student |
+----+-------+-----+---------+
```

可以看到，现在表中有 id=11 的记录。下面使用 DELETE 语句删除该记录：

```
mysql> DELETE FROM person WHERE id = 11;
Query OK, 1 row affected (0.02 sec)
```

语句执行完毕，查看执行结果：

```
mysql> SELECT * FROM person WHERE id=11;
Empty set (0.00 sec)
```

查询结果为空，说明删除操作成功。

【例 6.12】在 person 表中，使用 DELETE 语句同时删除多条记录。在前面 UPDATE 语句中将 age 字段值在 19~22 之间的记录的 info 字段值修改为 student，在这里删除这些记录。

执行删除操作前，使用 SELECT 语句查看一下的数据：

```
mysql> SELECT * FROM person WHERE age BETWEEN 19 AND 22;
+----+---------+-----+---------+
| id | name    | age | info    |
+----+---------+-----+---------+
|  1 | Green   |  20 | student |
|  2 | Suse    |  21 | student |
|  4 | Willam  |  22 | student |
|  7 | Dale    |  22 | student |
|  9 | Harry   |  21 | student |
| 10 | Harriet |  19 | student |
+----+---------+-----+---------+
```

可以看到，这些 age 字段值在 19~22 之间的记录存在表中。下面使用 DELETE 删除这些记录：

```
mysql> DELETE FROM person WHERE age BETWEEN 19 AND 22;
Query OK, 6 rows affected (0.00 sec)
```

语句执行完毕，查看执行结果：

```
mysql> SELECT * FROM person WHERE age BETWEEN 19 AND 22;
Empty set (0.00 sec)
```

查询结果为空，删除多条记录成功。

【例 6.13】删除 person 表中所有记录。

执行删除操作前，使用 SELECT 语句查看当前的数据：

```
mysql> SELECT * FROM person;
```

```
+----+---------+-----+-----------+
| id | name    | age | info      |
+----+---------+-----+-----------+
|  3 | Mary    |  24 | Musician  |
|  5 | Laura   |  25 | NULL      |
|  6 | Evans   |  27 | secretary |
| 12 | Beckham |  31 | police    |
+----+---------+-----+-----------+
```

结果显示 person 表中还有 4 条记录，执行 DELETE 语句删除这 4 条记录：

```
mysql> DELETE FROM person;
Query OK, 4 rows affected (0.00 sec)
```

语句执行完毕，查看执行结果：

```
mysql> SELECT * FROM person;
Empty set (0.00 sec)
```

查询结果为空，说明删除表中所有记录成功，现在 person 表中已经没有任何数据记录。

提示：如果想删除表中的所有记录，还可以使用 TRUNCATE TABLE 语句。TRUNCATE 将直接删除原来的表，并重新创建一个表，其语法结构为 TRUNCATE TABLE table_name。TRUNCATE 直接删除表而不是删除记录，因此执行速度比 DELETE 快。

6.4 为表增加计算列

什么叫计算列呢？简单来说就是某一列的值是通过别的列计算得来的。例如，a 列值为 1、b 列值为 2，c 列不需要手动插入，定义 a+b 的结果为 c 列的值，那么 c 就是计算列，它是通过别的列计算得来的。

增加计算列的语法格式如下：

```
col_name data_type [GENERATED ALWAYS] AS (expression)
    [VIRTUAL | STORED] [UNIQUE [KEY]] [COMMENT comment]
    [NOT NULL | NULL] [[PRIMARY] KEY]
```

在 MySQL 8.0 中，CREAE TABLE 和 ALTER TABLE 中都支持增加计算列。下面以 CREAE TABLE 为例进行讲解。

【例 6.14】定义数据表 tb1，然后定义字段 id、字段 a、字段 b 和字段 c，其中字段 c 为计算列，用于计算列 a+b 的值。

首先创建测试表 tb1，语句如下：

```
CREATE TABLE tb1(
id int(9) NOT NULL AUTO_INCREMENT,
a int(9) DEFAULT NULL,
b int(9) DEFAULT NULL,
c int(9) GENERATED ALWAYS AS ((a + b)) VIRTUAL,
PRIMARY KEY (`id`)
);
```

插入演示数据,语句如下:

```
insert into tb1(a,b) values (100,200);
```

查询数据表 tb1 中的数据,结果如下:

```
mysql> SELECT * FROM tb1;
+----+-----+-----+-----+
| id | a   | b   | c   |
+----+-----+-----+-----+
| 1  | 100 | 200 | 300 |
+----+-----+-----+-----+
```

更新数据中的数据,语句如下:

```
mysql> UPDATE tb1 SET a=500;
Query OK, 1 row affected (0.01 sec)
Rows matched: 1  Changed: 1  Warnings: 0
```

再次查看数据表中的数据,结果如下:

```
mysql> SELECT * FROM tb1;
+----+-----+-----+-----+
| id | a   | b   | c   |
+----+-----+-----+-----+
| 1  | 500 | 200 | 700 |
+----+-----+-----+-----+
```

从结果可以看出,字段 c 中的数据始终是字段 a 和字段 b 的和,并能随着字段 a 和字段 b 中数据的变化,自动重新计算 a+b 的值。

6.5　DDL 的原子化

在 MySQL 8.0 版本中,InnoDB 表的 DDL 支持事务完整性,即 DDL 操作要么成功要么回滚。DDL 操作回滚日志写入到 data dictionary 数据字典表 mysql.innodb_ddl_log(该表是隐藏的表,通过 show tables 无法看到)中,用于回滚操作。通过设置参数,可将 DDL 操作日志打印输出到 MySQL 错误日志中。

下面通过案例来对比不同的版本中 DDL 操作的区别。

分别在 MySQL 5.7 版本和 MySQL 8.0 版本中创建数据库和数据表,结果如下:

```
CREATE DATABASE mytest;
USE mytest;

CREATE TABLE bk1
(
bookid INT NOT NULL,
bookname VARCHAR(255)
);
mysql> SHOW TABLES;
+------------------+
| Tables_in_mytest |
```

```
+------------------+
| bk1              |
+------------------+
```

（1）在 MySQL 5.7 版本中，测试步骤如下：

删除数据表 bk1 和数据表 bk2，结果如下：

```
mysql> DROP TABLE BK1,BK2;
ERROR 1051 (42S02): Unknown table 'mytest.bk2'
```

再次查询数据库中的数据表名称，结果如下：

```
mysql> SHOW TABLES;
Empty set (0.00 sec)
```

从结果可以看出，虽然删除操作时报错了，但是仍然删除了数据表 bk1。

（2）在 MySQL 8.0 版本中，测试步骤如下：

删除数据表 bk1 和数据表 bk2，结果如下：

```
mysql> DROP TABLE bk1,bk2;
ERROR 1051 (42S02): Unknown table 'mytest.bk2'
```

再次查询数据库中的数据表名称，结果如下：

```
mysql> SHOW TABLES;
+------------------+
| Tables_in_mytest |
+------------------+
| bk1              |
+------------------+
```

从结果可以看出，数据表 bk1 并没有被删除。读者可以通过这两个版本操作的不同之处，体会一下 MySQL 8.0 版本 DDL 操作的原子化。

第 7 章

索引的设计和使用

索引用于快速找出在某个列中有一特定值的行。如果不使用索引，则 MySQL 必须从第 1 条记录开始读完整个表，直到找出相关的行。表越大，查询数据所花费的时间越多。如果表中查询的列有一个索引，MySQL 能快速到达某个位置去搜寻数据文件，而不必查看所有数据。本章将介绍与索引相关的内容，包括索引的含义和特点、索引的分类、索引的设计原则以及如何创建和删除索引。

7.1 索引简介

索引是对数据库表中一列或多列的值进行排序的一种结构，使用索引可提高数据库中特定数据的查询速度。本节将介绍索引的含义、分类和设计原则。

7.1.1 索引的含义和特点

索引是一个单独的、存储在磁盘上的数据库结构，包含着对数据表里所有记录的引用指针。使用索引可以快速找出在某个或多个列中有一特定值的行，所有 MySQL 列类型都可以被索引，对相关列使用索引是提高查询操作速度的最佳途径。

例如，数据库中有 2 万条记录，现在要执行一个查询"SELECT * FROM table where num=10000"，如果没有索引，就必须遍历整个表，直到 num 等于 10000 的这一行被找到为止；如果在 num 列上创建索引，MySQL 不需要任何扫描，直接在索引里面找 10000，就可以得知这一行的位置。可见，索引的建立可以提高数据库的查询速度。

索引是在存储引擎中实现的，因此，每种存储引擎的索引都不一定完全相同，并且每种存储引擎也不一定支持所有索引类型。根据存储引擎定义每个表的最大索引数和最大索引长度。所有存储引擎支持每个表至少 16 个索引，总索引长度至少为 256 字节。大多数存储引擎有更高的限制。MySQL 中索引的存储类型有两种，即 BTREE 和 HASH，具体和表的存储引擎相关；MyISAM 和 InnoDB 存储引擎只支持 BTREE 索引；MEMORY/HEAP 存储引擎可以支持 HASH 和 BTREE 索引。

索引的优点主要有以下几条：

（1）通过创建唯一索引，可以保证数据库表中每一行数据的唯一性。

（2）可以大大加快数据的查询速度，这也是创建索引的主要原因。

（3）在实现数据的参考完整性方面，可以加速表和表之间的连接。

（4）在使用分组和排序子句进行数据查询时，也可以显著减少查询中分组和排序的时间。

增加索引也有许多不利的方面，主要表现在如下几个方面：

（1）创建索引和维护索引要耗费时间，并且随着数据量的增加所耗费的时间也会增加。

（2）索引需要占用磁盘空间，除了数据表占数据空间之外，每一个索引还要占一定的物理空间。如果有大量的索引，索引文件可能比数据文件更快达到最大文件尺寸。

（3）当对表中的数据进行增加、删除和修改的时候，索引也要动态地维护，这样就降低了数据的维护速度。

7.1.2 索引的分类

MySQL 的索引可以分为以下几类。

1. 普通索引和唯一索引

普通索引是 MySQL 中的基本索引类型，允许在定义索引的列中插入重复值和空值。

唯一索引要求索引列的值必须唯一，但允许有空值。如果是组合索引，则列值的组合必须唯一。主键索引是一种特殊的唯一索引，不允许有空值。

2. 单列索引和组合索引

单列索引即一个索引只包含单个列，一个表可以有多个单列索引。

组合索引是指在表的多个字段组合上创建的索引，只有在查询条件中使用了这些字段的左边字段时，索引才会被使用。使用组合索引时遵循最左前缀集合。

3. 全文索引

全文索引类型为 FULLTEXT，在定义索引的列上支持值的全文查找，允许在这些索引列中插入重复值和空值。全文索引可以在 CHAR、VARCHAR 或者 TEXT 类型的列上创建。MySQL 中只有 MyISAM 存储引擎支持全文索引。

4. 空间索引

空间索引是对空间数据类型的字段建立的索引，MySQL 中的空间数据类型有 4 种，分别是 GEOMETRY、POINT、LINESTRING 和 POLYGON。MySQL 使用 SPATIAL 关键字进行扩展，使得能够用创建正规索引类似的语法创建空间索引。创建空间索引的列，必须将其声明为 NOT NULL，空间索引只能在存储引擎为 MyISAM 的表中创建。

7.1.3 索引的设计原则

索引设计不合理或者缺少索引都会对数据库和应用程序的性能造成障碍。高效的索引对于获得良好的性能非常重要。设计索引时，应该考虑以下准则：

（1）索引并非越多越好，一个表中如有大量的索引，不仅占用磁盘空间，还会影响 INSERT、DELETE、UPDATE 等语句的性能，因为在表中的数据更改时，索引也会进行调整和更新。

（2）避免对经常更新的表进行过多的索引，并且索引中的列要尽可能少。应该经常用于查询的字段创建索引，但要避免添加不必要的字段。

（3）数据量小的表最好不要使用索引，由于数据较少，查询花费的时间可能比遍历索引的时间还要短，索引可能不会产生优化效果。

（4）在条件表达式中经常用到的不同值较多的列上建立索引，在不同值很少的列上不要建立索引。比如，在学生表的"性别"字段上只有"男"与"女"两个不同值，因此就无须建立索引，如果对其建立索引不但不会提高查询效率，反而会严重降低数据更新速度。

（5）当唯一性是某种数据本身的特征时，指定唯一索引。使用唯一索引需能确保定义的列的数据完整性，以提高查询速度。

（6）在频繁进行排序或分组（即进行 GROUP BY 或 ORDER BY 操作）的列上建立索引，如果待排序的列有多个，可以在这些列上建立组合索引。

7.2 创建索引

MySQL 支持多种方法在单个或多个列上创建索引：在创建表的定义语句 CREATE TABLE 中指定索引列，使用 ALTER TABLE 语句在存在的表上创建索引，或者使用 CREATE INDEX 语句在已存在的表上添加索引。本节将详细介绍这 3 种方法。

7.2.1 创建表的时候创建索引

使用 CREATE TABLE 创建表时，除了可以定义列的数据类型，还可以定义主键约束、外键约束或者唯一性约束，而不论创建哪种约束，在定义约束的同时相当于在指定列上创建了一个索引。创建表时创建索引的基本语法格式如下：

```
CREATE TABLE table_name [col_name data_type]
[UNIQUE|FULLTEXT|SPATIAL] [INDEX|KEY] [index_name] (col_name [length]) [ASC | DESC]
```

UNIQUE、FULLTEXT 和 SPATIAL 为可选参数，分别表示唯一索引、全文索引和空间索引；INDEX 与 KEY 为同义词，两者作用相同，用来指定创建索引；col_name 为需要创建索引的字段列，该列必须从数据表中定义的多个列中选择；index_name 指定索引的名称，为可选参数，如果不指定，MySQL 默认 col_name 为索引值；length 为可选参数，表示索引的长度，只有字符串类型的字段才能指定索引长度；ASC 或 DESC 指定升序或者降序的索引值存储。

1. 创建普通索引

普通索引是最基本的索引类型，没有唯一性之类的限制，其作用只是加快对数据的访问速度。

【例 7.1】在 book 表中的 year_publication 字段上建立普通索引，SQL 语句如下：

```
CREATE TABLE book
(
bookid INT NOT NULL,
bookname VARCHAR(255) NOT NULL,
authors VARCHAR(255) NOT NULL,
info VARCHAR(255) NULL,
comment VARCHAR(255) NULL,
year_publication YEAR NOT NULL,
INDEX(year_publication)
);
```

该语句执行完毕之后，使用 SHOW CREATE TABLE 查看表结构：

```
mysql> SHOW CREATE table book \G
*************************** 1. row ***************************
       Table: book
Create Table: CREATE TABLE `book` (
  `bookid` int NOT NULL,
  `bookname` varchar(255) NOT NULL,
  `authors` varchar(255) NOT NULL,
  `info` varchar(255) DEFAULT NULL,
  `comment` varchar(255) DEFAULT NULL,
  `year_publication` year(4) NOT NULL,
  KEY `year_publication` (`year_publication`)
) ENGINE=InnoDB DEFAULT CHARSET=utf8mb4 COLLATE=utf8mb4_0900_ai_ci
```

由结果可以看到，book 表的 year_publication 字段成功建立了索引，其索引名称 year_publication 为 MySQL 自动添加。使用 EXPLAIN 语句查看索引是否正在使用：

```
mysql> EXPLAIN SELECT * FROM book WHERE year_publication=1990 \G
*************************** 1. row ***************************
           id: 1
  select_type: SIMPLE
        table: book
         type: ref
possible_keys: year_publication
          key: year_publication
      key_len: 1
          ref: const
         rows: 1
        Extra: Using index condition
```

EXPLAIN 语句输出结果的各个行解释如下：

（1）select_type 行指定所使用的 SELECT 查询类型，这里值为 SIMPLE，表示简单的 SELECT，不使用 UNION 或子查询。其他可能的取值有 PRIMARY、UNION、SUBQUERY 等。

（2）table 行指定数据库读取的数据表的名字，它们按被读取的先后顺序排列。

（3）type 行指定了本数据表与其他数据表之间的关联关系，可能的取值有 system、const、 eq_ref、

ref、range、index 和 All。

（4）possible_keys 行给出了 MySQL 在搜索数据记录时可选用的各个索引。

（5）key 行是 MySQL 实际选用的索引。

（6）key_len 行给出索引按字节计算的长度，key_len 数值越小，表示越快。

（7）ref 行给出了关联关系中另一个数据表里的数据列名。

（8）rows 行是 MySQL 在执行这个查询时预计会从这个数据表里读出的数据行的行数。

（9）Extra 行提供了与关联操作有关的信息。

可以看到，possible_keys 和 key 的值都为 year_publication，查询时使用了索引。

2. 创建唯一索引

创建唯一索引的主要原因是减少查询索引列操作的执行时间，尤其是对比较庞大的数据表。它与前面的普通索引类似，不同的就是：索引列的值必须唯一，但允许有空值。如果是组合索引，则列值的组合必须唯一。

【例 7.2】创建一个表 t1，在表中的 id 字段上使用 UNIQUE 关键字创建唯一索引。

```
CREATE TABLE t1
(
id INT NOT NULL,
name CHAR(30) NOT NULL,
UNIQUE INDEX UniqIdx(id)
);
```

该语句执行完毕之后，使用 SHOW CREATE TABLE 查看表结构：

```
mysql> SHOW CREATE table t1 \G
*************************** 1. row ***************************
       Table: t1
Create Table: CREATE TABLE `t1` (
  `id` int NOT NULL,
  `name` char(30) NOT NULL,
  UNIQUE KEY `UniqIdx` (`id`)
) ENGINE=InnoDB DEFAULT CHARSET=utf8mb4 COLLATE=utf8mb4_0900_ai_ci
```

由结果可以看到，id 字段上已经成功建立了一个名为 UniqIdx 的唯一索引。

3. 创建单列索引

单列索引是在数据表中的某一个字段上创建的索引，一个表中可以创建多个单列索引。前面两个例子中创建的索引都为单列索引。

【例 7.3】创建一个表 t2，在表中的 name 字段上创建单列索引。

表结构如下：

```
CREATE TABLE t2
(
id INT NOT NULL,
name CHAR(50) NULL,
INDEX SingleIdx(name(20))
);
```

该语句执行完毕之后，使用 SHOW CREATE TABLE 查看表结构：

```
mysql> SHOW CREATE table t2 \G
*************************** 1. row ***************************
       Table: t2
Create Table: CREATE TABLE `t2` (
  `id` int NOT NULL,
  `name` char(50) DEFAULT NULL,
  KEY `SingleIdx` (`name`(20))
) ENGINE=InnoDB DEFAULT CHARSET=utf8mb4 COLLATE=utf8mb4_0900_ai_ci
```

由结果可以看到，id 字段上已经成功建立了一个名为 SingleIdx 的单列索引，索引长度为 20。

4. 创建组合索引

组合索引是在多个字段上创建一个索引。

【例 7.4】创建表 t3，在表中的 id、name 和 age 字段上建立组合索引，SQL 语句如下：

```
CREATE TABLE t3
(
id INT NOT NULL,
name CHAR(30) NOT NULL,
age INT NOT NULL,
info VARCHAR(255),
INDEX MultiIdx(id, name, age)
);
```

该语句执行完毕之后，使用 SHOW CREATE TABLE 查看表结构：

```
mysql> SHOW CREATE table t3 \G
*** 1. row ***
       Table: t3
CREATE Table: CREATE TABLE `t3` (
  `id` int NOT NULL,
  `name` char(30) NOT NULL,
  `age` int NOT NULL,
  `info` varchar(255) DEFAULT NULL,
  KEY `MultiIdx` (`id`,`name`,`age`)
) ENGINE=InnoDB DEFAULT CHARSET=utf8mb4 COLLATE=utf8mb4_0900_ai_ci
```

由结果可以看到，id、name 和 age 字段上已经成功建立了一个名为 MultiIdx 的组合索引。

组合索引可起几个索引的作用，但是使用时并不是随便查询哪个字段都可以使用索引，而是遵从"最左前缀"：利用索引中最左边的列集来匹配行，这样的列集称为最左前缀。例如，这里由 id、name 和 age 3 个字段构成的索引，索引行中按 id、name、age 的顺序存放，索引可以搜索（id, name, age）、（id, name）或者 id 字段组合。如果列不构成索引最左面的前缀，那么 MySQL 不能使用局部索引，如（age）或者（name,age）组合则不能使用索引查询。

在 t3 表中，查询 id 和 name 字段，使用 EXPLAIN 语句查看索引的使用情况：

```
mysql> EXPLAIN SELECT * FROM t3 WHERE id=1 AND name='joe' \G
*** 1. row ***
          id: 1
  select_type: SIMPLE
        table: t3
```

```
        type: ref
possible_keys: MultiIdx
          key: MultiIdx
      key_len: 94
          ref: const,const
         rows: 1
        Extra: Using where
```

可以看到，查询 id 和 name 字段时，使用了名称 MultiIdx 的索引，如果查询（name,age）组合或者单独查询 name 和 age 字段，结果如下：

```
*** 1. row ***
           id: 1
  select_type: SIMPLE
        table: t3
         type: ALL
possible_keys: NULL
          key: NULL
      key_len: NULL
          ref: NULL
         rows: 1
        Extra: Using where
```

此时，possible_keys 和 key 值为 NULL，并没有使用在 t3 表中创建的索引进行查询。

5. 创建全文索引

FULLTEXT 全文索引可以用于全文搜索。只有 MyISAM 存储引擎支持 FULLTEXT 索引，并且只为 CHAR、VARCHAR 和 TEXT 列创建索引。索引总是对整个列进行，不支持局部（前缀）索引。

【例 7.5】创建表 t4，在表中的 info 字段上建立全文索引，SQL 语句如下：

```
CREATE TABLE t4
(
id INT NOT NULL,
name CHAR(30) NOT NULL,
age INT NOT NULL,
info VARCHAR(255),
FULLTEXT INDEX FullTxtIdx(info)
) ENGINE=MyISAM;
```

提示：因为 MySQL 8.0 中默认存储引擎为 InnoDB，在这里创建表时需要修改表的存储引擎为 MyISAM，否则创建索引会出错。

语句执行完毕之后，使用 SHOW CREATE TABLE 查看表结构：

```
mysql> SHOW CREATE table t4 \G
*************************** 1. row ***************************
       Table: t4
Create Table: CREATE TABLE `t4` (
  `id` int NOT NULL,
  `name` char(30) NOT NULL,
  `age` int NOT NULL,
  `info` varchar(255) DEFAULT NULL,
  FULLTEXT KEY `FullTxtIdx` (`info`)
```

```
) ENGINE=MyISAM DEFAULT CHARSET=utf8mb4 COLLATE=utf8mb4_0900_ai_ci
```

由结果可以看到，info 字段上已经成功建立了一个名为 FullTxtIdx 的 FULLTEXT 索引。全文索引非常适合于大型数据集，对于小的数据集，它的用处比较小。

6. 创建空间索引

空间索引必须在 MyISAM 类型的表中创建，且空间类型的字段必须为非空。

【例 7.6】创建表 t5，在空间类型为 GEOMETRY 的字段上创建空间索引，SQL 语句如下：

```
CREATE TABLE t5
(
g GEOMETRY NOT NULL,
SPATIAL INDEX spatIdx(g)
)ENGINE=MyISAM;
```

该语句执行完毕之后，使用 SHOW CREATE TABLE 查看表结构：

```
mysql> SHOW CREATE table t5 \G
*** 1. row ***
       Table: t5
CREATE Table: CREATE TABLE `t5` (
  `g` geometry NOT NULL,
  SPATIAL KEY `spatIdx` (`g`)
) ENGINE=MyISAM DEFAULT CHARSET=utf8mb4 COLLATE=utf8mb4_0900_ai_ci
```

可以看到，t5 表的 g 字段上创建了名称为 spatIdx 的空间索引。注意，创建时指定空间类型字段值的非空约束，并且表的存储引擎为 MyISAM。

7.2.2 在已经存在的表上创建索引

在已经存在的表中创建索引，可以使用 ALTER TABLE 语句或者 CREATE INDEX 语句。本节将介绍如何使用 ALTER TABLE 和 CREATE INDEX 语句在已知表字段上创建索引。

1. 使用 ALTER TABLE 语句创建索引

ALTER TABLE 创建索引的基本语法如下：

```
ALTER TABLE table_name  ADD [UNIQUE|FULLTEXT|SPATIAL]  [INDEX|KEY]
[index_name] (col_name[length],...) [ASC | DESC]
```

与创建表时创建索引的语法不同的是，这里使用了 ALTER TABLE 和 ADD 关键字，ADD 表示向表中添加索引。

【例 7.7】在 book 表中的 bookname 字段上建立名为 BkNameIdx 的普通索引。

添加索引之前，使用 SHOW INDEX 语句查看指定表中创建的索引：

```
mysql> SHOW INDEX FROM book \G
*** 1. Row ***
        Table: book
   Non_unique: 1
     Key_name: year_publication
 Seq_in_index: 1
```

```
        Column_name: year_publication
          Collation: A
        Cardinality: 0
           Sub_part: NULL
             Packed: NULL
               Null:
         Index_type: BTREE
            Comment:
      Index_comment:
```

其中各个主要参数的含义为:

(1) Table 表示创建索引的表。

(2) Non_unique 表示索引非唯一,1 代表是非唯一索引,0 代表唯一索引。

(3) Key_name 表示索引的名称。

(4) Seq_in_index 表示该字段在索引中的位置,单列索引该值为 1,组合索引为每个字段在索引定义中的顺序。

(5) Column_name 表示定义索引的列字段。

(6) Sub_part 表示索引的长度。

(7) Null 表示该字段是否能为空值。

(8) Index_type 表示索引类型。

可以看到,book 表中已经存在了一个索引,即前面已经定义的名称为 year_publication 索引,该索引为非唯一索引。

下面使用 ALTER TABLE 在 bookname 字段上添加索引,SQL 语句如下:

```
ALTER TABLE book ADD INDEX BkNameIdx( bookname(30) );
```

使用 SHOW INDEX 语句查看表中的索引:

```
mysql> SHOW INDEX FROM book \G
*** 1. Row ***
              Table: book
         Non_unique: 1
           Key_name: year_publication
       Seq_in_index: 1
        Column_name: year_publication
          Collation: A
        Cardinality: 0
           Sub_part: NULL
             Packed: NULL
               Null:
         Index_type: BTREE
            Comment:
      Index_comment:
*** 2. Row ***
              Table: book
         Non_unique: 1
           Key_name: BkNameIdx
       Seq_in_index: 1
        Column_name: bookname
          Collation: A
```

```
        Cardinality: 0
          Sub_part: 30
            Packed: NULL
              Null:
        Index_type: BTREE
           Comment:
     Index_comment:
```

可以看到,现在表中已经有了两个索引,一个是 year_publication,另一个为通过 ALTER TABLE 语句添加的、名称为 BkNameIdx 的索引,该索引为非唯一索引,长度为 30。

【例 7.8】在 book 表的 bookId 字段上建立名称为 UniqidIdx 的唯一索引,SQL 语句如下:

```
ALTER TABLE book ADD UNIQUE INDEX UniqidIdx ( bookId );
```

使用 SHOW INDEX 语句查看表中的索引:

```
mysql> SHOW INDEX FROM book \G
*** 1. Row ***
             Table: book
        Non_unique: 0
          Key_name: UniqidIdx
      Seq_in_index: 1
       Column_name: bookid
         Collation: A
       Cardinality: 0
          Sub_part: NULL
            Packed: NULL
              Null:
        Index_type: BTREE
           Comment:
     Index_comment:
```

可以看到 Non_unique 属性值为 0,表示名称为 UniqidIdx 的索引为唯一索引,创建唯一索引成功。

【例 7.9】在 book 表的 comment 字段上建立单列索引,SQL 语句如下:

```
ALTER TABLE book ADD INDEX BkcmtIdx ( comment(50) );
```

使用 SHOW INDEX 语句查看表中的索引:

```
*** 3. Row ***
             Table: book
        Non_unique: 1
          Key_name: BkcmtIdx
      Seq_in_index: 1
       Column_name: comment
         Collation: A
       Cardinality: 0
          Sub_part: 50
            Packed: NULL
              Null: YES
        Index_type: BTREE
           Comment:
     Index_comment:
```

可以看到,语句执行之后在 book 表的 comment 字段上建立了名称为 BkcmtIdx 的索引,长度为

50,在查询时,只需要检索前 50 个字符。

【例 7.10】在 book 表的 authors 和 info 字段上建立组合索引,SQL 语句如下:

```
ALTER TABLE book ADD INDEX BkAuAndInfoIdx ( authors(30),info(50) );
```

使用 SHOW INDEX 语句查看表中的索引:

```
mysql> SHOW INDEX FROM book \G
*** 4. Row ***
        Table: book
   Non_unique: 1
     Key_name: BkAuAndInfoIdx
 Seq_in_index: 1
  Column_name: authors
    Collation: A
  Cardinality: 0
     Sub_part: 30
       Packed: NULL
         Null:
   Index_type: BTREE
      Comment:
Index_comment:
*** 5. Row ***
        Table: book
   Non_unique: 1
     Key_name: BkAuAndInfoIdx
 Seq_in_index: 2
  Column_name: info
    Collation: A
  Cardinality: 0
     Sub_part: 50
       Packed: NULL
         Null: YES
   Index_type: BTREE
      Comment:
Index_comment:
```

可以看到名称为 BkAuAndInfoIdx 的索引由两个字段组成:authors 字段长度为 30,在组合索引中的序号为 1,该字段不允许空值 NULL;info 字段长度为 50,在组合索引中的序号为 2,该字段可以为空值 NULL。

【例 7.11】创建表 t6,在 t6 表上使用 ALTER TABLE 创建全文索引,SQL 语句如下:

首先创建表 t6,语句如下:

```
CREATE TABLE t6
(
id INT NOT NULL,
info CHAR(255)
) ENGINE=MyISAM;
```

注意,修改 ENGINE 参数为 MyISAM,MySQL 默认引擎 InnoDB 不支持全文索引。

使用 ALTER TABLE 语句在 info 字段上创建全文索引:

```
ALTER TABLE t6 ADD FULLTEXT INDEX infoFTIdx ( info );
```

使用 SHOW INDEX 语句查看索引：

```
mysql> SHOW index from t6 \G
*** 1. Row ***
        Table: t6
   Non_unique: 1
     Key_name: infoFTIdx
 Seq_in_index: 1
  Column_name: info
    Collation: NULL
  Cardinality: NULL
     Sub_part: NULL
       Packed: NULL
         Null: YES
   Index_type: FULLTEXT
      Comment:
ndex_comment:
```

可以看到，t6 表中已经创建了名称为 infoFTIdx 的索引，该索引在 info 字段上创建，类型为 FULLTEXT，允许空值。

【例 7.12】创建表 t7，在 t7 的空间数据类型字段 g 上创建名称为 spatIdx 的空间索引，SQL 语句如下：

```
CREATE TABLE t7 ( g GEOMETRY NOT NULL )ENGINE=MyISAM;
```

使用 ALTER TABLE 在表 t7 的 g 字段建立空间索引：

```
ALTER TABLE t7 ADD SPATIAL INDEX spatIdx(g);
```

使用 SHOW INDEX 语句查看索引：

```
mysql> SHOW index from t7 \G
*** 1. Row ***
        Table: t7
   Non_unique: 1
     Key_name: spatIdx
 Seq_in_index: 1
  Column_name: g
    Collation: A
  Cardinality: NULL
     Sub_part: 32
       Packed: NULL
         Null:
   Index_type: SPATIAL
      Comment:
Index_comment:
```

可以看到，t7 表的 g 字段上创建了名称为 spatIdx 的空间索引。

2. 使用 CREATE INDEX 创建索引

CREATE INDEX 语句可以在已经存在的表上添加索引。在 MySQL 中，CREATE INDEX 被映射到一个 ALTER TABLE 语句上，基本语法结构为：

```
CREATE [UNIQUE|FULLTEXT|SPATIAL] INDEX index_name
ON table_name (col_name[length],...) [ASC | DESC]
```

可以看到 CREATE INDEX 语句和 ALTER INDEX 语句的语法基本一样，只是关键字不同。

在这里，使用相同的表 book，假设该表中没有任何索引值，创建 book 表语句如下：

```
CREATE TABLE book
(
bookid INT NOT NULL,
bookname VARCHAR(255) NOT NULL,
authors VARCHAR(255) NOT NULL,
info VARCHAR(255) NULL,
comment VARCHAR(255) NULL,
year_publication YEAR NOT NULL
);
```

提示：读者可以将该数据库中的 book 表删除，按上面的语句重新建立，然后进行下面的操作。

【例 7.13】在 book 表中的 bookname 字段上建立名为 BkNameIdx 的普通索引，SQL 语句如下：

```
CREATE INDEX BkNameIdx ON book(bookname);
```

语句执行完毕之后，将在 book 表中创建名称为 BkNameIdx 的普通索引。读者可以使用 SHOW INDEX 或者 SHOW CREATE TABLE 语句查看 book 表中的索引，其索引内容与前面介绍的相同。

【例 7.14】在 book 表的 bookId 字段上建立名称为 UniqidIdx 的唯一索引，SQL 语句如下：

```
CREATE UNIQUE INDEX UniqidIdx ON book ( bookId );
```

语句执行完毕之后，将在 book 表中创建名称为 UniqidIdx 的唯一索引。

【例 7.15】在 book 表的 comment 字段上建立单列索引，SQL 语句如下：

```
CREATE INDEX BkcmtIdx ON book(comment(50) );
```

语句执行完毕之后，将在 book 表的 comment 字段上建立一个名为 BkcmtIdx 的单列索引，长度为 50。

【例 7.16】在 book 表的 authors 和 info 字段上建立组合索引，SQL 语句如下：

```
CREATE INDEX BkAuAndInfoIdx ON book ( authors(20),info(50) );
```

语句执行完毕之后，在 book 表的 authors 和 info 字段上建立了一个名为 BkAuAndInfoIdx 的组合索引，authors 的索引序号为 1、长度为 20，info 的索引序号为 2、长度为 50。

【例 7.17】先删除表 t6，再重新建立表 t6，在 t6 表中使用 CREATE INDEX 语句，在 CHAR 类型的 info 字段上创建全文索引，SQL 语句如下：

首先删除表 t6，并重新建立该表，可以输入下面的语句：

```
mysql> drop table t6;
Query OK, 0 rows affected (0.00 sec)

mysql> CREATE TABLE t6
    -> (
    -> id    INT NOT NULL,
```

```
    -> info CHAR(255)
    -> ) ENGINE=MyISAM;
Query OK, 0 rows affected (0.00 sec)
```

使用 CREATE INDEX 在 t6 表的 info 字段上创建名称为 infoFTIdx 的全文索引：

```
CREATE FULLTEXT INDEX infoFTIdx ON t6(info);
```

语句执行完毕之后，将在 t6 表中创建名称为 infoFTIdx 的索引，该索引在 info 字段上创建，类型为 FULLTEXT，允许空值。

【例 7.18】删除表 t7，重新创建表 t7，在 t7 表中使用 CREATE INDEX 语句，在空间数据类型字段 g 上创建名称为 spatIdx 的空间索引，操作如下：

首先删除表 t7，并重新建立该表，分别输入下面的语句：

```
mysql> drop table t7;
Query OK, 0 rows affected (0.00 sec)

mysql> CREATE TABLE t7 (g GEOMETRY NOT NULL) ENGINE=MyISAM;
Query OK, 0 rows affected (0.00 sec)
```

使用 CREATE INDEX 语句在表 t7 的 g 字段建立空间索引：

```
CREATE SPATIAL INDEX spatIdx ON t7 (g);
```

语句执行完毕之后，将在 t7 表中创建名称为 spatIdx 的空间索引，该索引在 g 字段上创建。

7.3 删除索引

MySQL 中删除索引使用 ALTER TABLE 或者 DROP INDEX 语句，两者可实现相同的功能，DROP INDEX 语句在内部被映射到一个 ALTER TABLE 语句中。

1. 使用 ALTER TABLE 删除索引

ALTER TABLE 删除索引的基本语法格式如下：

```
ALTER TABLE table_name DROP INDEX index_name;
```

【例 7.19】删除 book 表中名称为 UniqidIdx 的唯一索引，SQL 语句如下：

首先查看 book 表中是否有名称为 UniqidIdx 的索引，输入 SHOW 语句：

```
mysql> SHOW CREATE table book \G
*** 1. row ***
       Table: book
CREATE Table: CREATE TABLE `book` (
  `bookid` int NOT NULL,
  `bookname` varchar(255) NOT NULL,
  `authors` varchar(255) NOT NULL,
  `info` varchar(255) DEFAULT NULL,
  `year_publication` year(4) NOT NULL,
  UNIQUE KEY `UniqidIdx` (`bookid`),
```

```
  KEY `BkNameIdx` (`bookname`),
  KEY `BkAuAndInfoIdx` (`authors`(20),`info`(50))
) ENGINE=InnoDB DEFAULT CHARSET=utf8mb4 COLLATE=utf8mb4_0900_ai_ci
```

查询结果可以看到，book 表中有名称为 UniqidIdx 的唯一索引，该索引在 bookid 字段上创建。下面删除该索引，输入删除语句：

```
mysql> ALTER TABLE book DROP INDEX UniqidIdx;
```

语句执行完毕，使用 SHOW 语句查看索引是否被删除：

```
mysql> SHOW CREATE table book \G
*** 1. row ***
       Table: book
CREATE Table: CREATE TABLE `book` (
  `bookid` int NOT NULL,
  `bookname` varchar(255) NOT NULL,
  `authors` varchar(255) NOT NULL,
  `info` varchar(255) DEFAULT NULL,
  `year_publication` year(4) NOT NULL,
  KEY `BkNameIdx` (`bookname`),
  KEY `BkAuAndInfoIdx` (`authors`(20),`info`(50))
) ENGINE=InnoDB DEFAULT CHARSET=utf8mb4 COLLATE=utf8mb4_0900_ai_ci
```

由结果可以看到，book 表中已经没有名称为 uniqidIdx 的唯一索引，删除索引成功。

提示：具有 AUTO_INCREMENT 约束的字段的唯一索引不能被删除。

2. 使用 DROP INDEX 语句删除索引

DROP INDEX 删除索引的基本语法格式如下：

```
DROP INDEX index_name ON table_name;
```

【例 7.20】删除 book 表中名称为 BkAuAndInfoIdx 的组合索引，SQL 语句如下：

```
mysql> DROP INDEX BkAuAndInfoIdx ON book;
```

语句执行完毕，使用 SHOW 语句查看索引是否被删除：

```
mysql> SHOW CREATE table book \G
*** 1. row ***
       Table: book
CREATE Table: CREATE TABLE `book` (
  `bookid` int NOT NULL,
  `bookname` varchar(255) NOT NULL,
  `authors` varchar(255) NOT NULL,
  `info` varchar(255) DEFAULT NULL,
  `year_publication` year(4) NOT NULL,
  KEY `BkNameIdx` (`bookname`)
) ENGINE=InnoDB DEFAULT CHARSET=utf8
1 row in set (0.00 sec)
```

可以看到，book 表中已经没有名称为 BkAuAndInfoIdx 的组合索引，删除索引成功。

提示：删除表中的列时，如果要删除的列为索引的组成部分，则该列也会从索引中删除。如果组成索引的所有列都被删除，则整个索引将被删除。

7.4 统计直方图

MySQL 8.0 实现了统计直方图。利用直方图,用户可以对一张表的某一列做数据分布的统计,特别是针对没有索引的字段。这可以帮助查询优化器找到更优的执行计划。

7.4.1 直方图的优点

在数据库中,查询优化器负责将 SQL 转换成最有效的执行计划。有时候,查询优化器会找不到最优的执行计划,导致花费了更多不必要的时间。造成这种情况的主要原因是,查询优化器有时无法准确地知道以下几个问题的答案:

- 每个表有多少行?
- 每一列有多少不同的值?
- 每一列的数据分布情况如何?

例如,销售表 production 包括 id、tm、count 三个字段,分别表示编号、销售时间和销售数量。对比以下两个查询语句:

```
SELECT * FROM production WHERE tm BETWEEN "22:00:00" AND "23:59:00"
SELECT * FROM production WHERE tm BETWEEN "08:00:00" AND "12:00:00"
```

如果销售时间大部分集中在上午 8 点到 12 点,在查询销售情况时,第一个查询语句耗费的时间会远远大于第二个查询语句。

因为没有统计数据,优化器会假设 tm 的值是均匀分配的。如何才能使查询优化器知道数据的分布情况呢?一个解决方法就是在列上建立统计直方图。

直方图能近似获得一列的数据分布情况,从而让数据库知道它含有哪些数据。直方图有多种形式,MySQL 支持了两种:等宽直方图(singleton)和等高直方图(equi-height)。直方图的共同点是,它们都将数据分到了一系列的 buckets 中去。MySQL 会自动将数据划到不同的 buckets 中,也会自动决定创建哪种类型的直方图。

7.4.2 直方图的基本操作

创建直方图的语法格式如下:

```
ANALYZE TABLE table_name [UPDATE HISTOGRAM on col_name with N BUCKETS |DROP HISTOGRAM ON clo_name]
```

buckets 的默认值是 100。统计直方图的信息存储在数据字典表 column_statistcs 中,可以通过视图 information_schema.COLUMN_STATISTICS 访问。直方图以灵活的 JSON 格式存储。ANALYZE TABLE 会基于表大小自动判断是否要进行取样操作。ANALYZE TABLE 也会基于表中列的数据分布情况以及 bucket 的数量,来决定是否要建立等宽直方图(singleton)还是等高直方图(equi-height)。

创建用于测试的数据表 production,语句如下:

```
mysql> CREATE TABLE production (id int,tm TIME,count int);
```

在数据表 production 的字段 tm 上创建直方图，执行语句如下：

```
mysql> ANALYZE TABLE production UPDATE HISTOGRAM ON tm WITH 60 BUCKETS;
+-------------------+-----------+----------+------------------------------------------------+
| Table             | Op        | Msg_type | Msg_text                                       |
+-------------------+-----------+----------+------------------------------------------------+
| mytest.production | histogram | status   | Histogram statistics created for column 'tm'.  |
+-------------------+-----------+----------+------------------------------------------------+
```

buckets 的值必须指定，可以设置为 1 到 1024，默认值是 100。设置 buckets 值时，可以先设置低一些，如果没有满足需求，可以再往上增大。

对于不同的数据集合，buckets 的值取决于以下几个因素：

- 这列有多少不同的值。
- 数据的分布情况。
- 需要多高的准确性。

在数据表 production 的字段 tm 和字段 count 上创建直方图，执行语句如下：

```
mysql>ANALYZE TABLE production UPDATE HISTOGRAM ON tm,count WITH 60 BUCKETS;
+-------------------+-----------+----------+---------------------------------------------------+
| Table             | Op        | Msg_type | Msg_text                                          |
+-------------------+-----------+----------+---------------------------------------------------+
| mytest.production | histogram | status   | Histogram statistics created for column 'count'.  |
| mytest.production | histogram | status   | Histogram statistics created for column 'tm'.     |
+-------------------+-----------+----------+---------------------------------------------------+
```

再次创建直方图时，将会将上一个直方图重写。

如果需要删除已经创建的直方图，用 DROP HISTOGRAM 就可以实现：

```
mysql>ANALYZE TABLE production DROP HISTOGRAM ON tm,count;
+-------------------+-----------+----------+---------------------------------------------------+
| Table             | Op        | Msg_type | Msg_text                                          |
+-------------------+-----------+----------+---------------------------------------------------+
| mytest.production | histogram | status   | Histogram statistics removed for column 'count'.  |
| mytest.production | histogram | status   | Histogram statistics removed for column 'tm'.     |
+-------------------+-----------+----------+---------------------------------------------------+
```

直方图统计了表中某些字段的数据分布情况，为优化选择高效的执行计划提供参考。直方图与索引有着本质的区别：维护一个索引有代价，每一次的 INSERT、UPDATE、DELETE 都会需要更新索引，会对性能有一定的影响；而直方图一次创建永不更新，除非明确去更新它，所以不会影响 INSERT、UPDATE、DELETE 的性能。

建立直方图的时候，MySQL 服务器会将所有数据读到内存中，然后在内存中进行操作，包括排序。如果对一个很大的表建立直方图，可能会需要将几百兆的数据都读到内存中。为了规避这种风险，MySQL 会根据给定的 histogram_generation_max_mem_size 的值计算该将多少行数据读到内存中。

设置 histogram_generation_max_mem_size 值的方法如下：

```
mysql>SET histogram_generation_max_mem_size = 10000;
```

第 8 章

存储过程和函数

简单地说，存储过程就是一条或者多条 SQL 语句的集合，可视为批文件，但是其作用不仅限于批处理。本章将介绍如何创建存储过程和存储函数，如何使用变量，以及如何调用、查看、修改、删除存储过程和存储函数等内容。

8.1 创建存储过程和函数

存储程序可以分为存储过程和函数。在 MySQL 中，创建存储过程和函数使用的语句分别是 CREATE PROCEDURE 和 CREATE FUNCTION。使用 CALL 语句来调用存储过程，只能用输出变量返回值。函数可以从语句外调用（引用函数名），也能返回标量值。存储过程也可以调用其他存储过程。

8.1.1 创建存储过程

创建存储过程，需要使用 CREATE PROCEDURE 语句，基本语法格式如下：

```
CREATE PROCEDURE sp_name ( [proc_parameter] )
[characteristics ...] routine_body
```

CREATE PROCEDURE 为用来创建存储过程的关键字；sp_name 为存储过程的名称；proc_parameter 为指定存储过程的参数列表，列表形式如下：

```
[ IN | OUT | INOUT ] param_name type
```

其中，IN 表示输入参数，OUT 表示输出参数，INOUT 表示既可以输入也可以输出；param_name 表示参数名称；type 表示参数的类型，该类型可以是 MySQL 数据库中的任意类型。

characteristics 指定存储过程的特性，有以下取值：

- **LANGUAGE SQL**：说明 routine_body 部分是由 SQL 语句组成的，当前系统支持的语言为 SQL。SQL 是 LANGUAGE 特性的唯一值。
- **[NOT] DETERMINISTIC**：指明存储过程执行的结果是否正确。DETERMINISTIC 表示结果是确定的。每次执行存储过程时，相同的输入会得到相同的输出。NOT DETERMINISTIC 表示结果是不确定的，相同的输入可能得到不同的输出。如果没有指定任意一个值，默认为 NOT DETERMINISTIC。
- **{ CONTAINS SQL | NO SQL | READS SQL DATA | MODIFIES SQL DATA }**：指明子程序使用 SQL 语句的限制。CONTAINS SQL 表明子程序包含 SQL 语句，但是不包含读写数据的语句；NO SQL 表明子程序不包含 SQL 语句；READS SQL DATA 说明子程序包含读数据的语句；MODIFIES SQL DATA 表明子程序包含写数据的语句。默认情况下，系统会指定为 CONTAINS SQL。
- **SQL SECURITY { DEFINER | INVOKER }**：指明谁有权限来执行。DEFINER 表示只有定义者才能执行。INVOKER 表示拥有权限的调用者可以执行。默认情况下，系统指定为 DEFINER。
- **COMMENT 'string'**：注释信息，可以用来描述存储过程或函数。

routine_body 是 SQL 代码的内容，可以用 BEGIN…END 来表示 SQL 代码的开始和结束。

编写存储过程并不是一件简单的事情，可能存储过程中需要复杂的 SQL 语句，并且要有创建存储过程的权限；但是使用存储过程将简化操作，减少冗余的操作步骤，同时，还可以减少操作过程中的失误，提高效率，因此存储过程非常有用，而且应该尽可能地学会使用。

下面的代码演示了存储过程的内容，名称为 AvgFruitPrice，返回所有水果的平均价格，输入代码如下：

```
CREATE PROCEDURE AvgFruitPrice ()
BEGIN
    SELECT AVG(f_price) AS avgprice
    FROM fruits;
END;
```

上述代码中，此存储过程名为 AvgFruitPrice，使用 CREATE PROCEDURE AvgFruitPrice ()语句定义。此存储过程没有参数，但是后面的()仍然需要。BEGIN 和 END 语句用来限定存储过程体，过程本身仅是一个简单的 SELECT 语句（AVG 为求字段平均值的函数）。

【例 8.1】创建查看 fruits 表的存储过程，代码如下：

```
CREATE PROCEDURE Proc()
BEGIN
    SELECT * FROM fruits;
END ;
```

这行代码创建了一个查看 fruits 表的存储过程，每次调用这个存储过程的时候，都会执行 SELECT 语句查看表的内容，代码的执行过程如下：

```
MySQL> DELIMITER //
MySQL> CREATE PROCEDURE Proc()
    -> BEGIN
    -> SELECT * FROM fruits;
```

```
    -> END //
Query OK, 0 rows affected (0.00 sec)

MySQL> DELIMITER ;
```

这个存储过程和使用 SELECT 语句查看表的效果得到的结果是一样的，当然存储过程也可以是很多语句复杂的组合，就好像这个例子刚开始给出的那个语句一样，其本身也可以调用其他的函数来组成更加复杂的操作。

提示："DELIMITER //"语句的作用是将 MySQL 的结束符设置为"//"，因为 MySQL 默认的语句结束符号为英文分号";"。为了避免与存储过程中 SQL 语句结束符相冲突，需要使用 DELIMITER 改变存储过程的结束符，并以"END //"结束存储过程。存储过程定义完毕之后，再使用"DELIMITER ;"恢复默认结束符。DELIMITER 也可以指定其他符号作为结束符。

【例 8.2】创建名称为 CountProc 的存储过程，代码如下：

```
CREATE PROCEDURE CountProc (OUT param1 INT)
BEGIN
    SELECT COUNT(*) INTO param1 FROM fruits;
END;
```

上述代码的作用是创建一个获取 fruits 表记录条数的存储过程，名称是 CountProc，COUNT(*) 计算后把结果放入参数 param1 中。执行结果如下：

```
mysql> DELIMITER //
mysql> CREATE PROCEDURE CountProc(OUT param1 INT)
    -> BEGIN
    -> SELECT COUNT(*) INTO param1 FROM fruits;
    -> END;
    -> //
Query OK, 0 rows affected (0.00 sec)
mysql> DELIMITER ;
```

提示：当使用 DELIMITER 命令时，应该避免使用反斜杠（\）字符，因为反斜线是 MySQL 的转义字符。

8.1.2 创建存储函数

创建存储函数，需要使用 CREATE FUNCTION 语句，基本语法格式如下：

```
CREATE FUNCTION func_name ( [func_parameter] )
RETURNS type
[characteristic ...] routine_body
```

CREATE FUNCTION 为用来创建存储函数的关键字；func_name 表示存储函数的名称；func_parameter 为存储函数的参数列表，参数列表形式如下：

```
[ IN | OUT | INOUT ] param_name type
```

其中，IN 表示输入参数，OUT 表示输出参数，INOUT 表示既可以输入也可以输出；param_name 表示参数名称；type 表示参数的类型，该类型可以是 MySQL 数据库中的任意类型。

RETURNS type 语句表示函数返回数据的类型；characteristic 指定存储函数的特性，取值与创建存储过程时相同，这里不再赘述。

【例8.3】创建存储函数，名称为NameByZip，该函数返回SELECT语句的查询结果，数值类型为字符串型，代码如下：

```
CREATE FUNCTION NameByZip ()
 RETURNS CHAR(50)
 RETURN  (SELECT s_name FROM suppliers WHERE s_call='48075');
```

创建一个存储函数NameByZip，参数定义为空，返回一个CHAR类型的结果。代码的执行结果如下：

```
mysql> set global log_bin_trust_function_creators=TRUE;
mysql> DELIMITER //
mysql> CREATE FUNCTION NameByZip()
    -> RETURNS CHAR(50)
    -> RETURN  (SELECT s_name FROM suppliers WHERE s_call='48075');
    -> //

mysql> DELIMITER ;
```

如果在存储函数中的RETURN语句返回一个类型不同于函数的RETURNS子句中指定类型的值，返回值将被强制为恰当的类型。比如，如果一个函数返回一个ENUM或SET值，但是RETURN语句返回一个整数，对于SET成员集相应的ENUM成员，从函数返回的值是字符串。

提示：指定参数为IN、OUT或INOUT只对PROCEDURE是合法的（FUNCTION中总是默认为IN参数）。RETURNS子句只能对FUNCTION做指定，对函数而言这是强制的。它用来指定函数的返回类型，而且函数体必须包含一个RETURN value语句。

8.1.3 变量的使用

变量可以在子程序中声明并使用，这些变量的作用范围是在BEGIN…END程序中，本小节主要介绍如何定义变量和为变量赋值。

1. 定义变量

在存储过程中使用DECLARE语句定义变量，语法格式如下：

```
DECLARE var_name[,varname]... date_type [DEFAULT value];
```

var_name为局部变量的名称。DEFAULT value子句给变量提供一个默认值。值除了可以被声明为一个常数之外，还可以被指定为一个表达式。如果没有DEFAULT子句，则初始值为NULL。

【例8.4】定义名称为myparam的变量，类型为INT类型，默认值为100，代码如下：

```
DECLARE  myparam  INT  DEFAULT 100;
```

2. 为变量赋值

定义变量之后，为变量赋值可以改变变量的默认值。在MySQL中，使用SET语句为变量赋值，语法格式如下：

```
SET var_name = expr [, var_name = expr] ...;
```

在存储程序中的 SET 语句是一般 SET 语句的扩展版本。被参考变量可能是子程序内声明的变量，或者是全局服务器变量，如系统变量或者用户变量。

在存储程序中的 SET 语句作为预先存在的 SET 语法的一部分来实现，允许 SET a=x, b=y, …，这样的扩展语法。其中，不同的变量类型（局域变量和全局变量）可以被混合起来。这也允许把局部变量和一些只对系统变量有意义的选项合并起来。

【例 8.5】声明 3 个变量，分别为 var1、var2 和 var3，数据类型均为 INT，使用 SET 为变量赋值，代码如下：

```
DECLARE var1, var2, var3 INT;
SET var1 = 10, var2 = 20;
SET var3 = var1 + var2;
```

在 MySQL 中，还可以通过 SELECT…INTO 为一个或多个变量赋值，语法如下：

```
SELECT col_name[,...] INTO var_name[,...] table_expr;
```

这个 SELECT 语法把选定的列直接存储到对应位置的变量。col_name 表示字段名称；var_name 表示定义的变量名称；table_expr 表示查询条件表达式，包括表名称和 WHERE 子句。

【例 8.6】声明变量 fruitname 和 fruitprice，通过 SELECT…INTO 语句查询指定记录并为变量赋值，代码如下：

```
DECLARE fruitname CHAR(50);
DECLARE fruitprice DECIMAL(8,2);

SELECT f_name,f_price INTO fruitname, fruitprice
FROM fruits WHERE f_id ='a1';
```

8.1.4 定义条件和处理程序

特定条件需要特定处理。这些条件可以联系到错误以及子程序中的一般流程控制。定义条件是事先定义程序执行过程中遇到的问题，处理程序定义了在遇到这些问题时应当采取的处理方式，并且保证存储过程或函数在遇到警告或错误时能继续执行。这样可以增强存储程序处理问题的能力，避免程序异常停止运行。本节将介绍如何使用 DECLARE 关键字来定义条件和处理程序。

1．定义条件

定义条件使用 DECLARE 语句，语法格式如下：

```
DECLARE condition_name CONDITION FOR [condition_type]

[condition_type]:
SQLSTATE [VALUE] sqlstate_value | mysql_error_code
```

其中，condition_name 参数表示条件的名称；condition_type 参数表示条件的类型；sqlstate_value 和 MySQL_error_code 都可以表示 MySQL 的错误，sqlstate_value 为长度为 5 的字符串类型错误代码，MySQL_error_code 为数值类型错误代码。例如，在 ERROR 1142（42000）中，sqlstate_value 的值是 42000，MySQL_error_code 的值是 1142。

这个语句指定需要特殊处理的条件。它将一个名字和指定的错误条件关联起来。这个名字可以被用在随后定义处理程序的 DECLARE HANDLER 语句中。

【例 8.7】定义 "ERROR 1148（42000）" 错误，名称为 command_not_allowed。可以用两种不同的方法来定义，代码如下：

```
//方法一：使用 sqlstate_value
DECLARE  command_not_allowed CONDITION FOR SQLSTATE '42000';
//方法二：使用 mysql_error_code
DECLARE  command_not_allowed CONDITION  FOR  1148
```

2. 定义处理程序

定义处理程序时，使用 DECLARE 语句的语法如下：

```
DECLARE handler_type HANDLER FOR condition_value[,...] sp_statement
handler_type:
    CONTINUE | EXIT | UNDO

condition_value:
    SQLSTATE [VALUE] sqlstate_value
  | condition_name
  | SQLWARNING
  | NOT FOUND
  | SQLEXCEPTION
  | mysql_error_code
```

其中，handler_type 为错误处理方式，参数取 3 个值：CONTINUE、EXIT 和 UNDO。CONTINUE 表示遇到错误不处理，继续执行；EXIT 表示遇到错误马上退出；UNDO 表示遇到错误后撤回之前的操作，MySQL 中暂时不支持这样的操作。

condition_value 表示错误类型，可以有以下取值：

- SQLSTATE [VALUE] sqlstate_value：包含 5 个字符的字符串错误值。
- condition_name：表示 DECLARE CONDITION 定义的错误条件名称。
- SQLWARNING：匹配所有以 01 开头的 SQLSTATE 错误代码。
- NOT FOUND：匹配所有以 02 开头的 SQLSTATE 错误代码。
- SQLEXCEPTION：匹配所有没有被 SQLWARNING 或 NOT FOUND 捕获的 SQLSTATE 错误代码。
- MySQL_error_code：匹配数值类型错误代码。

sp_statement 参数为程序语句段，表示在遇到定义的错误时需要执行的存储过程或函数。

【例 8.8】定义处理程序的几种方式，代码如下：

```
//方法一：捕获 sqlstate_value
DECLARE CONTINUE HANDLER FOR SQLSTATE '42S02' SET @info='NO_SUCH_TABLE';
//方法二：捕获 mysql_error_code
DECLARE CONTINUE HANDLER FOR 1146 SET @info=' NO_SUCH_TABLE ';
//方法三：先定义条件，然后调用
DECLARE  no_such_table  CONDITION  FOR  1146;
DECLARE CONTINUE HANDLER FOR NO_SUCH_TABLE SET @info=' NO_SUCH_TABLE ';
//方法四：使用 SQLWARNING
DECLARE EXIT HANDLER FOR SQLWARNING SET @info='ERROR';
```

```
//方法五：使用 NOT FOUND
DECLARE EXIT HANDLER FOR NOT FOUND SET @info=' NO_SUCH_TABLE ';
//方法六：使用 SQLEXCEPTION
DECLARE EXIT HANDLER FOR SQLEXCEPTION SET @info='ERROR';
```

上述代码是 6 种定义处理程序的方法。

第一种方法是捕获 sqlstate_value 值。如果遇到 sqlstate_value 值为"42S02"，执行 CONTINUE 操作，并且输出"NO_SUCH_TABLE"信息。

第二种方法是捕获 MySQL_error_code 值。如果遇到 MySQL_error_code 值为 1146，执行 CONTINUE 操作，并且输出"NO_SUCH_TABLE"信息。

第三种方法是先定义条件，再调用条件。这里先定义 no_such_table 条件，遇到 1146 错误就执行 CONTINUE 操作。

第四种方法是使用 SQLWARNING。SQLWARNING 捕获所有以 01 开头的 sqlstate_value 值，然后执行 EXIT 操作，并且输出"ERROR"信息。

第五种方法是使用 NOT FOUND。NOT FOUND 捕获所有以 02 开头的 sqlstate_value 值，然后执行 EXIT 操作，并且输出"NO_SUCH_TABLE"信息。

第六种方法是使用 SQLEXCEPTION。SQLEXCEPTION 捕获所有没有被 SQLWARNING 或 NOT FOUND 捕获的 sqlstate_value 值，然后执行 EXIT 操作，并且输出"ERROR"信息。

【例 8.9】定义条件和处理程序，具体执行的过程如下：

```
mysql> CREATE TABLE test_db.t (s1 int,primary key (s1));

mysql> DELIMITER //

mysql> CREATE PROCEDURE handlerdemo ()
    -> BEGIN
    -> DECLARE CONTINUE HANDLER FOR SQLSTATE '23000' SET @x2 = 1;
    -> SET @x = 1;
    -> INSERT INTO test_db.t VALUES (1);
    -> SET @x = 2;
    -> INSERT INTO test_db.t VALUES (1);
    -> SET @x = 3;
    -> END;
    -> //

mysql> DELIMITER ;

/*调用存储过程*/
mysql> CALL handlerdemo();

/*查看调用过程结果*/
mysql> SELECT @x;
+------+
| @x   |
+------+
| 3    |
+------+
```

@x 是 1 个用户变量，执行结果 @x 等于 3，这表明 MySQL 被执行到程序的末尾。如果"DECLARE

CONTINUE HANDLER FOR SQLSTATE '23000' SET @x2 = 1;"这 1 行不在，第 2 个 INSERT 因 PRIMARY KEY 强制而失败之后，MySQL 可能已经采取默认（EXIT）路径，并且 SELECT @x 可能已经返回 2。

提示："@var_name"表示用户变量，使用 SET 语句为其赋值，用户变量与连接有关，一个客户端定义的变量不能被其他客户端看到或使用。当客户端退出时，该客户端连接的所有变量将自动释放。

8.1.5 光标的使用

查询语句可能返回多条记录，如果数据量非常大，需要在存储过程和储存函数中使用光标来逐条读取查询结果集中的记录。应用程序可以根据需要滚动或浏览其中的数据。本节将介绍如何声明、打开、使用和关闭光标。

光标必须在声明处理程序之前被声明，并且变量和条件还必须在声明光标或处理程序之前被声明。

1. 声明光标

在 MySQL 中，使用 DECLARE 关键字来声明光标，其语法的基本形式如下：

```
DECLARE cursor_name CURSOR FOR select_statement
```

其中，cursor_name 参数表示光标的名称；select_statement 参数表示 SELECT 语句的内容，返回一个用于创建光标的结果集。

【例 8.10】声明名称为 cursor_fruit 的光标，代码如下：

```
DECLARE cursor_fruit CURSOR FOR SELECT f_name, f_price FROM fruits ;
```

在上面的示例中，光标的名称为 cur_fruit，SELECT 语句部分从 fruits 表中查询出 f_name 和 f_price 字段的值。

2. 打开光标

打开光标的语法如下：

```
OPEN cursor_name{光标名称}
```

这个语句打开先前声明的名称为 cursor_name 的光标。

【例 8.11】打开名称为 cursor_fruit 的光标，代码如下：

```
OPEN  cursor_fruit ;
```

3. 使用光标

使用光标的语法如下：

```
FETCH cursor_name INTO var_name [, var_name] ...{参数名称}
```

其中，cursor_name 参数表示光标的名称；var_name 参数表示将光标中的 SELECT 语句查询出来的信息存入该参数中，var_name 必须在声明光标之前就定义好。

【例 8.12】 使用名称为 cursor_fruit 的光标，将查询出来的数据存入 fruit_name 和 fruit_price 这两个变量中，代码如下：

```
FETCH  cursor_fruit INTO fruit_name, fruit_price ;
```

上面的示例中，将光标 cursor_fruit 中用 SELECT 语句查询出来的信息存入 fruit_name 和 fruit_price 中。fruit_name 和 fruit_price 必须在前面已经定义。

4．关闭光标

关闭光标的语法如下：

CLOSE cursor_name{光标名称}

这个语句关闭先前打开的光标。

如果未被明确地关闭，光标在它被声明的复合语句的末尾关闭。

【例 8.13】 关闭名称为 cursor_fruit 的光标，代码如下：

```
CLOSE  cursor_fruit;
```

提示：MySQL 中光标只能在存储过程和函数中使用。

8.1.6　流程控制的使用

流程控制语句用来根据条件控制语句的执行。MySQL 中用来构造控制流程的语句有 IF 语句、CASE 语句、LOOP 语句、LEAVE 语句、ITERATE 语句、REPEAT 语句和 WHILE 语句。

每个流程中可能包含一个单独语句，或者是使用 BEGIN...END 构造的复合语句，构造可以被嵌套。本节将介绍这些控制流程语句。

1．IF 语句

IF 语句包含多个条件判断，根据判断的结果为 TRUE 或 FALSE 执行相应的语句，语法格式如下：

```
IF expr_condition THEN statement_list
    [ELSEIF expr_condition THEN statement_list] ...
    [ELSE statement_list]
END IF
```

IF 实现了一个基本的条件构造。如果 expr_condition 求值为真（TRUE），相应的 SQL 语句列表被执行；如果没有 expr_condition 匹配，则 ELSE 子句里的语句列表被执行。statement_list 可以包括一个或多个语句。

提示：MySQL 中还有一个 IF()函数，不同于这里描述的 IF 语句。

【例 8.14】 IF 语句的示例，代码如下：

```
IF val IS NULL
    THEN SELECT 'val is NULL';
    ELSE SELECT 'val is not NULL';
END IF;
```

该示例判断 val 值是否为空，如果 val 值为空，则输出字符串"val is NULL"，否则输出字符串"val is not NULL"。IF 语句都需要使用 END IF 来结束。

2. CASE 语句

CASE 是另一个进行条件判断的语句，有两种格式。

第 1 种格式如下：

```
CASE case_expr
    WHEN when_value THEN statement_list
    [WHEN when_value THEN statement_list] ...
    [ELSE statement_list]
END CASE
```

其中，case_expr 参数表示条件判断的表达式，决定了哪一个 WHEN 子句会被执行；when_value 参数表示表达式可能的值，如果某个 when_value 表达式与 case_expr 表达式结果相同，则执行对应 THEN 关键字后的 statement_list 中的语句；statement_list 参数表示不同 when_value 值的执行语句。

【例 8.15】使用 CASE 流程控制语句的第 1 种格式，判断 val 值等于 1、等于 2，或者两者都不等，语句如下：

```
CASE val
    WHEN 1 THEN SELECT 'val is 1';
    WHEN 2 THEN SELECT 'val is 2';
    ELSE SELECT 'val is not 1 or 2';
END CASE;
```

当 val 值为 1 时，输出字符串"val is 1"；当 val 值为 2 时，输出字符串"val is 2"；否则，输出字符串"val is not 1 or 2"。

CASE 语句的第 2 种格式如下：

```
CASE
    WHEN expr_condition THEN statement_list
    [WHEN expr_condition THEN statement_list] ...
    [ELSE statement_list]
END CASE
```

其中，expr_condition 参数表示条件判断语句；statement_list 参数表示不同条件的执行语句。该语句中，WHEN 语句将被逐个执行，直到某个 expr_condition 表达式为真，则执行对应 THEN 关键字后面的 statement_list 语句。如果没有条件匹配，则 ELSE 子句里的语句被执行。

提示：这里介绍的用在存储程序里的 CASE 语句与"控制流程函数"里描述的 SQL CASE 表达式的 CASE 语句稍微有点不同。这里的 CASE 语句不能有 ELSE NULL 子句，并且用 END CASE 替代 END 来终止。

【例 8.16】使用 CASE 流程控制语句的第 2 种格式，判断 val 是否为空、小于 0、大于 0 或者等于 0，语句如下：

```
CASE
    WHEN val is NULL THEN SELECT 'val is NULL';
    WHEN val < 0 THEN SELECT 'val is less than 0';
    WHEN val > 0 THEN SELECT 'val is greater than 0';
```

```
        ELSE SELECT 'val is 0';
END CASE;
```

当 val 值为空,输出字符串"val is NULL";当 val 值小于 0 时,输出字符串"val is less than 0";当 val 值大于 0 时,输出字符串"val is greater than 0";否则,输出字符串"val is 0"。

3. LOOP 语句

LOOP 循环语句用来重复执行某些语句,与 IF 和 CASE 语句相比,LOOP 只是创建一个循环操作的过程,并不进行条件判断。LOOP 内的语句一直重复执行直到循环被退出(使用 LEAVE 子句),跳出循环过程。LOOP 语句的基本格式如下:

```
[loop_label:] LOOP
    statement_list
END LOOP [loop_label]
```

其中,loop_label 表示 LOOP 语句的标注名称,该参数可以省略;statement_list 参数表示需要循环执行的语句。

【例 8.17】使用 LOOP 语句进行循环操作,id 值小于 10 时将重复执行循环过程,代码如下:

```
DECLARE id INT DEFAULT 0;
add_loop: LOOP
SET id = id + 1;
    IF id >= 10 THEN  LEAVE add_loop;
    END IF;
END LOOP add_ loop;
```

该示例循环执行 id 加 1 的操作。当 id 值小于 10 时,循环重复执行;当 id 值大于或者等于 10 时,使用 LEAVE 语句退出循环。LOOP 循环都以 END LOOP 结束。

4. LEAVE 语句

LEAVE 语句用来退出任何被标注的流程控制构造,基本格式如下:

```
LEAVE label
```

其中,label 参数表示循环的标志。LEAVE 和 BEGIN ... END 或循环一起被使用。

【例 8.18】使用 LEAVE 语句退出循环,代码如下:

```
add_num: LOOP
SET @count=@count+1;
IF @count=50 THEN LEAVE add_num ;
END LOOP add_num ;
```

该示例循环执行 count 加 1 的操作。当 count 的值等于 50 时,使用 LEAVE 语句跳出循环。

5. ITERATE 语句

ITERATE 语句将执行顺序转到语句段开头处,语句基本格式如下:

```
ITERATE label
```

ITERATE 只可以出现在 LOOP、REPEAT 和 WHILE 语句内。ITERATE 的意思为"再次循环",

label 参数表示循环的标志。ITERATE 语句必须跟在循环标志前面。

【例 8.19】ITERATE 语句示例，代码如下：

```
CREATE PROCEDURE doiterate()
BEGIN
DECLARE p1 INT DEFAULT 0;
my_loop: LOOP
    SET p1= p1 + 1;
    IF p1 < 10 THEN ITERATE my_loop;
    ELSEIF p1 > 20 THEN LEAVE my_loop;
    END IF;
    SELECT 'p1 is between 10 and 20';
END LOOP my_loop;
END
```

初始化 p1=0，如果 p1 的值小于 10 时，重复执行 p1 加 1 操作；当 p1 大于等于 10 并且小于等于 20 时，打印消息"p1 is between 10 and 20"；当 p1 大于 20 时，退出循环。

6. REPEAT 语句

REPEAT 语句创建一个带条件判断的循环过程，每次语句执行完毕之后，会对条件表达式进行判断，如果表达式为真，则循环结束；否则重复执行循环中的语句。REPEAT 语句的基本格式如下：

```
[repeat_label:] REPEAT
    statement_list
UNTIL expr_condition
END REPEAT [repeat_label]
```

repeat_label 为 REPEAT 语句的标注名称，该参数可以省略；REPEAT 语句内的语句或语句群被重复，直至 expr_condition 为真。

【例 8.20】REPEAT 语句示例，id 值小于 10 时将重复执行循环过程，代码如下：

```
DECLARE id INT DEFAULT 0;
REPEAT
SET id = id + 1;
UNTIL  id >= 10
END REPEAT;
```

该示例循环执行 id 加 1 的操作。当 id 值小于 10 时，循环重复执行；当 id 值大于或者等于 10 时，退出循环。REPEAT 循环都以 END REPEAT 结束。

7. WHILE 语句

WHILE 语句创建一个带条件判断的循环过程，与 REPEAT 不同，WHILE 在执行语句执行时，先对指定的表达式进行判断，如果为真，就执行循环内的语句，否则退出循环。WHILE 语句的基本格式如下：

```
[while_label:] WHILE expr_condition DO
    statement_list
END WHILE [while_label]
```

while_label 为 WHILE 语句的标注名称；expr_condition 为进行判断的表达式，如果表达式结果为真，WHILE 语句内的语句或语句群被执行，直至 expr_condition 为假，退出循环。

【例 8.21】WHILE 语句示例，i 值小于 10 时，将重复执行循环过程，代码如下：

```
DECLARE i INT DEFAULT 0;
WHILE i < 10 DO
    SET i = i + 1;
END WHILE;
```

8.2 调用存储过程和函数

存储过程已经定义好了，接下来需要知道如何调用这些过程和函数。存储过程和函数有多种调用方法。存储过程必须使用 CALL 语句调用，并且存储过程和数据库相关，如果要执行其他数据库中的存储过程，需要指定数据库名称，例如 CALL dbname.procname。存储函数的调用与 MySQL 中预定义的函数的调用方式相同。本节将介绍存储过程和存储函数的调用，主要包括调用存储过程的语法、调用存储函数的语法，以及存储过程和存储函数的调用实例。

8.2.1 调用存储过程

存储过程是通过 CALL 语句进行调用的，语法如下：

```
CALL sp_name([parameter[,...]])
```

CALL 语句调用一个先前用 CREATE PROCEDURE 创建的存储过程，其中 sp_name 为存储过程名称，parameter 为存储过程的参数。

【例 8.22】定义名为 CountProc1 的存储过程，然后调用这个存储过程。

定义存储过程：

```
mysql> use test_db;
mysql> DELIMITER //
mysql> CREATE PROCEDURE CountProc1 (IN sid INT, OUT num INT)
    -> BEGIN
    -> SELECT COUNT(*) INTO num FROM fruits WHERE s_id = sid;
    -> END //

mysql>  DELIMITER ;
```

调用存储过程：

```
mysql> CALL CountProc1 (101, @num);
```

查看返回结果：

```
mysql> SELECT @num;
+------+
| @num |
+------+
|   3  |
+------+
```

该存储过程返回了指定 s_id=101 的水果商提供的水果种类，返回值存储在 num 变量中，使用 SELECT 查看，返回结果为 3。

8.2.2 调用存储函数

在 MySQL 中，存储函数的使用方法与 MySQL 内部函数的使用方法是一样的。换言之，用户自己定义的存储函数与 MySQL 内部函数是一个性质的。区别在于，存储函数是用户自己定义的，而内部函数是 MySQL 的用户或开发者定义的。

【例 8.23】定义存储函数 CountProc2，然后调用这个函数，代码如下：

```
mysql> DELIMITER //
mysql> CREATE FUNCTION  CountProc2 (sid INT)
    -> RETURNS INT
    -> BEGIN
    -> RETURN (SELECT COUNT(*) FROM fruits WHERE s_id = sid);
    -> END;
    -> //

mysql>  DELIMITER ;
```

注意：如果在创建存储函数中报错"you *might* want to use the less safe log_bin_trust_function_creators variable"，需要执行以下代码：

```
mysql> SET GLOBAL log_bin_trust_function_creators = 1;
```

调用存储函数：

```
mysql> SELECT CountProc2(101);
+-----------------+
| Countproc(101)  |
+-----------------+
|               3 |
+-----------------+
1 row in set (0.00 sec)
```

可以看到，该例与上一个例子返回的结果相同。虽然存储函数和存储过程的定义稍有不同，但可以实现相同的功能，读者应该在实际应用中灵活选择。

8.3 查看存储过程和函数

MySQL 存储了存储过程和函数的状态信息，用户可以使用 SHOW STATUS 语句或 SHOW CREATE 语句来查看，也可直接从系统的 information_schema 数据库中查询。本节将通过实例来介绍这 3 种查看方法。

8.3.1 使用 SHOW STATUS 语句查看存储过程和函数的状态

SHOW STATUS 语句可以查看存储过程和函数的状态，其基本语法结构如下：

```
SHOW {PROCEDURE | FUNCTION} STATUS [LIKE 'pattern']
```

这个语句是一个 MySQL 的扩展，返回子程序的特征，如数据库、名字、类型、创建者及创建和修改日期。如果没有指定样式，那么根据使用的语句，所有存储程序或存储函数的信息都会被列出。其中，PROCEDURE 和 FUNCTION 分别表示查看存储过程和函数；LIKE 语句表示匹配存储过程或函数的名称。

【例 8.24】 SHOW STATUS 语句示例，代码如下：

```
SHOW PROCEDURE STATUS LIKE 'C%' \G
```

代码执行如下：

```
mysql> SHOW PROCEDURE STATUS LIKE 'C%' \G
*** 1. row ***
                  Db: test_db
                Name: CountProc
Type: PROCEDURE
             Definer: root@localhost
            Modified: 2018-11-21 13:52:28
             Created: 2018-11-21 13:52:28
       Security_type: DEFINER
             Comment:
character_set_client: gbk
collation_connection: gbk_chinese_ci
  Database Collation: utf8mb4_0900_ai_ci
```

"SHOW PROCEDURE STATUS LIKE 'C%'\G" 语句获取数据库中所有名称以字母 "C" 开头的存储过程的信息。通过上面的语句可以看到：这个存储函数所在的数据库为 test_db、存储函数的名称为 CountProc 等一些相关信息。

8.3.2 使用 SHOW CREATE 语句查看存储过程和函数的定义

除了 SHOW STATUS 之外，MySQL 还可以使用 SHOW CREATE 语句查看存储过程和函数的状态。

```
SHOW CREATE {PROCEDURE | FUNCTION} sp_name
```

这个语句是一个 MySQL 的扩展。类似于 SHOW CREATE TABLE，它返回一个可用来重新创建已命名子程序的确切字符串。PROCEDURE 和 FUNCTION 分别表示查看存储过程和函数；sp_name 参数表示匹配存储过程或函数的名称。

【例 8.25】 SHOW CREATE 语句示例，代码如下：

```
SHOW CREATE FUNCTION test_db.CountProc2 \G
```

代码执行如下：

```
mysql> SHOW CREATE FUNCTION test_db.CountProc2 \G
*************************** 1. row ***************************
       Function: CountProc2
        sql_mode: STRICT_TRANS_TABLES,NO_ENGINE_SUBSTITUTION
Create Function: CREATE DEFINER=`root`@`localhost` FUNCTION `CountProc2`(sid INT) RETURNS
```

```
int(11)
    BEGIN
    RETURN (SELECT COUNT(*) FROM fruits WHERE s_id = sid);
    END
character_set_client: gbk
collation_connection: gbk_chinese_ci
  Database Collation: utf8mb4_0900_ai_ci
```

执行上面的语句可以得到存储函数的名称为 CountProc2，sql_mode 为 sql 的模式，Create Function 为存储函数的具体定义语句，还有数据库设置的一些信息。

8.3.3 从 information_schema.Routines 表中查看存储过程和函数的信息

MySQL 中存储过程和函数的信息存储在 information_schema 数据库下的 Routines 表中。可以通过查询该表的记录来查询存储过程和函数的信息。其基本语法形式如下：

```
SELECT * FROM information_schema.Routines
WHERE ROUTINE_NAME='sp_name';
```

其中，ROUTINE_NAME 字段中存储的是存储过程和函数的名称；sp_name 参数表示存储过程或函数的名称。

【例 8.26】从 Routines 表中查询名称为 CountProc2 的存储函数的信息，代码如下：

```
SELECT * FROM information_schema.Routines WHERE ROUTINE_NAME='CountProc2'  AND  ROUTINE_TYPE = 'FUNCTION' \G
```

代码执行如下：

```
mysql> SELECT * FROM information_schema.Routines WHERE ROUTINE_NAME='CountProc2'  AND ROUTINE_TYPE = 'FUNCTION' \G
*************************** 1. row ***************************
           SPECIFIC_NAME: CountProc2
         ROUTINE_CATALOG: def
          ROUTINE_SCHEMA: test_db
            ROUTINE_NAME: CountProc2
            ROUTINE_TYPE: FUNCTION
               DATA_TYPE: int
CHARACTER_MAXIMUM_LENGTH: NULL
  CHARACTER_OCTET_LENGTH: NULL
       NUMERIC_PRECISION: 10
           NUMERIC_SCALE: 0
      DATETIME_PRECISION: NULL
      CHARACTER_SET_NAME: NULL
          COLLATION_NAME: NULL
          DTD_IDENTIFIER: int(11)
            ROUTINE_BODY: SQL
      ROUTINE_DEFINITION: BEGIN
    RETURN (SELECT COUNT(*) FROM fruits WHERE s_id = sid);
    END
           EXTERNAL_NAME: NULL
       EXTERNAL_LANGUAGE: SQL
          PARAMETER_STYLE: SQL
```

```
        IS_DETERMINISTIC: NO
        SQL_DATA_ACCESS: CONTAINS SQL
               SQL_PATH: NULL
          SECURITY_TYPE: DEFINER
                CREATED: 2018-11-21 16:57:09
           LAST_ALTERED: 2018-11-21 16:57:09
               SQL_MODE: STRICT_TRANS_TABLES,NO_ENGINE_SUBSTITUTION
        ROUTINE_COMMENT:
                DEFINER: root@localhost
   CHARACTER_SET_CLIENT: gbk
   COLLATION_CONNECTION: gbk_chinese_ci
     DATABASE_COLLATION: utf8mb4_0900_ai_ci
```

在 information_schema 数据库的 Routines 表中，存储着所有存储过程和函数的定义。使用 SELECT 语句查询 Routines 表中的存储过程和函数的定义时，一定要使用 ROUTINE_NAME 字段指定存储过程或函数的名称。否则，将查询出所有的存储过程或函数的定义。如果有存储过程和存储函数名称相同，就需要同时指定 ROUTINE_TYPE 字段，以表明查询的是哪种类型的存储程序。

8.4 修改存储过程和函数

使用 ALTER 语句可以修改存储过程或函数的特性，本节将介绍如何使用 ALTER 语句修改存储过程和函数。

```
ALTER {PROCEDURE | FUNCTION} sp_name [characteristic ...]
```

其中，sp_name 参数表示存储过程或函数的名称；characteristic 参数指定存储函数的特性，可能的取值有：

- CONTAINS SQL：表示子程序包含 SQL 语句，但不包含读或写数据的语句。
- NO SQL：表示子程序中不包含 SQL 语句。
- READS SQL DATA：表示子程序中包含读数据的语句。
- MODIFIES SQL DATA：表示子程序中包含写数据的语句。
- SQL SECURITY { DEFINER | INVOKER }：指明谁有权限来执行。
- DEFINER：表示只有定义者自己才能够执行。
- INVOKER：表示调用者可以执行。
- COMMENT 'string'：表示注释信息。

提示：修改存储过程使用 ALTER PROCEDURE 语句，修改存储函数使用 ALTER FUNCTION 语句。但是，这两个语句的结构是一样的，语句中的所有参数也是一样的。而且，它们与创建存储过程或函数的语句中的参数也是基本一样的。

【例 8.27】修改存储过程 CountProc 的定义。将读写权限改为 MODIFIES SQL DATA，并指明调用者可以执行，代码如下：

```
ALTER  PROCEDURE  CountProc  MODIFIES SQL DATA SQL SECURITY INVOKER ;
```

执行代码，并查看修改后的信息。结果显示如下：

```
//执行 ALTER PROCEDURE 语句
mysql> ALTER PROCEDURE CountProc MODIFIES SQL DATA SQL SECURITY INVOKER;
//查询修改后的 CountProc 表信息
mysql> SELECT SPECIFIC_NAME,SQL_DATA_ACCESS,SECURITY_TYPE
    -> FROM information_schema.Routines
    -> WHERE ROUTINE_NAME='CountProc' AND ROUTINE_TYPE='PROCEDURE';
+---------------+------------------+---------------+
| SPECIFIC_NAME | SQL_DATA_ACCESS  | SECURITY_TYPE |
+---------------+------------------+---------------+
| CountProc     | MODIFIES SQL DATA| INVOKER       |
+---------------+------------------+---------------+
```

结果显示，存储过程修改成功。从查询的结果可以看出，访问数据的权限（SQL_DATA_ACCESS）已经变成 MODIFIES SQL DATA，安全类型（SECURITY_TYPE）已经变成 INVOKER。

【例 8.28】修改存储函数 CountProc2 的定义，将读写权限改为 READS SQL DATA，并加上注释信息"FIND NAME"，代码如下：

```
ALTER FUNCTION CountProc2 READS SQL DATA COMMENT 'FIND NAME';
```

执行代码，并查看修改后的信息。结果显示如下：

```
//执行 ALTER FUNCTION 语句
mysql> ALTER FUNCTION CountProc2 READS SQL DATA COMMENT 'FIND NAME';
Query OK, 0 rows affected (0.00 sec)
//查看修改后的信息
mysql> SELECT SPECIFIC_NAME,SQL_DATA_ACCESS,ROUTINE_COMMENT
FROM information_schema.Routines
WHERE ROUTINE_NAME='CountProc2' AND ROUTINE_TYPE = 'FUNCTION';
+---------------+------------------+-----------------+
| SPECIFIC_NAME | SQL_DATA_ACCESS  | ROUTINE_COMMENT |
+---------------+------------------+-----------------+
| CountProc2    | READS SQL DATA   | FIND NAME       |
+---------------+------------------+-----------------+
```

存储函数修改成功。从查询的结果可以看出，访问数据的权限（SQL_DATA_ACCESS）已经变成 READS SQL DATA，函数注释（ROUTINE_COMMENT）已经变成 FIND NAME。

8.5 删除存储过程和函数

删除存储过程和函数，可以使用 DROP 语句，其语法结构如下：

```
DROP {PROCEDURE | FUNCTION} [IF EXISTS] sp_name
```

这个语句被用来移除一个存储过程或函数。sp_name 为要移除的存储过程或函数的名称。

IF EXISTS 子句是一个 MySQL 的扩展。如果程序或函数不存储，它可以防止发生错误，产生一个用 SHOW WARNINGS 查看的警告。

【例 8.29】删除存储过程和存储函数，代码如下：

```
DROP PROCEDURE CountProc;
DROP FUNCTION CountProc2;
```

上面语句的作用就是删除存储过程 CountProc 和存储函数 CountProc2。

8.6 全局变量的持久化

在 MySQL 数据库中，全局变量可以通过 SET GLOBAL 语句来设置。例如，设置服务器语句超时的限制，可以通过设置系统变量 max_execution_time 来实现：

```
SET GLOBAL MAX_EXECUTION_TIME=2000;
```

使用 SET GLOBAL 语句设置的变量值只会临时生效。数据库重启后，服务器又会从 MySQL 配置文件中读取变量的默认值。

MySQL 8.0 版本新增了 SET PERSIST 命令。例如，设置服务器的最大连接数为 1000：

```
SET PERSIST max_connections = 1000;
```

MySQL 会将该命令的配置保存到数据目录下的 mysqld-auto.cnf 文件中，下次启动时会读取该文件，用其中的配置来覆盖默认的配置文件。

下面通过一个案例来理解全部变量的持久化。

查看全局变量 max_connections 的值，结果如下：

```
mysql> show variables like '%max_connections%';
+------------------------+-------+
| Variable_name          | Value |
+------------------------+-------+
| max_connections        | 151   |
| mysqlx_max_connections | 100   |
+------------------------+-------+
```

设置全局变量 max_connections 的值：

```
mysql> set persist max_connections=1000;
```

重启 MySQL 服务器，再次查询 max_connections 的值：

```
mysql> show variables like '%max_connections%';
+------------------------+-------+
| Variable_name          | Value |
+------------------------+-------+
| max_connections        | 1000  |
| mysqlx_max_connections | 100   |
+------------------------+-------+
```

第 9 章

视 图

数据库中的视图是一个虚拟表。同真实的表一样，视图包含一系列带有名称的行和列数据。行和列数据来自由定义视图查询所引用的表，并且在引用视图时动态生成。本章将通过一些实例来介绍视图的含义、视图的作用、创建视图、查看视图、修改视图、更新视图和删除视图等 MySQL 视图的相关知识。

9.1 视图概述

视图是从一个或者多个表中导出的，视图的行为与表非常相似，但视图是一个虚拟表。在视图中用户可以使用 SELECT 语句查询数据，以及使用 INSERT、UPDATE 和 DELETE 修改记录。从 MySQL 5.0 开始提供了视图功能，视图可以使用户操作方便，而且可以保障数据库系统的安全。

9.1.1 视图的含义

视图是一个虚拟表，是从数据库中一个或多个表中导出来的表。视图还可以从已经存在的视图的基础上定义。

视图一经定义便存储在数据库中，与其相对应的数据并没有像表那样在数据库中再存储一份，通过视图看到的数据只是存放在基本表中的数据。对视图的操作与对表的操作一样，可以对其进行查询、修改和删除。当对通过视图看到的数据进行修改时，相应的基本表的数据也要发生变化；同时，若基本表的数据发生变化，则这种变化也可以自动反映到视图中。

下面有一个 student 表和 stu_info 表，在 student 表中包含了学生的 id 号和姓名，stu_info 表中包含了学生的 id 号、班级和家庭住址，而现在公布分班信息只需要 id 号、姓名和班级，该如何解决呢？通过学习后面的内容就可以找到完美的解决方案。

表设计如下：

```
CREATE TABLE student
(
s_id INT,
name VARCHAR(40)
);

CREATE TABLE stu_info
(
s_id INT,
glass VARCHAR(40),
addr VARCHAR(90)
);
```

通过 DESC 命令可以查看表的设计，可以获得字段、字段的定义、是否为主键、是否为空、默认值和扩展信息。

视图提供了一个很好的解决方法。创建一个视图，这些信息来自表的部分信息，其他的信息不取，这样既能满足用户要求也不破坏表原来的结构。

9.1.2 视图的作用

与直接从数据表中读取数据相比，视图有以下优点：

1. 简单化

看到的就是需要的。视图不仅可以简化用户对数据的理解，也可以简化它们的操作。那些被经常使用的查询可以被定义为视图，从而使得用户不必为以后的操作每次指定全部的条件。

2. 安全性

通过视图用户只能查询和修改他们所能见到的数据。数据库中的其他数据则既看不见也取不到。数据库授权命令可以使每个用户对数据库的检索限制到特定的数据库对象上，但不能授权到数据库特定行和特定的列上。通过视图，用户可以被限制在数据的不同子集上：

（1）使用权限可被限制在基表的行的子集上。
（2）使用权限可被限制在基表的列的子集上。
（3）使用权限可被限制在基表的行和列的子集上。
（4）使用权限可被限制在多个基表的连接所限定的行上。
（5）使用权限可被限制在基表中的数据的统计汇总上。
（6）使用权限可被限制在另一视图的一个子集上，或是一些视图和基表合并后的子集上。

3. 逻辑数据独立性

视图可帮助用户屏蔽真实表结构变化带来的影响。

9.2 创建视图

视图中包含了 SELECT 查询的结果，因此视图的创建基于 SELECT 语句和已存在的数据表。视图可以建立在一张表上，也可以建立在多张表上。本节将主要介绍创建视图的方法。

9.2.1 创建视图的语法形式

创建视图使用 CREATE VIEW 语句，基本语法格式如下：

```
CREATE [OR REPLACE] [ALGORITHM = {UNDEFINED | MERGE | TEMPTABLE}]
VIEW view_name [(column_list)]
AS SELECT_statement
[WITH [CASCADED | LOCAL] CHECK OPTION]
```

其中，CREATE 表示创建新的视图；REPLACE 表示替换已经创建的视图；ALGORITHM 表示视图选择的算法；view_name 为视图的名称，column_list 为属性列；SELECT_statement 表示 SELECT 语句；WITH [CASCADED | LOCAL] CHECK OPTION 参数表示视图在更新时保证在视图的权限范围之内。

ALGORITHM 的取值有 3 个，分别是 UNDEFINED、MERGE、TEMPTABLE。其中，UNDEFINED 表示 MySQL 将自动选择算法；MERGE 表示将使用的视图语句与视图定义合并起来，使得视图定义的某一部分取代语句对应的部分；TEMPTABLE 表示将视图的结果存入临时表，然后用临时表来执行语句。

CASCADED 与 LOCAL 为可选参数，CASCADED 为默认值，表示更新视图时要满足所有相关视图和表的条件；LOCAL 表示更新视图时满足该视图本身定义的条件即可。

该语句要求用户具有针对视图的 CREATE VIEW 权限，以及针对由 SELECT 语句选择的每一列上的某些权限。对于在 SELECT 语句中其他地方使用的列，必须具有 SELECT 权限。如果还有 OR REPLACE 子句，还必须在视图上具有 DROP 权限。

视图属于数据库对象。在默认情况下，将在当前数据库创建新视图。要想在给定数据库中明确创建视图，创建时应将名称指定为 db_name.view_name。

9.2.2 在单表上创建视图

MySQL 可以在单个数据表上创建视图。

【例 9.1】在 t 表上创建一个名为 view_t 的视图，代码如下：

首先创建基本表并插入数据，语句如下：

```
CREATE TABLE tv (quantity INT, price INT);
INSERT INTO tv VALUES(3, 50);
```

创建视图语句为：

```
CREATE VIEW view_t AS SELECT quantity, price, quantity *price FROM tv;
```

查询视图，执行如下：

```
mysql> SELECT * FROM view_t;
+----------+-------+----------------+
| quantity | price | quantity *price |
+----------+-------+----------------+
|        3 |    50 |            150 |
```

默认情况下创建的视图和基本表的字段是一样的，也可以通过指定视图字段的名称来创建视图。

【例 9.2】在 tv 表上创建一个名为 view_t2 的视图，代码如下：

mysql> CREATE VIEW view_t2(qty, price, total) AS SELECT quantity, price, quantity *price FROM tv;

语句执行成功后，查看 view_t2 视图中的数据：

```
mysql> SELECT * FROM view_t2;
+------+-------+-------+
| qty  | price | total |
+------+-------+-------+
|    3 |    50 |   150 |
+------+-------+-------+
```

可以看到，view_t2 和 view_t 两个视图中的字段名称不同，但数据却是相同的。因此，在使用视图的时候，可能用户根本就不需要了解基本表的结构，更接触不到实际表中的数据，从而保证了数据库的安全。

9.2.3 在多表上创建视图

MySQL 中也可以在两个或者两个以上的表上创建视图，可以使用 CREATE VIEW 语句实现。

【例 9.3】在表 student 和表 stu_info 上创建视图 stu_glass，代码如下：

首先向两个表中插入数据，输入语句如下：

```
mysql> INSERT INTO student VALUES(1,'wanglin1'),(2,'gaoli'),(3,'zhanghai');
mysql> INSERT INTO stu_info VALUES(1, 'wuban','henan'),(2,'liuban','hebei'),(3,'qiban','shandong');
```

创建视图 stu_glass，代码的执行如下：

```
mysql> CREATE VIEW stu_glass (id,name, glass) AS SELECT student.s_id,
student.name ,stu_info.glass FROM student ,stu_info WHERE student.s_id=stu_info.s_id;
```

```
mysql> SELECT * FROM stu_glass;
+----+----------+--------+
| id | name     | glass  |
+----+----------+--------+
|  1 | wanglin1 | wuban  |
|  2 | gaoli    | liuban |
|  3 | zhanghai | qiban  |
+----+----------+--------+
```

这个例子就解决了刚开始提出的那个问题，通过这个视图可以很好地保护基本表中的数据。这个视图中的信息很简单，只包含了 id、姓名和班级，id 字段对应 student 表中的 s_id 字段，name 字段对应 student 表中的 name 字段，glass 字段对应 stu_info 表中的 glass 字段。

9.3 查看视图

查看视图是查看数据库中已存在的视图的定义。查看视图必须要有 SHOW VIEW 的权限，MySQL 数据库下的 user 表中保存着这个信息。查看视图的方法包括 DESCRIBE、SHOW TABLE STATUS 和 SHOW CREATE VIEW，本节将介绍这些查看视图的方法。

9.3.1 使用 DESCRIBE 语句查看视图基本信息

DESCRIBE 可以用来查看视图，具体的语法如下：

```
DESCRIBE 视图名;
```

【例 9.4】通过 DESCRIBE 语句查看视图 view_t 的定义，代码如下：

```
DESCRIBE view_t;
```

执行结果如下：

```
mysql> DESCRIBE view_t;
+----------------+-----------+------+-----+---------+-------+
| Field          | Type      | Null | Key | Default | Extra |
+----------------+-----------+------+-----+---------+-------+
| quantity       | int       | YES  |     | NULL    |       |
| price          | int       | YES  |     | NULL    |       |
| quantity *price| bigint(21)| YES  |     | NULL    |       |
+----------------+-----------+------+-----+---------+-------+
```

结果显示出了视图的字段定义、字段的数据类型、是否为空、是否为主/外键、默认值和额外信息。

DESCRIBE 一般情况下都简写成 DESC，输入这个命令的执行结果和输入 DESCRIBE 的执行结果是一样的。

9.3.2 使用 SHOW TABLE STATUS 语句查看视图基本信息

查看视图的信息可以通过 SHOW TABLE STATUS 的方法完成，具体的语法如下：

```
SHOW TABLE STATUS LIKE '视图名';
```

【例 9.5】使用 SHOW TABLE STATUS 命令查看视图信息，代码如下：

```
SHOW TABLE STATUS LIKE 'view_t' \G
```

执行结果如下：

```
mysql> SHOW TABLE STATUS LIKE 'view_t' \G
*************************** 1. row ***************************
           Name: view_t
         Engine: NULL
        Version: NULL
     Row_format: NULL
           Rows: NULL
```

```
     Avg_row_length: NULL
       Data_length: NULL
   Max_data_length: NULL
      Index_length: NULL
         Data_free: NULL
    Auto_increment: NULL
       Create_time: 2022-03-17 09:06:24
       Update_time: NULL
        Check_time: NULL
         Collation: NULL
          Checksum: NULL
    Create_options: NULL
           Comment: VIEW
```

执行结果显示，表的说明 Comment 的值为 VIEW，说明该表为视图，其他的信息为 NULL，说明这是一个虚表。用同样的语句来查看一下数据表 tv 的信息，执行结果如下：

```
mysql> SHOW TABLE STATUS LIKE 'tv' \G
*************************** 1. row ***************************
              Name: tv
            Engine: InnoDB
           Version: 10
        Row_format: Dynamic
              Rows: 1
    Avg_row_length: 16384
       Data_length: 16384
   Max_data_length: 0
      Index_length: 0
         Data_free: 0
    Auto_increment: NULL
       Create_time: 2022-03-17 09:05:49
       Update_time: 2022-03-17 09:05:56
        Check_time: NULL
         Collation: utf8mb4_0900_ai_ci
          Checksum: NULL
    Create_options:
           Comment:
```

从查询的结果来看，这里的信息包含了存储引擎、创建时间等，Comment 信息为空，这就是视图和表的区别。

9.3.3 使用 SHOW CREATE VIEW 语句查看视图详细信息

使用 SHOW CREATE VIEW 语句可以查看视图的详细定义，语法如下：

```
SHOW CREATE VIEW 视图名;
```

【例 9.6】使用 SHOW CREATE VIEW 查看视图的详细定义，代码如下：

```
SHOW CREATE VIEW view_t \G
```

执行结果如下：

```
mysql> SHOW CREATE VIEW view_t \G
*************************** 1. row ***************************
```

```
            View: view_t
     Create View: CREATE ALGORITHM=UNDEFINED DEFINER=`root`@`localhost` SQL SECURITY
DEFINER VIEW `view_t` AS select `tv`.`quantity` AS `quantity`,`tv`.`price` AS
`price`,(`tv`.`quantity` * `tv`.`price`) AS `quantity*price` from `tv`
character_set_client: utf8mb4
collation_connection: utf8mb4_0900_ai_ci
```

执行结果显示视图的名称、创建视图的语句等信息。

9.3.4　在 views 表中查看视图详细信息

在 MySQL 中，information_schema 数据库的 views 表中存储了所有视图的定义。通过对 views 表的查询，可以查看数据库中所有视图的详细信息，查询语句如下：

```
SELECT * FROM information_schema.views;
```

【例 9.7】在 views 表中查看视图的详细定义，代码如下：

```
mysql> SELECT * FROM information_schema.views\G
*** 1. row ***
       TABLE_CATALOG: def
        TABLE_SCHEMA: chapter11db
          TABLE_NAME: stu_glass
     VIEW_DEFINITION: select `chapter11db`.`student`.`s_id` AS
`id`,`chapter11db`.`student`.`name` AS `name`,
  `chapter11db`.`stu_info`.`glass` AS `glass` from `chapter11db`.`student` join
`chapter11db`.`stu_info` where (
  `chapter11db`.`student`.`s_id` = `chapter11db`.`stu_info`.`s_id`)
        CHECK_OPTION: NONE
         IS_UPDATABLE: YES
             DEFINER: root@localhost
       SECURITY_TYPE: DEFINER
CHARACTER_SET_CLIENT: gbk
COLLATION_CONNECTION: gbk_chinese_ci
*** 2. row ***
       TABLE_CATALOG: def
        TABLE_SCHEMA: chapter11db
          TABLE_NAME: view_t
     VIEW_DEFINITION: select `chapter11db`.`t`.`quantity` AS
`quantity`,`chapter11db`.`t`.`price` AS `price`,(
  `chapter11db`.`t`.`quantity` * `chapter11db`.`t`.`price`) AS `quantity*price` from
`chapter11db`.`t`
        CHECK_OPTION: NONE
         IS_UPDATABLE: YES
             DEFINER: root@localhost
       SECURITY_TYPE: DEFINER
CHARACTER_SET_CLIENT: gbk
COLLATION_CONNECTION: gbk_chinese_ci
*** 3. row ***
       TABLE_CATALOG: def
        TABLE_SCHEMA: chapter11db
          TABLE_NAME: view_t2
     VIEW_DEFINITION: select `chapter11db`.`t`.`quantity` AS `qty`,`chapter11db`.`t`.`price`
AS `price`,(`chap
  ter11db`.`t`.`quantity` * `chapter11db`.`t`.`price`) AS `total` from `chapter11db`.`t`
```

```
         CHECK_OPTION: NONE
          IS_UPDATABLE: YES
               DEFINER: root@localhost
         SECURITY_TYPE: DEFINER
CHARACTER_SET_CLIENT: gbk
COLLATION_CONNECTION: gbk_chinese_ci
3 rows in set (0.03 sec)
```

查询的结果显示当前以及定义的所有视图的详细信息，在这里也可以看到前面定义的 3 个名称为 stu_glass、view_t 和 view_t2 视图的详细信息。

9.4 修改视图

修改视图是指修改数据库中存在的视图，当基本表的某些字段发生变化的时候，可以通过修改视图来保持与基本表的一致性。MySQL 中通过 CREATE OR REPLACE VIEW 语句和 ALTER 语句来修改视图。

9.4.1 使用 CREATE OR REPLACE VIEW 语句修改视图

在 MySQL 中修改视图，可使用 CREATE OR REPLACE VIEW 语句，语法如下：

```
CREATE [OR REPLACE] [ALGORITHM = {UNDEFINED | MERGE | TEMPTABLE}]
    VIEW view_name [(column_list)]
    AS SELECT_statement
    [WITH [CASCADED | LOCAL] CHECK OPTION]
```

可以看到，修改视图的语句和创建视图的语句是完全一样的。当视图已经存在时，修改语句对视图进行修改；当视图不存在时，创建视图。下面通过一个实例来说明。

【例 9.8】修改视图 view_t，代码如下：

```
CREATE OR REPLACE VIEW view_t AS SELECT * FROM tv;
```

首先通过 DESC 查看一下修改之前的视图，以便与修改之后的视图进行对比。执行的结果如下：

```
mysql> DESC view_t;
+----------------+-----------+------+-----+---------+-------+
| Field          | Type      | Null | Key | Default | Extra |
+----------------+-----------+------+-----+---------+-------+
| quantity       | int       | YES  |     | NULL    |       |
| price          | int       | YES  |     | NULL    |       |
| quantity*price | bigint(21)| YES  |     | NULL    |       |
+----------------+-----------+------+-----+---------+-------+

mysql> CREATE OR REPLACE VIEW view_t AS SELECT * FROM tv;

mysql> DESC view_t;
+----------+------+------+-----+---------+-------+
| Field    | Type | Null | Key | Default | Extra |
+----------+------+------+-----+---------+-------+
| quantity | int  | YES  |     | NULL    |       |
```

```
| price     | int       | YES  |     | NULL    |       |
+-----------+-----------+------+-----+---------+-------+
```

从执行的结果来看，相比原来的视图 view_t，新的视图 view_t 少了 1 个字段。

9.4.2 使用 ALTER 语句修改视图

ALTER 语句是 MySQL 提供的另外一种修改视图的方法，语法如下：

```
ALTER [ALGORITHM = {UNDEFINED | MERGE | TEMPTABLE}]
    VIEW view_name [(column_list)]
    AS SELECT_statement
    [WITH [CASCADED | LOCAL] CHECK OPTION]
```

这个语法中的关键字和前面视图的关键字是一样的，这里就不再详细介绍了。

【例 9.9】使用 ALTER 语句修改视图 view_t，代码如下：

```
ALTER VIEW view_t AS SELECT quantity FROM tv;
```

执行结果如下：

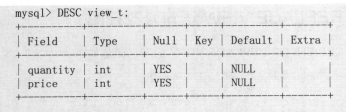

通过 ALTER 语句同样可以达到修改视图 view_t 的目的，从上面的执行过程来看，视图 view_t 只剩下一个 quantity 字段，修改成功。

9.5 更新视图

更新视图是指通过视图来插入、更新、删除表中的数据，因为视图是一个虚拟表，其中没有数据。通过视图所做的更新，都是转到基本表上进行更新的，如果对视图增加或者删除记录，实际上是对其基本表增加或者删除记录。本节将介绍视图更新的 3 种方法：INSERT、UPDATE 和 DELETE。

【例 9.10】使用 UPDATE 语句更新视图 view_t，代码如下：

```
UPDATE view_t SET quantity=5;
```

执行视图更新之前,查看基本表和视图的信息,执行结果如下:

```
mysql> SELECT * FROM view_t;
+----------+
| quantity |
+----------+
|        3 |
+----------+

mysql> SELECT * FROM tv;
+----------+-------+
| quantity | price |
+----------+-------+
|        3 |    50 |
+----------+-------+
```

使用 UPDATE 语句更新视图 view_t,执行过程如下:

```
mysql> UPDATE view_t SET quantity=5;
```

查看视图更新之后,基本表的内容如下:

```
mysql> SELECT * FROM tv;
+----------+-------+
| quantity | price |
+----------+-------+
|        5 |    50 |
+----------+-------+

mysql> SELECT * FROM view_t;
+----------+
| quantity |
+----------+
|        5 |
+----------+

mysql> SELECT * FROM view_t2;
+------+-------+-------+
| qty  | price | total |
+------+-------+-------+
|    5 |    50 |   250 |
+------+-------+-------+
```

对视图 view_t 更新后,基本表 tv 的内容也更新了;当对基本表 tv 更新后,另外一个视图 view_t2 中的内容也会更新。

【例 9.11】使用 INSERT 语句在基本表 tv 中插入一条记录,代码如下:

```
INSERT INTO tv VALUES (3,5);
```

执行结果如下:

```
mysql> INSERT INTO tv VALUES(3,5);
Query OK, 1 row affected (0.04 sec)

mysql> SELECT * FROM tv;
+----------+-------+
```

```
| quantity | price |
+----------+-------+
|    5     |  50   |
|    3     |   5   |
+----------+-------+
2 rows in set (0.00 sec)

mysql> SELECT * FROM view_t2;
+-----+-------+-------+
| qty | price | total |
+-----+-------+-------+
|  5  |  50   |  250  |
|  3  |   5   |   15  |
+-----+-------+-------+
2 rows in set (0.00 sec)
```

向表 tv 中插入一条记录，通过 SELECT 查看表 tv 和视图 view_t2，可以看到其中的内容也跟着更新，视图更新的不仅仅是数量和单价，总价也会更新。

【例 9.12】使用 DELETE 语句删除视图 view_t2 中的一条记录，代码如下：

```
DELETE FROM view_t2 WHERE price=5;
```

执行结果如下：

```
mysql> DELETE FROM view_t2 WHERE price=5;
Query OK, 1 row affected (0.03 sec)
mysql> SELECT * FROM view_t2;
+-----+-------+-------+
| qty | price | total |
+-----+-------+-------+
|  5  |  50   |  250  |
+-----+-------+-------+
1 row in set (0.00 sec)

mysql> SELECT * FROM tv;
+----------+-------+
| quantity | price |
+----------+-------+
|    5     |  50   |
+----------+-------+
1 row in set (0.02 sec)
```

在视图 view_t2 中删除 price=5 的记录，视图中的删除操作最终是通过删除基本表中相关的记录来实现的。查看删除操作之后的表 tv 和视图 view_t2，可以看到通过视图删除其所依赖的基本表中的数据。

当视图中包含有如下内容时，视图的更新操作将不能被执行：

（1）视图中不包含基表中被定义为非空的列。

（2）在定义视图的 SELECT 语句后的字段列表中使用了数学表达式。

（3）在定义视图的 SELECT 语句后的字段列表中使用聚合函数。

（4）在定义视图的 SELECT 语句中使用了 DISTINCT、UNION、TOP、GROUP BY 或 HAVING 子句。

9.6 删除视图

当视图不再需要时，可以将其删除。删除一个或多个视图可以使用 DROP VIEW 语句，语法如下：

```
DROP VIEW [IF EXISTS]
    view_name [, view_name] ...
    [RESTRICT | CASCADE]
```

其中，view_name 是要删除的视图名称，可以添加多个需要删除的视图名称，各个名称之间使用逗号分隔开。删除视图必须拥有 DROP 权限。

【例 9.13】删除 stu_glass 视图，代码如下：

```
DROP VIEW IF EXISTS stu_glass;
```

如果名称为 stu_glass 的视图存在，那么该视图将被删除。使用 SHOW CREATE VIEW 语句查看操作结果：

```
mysql> SHOW CREATE VIEW stu_glass;
ERROR 1146 (42S02): Table 'test_db.stu_glass' doesn't exist
```

可以看到，stu_glass 视图已经不存在，删除成功。

第 10 章

MySQL 触发器

MySQL 触发器和存储过程一样，都是嵌入到 MySQL 的一段程序。触发器是由事件来触发某个操作，这些事件包括 INSERT、UPDATAE 和 DELETE 语句。如果定义了触发程序，当数据库执行这些语句的时候，就会激发触发器执行相应的操作。触发程序是与表有关的命名数据库对象，当表上出现特定事件时，将激活该对象。本章将通过实例来介绍触发器的含义、创建触发器、查看触发器、触发器的使用方法以及删除触发器。

10.1 创建触发器

触发器（trigger）是一个特殊的存储过程，不同的是，执行存储过程要使用 CALL 语句来调用，而触发器的执行不需要使用 CALL 语句来调用，也不需要手工启动，只要当一个预定义的事件发生的时候，触发器就会被 MySQL 自动调用。比如，当对 fruits 表进行操作（INSERT、DELETE 或 UPDATE）时，就会激活它执行。

触发器可以查询其他表，而且可以包含复杂的 SQL 语句。它们主要用于满足复杂的业务规则或要求。例如，可以根据客户当前的账户状态控制是否允许插入新订单。本节将介绍如何创建触发器。

10.1.1 创建只有一个执行语句的触发器

创建一个触发器的语法如下：

```
CREATE TRIGGER trigger_name trigger_time trigger_event
ON tbl_name FOR EACH ROW trigger_stmt
```

其中，trigger_name 表示触发器名称，用户自行指定；trigger_time 表示触发时机，可以指定为 before 或 after；trigger_event 表示触发事件，包括 INSERT、UPDATE 和 DELETE；tbl_name 表示建立触发器的表名，即在哪张表上建立触发器；trigger_stmt 是触发器执行语句。

【例 10.1】 创建只有一个执行语句的触发器，代码如下：

```
CREATE TABLE account (acct_num INT, amount DECIMAL(10,2));
CREATE TRIGGER ins_sum BEFORE INSERT ON account
    FOR EACH ROW SET @sum = @sum + NEW.amount;
```

首先，创建一个 account 表，表中有两个字段，分别为 acct_num 字段（定义为 int 类型）和 amount 字段（定义成浮点类型）；其次，创建一个名为 ins_sum 的触发器，触发的条件是向数据表 account 插入数据之前，对新插入的 amount 字段值进行求和计算。

代码执行如下：

```
mysql> CREATE TABLE account (acct_num INT, amount DECIMAL(10,2));

mysql> CREATE TRIGGER ins_sum BEFORE INSERT ON account  FOR EACH ROW SET @sum = @sum + NEW.amount;

mysql> SET @sum =0;

mysql> INSERT INTO account VALUES(1,1.00), (2,2.00);

mysql> SELECT @sum;
+------+
| @sum |
+------+
| 3.00 |
+------+
```

首先，创建一个 account 表，在向表 account 插入数据之前，计算所有新插入的 account 表的 amount 值之和，触发器的名称为 ins_sum，条件是在向表插入数据之前触发。

10.1.2 创建有多个执行语句的触发器

创建具有多个执行语句的触发器的语法如下：

```
CREATE TRIGGER trigger_name trigger_time trigger_event
ON tbl_name FOR EACH ROW
BEGIN
    语句执行列表
END
```

其中，trigger_name 标识触发器的名称，用户自行指定；trigger_time 标识触发时机，可以指定为 before 或 after；trigger_event 标识触发事件，包括 INSERT、UPDATE 和 DELETE；tbl_name 标识建立触发器的表名，即在哪张表上建立触发器；触发器程序可以使用 BEGIN 和 END 作为开始和结束，中间包含多条语句。

【例 10.2】 创建一个包含多个执行语句的触发器，代码如下：

```
CREATE TABLE test1(a1 INT);
CREATE TABLE test2(a2 INT);
CREATE TABLE test3(a3 INT NOT NULL AUTO_INCREMENT PRIMARY KEY);
CREATE TABLE test4(
a4 INT NOT NULL AUTO_INCREMENT PRIMARY KEY,
b4 INT DEFAULT 0
);
```

```
DELIMITER //
CREATE TRIGGER testref BEFORE INSERT ON test1
  FOR EACH ROW BEGIN
    INSERT INTO test2 SET a2 = NEW.a1;
    DELETE FROM test3 WHERE a3 = NEW.a1;
    UPDATE test4 SET b4 = b4 + 1 WHERE a4 = NEW.a1;
  END
//
DELIMITER ;

INSERT INTO test3 (a3) VALUES
  (NULL), (NULL), (NULL), (NULL), (NULL),
  (NULL), (NULL), (NULL), (NULL), (NULL);

INSERT INTO test4 (a4) VALUES
  (0), (0), (0), (0), (0), (0), (0), (0), (0), (0);
```

上面的代码创建了一个名为 testref 的触发器。这个触发器的触发条件是在向表 test1 插入数据前执行触发器的语句，具体执行的代码如下：

```
mysql> INSERT INTO test1 VALUES (1), (3), (1), (7), (1), (8), (4), (4);
```

4 个表中的数据如下：

```
mysql> SELECT * FROM test1;
+------+
| a1   |
+------+
|    1 |
|    3 |
|    1 |
|    7 |
|    1 |
|    8 |
|    4 |
|    4 |
+------+
8 rows in set (0.00 sec)

mysql> SELECT * FROM test2;
+------+
| a2   |
+------+
|    1 |
|    3 |
|    1 |
|    7 |
|    1 |
|    8 |
|    4 |
|    4 |
+------+
```

```
8 rows in set (0.00 sec)

mysql> SELECT * FROM test3;
+----+
| a3 |
+----+
|  2 |
|  5 |
|  6 |
|  9 |
| 10 |
+----+
5 rows in set (0.00 sec)

mysql> SELECT * FROM test4;
+----+------+
| a4 | b4   |
+----+------+
|  1 |   3  |
|  2 |   0  |
|  3 |   1  |
|  4 |   2  |
|  5 |   0  |
|  6 |   0  |
|  7 |   1  |
|  8 |   1  |
|  9 |   0  |
| 10 |   0  |
+----+------+
10 rows in set (0.00 sec)
```

执行结果显示，在向表 test1 插入记录的时候，test2、test3、test4 都发生了变化。从这个例子看 INSERT 触发了触发器，向 test2 中插入了 test1 中的值，删除了 test3 中相同的内容，同时更新了 test4 中的 b4，即与插入的值相同的个数。

10.2 查看触发器

查看触发器是指查看数据库中已存在的触发器的定义、状态和语法信息等。可以通过命令来查看已经创建的触发器。本节将介绍两种查看触发器的方法，分别是 SHOW TRIGGERS 语句和在 TRIGGERS 表中查看触发器信息。

10.2.1 利用 SHOW TRIGGERS 语句查看触发器信息

通过 SHOW TRIGGERS 查看触发器的语句如下：

```
SHOW TRIGGERS;
```

【例 10.3】通过 SHOW TRIGGERS 语句查看一个触发器，代码如下：

```
SHOW TRIGGERS;
```

创建一个简单的触发器,名称为 trig_update,每次在 account 表更新数据之后都会向名称为 myevent 的数据表中插入一条记录,数据表 myevent 定义如下:

```
CREATE TABLE myevent
(
id int DEFAULT NULL,
evt_name char(20) DEFAULT NULL
) ;
```

创建触发器的执行代码如下:

```
mysql> CREATE TRIGGER trig_update AFTER UPDATE ON account
    -> FOR EACH ROW INSERT INTO myevent VALUES (1,'after update');
Query OK, 0 rows affected (0.00 sec)
```

使用 SHOW TRIGGERS 命令查看触发器:

```
mysql> SHOW TRIGGERS;
+------+------+-------+-----------+--------+---------+----------+---------+----------------------+--------------------+
|Trigger|Event|Table|Statement|Timing|Created|sql_mode|Definer|character_set_client|collatin_connection|Database Collation|
+------+------+-------+-----------+--------+---------+----------+---------+----------------------+--------------------+
|ins_sum|INSERT|account| SET @sum = @sum + NEW.amount| BEFORE | NULL| | root@localhost|latin1|latin1_wedish_ci  | latin1_swedish_ci |
| trig_update | UPDATE | account | INSERT INTO myevent VALUES (1,'after update') | AFTER| NULL| |root@localhost | utf8mb4  | utf8mb4_0900_ai_ci | utf8mb4_0900_ai_ci|
+------+------+-------+-----------+--------+---------+----------+---------+----------------------+--------------------+
2 rows in set (0.00 sec)
```

可以看到,信息显示比较混乱。如果在 SHOW TRIGGERS 命令的后面添加上"\G",显示信息会比较有条理,执行情况如下:

```
mysql> SHOW TRIGGERS \G
*************************** 1. row ***************************
             Trigger: ins_sum
               Event: INSERT
               Table: account
           Statement: SET @sum = @sum + NEW.amount
              Timing: BEFORE
             Created: 2022-03-17 09:27:50.92
            sql_mode: STRICT_TRANS_TABLES,NO_ENGINE_SUBSTITUTION
             Definer: root@localhost
character_set_client: utf8mb4
collation_connection: utf8mb4_0900_ai_ci
  Database Collation: utf8mb4_0900_ai_ci
*************************** 2. row ***************************
             Trigger: trig_update
               Event: UPDATE
               Table: account
           Statement: INSERT INTO myevent VALUES (1,'after update')
              Timing: AFTER
             Created: 2022-03-17 09:33:51.49
            sql_mode: STRICT_TRANS_TABLES,NO_ENGINE_SUBSTITUTION
```

```
              Definer: root@localhost
  character_set_client: utf8mb4
  collation_connection: utf8mb4_0900_ai_ci
    Database Collation: utf8mb4_0900_ai_ci
*************************** 3. row ***************************
               Trigger: testref
                 Event: INSERT
                 Table: test1
             Statement: BEGIN
  INSERT INTO test2 SET a2 = NEW.a1;
  DELETE FROM test3 WHERE a3 = NEW.a1;
  UPDATE test4 SET b4 = b4 + 1 WHERE a4 = NEW.a1;
END
                Timing: BEFORE
               Created: 2022-03-17 09:31:33.40
              sql_mode: STRICT_TRANS_TABLES,NO_ENGINE_SUBSTITUTION
               Definer: root@localhost
  character_set_client: utf8mb4
  collation_connection: utf8mb4_0900_ai_ci
    Database Collation: utf8mb4_0900_ai_ci
```

其中，Trigger 表示触发器的名称，在这里两个触发器的名称分别为 ins_sum 和 trig_update；Event 表示激活触发器的事件，这里的两个触发事件为插入操作 INSERT 和更新操作 UPDATE；Table 表示激活触发器的操作对象表，这里都为 account 表；Timing 表示触发器触发的时间，分别为插入操作之前（BEFORE）和更新操作之后（AFTER）；Statement 表示触发器执行的操作，还有一些其他信息，比如 SQL 的模式、触发器的定义账户和字符集等，这里不再一一介绍。

提示：SHOW TRIGGERS 语句查看当前创建的所有触发器信息，在触发器较少的情况下，使用该语句会很方便。如果要查看特定触发器的信息，可以直接从 INFORMATION_SCHEMA 数据库中的 TRIGGERS 表中查找。下面将介绍这种方法。

10.2.2 在 triggers 表中查看触发器信息

在 MySQL 中，所有触发器的定义都存在 INFORMATION_SCHEMA 数据库的 TRIGGERS 表中，可以通过查询命令 SELECT 查看，具体的语法如下：

```
SELECT * FROM INFORMATION_SCHEMA.TRIGGERS WHERE condition;
```

【例 10.4】通过 SELECT 命令查看触发器，代码如下：

```
SELECT * FROM INFORMATION_SCHEMA.TRIGGERS WHERE TRIGGER_NAME='trig_update'\G
```

上述命令通过 WHERE 来指定查看特定名称的触发器。指定触发器名称的执行情况如下：

```
*************************** 1. row ***************************
           TRIGGER_CATALOG: def
            TRIGGER_SCHEMA: test_db
              TRIGGER_NAME: trig_update
        EVENT_MANIPULATION: UPDATE
      EVENT_OBJECT_CATALOG: def
       EVENT_OBJECT_SCHEMA: test_db
        EVENT_OBJECT_TABLE: account
              ACTION_ORDER: 1
```

```
                  ACTION_CONDITION: NULL
                  ACTION_STATEMENT: INSERT INTO myevent VALUES (1,'after update')
                ACTION_ORIENTATION: ROW
                     ACTION_TIMING: AFTER
         ACTION_REFERENCE_OLD_TABLE: NULL
         ACTION_REFERENCE_NEW_TABLE: NULL
          ACTION_REFERENCE_OLD_ROW: OLD
          ACTION_REFERENCE_NEW_ROW: NEW
                           CREATED: 2022-03-17 09:33:51.49
                          SQL_MODE: STRICT_TRANS_TABLES,NO_ENGINE_SUBSTITUTION
                           DEFINER: root@localhost
              CHARACTER_SET_CLIENT: utf8mb4
              COLLATION_CONNECTION: utf8mb4_0900_ai_ci
                DATABASE_COLLATION: utf8mb4_0900_ai_ci
```

从上面的执行结果可以得知：TRIGGER_SCHEMA 表示触发器所在的数据库；TRIGGER_NAME 后面是触发器的名称；EVENT_OBJECT_TABLE 表示在哪个数据表上触发；ACTION_STATEMENT 表示触发器触发的时候执行的具体操作；ACTION_ORIENTATION 是 ROW，表示在每条记录上都触发；ACTION_TIMING 表示触发的时刻是 AFTER；剩下的是和系统相关的信息。

也可以不指定触发器名称，这样将查看所有的触发器，命令如下：

```
SELECT * FROM INFORMATION_SCHEMA.TRIGGERS \G
```

这个命令会显示 TRIGGERS 表中所有的触发器信息，读者自行测试一下。

10.3 触发器的使用

触发程序是与表有关的命名数据库对象，当表上出现特定事件时，将激活该对象。在某些触发程序的用法中，可用于检查插入到表中的值，或对更新涉及的值进行计算。

触发程序与表相关，当对表执行 INSERT、DELETE 或 UPDATE 语句时，将激活触发程序。可以将触发程序设置为在执行语句之前或之后激活。例如，可以在从表中删除每一行之前或在更新每一行之后激活触发程序。

【例 10.5】创建一个在 account 表插入记录之后更新 myevent 数据表的触发器，代码如下：

```
CREATE TRIGGER trig_insert AFTER INSERT ON account FOR EACH ROW INSERT INTO myevent VALUES (2,'after insert');
```

上面的代码创建了一个 trig_insert 触发器，在向表 account 插入数据之后，会向表 myevent 插入一组数据，代码执行如下：

```
mysql> CREATE TRIGGER trig_insert AFTER INSERT ON account
    -> FOR EACH ROW INSERT INTO myevent VALUES (2, 'after insert');

mysql> INSERT INTO account VALUES (1,1.00), (2,2.00);

mysql> SELECT * FROM myevent;
+------+--------------+
| id   | evt_name     |
```

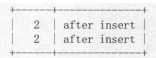

从执行的结果来看,创建了一个名称为 trig_insert 的触发器,在向 account 插入记录之后进行触发,执行的操作是向表 myevent 插入一条记录。

10.4 删除触发器

使用 DROP TRIGGER 语句可以删除 MySQL 中已经定义的触发器,删除触发器语句的基本语法格式如下:

```
DROP TRIGGER [schema_name.]trigger_name
```

其中,schema_name 表示数据库名称,是可选的。如果省略了 schema,将从当前数据库中舍弃触发程序;trigger_name 是要删除的触发器的名称。

【例 10.6】删除一个触发器,代码如下:

```
DROP TRIGGER test_db.ins_sum;
```

上面的代码中 test 是触发器所在的数据库,ins_sum 是一个触发器的名称。代码执行如下:

```
mysql> DROP TRIGGER test_db.ins_sum;
Query OK, 0 rows affected (0.00 sec)
```

触发器 ins_sum 删除成功。

第 11 章

存储引擎的选择

数据库存储引擎是数据库底层软件组件，数据库管理系统（DBMS）使用数据引擎进行创建、查询、更新和删除数据操作。不同的存储引擎提供不同的存储机制、索引技巧、锁定水平等功能，使用不同的存储引擎，还可以获得特定的功能。MySQL 的核心就是存储引擎。本章将讲解 MySQL 支持的各种存储引擎。

11.1 MySQL 的架构

MySQL 服务器由 SQL 层和存储引擎层构成。SQL 层主要功能包括权限判断、SQL 解析功能和查询缓存处理等，存储引擎层（Storage Engine Layer）完成底层数据库数据存储操作。MySQL 整体架构的 SQL 层和存储引擎层实际上各自都包含很多的小模块，各个模块的工作方式如图 11.1 所示。

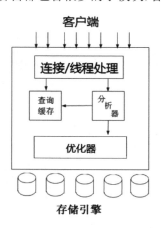

图 11.1　MySQL 各个模块的工作方式

从上图可以看出以下 3 点。

（1）客户端通过连接/线程处理层来连接 MySQL 数据库，连接/线程处理层主要用来处理客户端的请求、身份验证和数据库安全性验证等。

（2）查询缓存和查询分析器是 SQL 层的核心部分，其中主要涉及查询的解析、优化、缓存，以及所有内置的函数、存储过程、触发器、视图等功能。

（3）优化器主要负责存储和获取所有存储在 MySQL 中的数据。

可以把这三层统称为 MySQL 数据库的 SQL 层。

11.1.1　MySQL 物理文件的组成

MySQL 的物理文件包括日志文件、数据文件和其他文件，下面将详细介绍这些文件的含义和作用。

1．日志文件

在 MySQL 数据库中，日志文件主要记录了数据库操作信息和错误信息。常用的日志文件包括错误日志、二进制日志、查询日志、慢查询日志和 InnoDB 引擎在线 Redo 日志等。

（1）错误日志：Error Log

错误日志文件记录了 MySQL Server 运行过程中遇到的所有严重的错误信息，以及 MySQL 每次启动和关闭的详细信息。默认情况下错误日志功能是关闭的，启动时要重新配置---log-error[=file_name]选项，修改错误日志存放的目录和文件名称。

（2）二进制日志：Binary Log

二进制日志文件，就是常说的 binlog。二进制日志记录了 MySQL 所有修改数据库的操作，然后以二进制的形式记录在日志文件中，其中还包括每条语句所执行的时间和所消耗的资源，以及相关的事务信息。

默认情况下二进制日志功能是开启的，启动时可以重新配置--log-bin[=file_name]选项，修改二进制日志存放的目录和文件名称。

（3）查询日志：Query Log

默认的查询日志文件是 hostname.log。查询日志记录所有的查询操作，包括所有的 select 操作信息，体积比较大，开启后对性能有较大的影响，可以通过"--log[=file_name]"选项开启。如果需要跟踪某些特殊的 SQL 性能问题，可以短暂地打开该功能。

（4）慢查询日志：Slow Query Log

慢查询日志是指所有 SQL 执行的时间超过 long_query_time 变量的语句和达到 min_examined_row_limit 条的语句。用户可以针对这部分语句进行性能调优。慢查询日志通过设置--log-slow_queries[=file_name]选项开启后，将记录日志所在的路径和名称。MySQL 系统默认的慢查询日志的文件名是 hostname-slow.log，默认目录也是 data 目录。查看慢查询日志可以采用 mysqldumpslow 命令对慢查询日志进行分析。

（5）InnoDB 引擎在线 Redo 日志：InnoDB Redo Log

InnoDB 引擎在线 Redo 日志记录了 InnoDB 所做的所有物理变更和事务信息。通过 Redo 日志和 Undo 信息，InnoDB 大大地加强了事务的安全性。InnoDB 在线 Redo 日志默认存放在 data 目录下面，可以通过设置 innodb_log_group_home_dir 选项来更改日志的存放位置，通过 innodb_log_files_in_group 选项来设置日志的数量。

2．数据文件

MySQL 数据库会在 data 目录下面建立一个以数据库为名字的文件夹，用来存储数据库中的表文件数据。不同的数据库引擎，每个表的扩展名也不一样，例如，MyISAM 引擎用".MYD"作为扩展名，InnoDB 引擎可以用".ibd"，CSV 使用".csv"扩展名。

常见的数据文件如下：

（1）".frm"文件：无论是哪种存储引擎，创建表之后就一定会生成一个以表名命名的".frm"文件。frm 文件主要存放与表相关的数据信息，主要包括表结构的定义信息。当数据库崩溃时，用户可以通过 frm 文件来恢复数据表结构。

（2）".MYD"文件：MyISAM 存储引擎创建表的时候，每一个 MyISAM 类型的表都会有一个".MYD"文件与之对应。MYD 文件主要用来存放数据表的数据文件。

（3）".MYI"文件：每一个 MyISAM 类型的表都会有一个".MYD"文件和一个".MYI"文件，对于 MyISAM 存储引擎来说，可以被缓存的内容主要就是源于".MYI"文件中，文件中主要用来存储表数据文件中任何索引的数据树。

（4）".ibd"文件和".ibdata"文件：这两种文件主要是用来存储 InnoDB 存储引擎的数据，其中主要包括索引信息。InnoDB 存储引擎采用这两种数据文件，主要是因为 InnoDB 存储引擎的存储方式能够通过配置来决定是采用共享表空间，还是采用独享表空间的存储方式存储数据。

如果采用共享表空间的方式存储数据，则会采用 ibdata 文件来存储，所有的表共同使用一个或者多个 ibdata 文件。如果采用独享表空间的方式存储数据，则会采用 ibd 文件来存储。

共享表空间存储通过 innodb_data_home_dir 和 innodb_data_file_path 两个参数共同配置组成，innodb_data_home_dir 参数配置数据存放的总目录，innodb_data_file_path 参数配置每一个文件的路径及文件名称。如果需要添加新的 ibdata 文件，则需要在 innodb_data_file_path 参数后面配置，然后重新启动服务器才能够生效。

3．其他文件

MySQL 数据库系统除了日志文件、数据文件外，还包括其他的一些文件。例如系统配置文件、pid 文件、socket 文件等等。

MySQL 系统配置文件一般都在"etc/my.cnf"中。pid 文件类似于 Unix/Linux 操作系统下面的进程文件，MySQL 服务器的 pid 文件用来存放自己的进程 ID。MySQL 服务器启动后，socket 文件自动生成，该文件主要用来连接客户端。

11.1.2 MySQL 各逻辑块简介

MySQL 逻辑架构采用 SQL 层和存储引擎分离的方式，真正实现了数据存储和逻辑业务的分离，MySQL 的 SQL 层从宏观上可以分为三层，事实上 SQL 层包含了很多的子模块。下面就详细介绍 SQL 层各个子模块的功能。

1. 初始化模块

初始化模块就是在数据库启动的时候，对整个数据库做的一些初始化操作，例如，各种系统环境变量的初始化、各种缓存、存储引擎初始化设置等。

在 MySQL 初始化过程中，部分系统参数是通过 MySQL 数据库系统文件设置的。MySQL 系统参数可以通过 "mysqld --verbose –help" 命令来查看当前系统所有参数的设置。Linux 平台上 MySQL 数据库读取文件首先会读/etc/my.cnf 文件，该选项主要用来设置 MySQL 全局选项，许多初学者在 Linux 平台上安装 MySQL 失败就是因为/etc/my.cnf 的设置是系统默认的错误路径，对于初学者，可以将$MySQL_HOME/support_files/目录下面的配置文件复制到/etc/my.cnf 中，命令如下：

```
cp ./support_files/my_medium.cnf  /etc/my.cnf
```

MySQL 数据库读取完/etc/my.cnf 之后，接下来会解析$MySQL_HOME/my.cnf。在这个过程中，服务器会到 MySQL 安装目录下面解析数据库的相关配置。MySQL 启动初始化接着会解析 defaults-extra-file 附带选项，修改该参数可以指定系统配置文件，接下来数据库会解析有关用户的选项。

初始化模块是在 MySQL 数据库启动的时候，初始化数据库的各种环境变量、数据库初始化设置，以及初始化数据存储引擎的设置。

2. 核心 API

MySQL 数据库核心 API 主要实现了数据库底层操作的优化功能，其中主要包括 IO 操作、格式化输出、高性能存储数据结果算法的优化、字符串的处理，其中最重要的是内存管理。

3. 网络交互模块

MySQL 底层网络相互交互的模块抽象出来的接口，对外提供可以接收和发送数据的 API 接口，其他模块需要交互的时候，可以通过 API 接口调用。

4. 服务器客户端交互协议模块

MySQL 服务器采用 C/S 客户端的形式访问数据库，数据连接使用 MySQL C/S 客户端交互协议模块，实现了客户端与服务器端交互过程中所需要的一些独特的协议，这些协议都是建立在现有的网络协议之上。

5. 用户模块

用户模块主要功能是用于控制用户登录连接的权限和用户的授权管理。

6. 访问控制模块

访问控制模块主要用于监控用户的每一个操作。访问控制模块实现的功能就是根据用户模块中不同的用户授权，以及根据其数据库的各种约束，来控制用户对数据的访问。用户模块和访问控制模块结合起来，就组成了 MySQL 数据库的权限管理功能。

7. 连接管理、连接线程和线程管理

连接管理模块负责监听 MySQL Server 的各种请求，根据不同的请求，然后转发到线程管理模块，每个客户请求都会被数据库自动分配一个独立的线程为其单独服务，而连接线程的主要工作就是负责 MySQL Server 与客户端通信，线程管理模块负责管理这些生成的线程。

8. 转发模块

客户端连接 MySQL 之后会发送一些查询语句，在 MySQL Server 里面，连接线程接收到客户端的一个请求后，会直接将查询转发到各个对应的处理模块。转发模块主要就是根据查询语句语法分析，然后转发给不同的模块处理。

9. 缓存模块

查询缓存模块主要功能是将客户端查询的请求返回的结果集到缓存中，与查询的一个 HASH 值做一个对应。在查询的基表发生任何数据变化后，MySQL 会自动将其查询的缓存失效。在读写比例非常高的应用系统中，查询缓存对性能的提高是非常显著的。

10. 优化器模块

这个模块主要是将客户端发送的查询请求，在之前的算法的基础上分析，计算出一个最优的查询策略，优化之后会提高查询访问的速度，最后根据其最优策略返回查询语句。

11. 表变更管理模块

表变更管理模块主要负责完成 DML 和 DDL 语句的操作，例如，insert、update、delete、create table、alter table 等语句的处理。

12. 表维护模块

表维护模块主要用于检测表的状态，分析、优化表结构，以及修复表。

13. 系统状态管理模块

在客户端请求系统状态的时候，系统状态模块主要负责将各种状态的数据返回给用户。最常用的一些查询状态的命令包括 show status、show variables 等，都是通过这个模块负责返回的。

14. 表管理器

表管理器主要就是维护系统生成的表文件。例如，MyISAM 存储引擎类型表生成的是 frm 文件、MYD 文件以及 MYI 文件，表管理器的工作就是维护这些文件，将各个表结构的信息缓存起来。另外该模块还管理表级别的锁。

15. 日志记录模块

日志记录模块主要负责整个数据库逻辑层的日志文件，其中包含错误日志、二进制日志，以及慢查询日志等。

16. 复制模块

复制模块分为 Master 模块和 Slave 模块两部分。Master 模块主要负责复制环境中读取 Master 端的 binary 日志，以及 Slave 端的 I/O 线程交互等工作。

Slave 模块主要有两个线程，一个负责从 Master 请求和接收 binary 日志，并写入本地 I/O 线程，另一个从 relay log 读取日志事件，然后解析成可以在 Slave 端执行的命令，然后交给 Slave 端的 SQL 线程。

17. 存储引擎接口模块

MySQL 实现了其数据库底层存储引擎的插件式管理，将各种数据处理高度抽象化。

11.1.3 MySQL 各逻辑块协调工作

MySQL 各个逻辑模块协调工作的过程如图 11.2 所示。

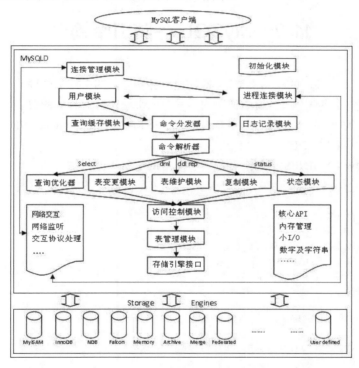

图 11.2　MySQL 各逻辑模块协调工作图

MySQL 启动后，MySQL 的初始化模块就从系统配置文件中读取系统参数和命令参数，并按照参数初始化整个系统，同时存储引擎也会启动。当初始化工作结束后，连接管理模块会监听并接收客户端的程序，连接管理模块会将连接请求转发给线程管理模块去请求一个连接线程。

线程模块接到请求之后会调用用户模块进行授权检查，通过检查后，会检测线程池里是否有空闲连接线程，如果有，就取出跟客户端连接上；如果没有，则建立一个新的线程与客户端建立连接。

MySQL 数据库中的请求有两种，一种是需要命令解析和分发模块解析才能执行请求的操作，另一种是不需要转发就可以直接执行的命令。如果 MySQL 初始阶段开启了日志功能，这时候日志模块会将请求记入日志，不管是哪种请求，都会记录到日志中。

Query 类型的请求，会将控制权交给 Query 解析器，Query 解析器会检查是否是 SELECT 类型的查询，如果是，则启动查询缓存模块，此时会查询缓存中已经存在结果集，如果存在，则将缓存中的数据返回给连接线程模块，之后连接线程会将数据传递到客户端。如果没有被缓存，或者不是一个可以缓存的查询，此时查询解析器进行相应的处理，通过查询分发器转给相关的处理模块。

如果解析器结果是 DML/DDL，则交给变更管理模块；如果是一些检查、修复类的查询，则交给表维护模块去处理；如果是一条没有被缓存的查询语句，则交给查询优化器模块。实际上表变更管理器又分为若干小的模块，例如：insert 处理器、delete 处理器、update 处理器、create 处理器，以及 alter 处理器，这些小模块负责不同的 DML 和 DDL。总之，查询优化器、表变更模块、表维护模块、复制模块、状态模块，都是根据命令解析器的结果不同而分发给不同类型的模块。

当一条命令执行完成之后，控制权都会还给连接线程模块，在上面各个模块处理过程中，各个模块都依赖于整个 MySQL 的核心 API 模块，比如内存管理、文件 I/O、字符串处理等。

11.2 MySQL 存储引擎简介

MySQL 提供了多个不同的存储引擎，包括处理事务安全表的引擎和处理非事务安全表的引擎。在 MySQL 中，不需要在整个服务器中使用同一种存储引擎，针对具体的要求，可以对每一个表使用不同的存储引擎。MySQL 支持的存储引擎有 InnoDB、MyISAM、Memory、NDB、Merge、Archive、Federated、CSV、BLACKHOLE 等。

可以使用 SHOW ENGINES 语句查看系统所支持的引擎类型，结果如下：

```
mysql> SHOW ENGINES \G
*** 1. row ***
      Engine: FEDERATED
     Support: NO
     Comment: Federated MySQL storage engine
Transactions: NULL
          XA: NULL
  Savepoints: NULL
*** 2. row ***
      Engine: MRG_MYISAM
     Support: YES
     Comment: Collection of identical MyISAM tables
Transactions: NO
          XA: NO
  Savepoints: NO
*** 3. row ***
      Engine: MyISAM
     Support: YES
     Comment: MyISAM storage engine
Transactions: NO
          XA: NO
  Savepoints: NO
*** 4. row ***
      Engine: BLACKHOLE
```

```
        Support: YES
        Comment: /dev/null storage engine (anything you write to it disappears)
   Transactions: NO
             XA: NO
      Savepoints: NO
*** 5. row ***
         Engine: CSV
        Support: YES
        Comment: CSV storage engine
   Transactions: NO
             XA: NO
      Savepoints: NO
*** 6. row ***
         Engine: MEMORY
        Support: YES
        Comment: Hash based, stored in memory, useful for temporary tables
   Transactions: NO
             XA: NO
      Savepoints: NO
*** 7. row ***
         Engine: ARCHIVE
        Support: YES
        Comment: Archive storage engine
   Transactions: NO
             XA: NO
      Savepoints: NO
*** 8. row ***
         Engine: InnoDB
        Support: DEFAULT
        Comment: Supports transactions, row-level locking, and foreign keys
   Transactions: YES
             XA: YES
      Savepoints: YES
*** 9. row ***
         Engine: PERFORMANCE_SCHEMA
        Support: YES
        Comment: Performance Schema
   Transactions: NO
             XA: NO
      Savepoints: NO
```

Support 列的值表示某种引擎是否能使用：YES 表示可以使用，NO 表示不能使用，DEFAULT 表示该引擎为当前默认的存储引擎。

查看当前默认的存储引擎，可以使用下面的命令：

```
mysql> show variables like '%storage_engine%';
+--------------------------------+-----------+
| Variable_name                  | Value     |
+--------------------------------+-----------+
| default_storage_engine         | InnoDB    |
| default_tmp_storage_engine     | InnoDB    |
| disabled_storage_engines       |           |
| internal_tmp_mem_storage_engine| TempTable |
+--------------------------------+-----------+
```

11.3　更改数据表的存储引擎

MySQL 数据库在创建表的时候，可以添加默认的存储引擎。例如下面的例子：

```
create table books(
 id int,
 name varchar(20) not null
)engine=MyISAM default charset=utf8mb4;
```

更改表的存储引擎的语法格式如下：

```
ALTER TABLE <表名> ENGINE=<更改后的存储引擎名>;
```

下面示例将数据表 books 的存储引擎修改为 InnoDB。

在修改存储引擎之前，先使用 SHOW create table 查看表 books 当前的存储引擎，结果如下：

```
mysql> SHOW create table books \G
*************************** 1. row ***************************
       Table: books
Create Table: create table `books` (
  `id` int DEFAULT NULL,
  `name` varchar(20) NOT NULL
) ENGINE=MyISAM DEFAULT CHARSET=utf8mb4 COLLATE=utf8mb4_0900_ai_ci
```

可以看到，表 books 当前的存储引擎为 ENGINE=MyISAM，接下来修改存储引擎类型，输入如下 SQL 语句并执行：

```
mysql> ALTER TABLE books ENGINE=InnoDB;
```

使用 SHOW create table 再次查看表 books 的存储引擎，发现表 books 的存储引擎变成了"InnoDB"，结果如下：

```
mysql> SHOW create table books \G
*************************** 1. row ***************************
       Table: books
Create Table: create table `books` (
  `id` int DEFAULT NULL,
  `name` varchar(20) NOT NULL
) ENGINE= InnoDB DEFAULT CHARSET=utf8mb4 COLLATE=utf8mb4_0900_ai_ci
```

11.4　各种存储引擎的特性

尽管 MySQL 数据库支持种类繁多的数据存储引擎，但是数据存储不管是使用哪种存储引擎，所有的存储数据都被记录到.frm 文件中，该文件记录了存储的数据，以及表的一些属性值。值得注意的是，不管是使用哪种数据存储引擎，都使用了高速缓存，数据库读取.frm 文件信息后会将表的信息缓存起来，提高了服务器下次读取数据的速度。

11.4.1 MyISAM

MyISAM 存储引擎在 MySQL 数据引擎中起源最早。MyISAM 的最初版本是 ISAM，最初的 MySQL 架构比较混乱，数据业务和数据存储没有很好地分离出来，MySQL 意识到需要更改架构，将前端的业务逻辑和后端数据存储以清晰的层次结构拆分开，于是 ISAM 被重构成 MyISAM，此时 MySQL 从架构上来讲真正实现了逻辑业务和数据存储的分离。

1. MyISAM 文件格式

MyISAM 在磁盘上存储三个文件，文件名和对应的表名是一致的。

（1）.frm 文件：存储表的定义数据。
（2）.MYD 文件：存储表具体记录数据。
（3）.MYI 文件：存储索引。

.frm 文件和.MYI 文件可以放在不同的目录下面，平均分配读取权限，可以获得更快的速度。MyISAM 存储引擎不支持事务，也不支持主键，对数据的存储和批量查询的速度比较快。

在实际应用中，对于不需要完整事务、主要以查询和增加记录为主的应用，往往采用 MyISAM 存储引擎。MyISAM 存储引擎只缓存索引，对数据文件采用操作系统缓存，如果索引数据超过系统所分配的缓存空间时候，也会采用操作系统来缓存索引。

2. MyISAM 文件修复

MyISAM 类型的表在数据存储的过程中可能会发生错误，此时可以通过使用 CHECK TABLE 语句检测 MyISAM 表的状态，然后用 REPAIR TABLE 语句修复损坏的表。

另外，还可以使用 myisamchk 工具修复损坏的表。通过 myisamchk 可以检测 MyISAM 类型表的健康状态和修复甚至优化表的存储。在 Linux 平台下，使用 myisamchk 检测某个表的时候，如果有用户同时也在操作这张表，即便这张表没有问题，也很有可能会提示该表已经被损坏，所以在检测数据库表的时候，应当停止 mysqld 服务。如果不想停掉 mysqld 服务，至少应该做一个 mysqladmin flush–tables 操作。

使用 myisamchk --help 命令可以获取 myisamchk 的用法，myisamchk 的命令格式如下：

```
myisamchk [options] tables[.MYI]
```

其中[options]用于指定 myisamchk 的操作，具体可选的参数含义如下：

（1）-c：表示用来检测表可能存在的错误。
（2）-d：表示打印出表的一些信息。
（3）-e：表示彻底地检查表。
（4）-i：表示打印有关被检查表的信息统计。
（5）-k #：一般与参数-r 一起使用，表示仅仅更新头#个索引，较高编号的索引被撤销。撤销的索引会通过 myisamchk –r 被重新激活。
（6）-q：与-r 一起可以使得修复速度更快。
（7）-r：表示可以修复几乎所有的损坏的问题。
（8）-o：相比-r 而言，-o 恢复表的能力要慢一些，但是能处理一些-r 不能处理的情况。

（9）-u：解开一个用 myisampack 压缩的表。

（10）-w：表示如果表被锁定，将会发生等待。

例如以下示例演示 myisamchk 操作：

```
[root@localhost books]# ls
books.frm   books.MYD   books.MYI   db.opt
[root@localhost books]# myisamchk -e books.MYI
Checking MyISAM file: books.MYI
Data records:          0    Deleted blocks:         0
- check file-size
- check record delete-chain
- check key delete-chain
- check index reference
- check data record references index: 1
- check record links
```

myisamchk 命令中 -O var=option 这个参数可以用来设置一些变量，-O 选项可能的变量如下：

```
key_buffer_size                 20971520
key_cache_block_size            1024
myisam_block_size               1024
read_buffer_size                2097152
write_buffer_size               2097152
sort_buffer_size                20971520
```

当在很大文件上运行 myisamchk 时候，往往需要较大的内存，缺省使用 3M 来修复。可以通过如下命令来设置内存空间：

```
[root@localhost books]# myisamchk -O sort_key_blocks=16 -e books.MYI
Checking MyISAM file: books.MYI
Data records:          0    Deleted blocks:         0
- check file-size
- check record delete-chain
- check key delete-chain
- check index reference
- check data record references index: 1
- check records and index references
```

3. MyISAM 表的存储格式

MyISAM 类型的表支持三种不同存储类型格式的表，分别如下：

（1）静态（固定长度）表。

（2）动态（可变长度）表。

（3）压缩表。

其中静态和动态表数据存储根据其表中数据列的类型自动选择，静态表是默认的存储格式，压缩表则只能通过 myisampack 工具创建。

静态表中的字段都是固定非变长的字段，这样每个记录都是固定长度的，这种存储方式的优势在于存储速度非常快，容易缓存，表发生缺损后容易恢复，缺点是静态表所占用的空间往往要比动态表多。

MyISAM 存储引擎采用动态表将会支持动态可变长度，字符型的列长是可变的，除了小于 4 个字符的。动态可变长字符通常比静态固定格式需要更少的存储空间，由于采用动态可变长度存储，所以出错的时候也比静态格式恢复更加困难，因为行变化如果很大的话，会被分成碎片。这个时候可以使用 myisamchk –ei 获取表的统计信息，并使用 myisamchk –r 来进行修复。

MyISAM 存储压缩表需要的磁盘存储空间最小，数据库系统提供了 myisampack 工具可以用来压缩 MyISAM 表，每行单独压缩，每列的压缩也不一样。

11.4.2 InnoDB 存储引擎

InnoDB 是事务型数据库的首选引擎，支持事务安全表（ACID），支持行锁定和外键。MySQL 5.5.5 之后，InnoDB 作为默认的存储引擎。InnoDB 写处理相对于 MyISAM 效率低一些，InnoDB 牺牲了存储和查询的效率，支持事务安全，支持自动增长列。对事务安全的支持，这是 InnoDB 成为 MySQL 最为流行的存储引擎之一的重要原因。下面重点介绍 InnoDB 存储引擎的特点。

1. 支持事务

MySQL 支持对 InnoDB 存储事务控制，实现了 SQL92 标准所定义的四个级别（read uncommitted、repeatable read、read committed 和 serializable）。MySQL 通过 commit、rollback、set autocommit、start transaction 等语法支持本地事务，具体语法如下所示：

```
START TRANSACTION | BEGIN [WORK]
COMMIT [WORK] [AND [NO] CHAIN] [[NO] RELEASE]
ROLLBACK [WORK] [AND [NO] CHAIN ][[NO] RELEASE]
SET AUTOCOMMIT = {0|1}
```

2. 自动增长列

InnoDB 表的自动增长需要在列的后面添加 auto_increment 属性，对表的添加数据过程中，可以插入空值，该列都可以自动增加数据。

以下示例可以实现自动增长列。

首先创建具有增长列属性的表，命令如下：

```
mysql> create table authors(
        id int primary key auto_increment,
        name varchar(10));
```

插入 id 列为空值的数据，命令如下：

```
mysql> insert into authors(name) values('ivan'),('susan'),('shark');
```

查询插入的结果，命令如下：

```
mysql> select * from authors;
+----+-------+
| id | name  |
+----+-------+
|  1 | ivan  |
|  2 | susan |
|  3 | shark |
+----+-------+
```

从结果可以看出，id 列自动插入了数值。

3. 外键约束

InnoDB 实现了外键这一数据库重要功能。从数据库性能上讲，数据库外键降低了数据库查询的效率，数据库表之间的耦合度更加紧密，但是对于不少用户来讲，采用外键约束可能是最低成本的选择方式。MySQL 支持外键的存储引擎只有 InnoDB。在创建外键的时候，要求父表必须有对应的索引，字表在创建外键的时候会添加相对应的索引。

下面是一个外键的例子，sclass 是主表，id 作为主表主键索引，st 表示子表，其中 class_id 作为外键对应 sclass 表的 id 值。

首先创建数据表 sclass，命令如下：

```
mysql> create table sclass(
       id int primary key auto_increment,
       cname varchar(20) not null,
       last_update timestamp not null default current_timestamp on update current_timestamp);
```

创建数据表 st，并添加外键约束，命令如下：

```
mysql> create table st(
       id integer primary key auto_increment,
       sname varchar(20) not null,
       class_id integer not null,
       last_update timestamp not null default current_timestamp on update current_timestamp,
       foreign key(class_id) references sclass(id) on delete restrict on update cascade);
```

对于上面创建的两个表，在做删除操作时，如果是删除主表的数据，子表对应的记录不会被删除；如果是更新主表，子表对应的记录会更新。

在物理存储方面，InnoDB 有自己独特的存储方式，数据也是存放在.frm 文件里面，但是表数据和索引数据是存放在一起的。InnoDB 的存储表和索引有以下两种方式：

（1）使用共享表空间存储，也就是所有表和索引数据存放在同一个表空间中，数据和索引在 innodb_data_home_dir 和 innodb_data_file_path 定义的表空间中，可以使一个或者多个数据文件。

（2）使用多表空间存储，这种存储方式创建的表结构存放在.frm 文件中，但是每个表的数据和索引被存放在一个单独的.ibd 文件中。如果是分区表，则每个分区对应单独的.ibd 文件，文件名称是"表名+分区名"，可以在创建分区的时候指定每个分区的数据文件的路径，这样的好处是可以将表的读取操作平均分布到若干个磁盘分区文件上，从而提高数据访问的效率。

如果要使用多表空间存储方式，需要设置 innodb_file_per_table 参数。该参数可以修改 InnoDB 为独立表空间模式，每个数据库的表都会生成一个数据空间。

可以使用如下方式查看多表存储空间模式是否已经开启：

```
mysql> SHOW VARIABLES LIKE '%per_table%';
+-----------------------+-------+
| Variable_name         | Value |
+-----------------------+-------+
| innodb_file_per_table | OFF   |
+-----------------------+-------+
```

在设置 innodb_file_per_table 参数之前，需要先关闭数据库，然后在 my.cnf 文件中设置或者添加该参数 innodb_file_per_table=1，并重启数据库才能生效。这时候，再查看多表存储空间模式是否已经开启：

```
mysql> show variables like '%per_table%';
+-----------------------+-------+
| Variable_name         | Value |
+-----------------------+-------+
| innodb_file_per_table | ON    |
+-----------------------+-------+
```

使用多表空间特性的表，可以比较方便地进行表备份和恢复操作，但是直接复制.idb 文件是不行的，可以通过以下命令：

```
ALTER TABLE TABLE_NAME DISCARD TABLESPACE;
ALTER TABLE TABLE_NAME IMPORT TABLESPACE;
```

InnoDB 在功能上跟 MyISAM 存储引擎有很大的不同，在参数配置上，InnoDB 也是单独处理的，InnoDB 所有的参数基本上都加了前缀"innodb_"。

如果想屏蔽 InnoDB 存储引擎，在 my.ini 配置文件中，将 skip-innodb 参数前的#去除，这样就无法创建 InnoDB 类型的表了。

11.4.3 MEMORY

MEMORY 存储引擎通过采用内存中的内容来创建表。每个 MEMORY 表实际上会和一个磁盘文件关联起来。文件名采用"表名.frm"的格式。MEMORY 类型的表访问速度非常快，因为数据来源于内存空间。MEMORY 存储引擎默认使用 HASH 索引，虽然 MEMORY 类型的表访问速度非常快，但是一旦数据库发生故障关闭，内存中的数据就会发生丢失。

下面创建一个 MEMORY 类型的表 grade，其执行效率非常快，命令如下：

```
mysql> create table grade (
       id int)engine= memory;
```

MEMORY 表的空间以小块来分配。创建 MEMORY 表的时候，可以通过添加一个索引指定是 HASH 索引或者 BTREE 索引。

例如，以下示例指定了索引类型为 HASH，命令如下：

```
mysql> create table table_memory(
       id int,
       index using hash(id)
       )engine=memory;
Query OK, 0 rows affected (0.08 sec)
```

查看索引类型的命令如下：

```
mysql> show index from table_memory \g;
Table:              table_memory
Non_unique:         1
Key_name:           id
Seq_in_index:       1
Column_name:        id
Collation:          NULL
```

```
Cardinality:            0
Sub_part:               NULL
Packed:                 NULL
NULL:                   YES
Index_type:             HASH
Comment:
```

可以通过如下命令删除 table_memory 的 HASH 索引：

```
mysql> drop index id on table_memory;
```

可以通过如下命令修改 table_memory 的索引：

```
mysql> create index id_index using btree on table_memory(id);
```

索引修改后，可以查看修改的效果，命令如下：

```
mysql> show index from table_memory \g;
Table:                  table_memory
Non_unique:             1
Key_name:               id_index
Seq_in_index:           1
Column_name:            id
Collation:              A
Cardinality:            NULL
Sub_part:               NULL
Packed:                 NULL
NULL:                   YES
Index_type:             BTREE
Comment:
```

MEMORY 表内容存储在内存中，如果一个内部表变得很大，服务器自动把它转换成为一个磁盘表。尺寸限制由 temp_table_size 系统变量的值来确定。每个 MEMORY 表容量的大小，可以通过设置 max_heap_table_size 变量的值来控制。max_heap_table_size 系统变量默认的值是 16MB。对于单个表，也可以在 create table 语句中指定一个 MAX_ROWS 表选项，参数 MAX_ROWS 指定表的最大行数。例如以下实例：

```
mysql> create table table_memory_1(
       id int primary key
       )engine=memory max_rows=10000;
Query OK, 0 rows affected (0.00 sec)
```

在 MySQL 服务器启动时候可以使用 init-file 选项。可以把 insert into... SELECT 或 LOAD DATA INFILE 这样的语句放入这个文件中，从而从数据源中装载数据表。

MEMORY 类型的存储引擎主要用于内容变化不频繁的代码表，对 MEMORY 存储引擎的表更新操作数据不会写入到磁盘文件中，所以在选择 MEMORY 存储引擎的时候需要考虑这一特性。

11.4.4　MERGE

MERGE 存储引擎是一组 MyISAM 表组合而成。将一组结构相同的 MyISAM 表组合成一个逻辑单元，通常也叫作 MRG_MYISAM 存储引擎。MERGE 表本身没有数据，对于 MERGE 类型表的插入操作，是通过 INSERT_METHOD 子句完成，可以使用 FIRST 或者 LAST 值，可以使其数据增加到第一个表,或者最后一个表上。其实,上述操作实际上是对内部 MyISAM 表进行操作,所以在创建 MERGE 表时候,MySQL 只会生成两个较小的文件,一个是.frm 的文件,用于存放数据;还有一个.MRG 文件,用于存放 MERGE 表的名称,包括 MERGE 表由哪些表组成。

下面是一个创建和使用 MERGE 表的例子:

步骤01 创建三个表 table_myisam_1、table_myisam_2、table_merge_12(MERGE 表)。

```
mysql> create table table_myisam_1(
       id int primary key,
       data datetime
       )engine=myisam;
Query OK, 0 rows affected (0.08 sec)

mysql> create table table_myisam_2(
       id int primary key,
       data datetime
       )engine=myisam;
Query OK, 0 rows affected (0.00 sec)

mysql> create table table_merge_12(
       id int primary key,
       data datetime
       )engine=merge union=(table_myisam_1,table_myisam_2) insert_method=first;
Query OK, 0 rows affected (0.03 sec)
```

步骤02 向前两个表 table_myisam_1、table_myisam_2 添加数据。

```
mysql> insert into table_myisam_1 values(1,'2022-1-2'),(2,'2022-1-3');

mysql> insert into table_myisam_2 values(1,'2022-1-2'),(2,'2022-1-3');
```

步骤03 查询 MERGE 表 table_merge_12 的数据。

```
mysql> select * from table_merge_12;
+----+---------------------+
| id | data                |
+----+---------------------+
|  1 | 2022-01-02 00:00:00 |
|  2 | 2022-01-03 00:00:00 |
|  1 | 2022-01-02 00:00:00 |
|  2 | 2022-01-03 00:00:00 |
+----+---------------------+
4 rows in set (0.00 sec)
```

步骤04 向 MERGE 表 table_merge_12 添加一条数据。

```
mysql> insert into table_merge_12 values(3,'2022-2-3');
Query OK, 1 row affected (0.00 sec)

mysql> select * from table_myisam_1;
+----+---------------------+
| id | data                |
```

```
+-----+---------------------+
|  1  | 2022-01-02 00:00:00 |
|  2  | 2022-01-03 00:00:00 |
|  3  | 2022-02-03 00:00:00 |
+-----+---------------------+
3 rows in set (0.00 sec)

mysql> select * from table_myisam_2;
+-----+---------------------+
| id  | data                |
+-----+---------------------+
|  1  | 2022-01-02 00:00:00 |
|  2  | 2022-01-03 00:00:00 |
+-----+---------------------+
2 rows in set (0.00 sec)
```

此时会发现 insert_method=first 起作用了，数据已经添加到前一个表中，merge 表还是前两个表合并的结果。通常如果没有指定 insert_method 参数，任何尝试往 MERGE 表中 INSERT 数据的操作，都会发生错误。通常使用 MERGE 表来透明地对多个表进行查询和更新操作。

11.5　选择合适的存储引擎

不同存储引擎都有各自的特点，以适应不同的需求。为了做出选择，首先需要考虑每一个存储引擎提供了哪些不同的功能。表 11.1 分析了常用存储引擎的对比情况。

表11.1　常用存储引擎的对比情况

特　　点	InnoDB	MyISAM	MEMORY	MERGE
存储限制	64TB	有	有	没有
事务安全	支持	-	-	-
锁机制	行锁	表锁	表锁	表锁
B 数索引	支持	支持	支持	支持
哈希索引	-	-	支持	-
全文索引	-	支持	-	-
集群索引	支持	-	-	-
数据缓存	支持	-	支持	-
索引缓存	支持	支持	支持	支持
数据可压缩	-	支持	-	-
空间使用	高	低	N/A	低
内存使用	高	低	中等	低
批量插入速度	低	高	高	高
支持外键	支持	-	-	-

如果要提供提交、回滚和崩溃恢复能力的事务安全（ACID 兼容）能力，并要求实现并发控制，InnoDB 是个很好的选择。如果数据表主要用来插入和查询记录，则 MyISAM 引擎能提供较高的处

理效率；如果只是临时存放数据，数据量不大，并且不需要较高的数据安全性，可以选择将数据保存在内存中的 Memory 引擎，MySQL 中使用该引擎作为临时表，存放查询的中间结果。如果只有 INSERT 和 SELECT 操作，可以选择 Archive 引擎，Archive 存储引擎支持高并发的插入操作，但是本身并不是事务安全的。Archive 存储引擎非常适合存储归档数据，如记录日志信息可以使用 Archive 引擎。

　　使用哪一种引擎要根据需要灵活选择，一个数据库中多个表可以使用不同引擎，以满足各种性能和实际需求。使用合适的存储引擎，将会提高整个数据库的性能。

第 12 章

MySQL 分区和事务控制

从 MySQL 5.1 版本开始支持数据分区表（partitioned table），分区表的使用大大增加了 MySQL 执行效率，本章将讲解 MySQL 分区表的使用。另外，本章还将讲解 MySQL 数据库事务和分布式事务，其中包括分布式事务的原理和语法。MySQL 分布式事务涉及多个事务性的活动。

12.1 合并表

合并表是将许多个 MyISAM 表合并成一个虚表，类似于使用 UNION 语句将多个表合并，合并表不是真的创造一个真正的表，它就像一个用于放置相似表的容器。

下面是一个合并表的例子。

步骤 01 创建数据存储引擎是 MyISAM 类型的表 mtable1 和 mtable2，命令如下：

```
mysql> create table mtable1(
    data int not null primary key
    )engine=myisam;

mysql> create table mtable2(
    data int not null primary key
    )engine=myisam;
```

步骤 02 向表 mtable1 和 mtable2 中插入数据，命令如下：

```
mysql> insert into mtable1 values(1),(2),(3);
mysql> insert into mtable2 values(2),(3),(4);
```

步骤 03 使用 UNION 语句创建表 mtable1 和 mtable2 的合并表 mergtable，命令如下：

```
mysql> create table mergtable(
    data int not null primary key
    )engine=merge union=(mtable1,mtable2) insert_method=last;
```

上述代码中的 insert_method=last 的含义如下：

如果表 mergtable 中插入一条记录，那么就将这条记录插入到合并表所合并的最后一个表里面，以上例子就是将记录插入到 mtable2 表中。

步骤01 查询合并表 mergtable 的信息，结果如下：

```
mysql> select * from mergtable;
+------+
| data |
+------+
|    1 |
|    2 |
|    2 |
|    3 |
|    3 |
|    4 |
+------+
```

值得注意的是，合并表所包含的表列的数量和类型跟所合并的表的列的数量和类型都是一样的。同时也可以看到每个表的列有主键，这会导致合并表有重复的行，这是合并表的一个局限。

步骤02 直接插入数据到 mergtable 表中，命令如下：

```
mysql> insert into mergtable values(5);
```

步骤03 查询 mtable1 表中的数据是否发生发生变化，命令如下：

```
mysql> select * from mtable1;
+------+
| data |
+------+
|    1 |
|    2 |
|    3 |
+------+
```

从结果可以看出，数据没有发生变化。

步骤04 查询 mtable2 表的数据是否发生变化，命令如下：

```
mysql> select * from mtable2;
+------+
| data |
+------+
|    2 |
|    3 |
|    4 |
|    5 |
+------+
```

从结果可以看出，插入到合并表 mergtable 的一条数据记录已经插入到 mtable2 表中了。

步骤05 删除表 mtable1 和 mtabl2，命令如下：

```
mysql> drop table mtable1,mtable2;
```

```
Query OK, 0 rows affected (0.00 sec)
```

步骤06 查询 mergtable 表，发生错误，命令如下：

```
mysql> select * from mergtable;
ERROR 1168 (HY000): Unable to open underlying table which is differently defined or of non-MyISAM type or doesn't exist
```

MySQL 的合并表对性能有一定的影响，下面是一些需要注意的事项：

（1）合并表看上去是一个表，事实上是逐个打开各个子表，这样的情况下，可能会因为缓存过多的表而导致超过 MySQL 缓存的最大设置。

（2）创建合并表的 CREATE 语句不会检查子表是否兼容，如果创建了一个有效的合并表后对某个表进行了修改，那么合并表也会发生错误。

12.2 分区表

从 MySQL 5.1 版本开始支持数据表分区，通俗地讲，表分区是将一张大表根据条件分割成若干小表。例如，某用户表的记录超过了 600 万条，那么就可以根据入库日期或者所在地将表分区。

12.2.1 认识分区表

分区表通过一些特殊的语句创建独立的空间，事实上创建分区表的每个分区都是有索引的独立表。分区看上去像一个单独的表。

数据库分区是一种物理数据库设计技术，分区的主要目的是为了让某些特定的查询操作减少响应时间，同时对于应用来讲说分区完全是透明的。MySQL 的分区主要有两种形式：水平分区（Horizontal Partitioning）和垂直分区（Vertical Partitioning）。

（1）水平分区：根据表的行进行分割，这种形式的分区一定是通过表的某个属性作为分割的条件。例如，某张表里面数据日期为 2011 年的数据和日期为 2012 年的数据分割开，就可以采用这种分区形式。

（2）垂直分区：通过对表的垂直划分来减少目标表的宽度，即某些特定的列被划分到特定的分区。

下面介绍 MySQL 各种分区表常用的操作案例。

12.2.2 RANGE 分区

RANGE 分区使用 values less than 操作符来进行定义，把连续且不相互重叠的字段分配给分区。

下面通过示例演示创建 RANGE 分区，命令如下：

```
mysql> create table emp(
    empno varchar(20) not null,
    empname varchar(20),
    deptno int,
```

```
        birthdate date,
        salary int
        )
        partition by range(salary)
        (
        partition p1 values less than(1000),
        partition p2 values less than(2000),
        partition p3 values less than(3000)
        );
Query OK, 0 rows affected (0.01 sec)

mysql> insert into emp values(1000,'kobe',12,'1888-08-08',1500);
Query OK, 1 row affected (0.01 sec)

mysql> insert into emp values(1000,'kobe',12,'1888-08-08',3500);
ERROR 1526 (HY000): Table has no partition for value 3500
```

此时，按照工资级别（字段 salary）进行表分区，partition by range 语法类似于 switch…case 语法，如果 salary 小于 1000，数据存储在 p1 分区；如果 salary 小于 2000，数据存储在 p2 分区；如果 salary 小于 3000，数据存储在 p3 分区。

上面插入的第二条数据工资级别（字段 salary）为 3500，此时没有分区用来存储该范围的数据，所以发生了错误。为了解决这种问题，加入 "partition p4 values less than maxvalue" 语句即可，命令如下：

```
mysql> drop table emp;
Query OK, 0 rows affected (0.00 sec)

mysql> create table emp(
        empno varchar(20) not null,
        empname varchar(20),
        deptno int,
        birthdate date,
        salary int
        )
        partition by range(salary)
        (
        partition p1 values less than(1000),
        partition p2 values less than(2000),
        partition p3 values less than(3000),
        partition p4 values less than maxvalue
        );
Query OK, 0 rows affected (0.01 sec)

mysql> insert into emp values(1000,'kobe',12,'1888-08-08',1000);
Query OK, 1 row affected (0.00 sec)

mysql> insert into emp values(1000,'durant',12,'1888-08-08',3500);
Query OK, 1 row affected (0.00 sec)
```

maxvalue 表示最大的可能的整数值。值得注意的是，values less than 子句中使用一个表达式也可以，不过表达式结果不能为 NULL，下面按照日期进行分区，命令如下：

```
mysql> drop table emp;
```

```
Query OK, 0 rows affected (0.00 sec)
mysql> create table emp(
       empno varchar(20) not null,
       empname varchar(20),
       deptno int,
       birthdate date,
       salary int
       )
       partition by range(year(birthdate))(
       partition p0 values less than(1980),
       partition p1 values less than(1990),
       partition p2 values less than(2000),
       partition p3 values less than maxvalue
       );
Query OK, 0 rows affected (0.01 sec)
```

该方案中，生日 1980 年以前的员工信息存储在 p0 分区中，生日 1990 年以前的员工信息存储在 p1 分区中，生日 2000 年以前的员工信息存储在 p2 分区中，2000 年以后出生的员工信息存储在 p3 分区中。

RANGE 分区很有用，常常使用在以下几种情况。

（1）如果要删除某个时间段的数据时，只需要删除分区即可。例如，要删除 1980 年以前出生员工的所有信息，此时执行"alter table emp drop partition p0"的效率要比执行"delete from emp where year(birthdate)<=1980"高效得多。

（2）如果使用包含日期或者时间的列，可以考虑用到 RANGE 分区。

（3）经常运行直接依赖于分割表的列的查询。比如，当执行某个查询，如"select count（*） from emp where year(birthdate) = 1999 group by empno"，此时 MySQL 数据库可以很迅速地确定只有分区 p2 需要扫描，这是因为查询条件对于其他分区不符合。

12.2.3 LIST 分区

LIST 分区类似 RANGE 分区，它们的区别主要在于，LIST 分区中每个分区的定义和选择是基于某列的值从属于一个集合，而 RANGE 分区是从属于一个连续区间值的集合。

下面通过示例演示创建 RANGE 分区，命令如下：

```
mysql> create table employees(
       empname varchar(20),
       deptno int,
       birthdate date not null,
       salary int
       )
       partition by list(deptno)
       (
       partition p1 values in (10,20),
       partition p2 values in (30),
       partition p3 values in (40)
       );
Query OK, 0 rows affected (0.01 sec)
```

以上示例以部门编号划分分区，10 号部门和 20 号部门的员工信息存储在 p1 分区，30 号部门的员工信息存储在 p2 分区，40 号部门的员工信息存储在 p3 分区。同 RANG 分区一样，如果插入数据的部门编号不在分区值列表中时，那么 INSERT 插入操作将失败并报错。

12.2.4　HASH 分区

　　HASH 分区是基于用户定义的表达式的返回值来进行选择的分区，该表达式使用将要插入到表中的这些行的列值进行计算。这个函数可以包含 MySQL 中有效的、产生非负整数值的任何表达式。

　　HASH 分区主要用来确保数据在预先确定数目的分区中平均分布。在 RANGE 和 LIST 分区中，必须明确指定一个给定的列值或列值集合应该保存在哪个分区中；而在 HASH 分区中，MySQL 自动完成这些工作，用户所要做的只是为将要被哈希的列值指定一个列值或表达式，以及指定被分区的表将要被分割成的分区数量。

　　看下面的例子：

```
mysql> create table htable(
       id int,
       name varchar(20),
       birthdate date not null,
       salary int
       )
       partition by hash(year(birthdate))
       partitions 4;
Query OK, 0 rows affected (0.00 sec)
```

　　当使用了"partition by hash"时，MySQL 将基于用户函数结果的模数，来确定使用哪个编号的分区。将要保存记录的分区编号为 N=MOD(表达式, num)。如果表 htable 中插入一条 birthdate 为 "2010-09-23" 的记录，可以通过如下方法计算该记录的分区。

　　mod(year('2010-09-23'),4)=mode(2010,4) =2

　　此时，该条记录的数据将会存储在分区编号为 2 的分区空间。

12.2.5　线性 HASH 分区

　　线性 HASH 分区和 HASH 分区的区别在于，线性哈希功能使用的一个线性的 2 的幂运算法则，而 HASH 分区使用的是哈希函数的模数。

　　先看下面的例子：

```
mysql> create table htable2(
       id int not null,
       name varchar(20),
       hired date not null default '1999-09-09',
       deptno int
       )
       partition by linear hash(year(hired))
       partitions 4;
Query OK, 0 rows affected (0.03 sec)
```

　　如果表 htable2 中插入一条 hired 为 "2010-09-23" 的记录，记录将要保存到的分区是 num 个分区中的分区 N。下列步骤通过哈希函数计算 N。

步骤01 找到下一个大于 num 的 2 的幂,把这个值称作 V,可以通过下面的公式得到。V=POWR(2,CEILING(LOG(2,num)))。假设 num 的值是 13,那么 LOG(2,13)就是 3.70043。CEILING(3.70043)就是 4,则 V=POWER(2,4),即等于 16。

步骤02 计算 N=F(column_list) & (V – 1) 此时当 N>=num 时,V=CEIL(V/2),此时 N=N & (V-1)。

下面示例通过线性哈希分区算法来计算分区 N 的值。线性哈希分区表 t1 通过下面的语句创建:

```
mysql> create table th1(
    col1 int,
    col2 char(5),
    col3 date
    )
    partition by linear hash( year(col3) )
    partitions 6;
Query OK, 0 rows affected (0.59 sec)
```

现在假设要插入两条记录到表 th1 中,其中一条记录 col3 列的值为 "2003-04-14",另一条记录 cols 列值为 "1998-10-19"。第一条记录要保存到的分区计算过程如下:

记录将要保存到 num 分区中的分区 N。假设 num 是 7 个分区,表 t1 使用线性 HASH 分区且有 4 个分区。

```
V = POWR(2,CEILING(LOG(2,num)))
V = POWR(2,CEILING(LOG(2,7))) = 8
N = YEAR('2003-04-14') & (8 - 1)
  = 2003 & 7
  = 3
```

N 的值是 3,很显然 3>=4 不成立,所以附件条件不执行,第一条记录的信息将存储在 3 号分区中。

第二条记录将要保存的分区序号计算如下:

```
V = POWR(2,CEILING(LOG(2,num)))
V = POWR(2,CEILING(LOG(2,7))) = 8
  = YEAR('1998-10-19') & (8 - 1)
  = 1998 & 7
  = 6
```

N 的值是 6,很显然 6>=4 成立,所以附件条件会执行。

```
V = CEIL(6/2) = 3
N = N & (V-1)
  = 6 & 2
  = 2
```

此时发现 2 >= 4 不成立,记录将被保存到#2 分区中。线性哈希分区的优点在于增加、删除、合并和拆分分区将变得更加快捷,有利于处理含有极其大量(1000G)数据的表。它的缺点在于,与使用常规 HASH 分区得到的数据分布相比,各个分区间数据的分布可能不大均衡。

12.2.6 KEY 分区

类似于 HASH 分区,区别在于 KEY 分区只支持计算一列或多列,且 MySQL 服务器提供其自身的哈希函数,这些函数是基于与 PASSWORD()一样的运算法则。

看下面的例子:

```
mysql> create table keytable(
    id int,
    name varchar(20) not null,
    deptno int,
    birthdate date not null,
    salary int
    )
    partition by key(birthdate)
    partitions 4;
Query OK, 0 rows affected (0.11 sec)
```

在 KEY 分区中使用关键字 LINEAR 和在 HASH 分区中使用具有同样的作用,分区的编号是通过 2 的幂算法得到,而不是通过模数算法。

12.2.7 复合分区

复合分区是分区表中每个分区的再次分割,子分区既可以使用 HASH 分区,也可以使用 KEY 分区。这也被称为子分区。

复合分区需要注意以下问题:

(1)如果一个分区中创建了复合分区,则其他分区也要有复合分区。
(2)如果创建了复合分区,则每个分区中的复合分区数必有相同。
(3)同一分区内的复合分区,名字不相同,不同分区内的复合分区名字可以相同。

下面通过示例演示不同的复合分区的创建方法。

1. RANGE-HASH 复合分区

创建 RANGE-HASH 复合分区的命令如下:

```
mysql> create table rhtable(
    no varchar(20) not null,
    name varchar(20),
    deptno int,
    birthdate date not null,
    salary int
    )
    partition by range(salary)
    subpartition by hash(year(birthdate))
    subpartitions 3
    (
    partition p1 values less than (2000),
    partition p2 values less than maxvalue
    );
Query OK, 0 rows affected (0.23 sec)
```

2. RANGE-KEY 复合分区

创建 RANGE-KEY 复合分区的命令如下:

```
mysql> create table rktable(
    no varchar(20) not null,
    name varchar(20),
    deptno int,
    birth date not null,
    salary int
    )
    partition by range(salary)
    subpartition by key(birth)
    subpartitions 3
    (
    partition p1 values less than (2000),
    partition p2 values less than maxvalue
    );
Query OK, 0 rows affected (0.07 sec)
```

3. LIST-HASH 复合分区

创建 LIST-HASH 复合分区的命令如下：

```
mysql> create table lhtable(
    no varchar(20) not null,
    name varchar(20),
    deptno int,
    birth date not null,
    salary int
    )
    partition by list(deptno)
    subpartition by hash( year(birth) )
    subpartitions 3
    (
    partition p1 values in (10),
    partition p2 values in (20)
    );
Query OK, 0 rows affected (0.08 sec)
```

4. LIST-KEY 复合分区

创建 LIST-KEY 复合分区的命令如下：

```
mysql> create table lktable(
    no varchar(20) not null,
    name varchar(20),
    deptno int,
    birthdate date not null,
    salary int
    )
    partition by list(deptno)
    subpartition by key(birthdate)
    subpartitions 3
    (
    partition p1 values in (10),
    partition p2 values in (20)
    );
Query OK, 0 rows affected (0.09 sec)
```

12.3　事务控制

MySQL 通过 SET AUTOCOMMIT、START TRANSACTION、COMMIT 和 ROLLBACK 等语句控制本地事务，具体语法如下：

```
START TRANSACTION | BEGIN [WORK];
COMMIT [WORK] [AND [NO] CHAIN] [[NO] RELEASE]
ROLLBACK [WORK] [AND [NO] CHAIN] [[NO] RELEASE]
SET AUTOCOMMIT = {0|1}
```

其中 START TRANSACTION 表示开启事务、COMMIT 表示提交事务、ROLLBACK 表示回滚事务、SET AUTOCOMMIT 用于设置是否自动提交事务。

默认情况下，MySQL 事务是自动提交的，如果需要通过明确的 COMMIT 和 ROLLBACK 在提交和回滚事务，那么需要通过明确的事务控制命令来开始事务，这是和 Oracle 的事务管理有明显不同的地方。如果应用从 Oracle 数据库迁移到 MySQL 数据库，则需要确保应用中是否对事务进行了明确的管理。

MySQL 的 AUTOCOMMIT（自动提交）默认是开启，对 MySQL 的性能有一定影响，举个例子来说，如果用户插入了 1000 条数据，MySQL 会提交事务 1000 次。这时可以把自动提交关闭掉，通过程序来控制，只要一次提交事务就可以了。

可以通过如下方式关闭自动提交功能，命令如下：

```
mysql> set @@autocommit=0;
Query OK, 0 rows affected (0.00 sec)
```

查看自动提交功能是否被关闭，命令如下：

```
mysql> show variables like "autocommit";
+---------------+-------+
| Variable_name | Value |
+---------------+-------+
| autocommit    | ON    |
+---------------+-------+
1 row in set (0.02 sec)
```

下面通过两个 Session（Session1 和 Session2）来理解事务控制的过程，具体操作步骤如下：

步骤 01 在 Session1 中，打开自动提交事务功能，然后创建表 ctable 并插入两条记录。命令如下：

```
mysql> set @@autocommit=1;
Query OK, 0 rows affected (0.00 sec)

mysql> create table ctable (data INT);

mysql> insert into ctable values(1);
Query OK, 1 row affected (0.02 sec)

mysql> insert into ctable values(2);
Query OK, 1 row affected (0.00 sec)
```

步骤 02 在 Session2 中，打开自动提交事务功能，然后查询表 ctable，命令如下：

```
mysql> set @@autocommit=1;
Query OK, 0 rows affected (0.00 sec)

mysql> select * from ctable;
+------+
| data |
+------+
|    1 |
|    2 |
+------+
2 rows in set (0.00 sec)
```

步骤03 在 Session1 中，关闭自动提交事务功能，然后向表 ctable 中插入两条记录，命令如下：

```
mysql> set @@autocommit=0;
Query OK, 0 rows affected (0.00 sec)

mysql> insert into ctable values(3);
Query OK, 1 row affected (0.00 sec)

mysql> insert into ctable values(4);
Query OK, 1 row affected (0.00 sec)
```

步骤04 在 Session2 中，查询表 ctable，命令如下：

```
mysql> select * from ctable;
+------+
| data |
+------+
|    1 |
|    2 |
+------+
2 rows in set (0.00 sec)
```

从结果可以看出，在 Session1 中新插入的两条记录没有查询出来。

步骤05 在 Session1 中，提交事务，命令如下：

```
mysql> commit;
Query OK, 0 rows affected (0.01 sec)
```

步骤06 在 Session2 中，查询表 ctable，命令如下：

```
mysql> select * from ctable;
+------+
| data |
+------+
|    1 |
|    2 |
|    3 |
|    4 |
+------+
4 rows in set (0.00 sec)
```

如果在表的锁定期间，如果使用 START TRANSACTION 命令开启一个新的事务，会造成一个隐含的 unlock tables 被执行，该操作存在一定的隐患。

下面通过一个案例来理解。

首先创建数据表 nbaplayer，命令如下：

```
create table nbaplayer(
        id int,
        name varchar(20),
        salary int);
```

步骤01 在 Session1 中，查询 nbaplayer 表，结果为空，命令如下：

```
mysql> select * from nbaplayer;
Empty set (0.00 sec)
```

步骤02 在 Session2 中，查询 nbaplayer 表，结果为空，命令如下：

```
mysql> select * from nbaplayer;
Empty set (0.00 sec)
```

步骤03 在 Session1 中，对表 nbaplayer 加写锁，命令如下：

```
mysql> lock table nbaplayer write;
Query OK, 0 rows affected (0.00 sec)
```

步骤04 在 Session2 中，向表 nbaplayer 中增加一条记录，命令如下：

```
mysql> insert into nbaplayer values(1,'kobe',10000);
等待
```

步骤05 在 Session1 中，插入一条记录，命令如下：

```
mysql> insert into nbaplayer values(2,'durant',40000);
Query OK, 1 row affected (0.02 sec)
```

步骤06 在 Session1 中，回滚刚才插入的记录，命令如下：

```
mysql> rollback;
Query OK, 0 rows affected (0.00 sec)
```

步骤07 在 Session1 中，开启一个新的事务，命令如下：

```
mysql> start transaction;
Query OK, 0 rows affected (0.00 sec)
```

步骤08 在 Session2 中，表锁被释放，此时成功增加该条记录，结果如下：

```
mysql> insert into nbaplayer values(1,'kobe',10000);
Query OK, 1 row affected (2 min 32.99 sec)
```

步骤09 在 Session2 中，查询 nbaplayer，命令如下：

```
mysql> select * from nbaplayer;
+------+--------+--------+
| id   | name   | salary |
+------+--------+--------+
|    2 | durant |  40000 |
|    1 | kobe   |  10000 |
+------+--------+--------+
2 rows in set (0.00 sec)
```

从结果可以看出，此时发现 Session1 的回滚操作并没有执行成功。

MySQL 提供的 LOCK IN SHARE MODE 锁，可以保证停止任何对它要读的数据行的更新或者删除操作。下面通过一个例子来理解。

步骤 01 在 Session1 中，开启一个新的事务，然后查询数据表 nbaplayer 的 salary 列的最大值，命令如下：

```
mysql> begin;
Query OK, 0 rows affected (0.00 sec)
mysql> select max(salary) from nbaplayer lock in share mode;
+-------------+
| max(salary) |
+-------------+
|       40000 |
+-------------+
1 row in set (0.00 sec)
```

步骤 02 在 Session2 中，尝试做更新操作，命令如下：

```
mysql> update nbaplayer set salary = 90000 where id = 1;
等待
```

步骤 03 在 Session1 中，提交事务，命令如下：

```
mysql> commit;
Query OK, 0 rows affected (0.00 sec)
```

步骤 04 在 Session2 中，等 Session1 的事务提交后，此时更新操作成功执行，结果如下：

```
mysql> update nbaplayer set salary = 90000 where id = 1;
Query OK, 1 row affected (16.25 sec)
Rows matched: 1  Changed: 1  Warnings: 0
```

12.4 MySQL 分布式事务

在 MySQL 中，使用分布式事务的应用程序涉及一个或多个资源管理器和一个事务管理器，分布式事务的事务参与者、资源管理器、事务管理器等位于不同的节点上，这些不同的节点相互协作共同完成一个具有逻辑完整性的事务。分布式事务的主要作用在于确保事务的一致性和完整性。

12.4.1 分布式事务的原理

资源管理器（Resource Manager，简称 RM）用于向事务提供资源，同时还具有管理事务提交或回滚的能力。数据库就是一种资源管理器。

事务管理器（Transaction Manager，简称 TM）用于和每个资源管理器通信，协调并完成事务的处理。一个分布式事务中各个事务均是分布式事务的"分支事务"。分布式事务和各分支通过一种命名方法进行标识。

MySQL 执行分布式事务，首先要考虑网络中涉及多少个事务管理器，MySQL 分布式事务管理，简单地讲就是同时管理若干事务管理器事务的一个过程，每个资源管理器的事务当执行到被提交或者被回滚的时候，根据每个资源管理器报告的有关情况，决定是否将这些事务作为一个原子性的操作执行全部提交或者全部回滚。因为 MySQL 分布式事务同时涉及多台 MySQL 服务器，所以在管理分布式事务的时候，必须考虑网络可能存在的故障问题。

用于执行分布式事务的过程使用两个阶段。

（1）第一阶段：所有的分支被预备。它们被事务管理器告知要准备提交，每个分支资源管理器记录分支的行动并指示任务的可行性。

（2）第二阶段：事务管理器告知资源管理器是否需要提交或者回滚。如果预备分支时，所有的分支指示它们将能够提交，那么所有的分支被告知提交。如果有一个分支出错，那么就要全部都要回滚。特性情况下，只要一个分支的时候，第二阶段则被省略。

分布式事务的主要作用在于确保事务的一致性和完整性。它利用分布式的计算机环境，将多个事务性的活动合并成一个事务单元，这些事务组合在一起构成原子操作，这些事务的活动要么一起执行并提交事务，要么回滚所有的操作，从而保证了多个活动之间的一致性和完整性。

12.4.2 分布式事务的语法

在 MySQL 中，执行分布式事务的语法格式如下：

```
XA {START|BEGIN} xid [JOIN|RESUME]
```

XA START xid 表示用于启动一个事务标识为 xid 的事务。xid 分布式事务表示的值，既可以由客户端提供，也可以由 MySQL 服务器生成。

结束分布式事务的语法格式如下：

```
XA END xid [SUSPEND [FOR MIGRATE]]
```

其中 xid 包括：gtrid [, bqual [, formatID]]，含义如下：

（1）gtrid 是一个分布式事务标识符。

（2）bqual 表示一个分支限定符，默认值是空字符串。对于一个分布式事务中的每个分支事务，bqual 值必须是唯一的。

（3）formatID 是一个数字，用于标识由 gtrid 和 bqual 值使用的格式，默认值为 1。

```
XA PREPARE xid
```

该命令使事务进入 PREPARE 状态，也就是两阶段提交的第一个阶段。

```
XA COMMIT xid [ONE PHASE]
```

该命令用来提交具体的分支事务。

```
XA ROLLBACK xid
```

该命令用来回滚具体的分支事务。也就是两阶段提交的第二个提交阶段，分支事务被实际的提交或者回滚。

```
XA RECOVER
```

该命令用于返回数据库中处于 PREPARE 状态的分支事务的详细信息。

分布式的关键在于如何确保分布式事务的完整性，以及在某个分支出现问题时如何解决故障。

分布式事务的相关命令就是提供给应用如何在多个独立的数据库之间进行分布式事务的管理，包括启动一个分支事务、使事务进入准备阶段，以及事务的实际提交、回滚操作等。

MySQL 分布式事务分为两类，内部分布式事务和外部分布式事务。内部分布式事务用于同一实例下跨多个数据引擎的事务，由二进制日志作为协调者；而外部分布式事务用于跨多个 MySQL 实例的分布式事务，需要应用层介入作为协调者，全局提交还是回滚，都是由应用层决定的，对应用层的要求比较高。

MySQL 分布式事务在某些特殊的情况下会存在一定的漏洞，当一个事务分支在 PREPARE 状态的时候失去了连接，在服务器重启以后，可以继续对分支事务进行提交或者回滚操作，没有写入二进制日志，这将导致事务部分丢失或者主从数据库不一致。

第 13 章

MySQL 性能优化

MySQL 性能优化就是通过合理安排资源，调整系统参数，使 MySQL 运行更快、更节省资源。MySQL 性能优化包括查询速度优化、数据库结构优化等。本章将为读者讲解以下几个内容：性能优化、查询优化、数据库结构优化、临时表性能优化、创建全局通用表空间、隐藏和显示索引。

13.1 优化简介

优化 MySQL 数据库是数据库管理员和数据库开发人员的必备技能。MySQL 优化，一方面是找出系统的瓶颈，提高 MySQL 数据库整体的性能；另一方面需要合理的结构设计和参数调整，以提高用户操作响应的速度；同时还要尽可能节省系统资源，以便系统可以提供更大负荷的服务。本节将为读者介绍 MySQL 数据库优化的基本知识。

MySQL 数据库优化是多方面的，原则是减少系统的瓶颈，减少资源的占用，增加系统的响应速度。例如，通过优化文件系统，提高磁盘 I\O 的读写速度；通过优化操作系统调度策略，提高 MySQL 在高负荷情况下的负载能力；优化表结构、索引、查询语句等使查询响应更快。

在 MySQL 中，可以使用 SHOW STATUS 语句查询一些 MySQL 数据库的性能参数。SHOW STATUS 语句的语法如下：

```
SHOW STATUS LIKE 'value';
```

其中，value 是要查询的参数值，一些常用的性能参数如下：

- Connections：连接 MySQL 服务器的次数。
- Uptime：MySQL 服务器的上线时间。
- Slow_queries：慢查询的次数。
- Com_select：查询操作的次数。
- Com_insert：插入操作的次数。

- Com_update：更新操作的次数。
- Com_delete：删除操作的次数。

查询 MySQL 服务器的连接次数，可以执行如下语句：

```
SHOW STATUS LIKE 'Connections';
```

查询 MySQL 服务器的慢查询次数，可以执行如下语句：

```
SHOW STATUS LIKE 'Slow_queries';
```

查询其他参数的方法和上面两个参数的查询方法相同。慢查询次数参数可以结合慢查询日志，找出慢查询语句，然后针对慢查询语句进行表结构优化或者查询语句优化。

13.2 优化查询

查询是数据库中最频繁的操作，提高查询速度可以有效地提高 MySQL 数据库的性能。本节将为读者介绍优化查询的方法。

13.2.1 分析查询语句

通过对查询语句的分析，可以了解查询语句的执行情况，找出查询语句执行的瓶颈，从而优化查询语句。MySQL 中提供了 EXPLAIN 语句和 DESCRIBE 语句，用来分析查询语句。本小节将为读者介绍使用 EXPLAIN 语句和 DESCRIBE 语句分析查询语句的方法。

EXPLAIN 语句的基本语法如下：

```
EXPLAIN [EXTENDED] SELECT select_options
```

使用 EXTENED 关键字，EXPLAIN 语句将产生附加信息。select_options 是 SELECT 语句的查询选项，包括 FROM WHERE 子句等。

执行该语句，可以分析 EXPLAIN 后面 SELECT 语句的执行情况，并且能够分析出所查询表的一些特征。

【例 13.1】使用 EXPLAIN 语句来分析一个查询语句，可执行如下语句：

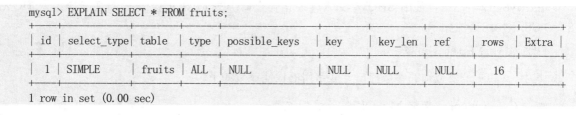

下面对查询结果进行解释。

- id：SELECT 识别符。这是 SELECT 的查询序列号。
- select_type：表示 SELECT 语句的类型。它可以是以下几种取值：SIMPLE 表示简单查询，其中不包括连接查询和子查询；PRIMARY 表示主查询，或者是最外层的查询语句；UNION 表示连接查询的第 2 个或后面的查询语句；DEPENDENT UNION，连接查询中的第 2 个或后面的 SELECT 语句，取决于外面的查询；UNION RESULT，连接查询的结果；SUBQUERY，子查询中的第 1 个 SELECT 语句；DEPENDENT SUBQUERY，子查询中的第 1 个 SELECT，取决于外面的查询；DERIVED，导出表的 SELECT（FROM 子句的子查询）。

- table：表示查询的表。
- type：表示表的连接类型。下面按照从最佳类型到最差类型的顺序给出各种连接类型。

（1）system

该表是仅有一行的系统表。这是 const 连接类型的一个特例。

（2）const

数据表最多只有一个匹配行，将在查询开始时被读取，并在余下的查询优化中作为常量对待。const 表查询速度很快，因为它们只读取一次。const 用于使用常数值比较 PRIMARY KEY 或 UNIQUE 索引的所有部分的场合。

在下面的查询中，tbl_name 可用于 const 表：

```
SELECT * from tbl_name WHERE primary_key=1;
SELECT * from tbl_name WHERE primary_key_part1=1AND primary_key_part2=2;
```

（3）eq_ref

对于每个来自前面的表的行组合，从该表中读取一行。当一个索引的所有部分都在查询中使用并且索引是 UNIQUE 或 PRIMARY KEY 时，即可使用这种类型。

eq_ref 可以用于使用"="操作符比较带索引的列。比较值可以为常量或一个在该表前面所读取的表的列的表达式。

在下面的例子中，MySQL 可以使用 eq_ref 连接来处理 ref_tables：

```
SELECT * FROM ref_table,other_table
WHERE ref_table.key_column=other_table.column;
SELECT * FROM ref_table,other_table
WHERE ref_table.key_column_part1=other_table.column
AND ref_table.key_column_part2=1;
```

（4）ref

对于来自前面的表的任意行组合，将从该表中读取所有匹配的行。这种类型用于索引既不是 UNIQUE 也不是 PRIMARY KEY 的情况，或者查询中使用了索引列的左子集，即索引中左边的部分列组合。ref 可以用于使用=或<=>操作符带索引的列。

在下面的例子中，MySQL 可以使用 ref 连接来处理 ref_tables：

```
SELECT * FROM ref_table WHERE key_column=expr;

SELECT * FROM ref_table,other_table
WHERE ref_table.key_column=other_table.column;

SELECT * FROM ref_table,other_table
WHERE ref_table.key_column_part1=other_table.column
AND ref_table.key_column_part2=1;
```

（5）ref_or_null

该连接类型如同 ref，但是添加了 MySQL 可以专门搜索包含 NULL 值的行。在解决子查询中经常使用该连接类型的优化。

在下面的例子中，MySQL 可以使用 ref_or_null 连接来处理 ref_tables：

```
SELECT * FROM ref_table
WHERE key_column=expr OR key_column IS NULL;
```

（6）index_merge

该连接类型表示使用了索引合并优化方法。在这种情况下，key 列包含了使用的索引的清单，key_len 包含了使用索引的最长关键元素。

（7）unique_subquery

该类型替换了下面形式的 IN 子查询的 ref：

```
value IN (SELECT primary_key FROM single_table WHERE some_expr)
```

unique_subquery 是一个索引查找函数，可以完全替换子查询，效率更高。

（8）index_subquery

该连接类型类似于 unique_subquery，可以替换 IN 子查询，但只适合下列形式的子查询中的非唯一索引：

```
value IN (SELECT key_column FROM single_table WHERE some_expr)
```

（9）range

只检索给定范围的行，使用一个索引来选择行。key 列显示使用了哪个索引。key_len 包含所使用索引的最长关键元素。

当使用=、<>、>、>=、<、<=、IS NULL、<=>、BETWEEN 或者 IN 操作符用常量比较关键字列时，类型为 range。

下面介绍几种检索指定行情况：

```
SELECT * FROM tbl_name
WHERE key_column = 10;

SELECT * FROM tbl_name
WHERE key_column BETWEEN 10 and 20;
SELECT * FROM tbl_name
WHERE key_column IN (10,20,30);

SELECT * FROM tbl_name
WHERE key_part1= 10 AND key_part2 IN (10,20,30);
```

（10）index

该连接类型与 ALL 相同，除了只扫描索引树。它通常比 ALL 快，因为索引文件通常比数据文件小。

（11）ALL

对于前面的表的任意行组合，进行完整的表扫描。如果表是第一个没标记 const 的表，这样不

- possible_keys：指出 MySQL 能使用哪个索引在该表中找到行。如果该列是 NULL，则没有相关的索引。在这种情况下，可以通过检查 WHERE 子句看它是否引用某些列或适合索引的列来提高查询性能。如果是这样，可以创建适合的索引来提高查询的性能。
- key：表示查询实际使用到的索引，如果没有选择索引，该列的值是 NULL。要想强制 MySQL 使用或忽视 possible_keys 列中的索引，在查询中使用 FORCE INDEX、USE INDEX 或者 IGNORE INDEX。参见 SELECT 语法。
- key_len：表示 MySQL 选择的索引字段按字节计算的长度，如果键是 NULL，则长度为 NULL。注意，通过 key_len 值可以确定 MySQL 将实际使用一个多列索引中的几个字段。
- ref：表示使用哪个列或常数与索引一起来查询记录。
- rows：显示 MySQL 在表中进行查询时必须检查的行数。
- Extra：表示 MySQL 在处理查询时的详细信息。

DESCRIBE 语句的使用方法与 EXPLAIN 语句是一样的，并且分析结果也是一样的。DESCRIBE 语句的语法形式如下：

```
DESCRIBE SELECT select_options
```

DESCRIBE 可以缩写成 DESC。

13.2.2 索引对查询速度的影响

MySQL 中提高性能的一个有效方式就是对数据表设计合理的索引。索引提供了高效访问数据的方法，并且可加快查询的速度，因此，索引对查询的速度有着至关重要的影响。使用索引可以快速地定位表中的某条记录，从而提高数据库查询的速度、提高数据库的性能。本小节将为读者介绍索引对查询速度的影响。

如果查询时没有使用索引，查询语句将扫描表中的所有记录。在数据量大的情况下，这样查询的速度会很慢。如果使用索引进行查询，查询语句可以根据索引快速定位到待查询记录，从而减少查询的记录数，达到提高查询速度的目的。

【例 13.2】下面是查询语句中不使用索引和使用索引的对比。首先，分析未使用索引时的查询情况，EXPLAIN 语句执行如下：

```
mysql> EXPLAIN SELECT * FROM fruits WHERE f_name='apple';
+----+-------------+--------+------+---------------+------+---------+------+------+-------------+
| id | select_type | table  | type | possible_keys | key  | key_len | ref  | rows | Extra       |
+----+-------------+--------+------+---------------+------+---------+------+------+-------------+
|  1 | SIMPLE      | fruits | ALL  | NULL          | NULL | NULL    | NULL |   15 | Using where |
+----+-------------+--------+------+---------------+------+---------+------+------+-------------+
1 row in set (0.00 sec)
```

可以看到，rows 列的值是 15，说明 "SELECT * FROM fruits WHERE f_name='apple';" 这个查询语句扫描了表中的 15 条记录。

然后，在 fruits 表的 f_name 字段上加上索引。执行添加索引的语句及结果如下：

```
mysql> CREATE INDEX index_name ON fruits(f_name);
Query OK, 0 rows affected (0.04 sec)
Records: 0  Duplicates: 0  Warnings: 0
```

现在，再分析上面的查询语句。执行的 EXPLAIN 语句及结果如下：

```
mysql> EXPLAIN SELECT * FROM fruits WHERE f_name='apple';
+----+-------------+--------+------+---------------+------------+---------+-------+------+-------------+
| id | select_type | table  | type | possible_keys | key        | key_len | ref   | rows | Extra       |
+----+-------------+--------+------+---------------+------------+---------+-------+------+-------------+
|  1 | SIMPLE      | fruits | ref  | index_name    | index_name | 255     | const |    1 | Using where |
+----+-------------+--------+------+---------------+------------+---------+-------+------+-------------+
1 row in set (0.00 sec)
```

结果显示，rows 列的值为 1，表示这个查询语句只扫描了表中的一条记录，其查询速度自然比扫描 15 条记录快；而且 possible_keys 和 key 的值都是 index_name，说明查询时使用了 index_name 索引。

13.2.3　使用索引查询

索引可以提高查询的速度，但并不是使用带有索引的字段查询时索引都会起作用。本小节将向读者介绍索引的使用。

使用索引有几种特殊情况，在这些情况下使用带有索引的字段查询时，有可能索引并没有起到提高查询速度的作用。下面重点介绍这几种特殊情况。

1. 使用 LIKE 关键字的查询语句

在使用 LIKE 关键字进行查询的查询语句中，如果匹配字符串的第一个字符为"%"，索引不会起作用。只有"%"不在第一个位置，索引才会起作用。下面将举例说明。

【例 13.3】查询语句中使用 LIKE 关键字，并且匹配的字符串中含有"%"字符，EXPLAIN 语句执行如下：

```
mysql> EXPLAIN SELECT * FROM fruits WHERE f_name like '%x';
+----+-------------+--------+------+---------------+------+---------+------+------+-------------+
| id | select_type | table  | type | possible_keys | key  | key_len | ref  | rows | Extra       |
+----+-------------+--------+------+---------------+------+---------+------+------+-------------+
|  1 | SIMPLE      | fruits | ALL  | NULL          | NULL | NULL    | NULL |   16 | Using where |
+----+-------------+--------+------+---------------+------+---------+------+------+-------------+
1 row in set (0.00 sec)

mysql> EXPLAIN SELECT * FROM fruits WHERE f_name like 'x%';
+----+-------------+--------+-------+---------------+------------+---------+------+------+-------------+
| id | select_type | table  | type  | possible_keys | key        | key_len | ref  | rows | Extra       |
+----+-------------+--------+-------+---------------+------------+---------+------+------+-------------+
|  1 | SIMPLE      | fruits | range | index_name    | index_name | 150     | NULL |    4 | Using where |
+----+-------------+--------+-------+---------------+------------+---------+------+------+-------------+
1 row in set (0.00 sec)
```

已知 f_name 字段上有索引 index_name。第 1 个查询语句执行后，rows 列的值为 16，表示这次查询过程中扫描了表中所有的 16 条记录；第 2 个查询语句执行后，rows 列的值为 4，表示这次查询过程扫描了 4 条记录。第 1 个查询语句中的索引没有起作用,因为第 1 个查询语句中 LIKE 关键字后的字符串以"%"开头,而第 2 个查询语句使用了索引 index_name。

2. 使用多列索引的查询语句

MySQL 可以为多个字段创建索引。一个索引可以包括 16 个字段。对于多列索引,只有查询条件中使用了这些字段中的第 1 个字段时,索引才会被使用。

【例 13.4】在表 fruits 中 f_id、f_price 字段上创建多列索引,验证多列索引的使用情况。

```
mysql> CREATE INDEX index_id_price ON fruits(f_id, f_price);
Query OK, 0 rows affected (0.39 sec)
Records: 0  Duplicates: 0  Warnings: 0
mysql> EXPLAIN SELECT * FROM fruits WHERE f_id='12';
+----+-------------+--------+-------+----------------------+---------+---------+-------+------+-------+
| id | select_type | table  | type  | possible_keys        | key     | key_len | ref   | rows | Extra |
+----+-------------+--------+-------+----------------------+---------+---------+-------+------+-------+
|  1 | SIMPLE      | fruits | const | PRIMARY,index_id_price | PRIMARY | 20      | const |    1 |       |
+----+-------------+--------+-------+----------------------+---------+---------+-------+------+-------+
1 row in set (0.00 sec)

mysql> EXPLAIN SELECT * FROM fruits WHERE f_price=5.2;
+----+-------------+--------+------+---------------+------+---------+------+------+-------------+
| id | select_type | table  | type | possible_keys | key  | key_len | ref  | rows | Extra       |
+----+-------------+--------+------+---------------+------+---------+------+------+-------------+
|  1 | SIMPLE      | fruits | ALL  | NULL          | NULL | NULL    | NULL |   16 | Using where |
+----+-------------+--------+------+---------------+------+---------+------+------+-------------+
1 row in set (0.00 sec)
```

从第 1 条语句查询结果可以看出,"f_id='12'"的记录有 1 条。第 1 条语句共扫描了 1 条记录,并且使用了索引 index_id_price。从第 2 条语句查询结果可以看出,rows 列的值是 16,说明查询语句共扫描了 16 条记录,并且 key 列值为 NULL,说明"SELECT * FROM fruits WHERE f_price=5.2;"语句并没有使用索引。因为 f_price 字段是多列索引的第 2 个字段,只有查询条件中使用了 f_id 字段,才会使 index_id_price 索引起作用。

3. 使用 OR 关键字的查询语句

查询语句的查询条件中只有 OR 关键字,且 OR 前后的两个条件中的列都是索引时,查询中才使用索引,否则查询将不使用索引。

【例 13.5】查询语句使用 OR 关键字的情况:

```
mysql> EXPLAIN SELECT * FROM fruits WHERE f_name='apple' or s_id=101 \G
*** 1. row ***
           id: 1
  select_type: SIMPLE
        table: fruits
         type: ALL
possible_keys: index_name
          key: NULL
      key_len: NULL
          ref: NULL
         rows: 16
```

```
            Extra: Using where
1 row in set (0.00 sec)

mysql> EXPLAIN SELECT * FROM fruits WHERE f_name='apple' or f_id='l2' \G
*** 1. row ***
           id: 1
  select_type: SIMPLE
        table: fruits
         type: index_merge
possible_keys: PRIMARY,index_name,index_id_price
          key: index_name,PRIMARY
      key_len: 510,20
          ref: NULL
         rows: 2
        Extra: Using union(index_name,PRIMARY); Using where
1 row in set (0.00 sec)
```

因为 s_id 字段上没有索引，所以第 1 条查询语句没有使用索引，总共查询了 16 条记录；第 2 条查询语句使用了 f_name 和 f_id 这两个索引，因为 id 字段和 name 字段上都有索引，所以查询的记录数为 2 条。

13.2.4 优化子查询

MySQL 从 4.1 版本开始支持子查询，使用子查询可以进行 SELECT 语句的嵌套查询，即一个 SELECT 查询的结果作为另一个 SELECT 语句的条件。子查询可以一次性完成很多逻辑上需要多个步骤才能完成的 SQL 操作。子查询虽然可以使查询语句很灵活，但执行效率不高。执行子查询时，MySQL 需要为内层查询语句的查询结果建立一个临时表。然后外层查询语句从临时表中查询记录。查询完毕后，再撤销这些临时表。因此，子查询的速度会受到一定的影响。如果查询的数据量比较大，这种影响就会随之增大。

在 MySQL 中，可以使用连接（JOIN）查询来替代子查询。连接查询不需要建立临时表，其速度比子查询要快，如果查询中使用索引，性能会更好。连接之所以更有效率，是因为 MySQL 不需要在内存中创建临时表来完成查询工作。

13.3 优化数据库结构

一个好的数据库设计方案对于提高数据库的性能常常会起到事半功倍的效果。合理的数据库结构不仅可以使数据库占用更小的磁盘空间，而且能够使查询速度更快。数据库结构的设计，需要考虑数据冗余、查询和更新的速度、字段的数据类型是否合理等多方面的内容。本节将为读者介绍优化数据库结构的方法。

13.3.1 将字段很多的表分解成多个表

对于字段较多的表，如果有些字段的使用频率很低，可以将这些字段分离出来，形成新表。因为当一个表的数据量很大时，会由于使用频率低的字段的存在而变慢。本小节将为读者介绍这种优化表的方法。

【例 13.6】假设会员表存储会员登录认证信息，该表中有很多字段，如 id、姓名、密码、地址、电话、个人描述字段。其中，地址、电话、个人描述等字段并不常用。可以将这些不常用字段分离出来组成另外一个表，将这个表取名叫 members_detail。表中有 member_id、address、telephone、description 等字段。其中，member_id 是会员编号，address 字段存储地址信息，telephone 字段存储电话信息，description 字段存储会员个人描述信息。这样就把会员表分成两个表，分别为 members 表和 members_detail 表。

创建这两个表的 SQL 语句如下：

```
CREATE TABLE members (
Id int NOT NULL AUTO_INCREMENT,
username varchar(255) DEFAULT NULL ,
password varchar(255) DEFAULT NULL ,
last_login_time datetime DEFAULT NULL ,
last_login_ip varchar(255) DEFAULT NULL ,
PRIMARY KEY (Id)
) ;
CREATE TABLE members_detail (
member_id int NOT NULL DEFAULT 0,
address varchar(255) DEFAULT NULL ,
telephone varchar(16) DEFAULT NULL ,
description text
) ;
```

这两个表的结构如下：

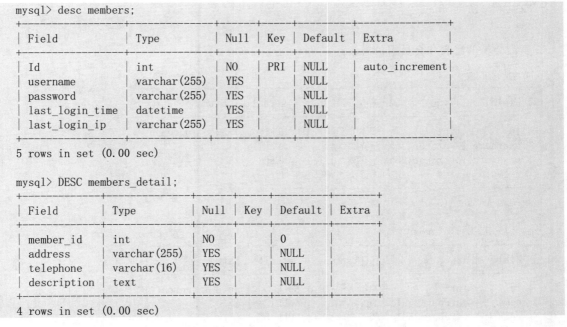

如果需要查询会员的详细信息，可以用会员的 id 来查询。如果需要将会员的基本信息和详细信息同时显示，可以将 members 表和 members_detail 表进行联合查询，查询语句如下：

```sql
SELECT * FROM members LEFT JOIN members_detail ON members.id=members_detail.member_id;
```

通过这种分解，可以提高表的查询效率。对于字段很多且有些字段使用不频繁的表，可以通过这种分解的方式来优化数据库的性能。

13.3.2　增加中间表

对于需要经常联合查询的表，可以建立中间表，以提高查询效率。通过建立中间表，把需要经常联合查询的数据插入到中间表中，然后将原来的联合查询改为对中间表的查询，以此来提高查询效率。本小节将为读者介绍增加中间表优化查询的方法。

首先，分析经常联合查询表中的字段。然后，使用这些字段建立一个中间表，并将原来联合查询的表的数据插入到中间表中。最后，使用中间表进行查询。

【例 13.7】会员信息表和会员组信息表的 SQL 语句如下：

```sql
CREATE TABLE vip(
Id int NOT NULL AUTO_INCREMENT,
username varchar(255) DEFAULT NULL,
password varchar(255) DEFAULT NULL,
groupId INT DEFAULT 0,
PRIMARY KEY (Id)
);
CREATE TABLE vip_group (
Id int NOT NULL AUTO_INCREMENT,
name varchar(255) DEFAULT NULL,
remark varchar(255) DEFAULT NULL,
PRIMARY KEY (Id)
);
```

查询会员信息表和会员组信息表：

```
mysql> DESC vip;
+----------+--------------+------+-----+---------+----------------+
| Field    | Type         | Null | Key | Default | Extra          |
+----------+--------------+------+-----+---------+----------------+
| Id       | int          | NO   | PRI | NULL    | auto_increment |
| username | varchar(255) | YES  |     | NULL    |                |
| password | varchar(255) | YES  |     | NULL    |                |
| groupId  | int          | YES  |     | 0       |                |
+----------+--------------+------+-----+---------+----------------+
4 rows in set (0.01 sec)

mysql> DESC vip_group;
+--------+--------------+------+-----+---------+----------------+
| Field  | Type         | Null | Key | Default | Extra          |
+--------+--------------+------+-----+---------+----------------+
| Id     | int          | NO   | PRI | NULL    | auto_increment |
| name   | varchar(255) | YES  |     | NULL    |                |
| remark | varchar(255) | YES  |     | NULL    |                |
+--------+--------------+------+-----+---------+----------------+
3 rows in set (0.01 sec)
```

已知现在有一个模块需要经常查询带有会员组名称、会员组备注（remark）、会员用户名信息等会员信息。根据这种情况可以创建一个 temp_vip 表。temp_vip 表中存储用户名（user_name）、会员组名称（group_name）和会员组备注（group_remark）信息。创建表的语句如下：

```
CREATE TABLE temp_vip (
Id int NOT NULL AUTO_INCREMENT,
user_name varchar(255) DEFAULT NULL,
group_name varchar(255) DEFAULT NULL,
group_remark varchar(255) DEFAULT NULL,
PRIMARY KEY (Id)
);
```

接下来，从会员信息表和会员组表中查询相关信息存储到临时表中：

```
mysql> INSERT INTO temp_vip(user_name, group_name, group_remark)
    SELECT v.username,g.name,g.remark
    FROM vip as v ,vip_group as g
    WHERE v.groupId =g.Id;
Query OK, 0 rows affected (0.95 sec)
Records: 0  Duplicates: 0  Warnings: 0
```

以后，可以直接从 temp_vip 表中查询会员名、会员组名称和会员组备注，而不用每次都进行联合查询。这样可以提高数据库的查询速度。

13.3.3 增加冗余字段

设计数据库表时应尽量遵循范式理论的规约，尽可能减少冗余字段，让数据库设计看起来精致、优雅。但是，合理地加入冗余字段可以提高查询速度。本小节将为读者介绍通过增加冗余字段来优化查询速度的方法。

表的规范化程度越高，表与表之间的关系就越多，需要连接查询的情况也就越多。例如，员工的信息存储在 staff 表中，部门信息存储在 department 表中。通过 staff 表中的 department_id 字段与 department 表建立关联关系。如果要查询一个员工所在部门的名称，必须从 staff 表中查找员工所在部门的编号（department_id），然后根据这个编号去 department 表查找部门的名称。如果经常需要进行这个操作，连接查询会浪费很多时间。可以在 staff 表中增加一个冗余字段 department_name，该字段用来存储员工所在部门的名称，这样就不用每次都进行连接操作了。

提示：冗余字段会导致一些问题。比如，冗余字段的值在一个表中被修改了，就要想办法在其他表中更新该字段，否则就会使原本一致的数据变得不一致。分解表、增加中间表和增加冗余字段都浪费了一定的磁盘空间。从数据库性能来看，为了提高查询速度而增加少量的冗余，大部分情形下是可以接受的。是否通过增加冗余来提高数据库性能，要根据实际需求来综合分析。

13.3.4 优化插入记录的速度

插入记录时，影响插入速度的主要因素是索引、唯一性校验、一次插入记录条数等。根据这些情况，可以分别进行优化。本小节将为读者介绍优化插入记录速度的几种方法。

1. MyISAM 引擎的表优化

对于 MyISAM 引擎的表，常见的优化方法如下：

（1）禁用索引

对于非空表，插入记录时，MySQL 会根据表的索引对插入的记录建立索引。如果插入大量数据，建立索引会降低插入记录的速度。为了解决这种情况，可以在插入记录之前禁用索引，数据插入完毕后再开启索引。禁用索引的语句如下：

```
ALTER TABLE table_name DISABLE KEYS;
```

其中，table_name 是禁用索引的表的表名。

重新开启索引的语句如下：

```
ALTER TABLE table_name ENABLE KEYS;
```

对于空表批量导入数据，则不需要进行此操作，因为 MyISAM 引擎的表是在导入数据之后才建立索引的。

（2）禁用唯一性检查

插入数据时，MySQL 会对插入的记录进行唯一性校验。这种唯一性校验也会降低插入记录的速度。为了降低这种情况对查询速度的影响，可以在插入记录之前禁用唯一性检查，等到记录插入完毕后再开启。禁用唯一性检查的语句如下：

```
SET UNIQUE_CHECKS=0;
```

开启唯一性检查的语句如下：

```
SET UNIQUE_CHECKS=1;
```

（3）使用批量插入

插入多条记录时，可以使用一条 INSERT 语句插入一条记录；也可以使用一条 INSERT 语句插入多条记录。插入一条记录的 INSERT 语句情形如下：

```
INSERT INTO fruits VALUES('x1', '101', 'mongo2', '5.7');
INSERT INTO fruits VALUES('x2', '101', 'mongo3', '5.7');
INSERT INTO fruits VALUES('x3', '101', 'mongo4', '5.7');
```

使用一条 INSERT 语句插入多条记录的情形如下：

```
INSERT INTO fruits VALUES
('x1', '101', 'mongo2', '5.7'),
('x2', '101', 'mongo3', '5.7'),
('x3', '101', 'mongo4', '5.7');
```

第 2 种情形的插入速度要比第 1 种情形要快。

（4）使用 LOAD DATA INFILE 批量导入

当需要批量导入数据时，如果能用 LOAD DATA INFILE 语句，就尽量使用。因为 LOAD DATA INFILE 语句导入数据的速度比 INSERT 语句快。

2. InnoDB 引擎的表优化

对于 InnoDB 引擎的表，常见的优化方法如下：

（1）禁用唯一性检查

插入数据之前执行 set unique_checks=0 来禁止对唯一索引的检查，数据导入完成之后再运行 set unique_checks=1。这个和 MyISAM 引擎的使用方法一样。

（2）禁用外键检查

插入数据之前执行禁止对外键的检查，数据插入完成之后再恢复对外键的检查。禁用外键检查的语句如下：

```
SET foreign_key_checks=0;
```

恢复对外键的检查语句如下：

```
SET foreign_key_checks=1;
```

（3）禁止自动提交

插入数据之前禁止事务的自动提交，数据插入完成之后，执行恢复自动提交操作。禁止自动提交的语句如下：

```
set autocommit=0;
```

恢复自动提交的语句如下：

```
set autocommit=1;
```

13.3.5 分析表、检查表和优化表

MySQL 提供了分析表、检查表和优化表的语句。分析表主要是分析关键字的分布；检查表主要是检查表中是否存在错误；优化表主要是消除删除或者更新造成的空间浪费。本小节将为读者介绍分析表、检查表和优化表的方法。

1. 分析表

MySQL 中提供了 ANALYZE TABLE 语句来分析表。ANALYZE TABLE 语句的基本语法如下：

```
ANALYZE [LOCAL | NO_WRITE_TO_BINLOG] TABLE tbl_name[,tbl_name]...
```

LOCAL 关键字是 NO_WRITE_TO_BINLOG 关键字的别名，二者都是执行过程不写入二进制日志，tbl_name 为分析表的表名，可以有一个或多个。

使用 ANALYZE TABLE 分析表的过程中，数据库系统会自动对表加一个只读锁。在分析期间，只能读取表中的记录，不能更新和插入记录。ANALYZE TABLE 语句能够分析 InnoDB、BDB 和 MyISAM 类型的表。

【例 13.8】使用 ANALYZE TABLE 来分析 message 表，执行的语句及结果如下：

```
mysql> ANALYZE TABLE members;
+------------------+---------+----------+----------+
| Table            | Op      | Msg_type | Msg_text |
+------------------+---------+----------+----------+
```

```
| test_db.members  | analyze   | status      | OK           |
+------------------+-----------+-------------+--------------+
```

上面结果显示的信息说明如下：

- Table：表示分析的表的名称。
- Op：表示执行的操作。analyze 表示进行分析操作。
- Msg_type：表示信息类型，其值通常是状态（status）、信息（info）、注意（note）、警告（warning）和错误（error）之一。
- Msg_text：显示信息。

2．检查表

MySQL 中可以使用 CHECK TABLE 语句来检查表。CHECK TABLE 语句能够检查 InnoDB 和 MyISAM 类型的表是否存在错误。对于 MyISAM 类型的表，CHECK TABLE 语句还会更新关键字统计数据。而且，CHECK TABLE 也可以检查视图是否有错误，比如在视图定义中被引用的表已不存在。该语句的基本语法如下：

```
CHECK TABLE tbl_name [, tbl_name] ... [option] ...
option = {QUICK | FAST | MEDIUM | EXTENDED | CHANGED}
```

其中，tbl_name 是表名。option 参数有 5 个取值，分别是 QUICK、FAST、MEDIUM、EXTENDED 和 CHANGED，各个选项的意义如下：

- QUICK：不扫描行，不检查错误的连接。
- FAST：只检查没有被正确关闭的表。
- CHANGED：只检查上次检查后被更改的表和没有被正确关闭的表。
- MEDIUM：扫描行，以验证被删除的连接是有效的。也可以计算各行的关键字校验和，并使用计算出的校验和验证这一点。
- EXTENDED：对每行的所有关键字进行一个全面的关键字查找。这可以确保表是 100% 一致的，但是花的时间较长。

option 只对 MyISAM 类型的表有效，对 InnoDB 类型的表无效。CHECK TABLE 语句在执行过程中也会给表加上只读锁。

3．优化表

MySQL 中使用 OPTIMIZE TABLE 语句来优化表。该语句对 InnoDB 和 MyISAM 类型的表都有效。但是，OPTILMIZE TABLE 语句只能优化表中 VARCHAR、BLOB 或 TEXT 类型的字段。OPTILMIZE TABLE 语句的基本语法如下：

```
OPTIMIZE [LOCAL | NO_WRITE_TO_BINLOG] TABLE tbl_name [, tbl_name] ...
```

其中，LOCAL | NO_WRITE_TO_BINLOG 关键字的意义和分析表相同，都是指定不写入二进制日志；tbl_name 是表名。

通过 OPTIMIZE TABLE 语句可以消除删除和更新造成的文件碎片。OPTIMIZE TABLE 语句在执行过程中也会给表加上只读锁。

提示：一个表使用了 TEXT 或者 BLOB 这样的数据类型，如果已经删除了表的一大部分，或者已经对含有可变长度行的表（含有 VARCHAR、BLOB 或 TEXT 列的表）进行了很多更新，则应使用 OPTIMIZE TABLE 来重新利用未使用的空间，并整理数据文件的碎片。在多数的设置中，根本不需要运行 OPTIMIZE TABLE。即使对可变长度的行进行了大量的更新，也不需要经常运行，每周一次或每月一次即可，并且只需要对特定的表运行。

13.4 临时表性能优化

在 MySQL 8.0 中，用户可以把数据库和表归组到逻辑和物理表空间中，这样做可以提高资源的利用率。

MySQL 8.0 使用 CREATE TABLESPACE 语句来创建一个通用表空间。这个功能可以让用户自由地选择表和表空间之间的映射。例如，创建表空间和设置这个表空间应该含有什么样的表。这也让在同一个表空间的用户对所有的表分组，因此在文件系统一个单独的文件内持有他们所有的数据，同时为通用表空间实现了元数据锁。

优化普通 SQL 临时表性能是 MySQL 8.0 的目标之一。首先，通过优化临时表在磁盘中的不必要步骤，临时表的创建和移除成为一个轻量级的操作。将临时表移动到一个单独的表空间中，恢复临时表的过程就变得非常简单，就是在启动时重新创建临时表的单一过程。

MySQL 8.0 去掉了临时表中不必要的持久化。临时表仅仅在连接和会话内被创建，然后通过服务的生命周期绑定它们。通过移除不必要的 UNDO 和 REDO 日志，改变缓冲和锁，为临时表做了优化操作。

MySQL 8.0 增加了 UNDO 日志一个额外的类型，这个类型的日志被保存在一个单独的临时表空间中，在恢复期间不会被调用，而是在回滚操作中才会被调用。

MySQL 8.0 为临时表设定了一个特别类型，称之为"内在临时表"。内在临时表和普通临时表很像，只是内在临时表使用宽松的 ACID 和 MVCC 语义。

MYSQL 8.0 为了提高临时表相关的性能，对临时表相关的部分进行了大幅修改，包括引入新的临时表空间（ibtmp1）；对于临时表的 DDL，不持久化相关表定义；对于临时表的 DML，不写 redo、关闭 change buffer 等。

InnoDB 临时表元数据不再存储于 InnoDB 系统表，而是存储在 INNODB_TEMP_TABLE_INFO 中，包含所有用户和系统创建的临时表信息。该表在第一次运行 select 时被创建，下面举例说明：

```
mysql> SELECT * FROM information_schema.innodb_temp_table_info;
Empty set (0.00 sec)

mysql> create temporary table temp_1(id int,name varchar(100))default charset utf8;
Query OK, 0 rows affected (0.00 sec)

mysql> select * from information_schema.innodb_temp_table_info;
+----------+--------------+--------+------------+
| TABLE_ID | NAME         | N_COLS | SPACE      |
+----------+--------------+--------+------------+
|     1065 | #sql1060_8_8 |      5 | 4294566162 |
+----------+--------------+--------+------------+
```

MySQL 8.0 使用了独立的临时表空间来存储临时表数据，但不能是压缩表。临时表空间在实例启动的时候进行创建、shutdown 的时候进行删除，即为所有非压缩的 innodb 临时表提供一个独立的表空间。默认的临时表空间文件为 ibtmp1，位于数据目录中。通过 innodb_temp_data_file_path 参数可指定临时表空间的路径和大小，默认为 12MB。只有重启实例才能回收临时表空间文件 ibtmp1 的大小。create temporary table 和 using temporary table 将共用这个临时表空间。

```
mysql> show variables like 'innodb_temp_data_file_path';
+----------------------------+-----------------------+
| Variable_name              | Value                 |
+----------------------------+-----------------------+
| innodb_temp_data_file_path | ibtmp1:12M:autoextend |
+----------------------------+-----------------------+
```

在 MySQL 8.0 中，临时表在连接断开或者数据库实例关闭的时候会进行删除，从而提高了性能。只有临时表的元数据使用了 redo 保护，保护元数据的完整性，以便异常启动后进行清理工作。

临时表的元数据在 MySQL 8.0 之后使用了一个独立的表（innodb_temp_table_info）进行保存，不用使用 redo 保护，元数据也只保存在内存中。但这有一个前提，即必须使用共享的临时表空间，如果使用 file-per-table，仍然需要持久化元数据，以便异常恢复清理。临时表需要 undo log，用于 MySQL 运行时的回滚。

在 MySQL 8.0 中，新增一个系统选项 internal_tmp_disk_storage_engine，可定义磁盘临时表的引擎类型，默认为 InnoDB，可选 MyISAM。在这以前，只能使用 MyISAM。在 MySQL 5.6.3 以后新增的参数 default_tmp_storage_engine 是控制 create temporary table 创建的临时表存储引擎，在以前默认是 MEMORY。

查看结果如下：

```
mysql> show variables like '%engine%';
+----------------------------------+-----------+
| Variable_name                    | Value     |
+----------------------------------+-----------+
| default_storage_engine           | InnoDB    |
| default_tmp_storage_engine       | InnoDB    |
| disabled_storage_engines         |           |
| internal_tmp_disk_storage_engine | InnoDB    |
| internal_tmp_mem_storage_engine  | TempTable |
| use_secondary_engine             | ON        |
+----------------------------------+-----------+
```

13.5　创建全局通用表空间

MySQL 8.0 支持创建全局通用表空间，全局表空间可以被所有数据库的表共享，而且相比于独享表空间，手动创建共享表空间可以节约元数据方面的内存。可以在创建表的时候指定属于哪个表空间，也可以对已有表进行表空间修改。

下面创建名为 dxy 的共享表空间，SQL 语句如下：

```
mysql> CREATE TABLESPACE dxy ADD datafile 'dxy.ibd' file_block_size=16k;
```

```
Query OK, 0 rows affected (0.10 sec)
```

指定表空间，SQL 语句如下：

```
mysql> CREATE TABLE t1(id int,name varchar(10))engine = innodb default charset utf8mb4 tablespace dxy;
Query OK, 0 rows affected (0.01 sec)
```

也可以通过 ALTER TABLE 语句指定表空间，SQL 语句如下：

```
mysql> alter table t1 tablespace dxy;
```

如何删除共享表空间？因为是共享表空间，所以不能直接通过 drop table tablename 来删除，也不能回收空间。当确定共享表空间的数据都没用，并且依赖该表空间的表均已经删除时，可以通过 drop tablespace 来删除共享表空间，以释放空间，如果依赖该共享表空间的表存在，就会删除失败。

首先，删除依赖该表空间的数据表，SQL 语句如下：

```
mysql> DROP TABLE t1;
Query OK, 0 rows affected (0.16 sec)
```

最后，删除表空间，SQL 语句如下：

```
mysql> DROP TABLESPACE dxy;
Query OK, 0 rows affected (0.01 sec)
```

13.6　隐藏和显示索引

不可见索引的特性对于性能调试非常有用。在 MySQL 8.0 中，索引可以被"隐藏"和"显示"。当一个索引被隐藏时，它不会被查询优化器所使用。也就是说，管理员可以隐藏一个索引，然后观察对数据库的影响。如果数据库性能有所下降，就说明这个索引是有用的，于是将其"恢复显示"即可；如果数据库性能看不出变化，说明这个索引是多余的，可以删掉了。

下面通过一个案例来了解如何隐藏和显示索引。

创建不可见索引，执行语句如下：

```
mysql> CREATE TABLE test10(a int,b int,index idx_a_b(a,b desc) invisible );
Query OK, 0 rows affected (0.18 sec)
```

查看索引 idx_a_b 的属性 Visible 的值，执行语句如下：

```
mysql>SHOW INDEX FROM test10 \G
*************************** 1. row ***************************
        Table: test10
   Non_unique: 1
     Key_name: idx_a_b
 Seq_in_index: 2
  Column_name: b
    Collation: D
  Cardinality: 0
     Sub_part: NULL
       Packed: NULL
         Null: YES
```

```
       Index_type: BTREE
         Comment:
   Index_comment:
         Visible: NO
      Expression: NULL
```

从结果可以看出，Visible 的属性值为 NO。

显示不可见索引，执行语句如下：

```
mysql>ALTER TABLE test10 ALTER index idx_a_b visible;
Query OK, 0 rows affected (0.12 sec)
Records: 0  Duplicates: 0  Warnings: 0
```

再次查看索引 idx_a_b 的属性 Visible 的值，执行语句如下：

```
mysql>SHOW INDEX FROM test1 \G
*************************** 1. row ***************************
           Table: test1
      Non_unique: 1
        Key_name: idx_a_b
    Seq_in_index: 2
     Column_name: b
       Collation: D
     Cardinality: 0
        Sub_part: NULL
          Packed: NULL
            Null: YES
      Index_type: BTREE
         Comment:
   Index_comment:
         Visible: YES
      Expression: NULL
```

从结果可以看出，Visible 的属性值为 YES。

注意：当索引被隐藏时，它的内容仍然是和正常索引一样实时更新的。如果一个索引需要长期被隐藏，那么可以将其删除，因为索引的存在会影响插入、更新和删除的性能。

再次显示索引，执行语句如下：

```
mysql>ALTER TABLE test10 ALTER index idx_a_b invisible;
Query OK, 0 rows affected (0.07 sec)
Records: 0  Duplicates: 0  Warnings: 0
```

注意：数据表中的主键不能被设置为 invisible。

第 14 章

MySQL 的锁定机制

MySQL 与其他数据库在锁定机制方面最大的不同之处在于，对于不同的存储引擎支持不同的锁定机制。例如，InnoDB 存储引擎支持行级锁（row-level locking），也支持表级锁（table-level locking），默认的情况下是采用行级锁；MyISAM 和 MEMORY 存储引擎采用的是表级锁。本章将详细讲解 MySQL 中不同的存储引擎的锁定机制。

14.1 认识 MySQL 的锁定机制

MySQL 与其他数据库在锁定机制方面最大的不同之处在于，对于不同的存储引擎支持不同的锁定机制。例如，InnoDB 存储引擎支持行级锁（row-level locking），也支持表级锁（table-level locking），默认的情况下是采用行级锁；MyISAM 和 MEMORY 存储引擎采用的是表级锁。BDB 存储引擎采用的是页面锁（page-level locking），同时也支持表级锁。

总的来说，MySQL 各存储引擎使用了三种级别的锁定机制：行级锁定、页级锁定和表级锁定。下面分析一下这三种级别的锁机制的特点。

1. 行级锁定

行级锁最大的特点是锁定对象的颗粒度很小，发生锁定资源争用的概率也很小，能够给予应用程序尽可能大的并发处理能力，从而提高一些需要高并发的应用系统的整体性能。

虽然能够在并发处理能力上有较大的优势，但是行级锁也存在不少弊端。由于行级锁的颗粒度比较小，所以每次获取锁和释放锁会消耗比较大，加锁比较慢，很容易发生死锁。

行级锁定不是 MySQL 自己实现的锁定方式，而是由其他存储引擎所实现的，比如 InnoDB 存储引擎。InnoDB 实现了两种类型的行级锁，包括共享锁和排他锁，而在锁定机制的实现过程中为了让行级锁定和表级锁定共存，InnoDB 使用了两种内部使用的意向锁，也就是意向共享锁和意向排他锁。各个锁的含义如下：

（1）共享锁（S）：允许一个事务读取一行数据时，阻止其他事务读取相同数据的排他锁。

（2）排他锁（X）：允许获得排他锁的事务更新数据，阻止其他事务读取相同数据的共享锁和排他锁。

（3）意向共享锁（IS）：事务打算给数据行加行共享锁。事务在给一个数据行加共享锁前必须先取得该表的 IS 锁。

（4）意向排他锁（IX）：事务打算给数据行加行排他锁。事务在给一个数据行加排他锁前必须先取得该表的 IX 锁。

上面这 4 种锁的共存逻辑关系如表 14.1 所示。

表14.1　4种锁的共存逻辑关系表

共享锁（S）	排他锁（X）	意向共享锁（IS）	意向排他锁（IX）
兼容	冲突	兼容	冲突
冲突	冲突	冲突	冲突
兼容	冲突	兼容	兼容
冲突	冲突	兼容	兼容

如果一个事务请求的锁模式与当前的锁模式兼容，InnoDB 就将请求的锁授予该事务；如果两者不兼容，那么该事务就要等待锁释放。

意向锁是 InnoDB 存储引擎自动加的，对于普通 SELECT 语句，InnoDB 不会加任何锁，对于 INSERT、UPDATE、DELETE 语句，InnoDB 会自动给涉及的数据加排他锁，InnoDB 可以通过以下语句添加共享锁和排他锁。

（1）添加共享锁（S）的语句如下：

```
SELECT * FROM table_name WHERE ... LOCK IN SHARE MODE
```

（2）添加排他锁（X）的语句如下：

```
SELECT * FROM table_name WHERE ... FOR UPDATE
```

共享锁和排他锁的详细用法将在本章后面的小节中详细讲解。

2. 表级锁定

与行级锁不同的是，表级锁的锁定机制的颗粒度最大，该锁定机制的最大特点是系统开销比较小，由于实现逻辑非常简单，所以带来的系统的负面影响也最小。由于表级锁一次性将整个表锁定，因此可以很好地避免死锁的问题。

同时，表级锁定机制也存在一定的缺陷，由于表级锁的锁定机制颗粒很大，所以发生锁冲突的概率也最高，表级锁定机制下并发度也最低。

MySQL 数据库的表级锁定主要分为两种类型，一种是读锁定，另一种是写锁定。MySQL 数据库提供了 4 种队列来维护这两种锁，这 4 种队列间接地说明了数据库表级锁的四种状态，这 4 种队列如下：

（1）Current read lock queue (lock read)。

（2）Padding read lock queue (lock read wait)。

（3）Current write lock queue (lock write)。

（4）Padding write lock queue (lock write wait)。

其中 Current read lock queue 中存放的是当前持有读锁的所有线程，而正在等待资源的信息则存放在 Padding read lock queue 中。同样，Current write lock queue 中存放的是当前持有写锁的所有线程，而正在等待对资源写操作的信息则存放在 Padding write lock queue 中。

MySQL 内部实现读锁和写锁有多达 11 种具体的锁定类型，由系统中一个枚举类型变量（thr_lock_type）定义，具体各种锁定类型如表 14.2 所示。

表14.2　各种锁定类型的含义

锁定类型	含　　义
IGNORE	当发生锁请求的时候内部交互使用，在锁定结构和队列中并不会有任何信息存储
UNLOCK	释放锁定请求的交互用锁类型
READ	普通读锁定
WRITE	普通写锁定
READ_WITH_SHARED_LOCKS	在 InnoDB 中使用到，语法为：SELECT…LOCK IN SHARE MODE
READ_HIGH_PRIORITY	高优先级读锁定
READ_NO_INSERT	不允许 Concurrent Insert 的锁定
WRITE_ALLOW_WRITE	这个类型实际上就是由存储引擎自行处理锁定的时候，MySQL 允许其他线程再获取读或者写锁定，因为即使资源冲突，存储引擎自行处理
WRITE_ALLOW_READ	这种锁定发生在对表 DDL 操作的时候，MySQL 可以允许其他线程获取读锁定，因为 MySQL 是通过重建整个表然后再 RENAME 而实现该功能的，所在整个过程中表依然可以提供读服务
WRITE_CONCURRENT_INSERT	正在进行 Concurrent Insert 时锁使用的锁定方式，该锁定进行的时候，除了 READ_NO_INSERT 之外的其他任何读锁定请求都不会被阻塞
WRITE_DELAYED	在使用 INSERT DELAYED 时的锁定类型
WRITE_LOW_PRIORITY	声明的低级别锁定方式,通过设置 LOW_PRIORITY_UPDATE=1 而产生
WRITE_ONLY	当在操作过程中某个锁定异常中断之后，系统内部需要进行 CLOSE TABLE 操作，在这个过程中出现的锁定类型就是 WRITE_ONLY

对于 MySQL 数据库读锁和写锁的加锁方式，通常使用 LOCK TABLE 和 UNLOCK TABLE 实现对表的加锁和解锁，表 14.3 是一个获得表锁和释放表锁的详细过程。

表14.3　一个获得表锁和释放表锁的例子

Session1	Session2

创建数据表 lock_table_test 并插入演示数据，命令如下：
`mysql>create table lock_table_test(`
`        id int,`
`        data varchar(20));`
`mysql>insert into lock_table_test`
`values(1,'t');`
获得表 lock_table_test 的 READ 锁，命令如下：
`mysql> lock table lock_table_test read;`
`Query OK, 0 rows affected (0.00 sec)`	

(续表)

Session1	Session2
当前 Session1 可以查询出该表的记录，命令如下： mysql> select * from lock_table_test limit 0,1; +------+------+ | id | data | +------+------+ | 1 | t | +------+------+ 1 row in set (0.00 sec)	Session2 也查询出该表的记录，命令如下： mysql> select *from lock_table_test limit 0,1; +------+------+ | id | data | +------+------+ | 1 | t | +------+------+ 1 row in set (0.00 sec)
	Session2 更新锁定表将会发生等待以获得锁，命令如下： mysql> update lock_table_test set data='test' where id = 1; 等待
释放锁，命令如下： mysql> unlock tables; Query OK, 0 rows affected (0.00 sec)	等待
	Session2 获得锁，更新操作执行完成，命令如下： mysql> update lock_table_test set data='test' where id = 1; Query OK, 1 rows affected (1 min 1.77 sec) Rows matched: 1 Changed: 1 Warnings: 0

3. 页级锁定

页级锁定在 MySQL 中是比较特殊的一种锁定机制，页级锁定的特点是颗粒度介于行级锁定与表级锁定之间，所以获取锁定所需要的资源开销，以及锁提供的并发处理的能力也介于表级锁定和行级锁定之间。

在数据库实现资源锁定的过程中，锁定机制的粒度越小，数据库实现的算法越复杂，数据库所消耗的内存也越大。不过，随着锁机制粒度越来越小，应用的并发发生锁等待的机率也越来越小，系统整体性能也随之增高。

MySQL 使用写队列和读队列来完成数据库的读和写操作，所以说，MySQL 数据库存在读锁和写锁的概念。对于写锁而言，如果表没有加入写锁，那么对其表加写锁；如果表已经加了写锁，此时会将写操作的请求放入写锁的队列中。而对于读锁而言，如果没有加入读锁，那么请求会加入一个读操作的锁，其他读操作的请求会放到读锁的队列中。

下面通过一个简单的例子来说明读写操作，具体操作步骤如下：

步骤 01 首先创建表 content 并插入数据，语句如下：

```
mysql> create table content(id int,content varchar(20));
Query OK, 0 rows affected (0.53 sec)
mysql> insert into content values(1,'wangfei');
```

步骤 02 向表 content 里面添加大量的数据，数据越多，效果越明显。重复多次执行以下语句即可：

```
mysql> insert into content select * from content;
```

```
Query OK, 16384 rows affected (0.06 sec)
Records: 16384  Duplicates: 0  Warnings: 0

mysql> insert into content select * from content;
Query OK, 32768 rows affected (0.12 sec)
Records: 32768  Duplicates: 0  Warnings: 0
```

步骤03 此时，准备工作已经完成，可以根据表14.4所示运行读写操作，理解 MySQL 读写队列运行的过程。

表14.4 运行MySQL读写队列的过程

Session1	Session2
mysql>select count(*) from content;	
	此时执行： mysql> update content set content = 'test2' where id = 1;
	等待
查询结束	
	mysql> update content set content = 'test2' where id = 1; Query OK, 0 rows affected (5.16 sec) Rows matched: 1048576 Changed: 0 Warnings: 0
Session1 关闭	
	mysql> update content set content = 'test2' where id = 1; Query OK, 0 rows affected (3.14 sec) Rows matched: 1048576 Changed: 0 Warnings: 0

对于 Session1，此时表没有加锁，那么此时 Session1 操作会对表加上一个读锁；对于 Session2，此时表 content 因为已经加上了读锁，所以会将 update 请求放到锁定队列中。

从上述特点可见，很难笼统地说哪种锁定机制好，只能根据具体应用的特点来选择哪种锁定机制更合适。仅从锁的角度来看，表级锁更适合以查询为主，只有少量按索引条件更新数据的应用。而行级锁更适合有大量按索引条件并发更新少量不同数据，同时又有并发查询的应用。

14.2 MyISAM 的锁定机制

MyISAM 基本上可以说是对 MySQL 所提供的锁定机制所实现的表级锁定依赖最大的一种存储引擎了。MyISAM 的表锁是使用最为广泛的锁类型，本节详细介绍 MyISAM 表锁的使用。

14.2.1 MyISAM 表级锁的锁模式

MySQL 的表级锁有两种模式：表共享读锁（table read lock）和表独占写锁（table write lock）。

锁模式的兼容性如表 14.5 所示。

表14.5　锁模式的兼容性

None	读　锁	写　锁
兼容	兼容	冲突
兼容	冲突	冲突

对于 MyISAM 表的读操作不会因为不同进程访问资源而发生阻塞，而对于 MyISAM 表的写操作会阻塞其他用户对同一表的读和写操作。

下面通过一个例子来看一下 MySQL 表级锁的读锁过程，如表 14.6 所示。

注意：本章节的数据表操作全部在数据库 test 下，读者可以自己新建一个数据库 test 进行试验。

表14.6　MySQL表锁的读锁的过程

Session1	Session2

首先新建表 read_lock，然后添加一条数据，命令如下：
`mysql> create table read_lock(`
　　　`id int,`
　　　`data varchar(20)`
　　`)engine=myisam;`
`Query OK, 0 rows affected (0.01 sec)`
`mysql>insert into read_lock`
`values(102,'data');`
`Query OK, 1 row affected (0.01 sec)`	

获得 readk_lock 表的读锁，命令如下： `mysql> lock table read_lock read;` `Query OK, 0 rows affected (0.00 sec)` 查询 read_lock 表的数据，命令如下： `mysql> select * from read_lock;` `+------+------+` `| id	data |` `+------+------+` `| 102	data |` `+------+------+` `1 row in set (0.00 sec)` 执行更新操作，发现不能够执行，命令如下： `mysql> update read_lock set data='test' where id = 102;` `ERROR 1099 (HY000): Table 'read_lock' was locked with a READ lock and can't be updated` 插入数据时不能执行，执行命令如下： `mysql>insert into read_lock` `values(13,'data2');` `ERROR 1099 (HY000): Table 'read_lock' was locked with a READ lock and can't be updated`	其他 Session 是可以查询数据的，如下： `mysql> select *from read_lock;` `+------+------+` `| id	data |` `+------+------+` `| 102	data |` `+------+------+` `1 row in set (0.00 sec)`  　　`mysql> insert into read_lock values(13,'data3');` 等待

释放读锁，命令如下：
`mysql> unlock tables;`
`Query OK, 0 rows affected (0.00 sec)`	

	成功添加一条数据，命令如下：
	`mysql> insert into read_lock`
	`values(103,'data3');`
	`Query OK, 1 row affected (3 min 26.30 sec)`

接下来再看一下 MySQL 表级锁的写锁测试的例子，如表 14.7 所示。

表14.7　一个MySQL表锁的写锁测试的例子

Session1	Session2

获得表 read_lock 表的写锁，命令如下：
`mysql> lock table read_lock write;`
`Query OK, 0 rows affected (0.00 sec)`	

当前 Session 对锁定表的查询、更新、插入操作都是可以执行的，执行命令如下： `mysql> select * from read_lock;` `+------+-------+` `| id	data |` `+------+-------+` `| 102	data |` `| 103	data3 |` `+------+-------+` `2 rows in set (0.00 sec)`  `mysql> update read_lock set data='tt' where id = 103;` `Query OK, 1 row affected (0.05 sec)` `Rows matched: 1 Changed: 1 Warnings: 0`  `mysql> insert into read_lock values(104,'data4');` `Query OK, 1 row affected (0.00 sec)`	Session2 对锁定表的查询被阻塞，需要等待锁释放，执行命令如下： `mysql> select * from read_lock;` 等待

释放锁，命令如下：
`mysql> unlock tables;`
`Query OK, 0 rows affected (0.00 sec)`	

	Session2 获得锁，查询返回结果，命令如下： `mysql> select * from read_lock;` `+------+-------+` `| id	data |` `+------+-------+` `| 102	data |` `| 103	tt |` `| 104	data4 |` `+------+-------+` `3 rows in set (54.45 sec)`

14.2.2　获取 MyISAM 表级锁的争用情况

MyISAM 存储引擎只支持表锁，MySQL 数据库可以通过检查 table_locks_waited 和 table_locks_immediate 状态变量来分析系统上表锁的争夺情况，命令如下：

```
mysql> show status like 'table%';
```

```
+------------------------+-------+
| Variable_name          | Value |
+------------------------+-------+
| Table_locks_immediate  | 120   |
| Table_locks_waited     | 0     |
+------------------------+-------+
2 rows in set (0.25 sec)
```

这里有两个状态变量记录了 MySQL 内部表级锁定的情况，两个变量的含义如下：

（1）Table_locks_immediate：产生表级锁定的次数。

（2）Table_locks_waited：出现比较锁定争用而发生等待的次数。

如果 Table_locks_waited 的值比较高，那么说明了存在着比较严重的表级锁争用的情况。MyISAM 在读操作占主导的情况下是高效的，但是一旦出现大量读写操作并发，同 InnoDB 相比，MyISAM 的执行效率就会直线下降。对于 MyISAM 存储引擎表，新的数据会被附加到数据文件的结尾，如果经常做一些 UPDATE 和 DELETE 操作，数据将不会是连续的，数据文件中就会出现很多空洞，此时再插入新的数据时，默认情况下会先看这些空洞是否可以容纳新数据，如果可以容纳新的数据，那么会将数据保存到空洞里面去，反之，会将新的数据保存到数据文件的结尾。这样做是为了减少文件的大小，并减少文件碎片的产生。

MyISAM 存储引擎表往往因为读表请求的增加，会出现比较严重的读写锁的问题，所以经常在实际应用中采用主从分离，主从服务器把读写操作分离出来，主服务器执行写操作，而从服务器负责查询操作，此时往往因为主服务器执行完了写入的操作，但从服务器有大量的查询操作，会被这些来自主服务器和从服务器同步的 UPDATE 和 INSERT 操作严重堵塞，最后造成所有的 MySQL 从库负载迅速上升。在解决 MyISAM 读写互斥问题中，由于没有办法在短期内增加读的服务器，所以通过对 MySQL 进行一些配置，以牺牲数据实时性为代价，来换取所有服务器的安全。具体配置如下：

（1）当对于同一个 MyISAM 表进行查询和插入操作时，为了降低锁竞争频率，可以把 concurrent_insert 的值设置为 2，此时不管表有没有空洞，都允许数据文件结尾并发插入数据，至于产生的文件碎片，可以定期使用 OPTIMIZE TABLE 语法进行优化。

（2）在默认情况下，写操作的优先级要高于读操作的优先级别，即便先发送的是读操作请求，后发送的是写操作请求，此时也会优先处理写请求，然后处理读请求。可以考虑设置 max_write_lock_count=1，此时当系统处理一个写操作后，就会暂停写操作，给读操作执行的机会。

（3）降低写操作的优先级，给读操作更高的优先级别，可以将 low-priority-updates 设置为 1。

14.2.3　MyISAM 表级锁加锁方法

MyISAM 在执行查询语句的时候，会自动给查询语句涉及数据库表记录添加读锁，而在执行数据更新操作，比如执行 UPDATE、INSERT、DELETE 等语句的时候，MySQL 数据库会自动为涉及的表记录添加写锁。

对于 MySQL 数据库读锁和写锁的加锁方式，通常使用命令 LOCK TABLE 和 UNLOCK TABLE 实现手动加表级锁，LOCK TABLE、UNLOCK TABLE 可以用来锁定和释放当前线程的表，具体语法如下：

第 14 章 MySQL 的锁定机制

```
LOCK TABLE
    tbl_name [As alias] {READ[LOCAL] | [LOW_PRIORITY] WRITE}
    [, tbl_name [AS alias] {READ [LOCAL] | [LOW_PRIORITY] WRITE}] ...
UNLOCK TABLES
```

要想使用 LOCK TABLE，必须拥有有关表的 LOCK TABLE 权限和 SELECT 权限，如果一个线程获得对一个表的读锁，该线程只能从该表中读取数据；如果一个线程获得对一个表的写锁，那么该线程只可以对表进行写数据，此时其他线程会被阻塞，直至写被释放为止。

下面详细讲解一下 LOCK TABLE 的用法。

当对表以别名的形式锁定时，不能在一次查询中多次使用一个已锁定的表使用别名代替，在此情况下，必须分别获得对每个别名的锁定，举例如下：

```
mysql> create table c_table (
        id int,
        data varchar(20)
        )engine=myisam;
Query OK, 0 rows affected (0.01 sec)
mysql>insert into c_table values(1,'test');
mysql>insert into c_table values(2,'test2');
mysql>insert into c_table values(3,'test3');
mysql>insert into c_table values(4,'test4');
mysql> lock table c_table write, c_table as c write;
Query OK, 0 rows affected (0.22 sec)

mysql> insert into c_table select * from c_table;
ERROR 1100 (HY000): Table 'c_table' was not locked with LOCK TABLES

mysql> insert into c_table select *from c_table as c;
Query OK, 4 rows affected (0.06 sec)
Records: 4  Duplicates: 0  Warnings: 0
```

如果使用一个别名来引用一个表，那么必须使用相同的别名锁定该表；如果没有别名，则不会锁定该表，举例如下：

```
mysql> lock table c_table read;
Query OK, 0 rows affected (0.00 sec)

mysql> select * from c_table as c;
ERROR 1100 (HY000): Table 'c' was not locked with LOCK TABLES
```

如果使用一个别名锁定一个表，那么必须使用该别名在查询中引用该表。举例如下：

```
mysql> unlock tables;
Query OK, 0 rows affected (0.00 sec)

mysql> lock table c_table as c read;
Query OK, 0 rows affected (0.00 sec)

mysql> select * from c_table;
ERROR 1100 (HY000): Table 'c_table' was not locked with LOCK TABLES

mysql> select * from c_table as c;
+------+-------+
| id   | data  |
```

```
+------+-------+
|   1  | test  |
|   2  | data2 |
|   3  | data3 |
|   4  | data4 |
|   1  | test  |
|   2  | data2 |
|   3  | data3 |
|   4  | data4 |
+------+-------+
8 rows in set (0.01 sec)
```

READ LOCAL 和 READ 之间的区别是：READ LOCAL 允许在锁定被保持时，执行非冲突性 INSERT 语句同时插入，其作用就是在满足 MyISAM 表并发插入条件的情况下，允许其他用户在表尾并发插入记录。

如果对一个表使用 LOW_PRIORITY WRITE 锁定，这意味着 MySQL 等待特定的锁定，直到没有申请 READ 锁定的线程时为止。当线程已经获得 WRITE 锁定，并在等待得到锁定表清单中的用于下一个表的锁定时，所有其他线程会等待 WRITE 锁定被释放。

14.2.4　MyISAM Concurrent Insert 的特性

MyISAM 存储引擎有个系统变量 concurrent_insert，专门用以控制其并发插入的行为，其值分别是 0、1、2。

（1）当 concurrent_insert 的值为 0 的时候，不允许并发插入。

（2）当 concurrent_insert 的值为 1 的时候，如果表中没有被删除的行，MyISAM 允许在一个进程读表的同时，另一个进程从表尾插入记录。

（3）当 concurrent_insert 的值为 2 的时候，无论表中有没有被删除的行，都允许在表尾并发插入记录。

MySQL 数据库默认 concurrent_insert 的值为 1，即 MySQL 允许一个进程读表的同时，另一个进程从表尾插入记录。

在如表 14.8 所示的例子中，Session1 获得表 c_table2 的 READ LOCAL 锁，此时 Session2 可以插入数据记录，但是更新操作发生阻塞，此时 Session1 不能对表进行插入和更新数据操作。等到 Session1 释放锁资源后，Session2 完成更新操作。

表14.8　MyISAM存储引擎并发操作的例子

Session1	Session2

新建数据表 c_table2 并插入演示数据，命令如下：
mysql> create table c_table2(
　　　id int,
　　　data varchar(20)
　　　)engine=myisam;
Query OK, 0 rows affected (0.09 sec)
mysql> insert into c_table2 values(1,'data');
mysql> insert into c_table2 values(2,'data2');	

(续表)

Session1	Session2

获得表 c_table2 的 READ LOCAL 锁，命令如下：
mysql> lock table c_table2 read local;
Query OK, 0 rows affected (0.02 sec)
当前 Session 不能对表进行插入数据操作，执行结果如下：
mysql> insert into c_table2 values(3,'data3');
ERROR 1099 (HY000): Table 'c_table2' was locked with a READ lock and can't be updated
当前 Session 不能对表进行更新数据操作，执行结果如下：
mysql> update c_table2 set data='test' where id = 1;
ERROR 1099 (HY000): Table 'c_table2' was locked with a READ lock and can't be updated
	Session2 可以进行插入操作，但是更新会等待，执行结果如下：
	mysql> insert into c_table2 values(3,'data3');
	Query OK, 1 row affected (0.00 sec)
	mysql> update c_table2 set data='test' where id = 1;
	等待

Session1 查询不能访问其他 Session 插入的数据，执行结果如下： mysql> select * from c_table2; +------+-------+ | id | data | +------+-------+ | 1 | test | | 2 | data2 | +------+-------+ 2 rows in set (0.00 sec)	
释放锁，执行结果如下： mysql> unlock tables; Query OK, 0 rows affected (0.00 sec) 当前 Session 解锁后，可以查询到其他 Session 插入的记录，执行结果如下： mysql> select * from c_table2; +------+-------+ | id | data | +------+-------+ | 1 | test | | 2 | data2 | | 3 | data3 | +------+-------+ 3 rows in set (0.00 sec)	Session2 获得锁，更新操作完成，执行结果如下： mysql> update c_table2 set data='test' where id = 1; Query OK, 1 row affected (2 min 3.42 sec) Rows matched: 1 Changed: 1 Warnings: 0

14.2.5　MyISAM 表锁优化建议

对于 MyISAM 存储引擎，使用表级锁虽然在实现过程中，比行级锁和页级锁所带来的附加成本要小，所消耗的资源也是最小的，但是，MyISAM 表级锁的颗粒比较大，在数据库并发处理过程中产生的数据资源争用的情况也会比其他的锁定级别都要多，从而在较大程度上会降低并发处理能力。MyISAM 表锁的优化建议如下：

（1）MyISAM 表锁的锁定级别是固定的，所以在考虑 MyISAM 表级锁优化时，重点考虑如何提高并发的效率。

（2）减少锁定的时间，尽量让查询的时间更短。减少比较复杂的查询语句，可以考虑将复杂的查询分解成多个小的查询。尽可能建立足够高效的索引，让数据检索更迅速。尽量让 MyISAM 存储引擎的表控制字段类型。利用合理的机会优化 MyISAM 表数据文件。

（3）MyISAM 表级锁可以考虑分离能并行的操作，对于读锁互相阻塞的表级锁，可能会觉得在存储引擎的表上就只能是完全的串行化，没有办法再并行了。但是，MyISAM 的存储引擎还有个非常有用的特性，就是 Concurrent Insert 的特性。可以考虑设置 Concurrent Insert 值为 2，此时无论 MyISAM 存储引擎的数据文件的中间部分是否存在空洞（因为删除数据而留下的空闲空间），都允许数据文件尾部进行插入操作。

（4）MyISAM 的表级锁定对于读和写是有不同优先级别设定的。默认情况下，写操作的优先级别高于读操作的优先级别，可以考虑根据应用的实际情况来设置读锁和写锁的优先级别，通过设置系统参数 low_priority_updates=1，可以将写的优先级别设置为低于读的优先级别。

14.3　InnoDB 的锁定机制

MySQL 的 MyISAM 存储引擎只支持表级锁，随着应用对事务的完整性和并发性要求越来越高，MySQL 出现了基于支持事务的存储引擎，随后又出现了支持行级锁的 InnoDB 存储引擎。

MySQL 数据库最常见的两种存储引擎是 MyISAM 和 InnoDB，这两种存储引擎各有优缺点，MyISAM 类型不支持事务处理，但执行的效率比 InnoDB 类型的存储引擎更快；而 InnoDB 类型的存储引擎支持事务特性，并且 InnoDB 提供外键等数据库高级功能。在处理数据量上，InnoDB 可以处理海量数据，并且在具有良好索引的基础上，InnoDB 的查询速度要比 MyISAM 快。另外，InnoDB 存储引擎采用了行级锁，本节将会详细介绍 InnoDB 存储引擎的行级锁。

14.3.1　InnoDB 行级锁模式

InnoDB 存储引擎支持行级锁，支持事务处理。事务是由一组 SQL 语句组成的逻辑处理单元，它的 ACID 特性如下：

（1）原子性（Atomicity）：事务具有原子不可分割的特性，要么一起执行，要么都不执行。

（2）一致性（Consistency）：在事务开始和事务结束时，数据都保持一致状态。

（3）隔离性（Isolation）：在事务开始和结束过程中，事务保持着一定的隔离特性，保证事务不受外部并发数据操作的影响。

（4）持久性（Durability）：事务完成后，数据将会被持久化到数据库中。

InnoDB 存储引擎并发事务处理能力大大增加了数据库资源的利用率，提高了数据库系统的事务吞吐量，但并发事务同时也存在一些问题，主要包括：更新丢失（Lost Update）、脏读（Dirty Reads）、不可重复读（Non-Repeatable Reads）、幻读（Phantom Reads）。它们的具体含义如下：

（1）更新丢失：两个事务更新同一行数据，但是第二个事务却中途失败退出了，导致对两个事务的修改都失效了，这时系统没有执行任何锁操作，因此并发事务并没有被隔离。

（2）脏读：一个事务读了某行数据，但是另一个事务已经更新了这行数据，这是非常危险的，很可能所有的操作都会被回滚。

（3）不可重复读：一个事务对一行数据重复读取两次，可是得到了不同的结果。在两次读取数据的中途，有可能存在另外一个事务对数据进行了修改的情况。

（4）幻读：事务在操作过程中进行两次查询，第二次查询结果包含了第一次没有出现的数据，出现幻读的主要原因是，两次查询过程中另一个事务插入了新的数据。

数据库并发中的"更新丢失"通常是应该完全避免的。但防止更新丢失数据，并不能单靠数据库事务控制来解决，需要应用程序对要更新的数据加必要的锁来解决，而以上出现的数据库脏读、不可重复读、幻读，都必须由数据库提供一定的事务隔离机制来解决。为了避免数据库事务带来的问题，在标准 SQL 规范中定义了 4 个事务的隔离级别，不同的隔离级别对事务处理不一样。

数据库隔离级别包括：未提交读（Read uncommitted）、已提交读（Read committed）、可重复读（Repeatable read）、可序列化（Serializable）。事务的隔离级别越高，并发副作用就越小，但付出的代价也越大。这 4 种数据库隔离级别的比较如表 14.9 所示。

表14.9 数据库隔离级别的比较

隔离级别	读数据一致性及允许的并发副作用			
	读数据一致性	脏读	不可重复读	幻读
未提交读（Read uncommitted）	最低级别，只能保证不读取物理上损坏的数据	是	是	是
已提交读（Read committed）	语句级	否	是	是
可重复读（Repeatable read）	事务级	否	否	是
可序列化（Serializable）	最高级别，事务级	否	否	否

InnoDB 存储引擎实现了四种行锁，分别是：共享锁（S）、排他锁（X）、意向共享锁（IS）、意向排他锁（IX），下面进一步学习这四种行锁的知识。

首先，使用共享锁和排他锁必须要满足以下几个条件：

（1）设置 autocommit 的值是 OFF 或者 0。

（2）表的数据引擎是支持事务的，比如 InnoDB 数据引擎。

（3）如果不管 autocommit，手动在事务里执行操作，这个时候要使用 begin 或者 start transaction 开始事务。

（4）不要在锁定事务规定的时间之外使用共享锁和排他锁。

下面通过一个例子来理解。首先建立一个表，然后添加记录，命令如下：

```
mysql> create table s_table(
    id int,
    data varchar(20)
    )engine=innodb;
Query OK, 0 rows affected (0.13 sec)

mysql> insert into s_table values(120,'test');
Query OK, 1 row affected (0.00 sec)
```

使用 InnoDB 存储引擎共享锁的过程如表 14.10 所示。

表14.10 一个InnoDB存储引擎共享锁的例子

Session1	Session2

关闭自动事务提交，查看数据表 s_table 中 id 为 120 的数据，命令如下：
mysql> set autocommit=0;
Query OK, 0 rows affected (0.00 sec)
mysql> select id,data from s_table where id = 120;
+------+------+
| id | data |
+------+------+
| 120 | test |
+------+------+
1 row in set (0.00 sec)
	关闭自动事务提交，查看数据表 s_table 中 id 为 120 的数据，命令如下：
	mysql> set autocommit=0;
	Query OK, 0 rows affected (0.00 sec)
	mysql> select id,data from s_table where id = 120;
	+------+------+
	| id | data |
	+------+------+
	| 120 | test |
	+------+------+
	1 row in set (0.00 sec)

当前 Session 对 id=120 的这条记录添加共享锁，命令如下： mysql> select id,data from s_table where id = 120 lock in share mode; +------+------+ | id | data | +------+------+ | 120 | test | +------+------+ 1 row in set (0.00 sec)	
	Session2 可以查询到 id=120 的该条记录，同时也可以对该条记录加共享锁，命令如下： mysql> select id,data from s_table where id = 120 lock in share mode; +------+------+ | id | data | +------+------+ | 120 | test | +------+------+ 1 row in set (0.00 sec)
当前 Session 对该条记录进行更新操作，此时会发生等待，命令如下： mysql> update s_table set data='test2' where id = 120; 等待	
	当前 Session 同样使用 update 语句进行更新操作，则会导致死锁退出，命令如下： mysql> update s_table set data='test2' where id=120; ERROR 1213 (40001): Deadlock found when trying to get lock; try restarting transaction
获得锁之后，可以更新成功，命令如下： mysql> update s_table set data='test2' where id = 120; Query OK, 1 row affected (10.38 sec) Rows matched: 1 Changed: 1 Warnings: 0	

下面接着学习 InnoDB 存储引擎使用排他锁的例子，具体过程如表 14.11 所示。

表14.11　一个InnoDB存储引擎排他锁的例子

Session1	Session2

关闭自动事务提交，查看数据表 s_table 中 id 为 120 的数据，命令如下：
mysql> set autocommit=0;
Query OK, 0 rows affected (0.00 sec)
 	
mysql> select id,data from s_table where id = 120;
+------+------+
| id | data |
+------+------+
| 120 | test |
+------+------+
1 row in set (0.00 sec)
	关闭自动事务提交，查看数据表 s_table 中 id 为 120 的数据，命令如下：
	mysql> set autocommit=0;
	Query OK, 0 rows affected (0.00 sec)
	
	mysql> select id,data from s_table where id = 120;
	+------+------+
	| id | data |
	+------+------+
	| 120 | test |
	+------+------+
	1 row in set (0.00 sec)

当前 Session 对 id=120 的这条记录加 for update 的排他锁，命令如下： mysql> select id,data from s_table where id = 120 for update; +------+------+ | id | data | +------+------+ | 120 | test | +------+------+ 1 row in set (0.00 sec)	

	当前 Session，直接查询 select id,data from s_table where id = 120 该条记录，命令如下：
	mysql> select id,data from s_table where id = 120;
	+------+------+
	| id | data |
	+------+------+
	| 120 | test |
	+------+------+
	1 row in set (0.00 sec)
	如果当前 Session 对 id=120 的这条记录加 for update 的排他锁，会发生等待，命令如下：
	mysql> select id,data from s_table where id = 120 for update;
	等待

接下来当前 Session 更新该条记录，然后提交事务，事务提交后会释放锁，命令如下： mysql> update s_table set data = 'test4' where id = 120; Query OK, 1 row affected (0.00 sec) Rows matched: 1 Changed: 1 Warnings: 0 mysql> commit; Query OK, 0 rows affected (0.01 sec)	

(续表)

Session1	Session2
	当前 Session2 得到 Session1 提交的记录，执行命令如下： `mysql> select id,data from s_table where id = 120 for update;` `+------+-------+` `| id	data |` `+------+-------+` `| 120	test4 |` `+------+-------+` `1 row in set (12.11 sec)`

14.3.2 获取 InnoDB 行级锁的争用情况

对于 InnoDB 所使用的行级锁定，MySQL 系统是通过另外一组更为详细的状态来记录的，查看命令如下：

```
mysql> show status like '%innodb_row_lock%';
+-------------------------------+--------+
| Variable_name                 | Value  |
+-------------------------------+--------+
| Innodb_row_lock_current_waits | 0      |
| Innodb_row_lock_time          | 211656 |
| Innodb_row_lock_time_avg      | 26457  |
| Innodb_row_lock_time_max      | 50968  |
| Innodb_row_lock_waits         | 8      |
+-------------------------------+--------+
5 rows in set (0.00 sec)
```

InnoDB 的行级锁定状态不仅记录了锁定等待次数，还记录了锁定总时间长，每次平均时长，以及最大的时长，各个状态变量的说明如下：

（1）Innodb_row_lock_current_waits：当前正在等待锁定的数量。

（2）Innodb_row_lock_time：从系统启动到现在锁定总时间长度。

（3）Innodb_row_lock_time_avg：每次等待锁花费的平均时间。

（4）Innodb_row_lock_time_max：从系统启动到现在等待最长的一次所花的时间。

（5）Innodb_row_lock_waits：系统启动后到现在总共等待的次数。

根据 Innodb 提供的这些系统状态的分析，制定相应的优化计划，尤其是当等待次数比较高的时候，而且每次等待时间也比较大的时候，就需要分析系统出现这种情况的原因。

此外，还可以通过如下操作步骤来监控 InnoDB 行级锁并发争用的情况：

步骤 01 创建 Innodb Monitor 表，从而打开 InnoDB 的监控功能，命令如下：

```
mysql> create table innodb_monitor(a int)engine=innodb;
Query OK, 0 rows affected (0.01 sec)
```

步骤 02 使用 "SHOW ENGINE INNODB STATUS\G" 语句查看 InnoDB 行级锁并发争用的情况，命令如下：

```
mysql> SHOW ENGINE INNODB STATUS\G
```

```
*************************** 1. row ***************************
  Type: InnoDB
  Name:
Status:
=====================================
2022-03-25 09:25:03 0x1f28 INNODB MONITOR OUTPUT
=====================================
Per second averages calculated from the last 36 seconds
-----------------
BACKGROUND THREAD
-----------------
srv_master_thread loops: 324 srv_active, 0 srv_shutdown, 221817 srv_idle
srv_master_thread log flush and writes: 0
----------
SEMAPHORES
----------
OS WAIT ARRAY INFO: reservation count 3503
OS WAIT ARRAY INFO: signal count 2977
RW-shared spins 0, rounds 0, OS waits 0
RW-excl spins 0, rounds 0, OS waits 0
RW-sx spins 0, rounds 0, OS waits 0
Spin rounds per wait: 0.00 RW-shared, 0.00 RW-excl, 0.00 RW-sx
------------------------
LATEST DETECTED DEADLOCK
------------------------
2022-03-24 19:26:58 0x2058
*** (1) TRANSACTION:
TRANSACTION 3516, ACTIVE 135 sec starting index read
mysql tables in use 1, locked 1
LOCK WAIT 4 lock struct(s), heap size 1128, 3 row lock(s)
MySQL thread id 35, OS thread handle 7976, query id 7555 localhost ::1 root updating
update s_table set data='test2' where id = 120

*** (1) HOLDS THE LOCK(S):
RECORD LOCKS space id 175 page no 4 n bits 72 index GEN_CLUST_INDEX of table `test`.`s_table`
trx id 3516 lock mode S
Record lock, heap no 1 PHYSICAL RECORD: n_fields 1; compact format; info bits 0
 0: len 8; hex 73757072656d756d; asc supremum;;

Record lock, heap no 2 PHYSICAL RECORD: n_fields 5; compact format; info bits 0
 0: len 6; hex 00000001027c; asc     |;;
 1: len 6; hex 000000000dbb; asc      ;;
 2: len 7; hex 82000000db0110; asc        ;;
 3: len 4; hex 80000078; asc    x;;
 4: len 4; hex 74657374; asc test;;

*** (1) WAITING FOR THIS LOCK TO BE GRANTED:
RECORD LOCKS space id 175 page no 4 n bits 72 index GEN_CLUST_INDEX of table `test`.`s_table`
trx id 3516 lock_mode X waiting
Record lock, heap no 2 PHYSICAL RECORD: n_fields 5; compact format; info bits 0
 0: len 6; hex 00000001027c; asc     |;;
 1: len 6; hex 000000000dbb; asc      ;;
 2: len 7; hex 82000000db0110; asc        ;;
 3: len 4; hex 80000078; asc    x;;
 4: len 4; hex 74657374; asc test;;


*** (2) TRANSACTION:
TRANSACTION 3517, ACTIVE 109 sec starting index read
mysql tables in use 1, locked 1
```

```
LOCK WAIT 4 lock struct(s), heap size 1128, 3 row lock(s)
MySQL thread id 36, OS thread handle 920, query id 7556 localhost ::1 root updating
update s_table set data='test2' where id = 120

*** (2) HOLDS THE LOCK(S):
RECORD LOCKS space id 175 page no 4 n bits 72 index GEN_CLUST_INDEX of table `test`.`s_table`
trx id 3517 lock mode S
Record lock, heap no 1 PHYSICAL RECORD: n_fields 1; compact format; info bits 0
 0: len 8; hex 73757072656d756d; asc supremum;;

Record lock, heap no 2 PHYSICAL RECORD: n_fields 5; compact format; info bits 0
 0: len 6; hex 00000001027c; asc     |;;
 1: len 6; hex 000000000dbb; asc     ;;
 2: len 7; hex 82000000db0110; asc       ;;
 3: len 4; hex 80000078; asc     x;;
 4: len 4; hex 74657374; asc test;;


*** (2) WAITING FOR THIS LOCK TO BE GRANTED:
RECORD LOCKS space id 175 page no 4 n bits 72 index GEN_CLUST_INDEX of table `test`.`s_table`
trx id 3517 lock_mode X waiting
Record lock, heap no 2 PHYSICAL RECORD: n_fields 5; compact format; info bits 0
 0: len 6; hex 00000001027c; asc     |;;
 1: len 6; hex 000000000dbb; asc     ;;
 2: len 7; hex 82000000db0110; asc       ;;
 3: len 4; hex 80000078; asc     x;;
 4: len 4; hex 74657374; asc test;;

*** WE ROLL BACK TRANSACTION (2)
------------
TRANSACTIONS
------------
Trx id counter 3529
Purge done for trx's n:o < 3529 undo n:o < 0 state: running but idle
History list length 0
LIST OF TRANSACTIONS FOR EACH SESSION:
---TRANSACTION 283175825169536, not started
0 lock struct(s), heap size 1128, 0 row lock(s)
---TRANSACTION 283175825168760, not started
0 lock struct(s), heap size 1128, 0 row lock(s)
---TRANSACTION 283175825167984, not started
0 lock struct(s), heap size 1128, 0 row lock(s)
--------
FILE I/O
--------
I/O thread 0 state: wait Windows aio (insert buffer thread)
I/O thread 1 state: wait Windows aio (log thread)
I/O thread 2 state: wait Windows aio (read thread)
I/O thread 3 state: wait Windows aio (read thread)
I/O thread 4 state: wait Windows aio (read thread)
I/O thread 5 state: wait Windows aio (read thread)
I/O thread 6 state: wait Windows aio (write thread)
I/O thread 7 state: wait Windows aio (write thread)
I/O thread 8 state: wait Windows aio (write thread)
I/O thread 9 state: wait Windows aio (write thread)
Pending normal aio reads: [0, 0, 0, 0] , aio writes: [0, 0, 0, 0] ,
 ibuf aio reads:, log i/o's:, sync i/o's:
Pending flushes (fsync) log: 0; buffer pool: 0
28172 OS file reads, 30488 OS file writes, 10597 OS fsyncs
0.00 reads/s, 0 avg bytes/read, 0.00 writes/s, 0.00 fsyncs/s
```

```
----------------------------------
INSERT BUFFER AND ADAPTIVE HASH INDEX
----------------------------------
Ibuf: size 1, free list len 0, seg size 2, 0 merges
merged operations:
 insert 0, delete mark 0, delete 0
discarded operations:
 insert 0, delete mark 0, delete 0
Hash table size 2267, node heap has 1 buffer(s)
Hash table size 2267, node heap has 1 buffer(s)
Hash table size 2267, node heap has 1 buffer(s)
Hash table size 2267, node heap has 1 buffer(s)
Hash table size 2267, node heap has 1 buffer(s)
Hash table size 2267, node heap has 1 buffer(s)
Hash table size 2267, node heap has 1 buffer(s)
Hash table size 2267, node heap has 1 buffer(s)
0.00 hash searches/s, 0.00 non-hash searches/s
---
LOG
---
Log sequence number           42151327
Log buffer assigned up to     42151327
Log buffer completed up to    42151327
Log written up to             42151327
Log flushed up to             42151327
Added dirty pages up to       42151327
Pages flushed up to           42151327
Last checkpoint at            42151327
13848 log i/o's done, 0.00 log i/o's/second
----------------------
BUFFER POOL AND MEMORY
----------------------
Total large memory allocated 0
Dictionary memory allocated 1795369
Buffer pool size   511
Free buffers       109
Database pages     394
Old database pages 0
Modified db pages  0
Pending reads      0
Pending writes: LRU 0, flush list 0, single page 0
Pages made young 0, not young 0
0.00 youngs/s, 0.00 non-youngs/s
Pages read 28147, created 2053, written 11918
0.00 reads/s, 0.00 creates/s, 0.00 writes/s
No buffer pool page gets since the last printout
Pages read ahead 0.00/s, evicted without access 0.00/s, Random read ahead 0.00/s
LRU len: 394, unzip_LRU len: 0
I/O sum[0]:cur[0], unzip sum[0]:cur[0]
--------------
ROW OPERATIONS
--------------
0 queries inside InnoDB, 0 queries in queue
0 read views open inside InnoDB
Process ID=1448, Main thread ID=1040 , state=sleeping
Number of rows inserted 118378, updated 65557, deleted 46, read 198464
0.00 inserts/s, 0.00 updates/s, 0.00 deletes/s, 0.00 reads/s
Number of system rows inserted 3736, updated 4152, deleted 837, read 22291
0.00 inserts/s, 0.00 updates/s, 0.00 deletes/s, 0.00 reads/s
----------------------------
```

```
END OF INNODB MONITOR OUTPUT
============================
```

14.3.3 InnoDB 行级锁的实现方法

InnoDB 行级锁是通过给索引上的索引项加锁来实现的。InnoDB 行级锁只有通过索引条件检索数据，InnoDB 才使用行级锁，否则，InnoDB 将使用表锁。

在不通过索引条件查询的时候，InnoDB 使用的是表锁，而不是行锁。例如，在表 14.12 所示的案例中，表 t_innodb_no_index 没有索引，此时使用的是表锁定。

表14.12 InnoDB行锁定机制的例子

Session1	Session2

创建表 t_innodb_no_index，然后添加数据，命令如下：
mysql> create table t_innodb_no_index(
    id int,
    name varchar(20)
)engine=innodb;	

向表 t_innodb_no_index 表中添加两条数据，命令如下：
mysql>insert into t_innodb_no_index values(1,'data1');
Query OK, 1 row affected (0.03 sec)
 	
mysql> insert into t_innodb_no_index values(2,'data2');
Query OK, 1 row affected (0.00 sec)	

关闭事务自动提交，查询数据，命令如下：
mysql> set autocommit=0;
Query OK, 0 rows affected (0.00 sec)
 	
mysql> select * from t_innodb_no_index where id = 1;
+------+-------+
| id | name |
+------+-------+
| 1 | data1 |
+------+-------+
1 row in set (0.06 sec)
	关闭事务自动提交，查询数据，命令如下：
	mysql> set autocommit=0;
	Query OK, 0 rows affected (0.00 sec)
	
	mysql> select * from t_innodb_no_index where id = 2;
	+------+-------+
	| id | name |
	+------+-------+
	| 2 | data2 |
	+------+-------+
	1 row in set (0.00 sec)

查询数据，命令如下： mysql> select * from t_innodb_no_index where id = 1 for update; +------+-------+ | id | name | +------+-------+ | 1 | data1 | +------+-------+ 1 row in set (0.00 sec)	
	查询数据，命令如下： mysql> select * from t_innodb_no_index where id = 2 for update; 等待

从结果可以看出，在表 t_innodb_no_index 没有索引的情况下，InnoDB 使用的只是表级锁。如

果使用的表包含索引，此时 InnoDB 将会锁定符合条件的行。我们通过例子来理解包含索引的表锁定行的操作过程，具体如表 14.13 所示。

表14.13 InnoDB行锁定机制的例子

Session1	Session2

创建表 t_innodb_lock，命令如下：
mysql> create table t_innodb_lock(a int,b varchar(20)) engine=innodb;
Query OK, 0 rows affected (0.11 sec)
给表 t_innodb_lock 给 a 列添加索引，命令如下：
mysql> create index index_innodb_t on t_innodb_lock(a);
Query OK, 0 rows affected (0.05 sec)
Records: 0　Duplicates: 0　Warnings: 0
给表 t_innodb_lock 给 b 列添加索引，命令如下：
mysql> create index index_innodb_t2 on t_innodb_lock(b);
Query OK, 0 rows affected (0.05 sec)
Records: 0　Duplicates: 0　Warnings: 0
插入数据，命令如下：
mysql>insert into t_innodb_lock values(1,'b1');
Query OK, 1 row affected (0.00 sec)
 	
mysql>insert into t_innodb_lock values(2,'b2');
Query OK, 1 row affected (0.00 sec)	

关闭事务自动提交，查询数据，命令如下：
mysql> set autocommit=0;
Query OK, 0 rows affected (0.06 sec)
mysql> select * from t_innodb_lock where a = 1 for update;
+------+------+
| a　　| b　　|
+------+------+
|　1　| b1　|
+------+------+
1 row in set (0.00 sec)
	关闭事务自动提交，命令如下：
	mysql> set autocommit=0;
	Query OK, 0 rows affected (0.06 sec)

	查询数据，命令如下：
	mysql> select * from t_innodb_lock where a = 2 for update;
	ERROR 3024 (HY000): Query execution was interrupted, maximum statement execution time exceeded
	
	mysql> select * from t_innodb_lock where a = 1 for update;
	ERROR 3024 (HY000): Query execution was interrupted, maximum statement execution time exceeded

(续表)

Session1	Session2
提交事务，命令如下： mysql> commit; Query OK, 0 rows affected (0.00 sec)	
	查询数据，命令如下： mysql> select * from t_innodb_lock where a = 1 for update; +------+------+ | a | b | +------+------+ | 1 | b1 | +------+------+
	提交事务，命令如下： mysql> commit; Query OK, 0 rows affected (0.00 sec)
开始事务，然后更新数据，命令如下： mysql> start transaction; Query OK, 0 rows affected (0.00 sec) mysql> update t_innodb_lock set b='test' where a = 1; Query OK, 1 row affected (0.13 sec) Rows matched: 1 Changed: 1 Warnings: 0	mysql> start transaction; Query OK, 0 rows affected (0.00 sec)
	更新数据，命令如下： mysql> update t_innodb_lock set b='test' where a = 1; 被阻塞，等待
提交事务，命令如下： mysql> commit; Query OK, 0 rows affected (0.00 sec)	
	阻塞解除，更新操作执成功，命令如下： mysql> update t_innodb_lock set b='test2' where a=1; Query OK, 1 row affected (8.42 sec) Rows matched: 1 Changed: 1 Warnings: 0

当表中锁定其中某几行的时候，此时不同的事务可以使用不同的索引锁定不同的行。另外，不论使用主键索引、唯一索引或者普通索引，InnoDB 都会使用行级锁来对数据加锁。下面通过例子来理解 InnoDB 行锁定机制，操作过程如表 14.14 所示。

表14.14　InnoDB行锁定机制的例子

Session1	Session2
提交事务，命令如下： mysql> commit; Query OK, 0 rows affected (0.00 sec) 开启事务，命令如下： mysql> start transaction; Query OK, 0 rows affected (0.00 sec)	提交事务，命令如下： mysql> commit; Query OK, 0 rows affected (0.00 sec) 开启事务，命令如下： mysql> start transaction; Query OK, 0 rows affected (0.00 sec)

(续表)

Session1	Session2
查询数据，命令如下： mysql> select * from t_innodb_lock where a = 1 for update; +------+------+ | a | b | +------+------+ | 1 | test | +------+------+	
	使用 b 列的索引访问记录，可以获得该条记录的行级锁，命令如下： mysql> select *from t_innodb_lock where a = 2 for update; +------+------+ | a | b | +------+------+ | 2 | b2 | +------+------+
	由于 b='test'，a=1 的记录已经被 Session1 锁定，查询该记录时会出现错误提示，命令如下： mysql> select *from t_innodb_lock where b = 'test' for update; ERROR 3024 (HY000): Query execution was interrupted, maximum statement execution time exceeded
提交事务，命令如下： mysql> commit; Query OK, 0 rows affected (0.02 sec)	
	查询 b='test' 的记录，命令如下： mysql> select *from t_innodb_lock where b = 'test' for update; +------+------+ | a | b | +------+------+ | 1 | test | +------+------+

由此可以看出，Session2 查询其他被 Session1 锁定的行，同样会发生阻塞等待锁资源。

14.3.4　间隙锁（Net-Key 锁）

在更新 InnoDB 存储引擎表中某个区间的数据时，此时将会锁定这个区间的所有记录。例如，update xxx where id between 1 and 100，此时它会锁住 id 从 1 到 100 之间所有的记录。值得注意的是，在这个区间中假设某条记录并不存在，该条记录也会被锁住，这个时候，如果另外一个 Session 往这个表中添加一条记录时，此时必须要等到上一个事务释放锁资源。

InnoDB 使用间隙锁的目的，一方面是为了防止幻读，如果没有添加间隙锁，如果其他事务中添加 id 在 1 到 100 之间的某条记录，此时会发生幻读；另一方面是为了满足其恢复和赋值的需求。

下面看一个 InnoDB 存储引擎的间隙锁阻塞的例子，操作过程如表 14.15 所示。

表14.15　InnoDB存储引擎的间隙锁阻塞例子

Session1	Session2

创建数据表 n_innodb_lock 并插入演示数据，命令如下：
`mysql> create table n_innodb_lock(id int) engine=innodb;`
 	
`mysql> insert into n_innodb_lock values(1),(2),(3),(4),(5),(6),(7),(8),(9),(10);`	

查看数据库的隔离级别，命令如下：
`mysql>select @@transaction_isolation;`
`+-------------------------+`
`| @@transaction_isolation |`
`+-------------------------+`
`| REPEATABLE-READ         |`
`+-------------------------+`
`1 row in set (0.00 sec)`
关闭事务的自动提交功能，命令如下：
`mysql> set autocommit = 0;`
`Query OK, 0 rows affected (0.00 sec)`
	查看数据库的隔离级别，命令如下：
	`mysql>select @@transaction_isolation;`
	`+-------------------------+`
	`| @@transaction_isolation |`
	`+-------------------------+`
	`| REPEATABLE-READ         |`
	`+-------------------------+`
	`1 row in set (0.00 sec)`
	关闭事务的自动提交功能，命令如下：
	`mysql> set autocommit = 0;`
	`Query OK, 0 rows affected (0.00 sec)`

对不存在的记录加 for update，命令如下：
`mysql> select max(id) from n_innodb_lock ;`
`+---------+`
`| max(id) |`
`+---------+`
`|      10 |`
`+---------+`
`1 row in set (0.00 sec)`
 	
`mysql> select * from n_innodb_lock where id=100 for update;`
`Empty set (0.00 sec)`	

	此时，插入一条新的记录，命令如下： `mysql>insert into n_innodb_lock(id) values(11);` 等待

执行回滚操作，命令如下：
`mysql> rollback;`
`Query OK, 0 rows affected (13.04 sec)`	

	由于 Session1 回滚操作释放了间隙锁，当前成功添加记录，结果如下：
	　　`mysql> insert into n_innodb_lock(id) values(11);`
	`Query OK, 1 row affected (15.51 sec)`

14.3.5　InnoDB 在不同隔离级别下加锁的差异

在不同的隔离级别下，InnoDB 处理 SQL 语句时所采用的一致性和需要的锁是不同的。

对于 SQL 语句而言，隔离级别越高，InnoDB 存储引擎给记录添加的锁就越严格，产生锁冲突的可能性就越高，对并发的性能影响就越大。因此，应该尽量使用较低的隔离级别，以降低并发中锁争用出现的机率。

对于一些需要使用较高隔离级别的情况，可以通过如下操作更换隔离级别：

```
mysql> set session transaction isolation level repeatable read;
Query OK, 0 rows affected (0.48 sec)

mysql> set session transaction isolation level serializable;
Query OK, 0 rows affected (0.00 sec)
```

14.3.6　InnoDB 存储引擎中的死锁

一般情况下，如果 InnoDB 存储引擎发生了死锁状况，通常是一个事务释放锁并回滚，另一个事务获得锁，继续完成事务。但在涉及外部锁，或涉及表锁情况下，InnoDB 并不能完全自动检测到死锁，此时，需要通过设置锁等待时间（innodb_lock_wait_timeout）来解决。通常情况下，死锁都是应用设计的问题，通过调整业务流程、事务大小、数据库访问的 SQL 语句，大多数死锁都可以避免。下面来看一个 InnoDB 存储引擎发生死锁的例子，操作过程如表 14.16 所示。

表14.16　InnoDB存储引擎发生死锁的例子

Session1	Session2
关闭事务的自动提交功能，命令如下： mysql> set autocommit=0; Query OK, 0 rows affected (0.05 sec)	关闭事务的自动提交功能，命令如下： mysql> set autocommit=0; Query OK, 0 rows affected (0.05 sec)
更新数据，命令如下： mysql> update t_innodb_lock set b='test' where a = 1; Query OK, 1 row affected (0.08 sec) Rows matched: 2 Changed: 1 Warnings: 0	
	更新数据，命令如下： mysql> update t_innodb_lock set b='test' where a = 2; Query OK, 1 row affected (0.03 sec) Rows matched: 1 Changed: 1 Warnings: 0
等待 Session2 释放资源，被阻塞，此时发生死锁，命令如下： mysql> update t_innodb_lock set b='test' where a = 2; ERROR 1213 (40001): Deadlock found when trying to get lock; try restarting transaction	
	获得锁资源，更新操作成功执行，结果如下： mysql> update t_innodb_lock set b='test' where a = 2; Query OK, 1 row affected (38.66 sec) Rows matched: 1 Changed: 1 Warnings: 0

通常在应用中，在 REPEATABLE-READ 隔离级别下，如果两个线程同时以相同的机率使用了排他锁，在没有符合该条件的记录情况下，两个线程都会加锁成功。程序发现记录尚不存在，就试图插入一条新记录，如果两个线程都这么做，就会发生死锁。这种情况下，将隔离级别改成 READ COMMIT，就可以避免死锁。下面通过案例来学习如何避免死锁。

步骤01 创建测试表 innodb_dead_lock，插入测试数据，然后添加索引 dead_index_id，命令如下：

```
mysql> create table innodb_dead_lock(
    id int,
    data varchar(20)
    )engine=innodb;
Query OK, 0 rows affected (0.14 sec)
mysql> insert into innodb_dead_lock values(1,'data');
Query OK, 1 row affected (0.05 sec)
mysql> insert into innodb_dead_lock values(2,'data2');
Query OK, 1 row affected (0.00 sec)
mysql> create index dead_index_id on innodb_dead_lock(id);
Query OK, 0 rows affected (0.08 sec)
Records: 0  Duplicates: 0  Warnings: 0
```

步骤02 接着执行过程如表 14.17 所示。

表14.17 死锁的执行过程

Session1	Session2

关闭事务的自动提交功能，命令如下：
`mysql> set autocommit=0;`
`Query OK, 0 rows affected (0.05 sec)`
查看数据库的隔离级别，命令如下：
`mysql> select @@transaction_isolation;`
`+-------------------------+`
`| @@transaction_isolation |`
`+-------------------------+`
`| REPEATABLE-READ         |`
`+-------------------------+`
`1 row in set (0.00 sec)`
启动事务，命令如下：
`mysql> start transaction;`
`Query OK, 0 rows affected (0.00 sec)`
	关闭事务的自动提交功能，命令如下：
	`mysql> set autocommit=0;`
	`Query OK, 0 rows affected (0.05 sec)`
	查看数据库的隔离级别，命令如下：
	`mysql> select @@transaction_isolation;`
	`+-------------------------+`
	`| @@transaction_isolation |`
	`+-------------------------+`
	`| REPEATABLE-READ         |`
	`+-------------------------+`
	`1 row in set (0.00 sec)`
	启动事务，命令如下：
	`mysql> start transaction;`
	`Query OK, 0 rows affected (0.00 sec)`

查询数据，命令如下：
`mysql> select * from innodb_dead_lock where id=102 for update;`
`Empty set (0.00 sec)`	

	查询数据，命令如下：
	`mysql> select * from innodb_dead_lock where id=102 for update;`
	`Empty set (0.00 sec)`

插入数据，命令如下： `mysql> insert into innodb_dead_lock values(102,'data102');` 等待	

Session1	Session2
	其他 Session 已经对记录进行了更新，此时再插入此条记录就会发生死锁并退出，结果如下： mysql> insert into innodb_dead_lock values(102,'data102'); ERROR 1213 (40001): Deadlock found when trying to get lock; try restarting transaction
由于其他 Session 已经退出，当前 Session 可以获得锁并成功插入记录，结果如下： mysql> insert into innodb_dead_lock values(102,'data102'); Query OK, 1 row affected (29.95 sec)	

14.3.7　InnoDB 行级锁优化建议

InnoDB 存储引擎由于实现了行级锁，很显然，在锁定方面行级锁的颗粒更小，实现更为复杂，所带来的性能损耗也比表级锁更高，但是 InnoDB 行锁在并发性能上远远要高于表级锁定。当系统并发量较高的时候，InnoDB 的整体性能和 MyISAM 相比优势就比较明显了，所以说在选择使用哪种锁的时候，应该考虑到应用的是否有很大的并发量。想要合理使用 InnoDB 的行锁，就要做到扬长避短，具体表现为以下几点：

（1）尽量控制事务的大小，减少锁定的资源量和锁定时间长度。

（2）尽可能让所有的数据检索都通过索引来完成，从而避免因为无法通过索引加锁而升级为表级锁定。

（3）尽可能减少基于范围的数据检索过滤条件，避免因为间隙锁带来的负面影响而锁定了不该锁定的记录。

（4）在业务环节允许的情况下，尽量使用较低级别的事务隔离，以减少因为事务隔离级别锁带来的附加成本。

（5）合理使用索引，让 InnoDB 在索引上面加锁的时候更加准确。

（6）在应用中，尽可能按照相同的访问顺序来访问，防止产生死锁。

（7）在同一个事务中，尽可能做到一次锁定所需的所有资源，减少产生死锁的机率。

（8）对于容易产生死锁的业务，可以放弃使用 InnoDB 行级锁定，尝试使用表级锁定来减少死锁产生的概率。

（9）不要申请超过实际需要的锁级别。

14.4　跳过锁等待

在 MySQL 5.7 版本中，SELECT...FOR UPDATE 语句在执行的时候，如果获取不到锁，会一直等待，直到超时 innodb_lock_wait_timeout。

在 MySQL 8.0 版本中，通过添加 NOWAIT 和 SKIP LOCKED 语法，能够立即返回。如果查询的行已经加锁，那么 NOWAIT 会立即报错返回，而 SKIP LOCKED 也会立即返回，只是返回的结果中不包含被锁定的行。

下面通过表 14.18 来理解 MySQL 8.0 版本中如何跳过锁等待。

表14.18　跳过锁等待的例子

Session1	Session2
创建表 bbs1，然后添加数据，命令如下： mysql> CREATE TABLE bbs1(　　　id int, 　　　name varchar(20)); mysql>INSERT INTO bbs1 (id,name) 　　VALUES(101,'lili'), 　　　(102,'zhangfeng'), 　　　(103,'wangming'), 　　　(104,'wangxiao');	
向数据表 bbs1 中添加排他锁，命令如下： mysql> BEGIN; mysql> SELECT * FROM bbs1 WHERE id = 102 FOR UPDATE; +------+-----------+ | id | name | +------+-----------+ | 102 | zhangfeng | +------+-----------+	
	添加 NOWAIT 语法： 　　mysql> SELECT * FROM bbs1 WHERE id = 102 FOR UPDATE NOWAIT; ERROR 3572 (HY000): Statement aborted because lock(s) could not be acquired immediately and NOWAIT is set.
	添加 SKIP LOCKED 语法： 　　mysql> SELECT * FROM bbs1 WHERE id = 102 FOR UPDATE SKIP LOCKED; Empty set (0.00 sec)

第 15 章

MySQL 服务器性能优化

和大多数数据库一样，MySQL 提供了很多参数来进行服务器的优化设置。MySQL 数据库服务器第一次启动的时候，很多参数都是默认设置的，这并不能完全满足实际生产环境的需求，为此数据库管理员需要进行必要的设置。本章将主要讲解 MySQL 服务器优化方面的一些知识和技巧，包括 MySQL 服务器配置优化、I/O 调优，以及其他优化技巧。

15.1 优化 MySQL 服务器简介

MySQL 服务器主要从两方面来优化：一方面是对硬件进行优化；另一方面是对 MySQL 服务的参数进行优化。这部分的内容需要较全面的知识，一般只有专业的数据库管理员才能进行这一类的优化。对于可以定制参数的操作系统，也可以针对 MySQL 进行操作系统优化。本节将为读者介绍优化 MySQL 服务器的方法。

15.1.1 优化服务器硬件

服务器的硬件性能直接决定着 MySQL 数据库的性能，硬件的性能瓶颈直接决定 MySQL 数据库的运行速度和效率。针对性能瓶颈，提高硬件配置，可以提高 MySQL 数据库查询、更新的速度。本小节将为读者介绍以下优化服务器硬件的方法：

（1）配置较大的内存。足够大的内存是提高 MySQL 数据库性能的方法之一。内存的速度比磁盘 I/O 快得多，可以通过增加系统的缓冲区容量，使数据在内存停留的时间更长，以减少磁盘 I/O。

（2）配置高速磁盘系统，以减少读盘的等待时间，提高响应速度。

（3）合理分布磁盘 I/O，把磁盘 I/O 分散在多个设备上，以减少资源竞争，提高并行操作能力。

（4）配置多处理器。MySQL 是多线程的数据库，多处理器可同时执行多个线程。

15.1.2 优化 MySQL 的参数

通过优化 MySQL 的参数可以提高资源利用率，从而达到提高 MySQL 服务器性能的目的。本小节将为读者介绍这些配置参数。

MySQL 服务的配置参数都在 my.cnf 或者 my.ini 文件的[MySQLd]组中。下面针对几个对性能影响比较大的参数进行详细介绍。

- key_buffer_size：表示索引缓冲区的大小。索引缓冲区所有的线程共享。增加索引缓冲区可以得到更好处理的索引（对所有读和多重写）。当然，这个值也不是越大越好，它的大小取决于内存的大小。如果这个值太大，导致操作系统频繁换页，也会降低系统性能。
- table_cache：表示同时打开的表的个数。这个值越大，能够同时打开的表的个数越多。这个值不是越大越好，因为同时打开的表太多会影响操作系统的性能。
- query_cache_size：表示查询缓冲区的大小。该参数需要和 query_cache_type 配合使用。当 query_cache_type 值是 0 时，所有的查询都不使用查询缓冲区。但是，query_cache_type=0 并不会导致 MySQL 释放 query_cache_size 所配置的缓冲区内存。当 query_cache_type=1 时，所有的查询都将使用查询缓冲区，除非在查询语句中指定 SQL_NO_CACHE，如 SELECT SQL_NO_CACHE * FROM tbl_name。当 query_cache_type=2 时，只有在查询语句中使用 SQL_CACHE 关键字，查询才会使用查询缓冲区。使用查询缓冲区可以提高查询的速度，这种方式只适用于修改操作少且经常执行相同的查询操作的情况。
- sort_buffer_size：表示排序缓存区的大小。这个值越大，进行排序的速度越快。
- read_buffer_size：表示每个线程连续扫描时为扫描的每个表分配的缓冲区的大小（字节）。当线程从表中连续读取记录时需要用到这个缓冲区。SET SESSION read_buffer_size=n 可以临时设置该参数的值。
- read_rnd_buffer_size：表示为每个线程保留的缓冲区的大小，与 read_buffer_size 相似。但主要用于存储按特定顺序读取出来的记录。也可以用 SET SESSION read_rnd_buffer_size=n 来临时设置该参数的值。如果频繁进行多次连续扫描，可以增加该值。
- innodb_buffer_pool_size：表示 InnoDB 类型的表和索引的最大缓存。这个值越大，查询的速度就会越快，但是这个值太大会影响操作系统的性能。
- max_connections：表示数据库的最大连接数。这个连接数不是越大越好，因为这些连接会浪费内存的资源。过多的连接可能会导致 MySQL 服务器无法正常运行。
- innodb_flush_log_at_trx_commit：表示何时将缓冲区的数据写入日志文件，并且将日志文件写入磁盘中。该参数对于 innoDB 引擎非常重要。该参数有 3 个值，分别为 0、1 和 2。值为 0 时，表示每隔 1 秒将数据写入日志文件并将日志文件写入磁盘；值为 1 时，表示每次提交事务时将数据写入日志文件并将日志文件写入磁盘；值为 2 时，表示每次提交事务时将数据写入日志文件，每隔 1 秒将日志文件写入磁盘。该参数的默认值为 1。默认值 1 安全性最高，但是每次事务提交或事务外的指令都需要把日志写入（flush）硬盘，是比较费时的；0 值更快一点，但安全方面比较差；2 值日志仍然会每秒写入到硬盘，所以即使出现故障，一般也不会丢失超过 2 秒的更新。
- back_log：表示在 MySQL 暂时停止回答新请求之前的短时间内，多少个请求可以被存在堆栈中。换句话说，该值表示对到来的 TCP/IP 连接的侦听队列的大小。只有期望在一个短时间内有很多连接时，才需要增加该参数的值。操作系统在这个队列大小上也有限制，设定 back_log 高于操作系统的限制将是无效的。
- interactive_timeout：表示服务器在关闭连接前等待行动的秒数。
- sort_buffer_size：表示每个需要进行排序的线程分配的缓冲区大小。增加这个参数的值可以提高 ORDER BY 或 GROUP BY 操作的速度，默认数值是 2097144 字节（约 2MB）。
- thread_cache_size：表示可以复用的线程数量。如果有很多新的线程，为了提高性能，可以增大该参数的值。
- wait_timeout：表示服务器在关闭一个连接时等待行动的秒数，默认数值是 28800。

合理地配置这些参数可以提高 MySQL 服务器的性能。除上述参数以外，还有 innodb_log_buffer_size、innodb_log_file_size 等参数。配置完参数以后，需要重新启动 MySQL 服务才会生效。

15.2　影响 MySQL 性能的重要参数

在 MySQL 数据库系统中，有些参数直接影响系统的整体性能。MySQL 默认的参数设置往往不能达到实际工作的要求，此时需要通过调整服务器的一些参数来满足实际应用的需求。本小节将介绍一些 MySQL 服务器的参数调整方法，用来优化 MySQL 服务器的性能，这里是一些通用的方法，不涉及具体的存储引擎。

15.2.1　查看性能参数的方法

MySQL 服务启动后，可以使用 SHOW VARIABLES 语句来查询服务器的一些静态参数，比如，缓冲区大小、字符集、数据文件名称等信息，命令如下：

```
mysql> show variables;
+--------------------------+-------------------------------------------+
| Variable_name            | Value                                     |
+--------------------------+-------------------------------------------+
| auto_increment_increment | 1                                         |
| auto_increment_offset    | 1                                         |
| autocommit               | ON                                        |
| automatic_sp_privileges  | ON                                        |
| back_log                 | 50                                        |
| basedir                  | /root/tools/mysql-5.1.30-linux-i686-glibc23/ |
| big_tables               | OFF                                       |
| binlog_cache_size        | 32768                                     |
| binlog_format            | STATEMENT                                 |
| bulk_insert_buffer_size  | 8388608                                   |
| character_set_client     | utf8                                      |
| ...                      | ...                                       |
+--------------------------+-------------------------------------------+
267 rows in set (0.08 sec)
```

SHOW RARIABLES 可以查看 MySQL 启动之前已经配置好的一些系统静态参数。SHOW STATUS 命令可以查询服务器运行中的状态信息，比如，当前连接数、锁等待状态信息，如下所示：

```
mysql> show status;
+-----------------------------+---------+
| Variable_name               | Value   |
+-----------------------------+---------+
| Aborted_clients             | 0       |
| Aborted_connects            | 0       |
| Binlog_cache_disk_use       | 0       |
| Binlog_cache_use            | 0       |
| Bytes_received              | 1567    |
| Bytes_sent                  | 24957   |
| Com_admin_commands          | 0       |
| Com_assign_to_keycache      | 0       |
| Com_alter_db                | 0       |
| Com_alter_db_upgrade        | 0       |
| Com_alter_event             | 0       |
| Com_alter_function          | 0       |
| Com_alter_procedure         | 0       |
| Com_alter_server            | 0       |
| Com_alter_table             | 0       |
| Com_alter_tablespace        | 0       |
| Com_analyze                 | 0       |
| Com_backup_table            | 0       |
| Com_begin                   | 0       |
| Com_binlog                  | 0       |
| Com_call_procedure          | 0       |
| Com_change_db               | 0       |
| Com_change_master           | 0       |
| ...                         | ...     |
+-----------------------------+---------+
290 rows in set (0.80 sec)
```

同时，可以在操作系统下直接查询数据库的状态，命令如下：

```
[root@localhost ~]#mysqladmin -uroot variables
+---------------------------+-------------------------------------------+
| Variable_name             | Value                                     |
+---------------------------+-------------------------------------------+
| auto_increment_increment  | 1                                         |
| auto_increment_offset     | 1                                         |
| autocommit                | ON                                        |
| automatic_sp_privileges   | ON                                        |
| back_log                  | 50                                        |
| basedir                   | /root/tools/mysql-8.0.28-linux-i686-glibc23/ |
| big_tables                | OFF                                       |
| binlog_cache_size         | 32768                                     |
| binlog_format             | STATEMENT                                 |
| bulk_insert_buffer_size   | 8388608                                   |
| character_set_client      | utf8                                      |
| ...                       | ...                                       |
+---------------------------+-------------------------------------------+
267 rows in set (0.08 sec)
```

MySQL 服务器参数比较多，如果需要了解某个参数的含义，可以通过如下命令查询：

```
[root@localhost ~]#mysqld -verbose -help|more
mysqld  Ver 8.0.28-log for pc-linux-gnu on i686 (MySQL Community Server (GPL))
Copyright (C) 2022 MySQL AB, by Monty and others
This software comes with ABSOLUTELY NO WARRANTY. This is free software,
and you are welcome to modify and redistribute it under the GPL license

Starts the MySQL database server

Usage: mysqld [OPTIONS]

Default options are read from the following files in the given order:
/etc/my.cnf /usr/local/mysql/etc/my.cnf ~/.my.cnf
The following groups are read: mysql_cluster mysqld server mysqld-5.0
The following options may be given as the first argument:
--print-defaults    Print the program argument list and exit
--no-defaults       Don't read default options from any options file
--defaults-file=#   Only read default options from the given file #
--defaults-extra-file=# Read this file after the global files are read
 -?, --help         Display this help and exit.
 --abort-slave-event-count=#
                    Option used by mysql-test for debugging and testing of
                    replication.
 --allow-suspicious-udfs
                    Allows use of UDFs consisting of only one symbol xxx()
                    without corresponding xxx_init() or xxx_deinit(). That
                    also means that one can load any function from any
                    library, for example exit() from libc.so
 -a, --ansi         Use ANSI SQL syntax instead of MySQL syntax. This mode
                    will also set transaction isolation level 'serializable'.
 --auto-increment-increment[=#]
                    Auto-increment columns are incremented by this
 --auto-increment-offset[=#]
                    Offset added to Auto-increment columns. Used when
                    auto-increment-increment != 1
 --automatic-sp-privileges
                    Creating and dropping stored procedures alters ACLs.
                    Disable with --skip-automatic-sp-privileges.
 -b, --basedir=name Path to installation directory. All paths are usually
                    resolved relative to this.
 --bdb              Enable Berkeley DB (if this version of MySQL supports
                    it). Disable with --skip-bdb (will save memory).
...
sync-frm                                TRUE
table_cache                             64
table_lock_wait_timeout                 50
thread_cache_size                       0
thread_concurrency                      10
thread_stack                            196608
time_format                             (No default value)
tmp_table_size                          33554432
transaction_alloc_block_size            8192
transaction_prealloc_size               4096
updatable_views_with_limit              1
```

```
wait_timeout                    28800
To see what values a running MySQL server is using, type
'mysqladmin variables' instead of 'mysqld --verbose --help'.
```

从查询的结果可以看出，包含了当前服务器参数的介绍。如果需要查询其中某些参数的含义，可以通过下面的命令过滤查询，查询结果如下：

```
[root@localhost ~]# mysqld --verbose --help|grep thread
  --log-slave-updates Tells the slave to log the updates from the slave thread
                      The number of seconds the slave thread will sleep before
                      set, the slave thread will not be started. Note that the
                      master and where the I/O replication thread is in the
                      The password the slave thread will authenticate with when
  --master-user=name  The username the slave thread will use for authentication
                      for other threads.
                      A dedicated thread is created to, at the given
                      the SQL replication thread is in the relay logs.
                      Tells the slave thread to restrict replication to the
                      Tells the slave thread to restrict replication to the
                      Tells the slave thread to not replicate to the specified
                      Tells the slave thread to not replicate to the specified
                      Tells the slave thread to restrict replication to the
                      Tells the slave thread to not replicate to the tables
  --skip-thread-priority
                      Don't give threads different priorities.
                      Tells the slave thread to continue replication when a
                      have. This comes into play when the main MySQL thread
                      that this is a limit per thread!
                      How long a INSERT DELAYED thread should wait for INSERT
  --flush_time=#      A dedicated thread is created to flush all tables at the
                      Number of times a thread is allowed to enter InnoDB
  --innodb_file_io_threads=#
                      Number of file I/O threads in InnoDB.
  --innodb_thread_concurrency=#
                      environments. Sets the maximum number of threads allowed
                      inside InnoDB. Value 0 will disable the thread
  --innodb_thread_sleep_delay=#
                      Time of innodb thread sleeping before joining InnoDB
  --max_delayed_threads=#
                      Don't start more than this number of threads to handle
  --myisam_repair_threads=#
                      Number of threads to use when repairing MyISAM tables.
                      Each thread that does a sequential scan allocates a
                      exception for replication (slave) threads and users with
                      Number of times the slave SQL thread will retry a
                      If creating the thread takes longer than this value (in
                      seconds), the Slow_launch_threads counter will be
                      Each thread that needs to do a sort allocates a buffer of
  --table_cache=#     The number of open tables for all threads.
  --thread_cache_size=#
                      How many threads we should keep in a cache for reuse.
  --thread_concurrency=#
                      Permits the application to give the threads system a hint
                      for the desired number of threads that should be run at
```

```
    --thread_stack=#         The stack size for each thread.
innodb_file_io_threads              4
innodb_thread_concurrency           8
innodb_thread_sleep_delay           10000
max_delayed_threads                 20
myisam_repair_threads               1
thread_cache_size                   0
thread_concurrency                  10
thread_stack                        196608
```

mysqld --verbose --help|grep thread 命令会把包含 thread 信息查询出来，了解查询 MySQL 服务器参数的方法之后，接下来介绍 MySQL 数据库性能有重要影响的参数。在 MySQL 数据库中，key_buffer_size 和 table_cache 参数仅仅适用于 MyISAM 存储引擎。

15.2.2 key_buffer_size 的设置

在 MySQL 数据库中，key_buffer_size 参数是对 MyISAM 表性能影响最大的一个参数，先来看看 mysqld 中如何定义 key_buffer_size 参数的，如下所示：

```
[root@localhost ~]# mysqld --verbose --help|grep "\-\-key_buffer_size" -A 5
  --key_buffer_size=#  The size of the buffer used for index blocks for MyISAM
                       tables. Increase this to get better index handling (for
                       all reads and multiple writes) to as much as you can
                       afford; 64M on a 256M machine that mainly runs MySQL is
                       quite common.
```

该参数用来设置索引块缓存的大小，只适用于 MyISAM 存储引擎，MySQL 5.1 以后提供了多个 key_buffer，可以将制定的表索引缓存到指定的 key_buffer，这样可以更好地降低线程之间的竞争。下面查询一下 key_buffer_size 的值：

```
mysql> show variables like 'key_buffer_size';
+-----------------+----------+
| Variable_name   | Value    |
+-----------------+----------+
| key_buffer_size | 16777216 |
+-----------------+----------+
1 row in set (0.09 sec)
```

系统分配了索引缓冲区 16MB 的内存，下面修改该参数值到 200MB：

```
mysql> set global key_buffer_size=204800;
Query OK, 0 rows affected (0.64 sec)

mysql> show variables like 'key_buffer_size';
+-----------------+--------+
| Variable_name   | Value  |
+-----------------+--------+
| key_buffer_size | 204800 |
+-----------------+--------+
1 row in set (0.06 sec)
```

上面语句设置默认的 key_buffer。下面介绍如何设置多个 key_buffer，先建立一个索引缓存：

```
mysql> set global hot_cache2.key_buffer_size=128*1024;
Query OK, 0 rows affected (0.05 sec)
```

将相关表的索引放到指定的索引缓存中，下面将表 t 和表 t2 的索引放到 hot_cache2 缓存中：

```
mysql> cache index t,t2 in hot_cache2;
+---------+--------------------+----------+----------+
| Table   | Op                 | Msg_type | Msg_text |
+---------+--------------------+----------+----------+
| test.t  | assign_to_keycache | status   | OK       |
| test.t2 | assign_to_keycache | status   | OK       |
+---------+--------------------+----------+----------+
2 rows in set (0.00 sec)
```

如果想要表 t 的索引加载到默认的缓存（key_buffer）中，可以使用如下语句：

```
mysql> load index into cache t;
+--------+--------------+----------+----------+
| Table  | Op           | Msg_type | Msg_text |
+--------+--------------+----------+----------+
| test.t | preload_keys | status   | OK       |
+--------+--------------+----------+----------+
1 row in set (0.00 sec)
```

如果要删除索引缓存，则只需要设置该缓存区大小为 0 即可，如下所示：

```
mysql> set global hot_cache2.key_buffer_size=0;
Query OK, 0 rows affected (0.00 sec)
```

值得注意的是，不能删除默认的索引缓存区，下面看一下删除后的情况：

```
mysql> show variables like 'key_buffer_size';
+-----------------+--------+
| Variable_name   | Value  |
+-----------------+--------+
| key_buffer_size | 204800 |
+-----------------+--------+
1 row in set (0.02 sec)
mysql> set global key_buffer_size=0;
ERROR 1438 (HY000): Cannot drop default keycache
```

接下来查询 key_buffer_size 的大小，可以发现并没有删除成功，如下所示：

```
mysql> show variables like 'key_buffer_size';
+-----------------+--------+
| Variable_name   | Value  |
+-----------------+--------+
| key_buffer_size | 204800 |
+-----------------+--------+
1 row in set (0.00 sec)
```

Cache index 命令可以将多个表的索引加载到指定的索引缓冲区中，但每次数据库重启后，索引缓冲区中的数据会清空，此时可以考虑在配置文件/etc/my.cnf 中添加执行 init-file 选项，每次服务器启动的时候，自动将指定表的索引加载到缓冲区中，如下所示：

```
[root@localhost ~]# more /etc/my.cnf
[mysqld]
```

```
port     = 3309
...
key_buffer_size = 1G
hot_cache.key_buffer_size = 512M
init_file = /root/initSQL/mysqld_init.sql
...
```

数据库每次启动时，执行/root/initSQL/mysqld_init.sql 脚本文件，该文件可以将多个表索引加载到缓存冲区中，下面的例子将表 t 和表 t1 的索引加载到 hot_cache 缓冲区中：

```
cache index t,t1 in hot_cache;
```

下面语句可以验证下配置是否生效：

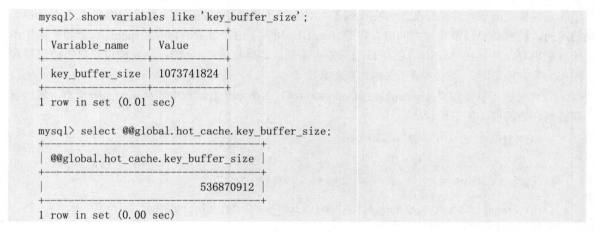

15.2.3　内存参数的设置

对于数据库系统，内存设置都会直接影响到系统的性能，因为内存的读写速度要远远高于磁盘的速度，所以应尽量让数据的读写直接在内存中进行。然而由于系统内存资源的限制，过多使用内存空间必将降低系统系统的整体性能，所以内存的设置需要兼顾服务器的配置。

1. innodb_buffer_pool_size 的设置

在 mysqld 中，对 innodb_buffer_pool_size 参数的定义如下：

```
[root@localhost ~]# mysqld --verbose --help|grep "\-\-innodb_buffer_pool_size" -A 2
 --innodb_buffer_pool_size=#
          The size of the memory buffer InnoDB uses to cache data
          and indexes of its tables.
```

该参数的作用是设置缓存 innodb 表和索引数据。这个值设置越高，访问表中数据需要的磁盘 I/O 越少。该参数的默认值是 128MB，建议不要将该参数设置过大，设置该参数应该遵循以下的分配原则。

（1）如果数据库服务器是一个独立的服务器，可以考虑使用物理内存大小的 70%~80%。
（2）由于该参数不能动态更改，想要修改这个值，需要重启 mysqld 服务。
（3）不要把该参数设置太大，如果分配过大，会导致操作系统的虚拟空间被占用，操作系统变慢，从而减低 SQL 查询的效率。

2. innodb_additional_mem_pool 的设置

在 mysqld 中，对 innodb_additional_mem_pool 参数的定义如下：

```
[root@localhost~]# mysqld --verbose --help|grep "\-\-innodb_additional_mem_pool" -A 3
  --innodb_additional_mem_pool_size=#
                    Size of a memory pool InnoDB uses to store data
                    dictionary information and other internal data
                    structures.
```

这个参数用来设置 InnoDB 存储的数据目录信息和其他内部数据结构的内存池大小。应用程序里的表越多，需要在这里分配的内存就越多。对于一个相对稳定的应用，这个参数的大小也是相对稳定的，没有必要预留非常大的值。如果 InnoDB 用光了这个池内的内存，InnoDB 开始从操作系统分配内存，并且往 MySQL 错误日志写警告信息。其默认值是 1MB，当发现错误日志中已经有相关的警告信息时，就应该适当的增加该参数的大小。

管理员也可以使用"show engine innodb status\G;"命令查询运行中数据库的当前状态，然后再调整到合适的值，如下所示：

```
mysql> show engine innodb status\G;
...
----------------------
BUFFER POOL AND MEMORY
----------------------
Total memory allocated 17445316; in additional pool allocated 869760
Dictionary memory allocated 25752
Buffer pool size   512
Free buffers       488
Database pages     24
Modified db pages  0
Pending reads 0
Pending writes: LRU 0, flush list 0, single page 0
Pages read 24, created 0, written 0
0.00 reads/s, 0.00 creates/s, 0.00 writes/s
No buffer pool page gets since the last printout
...
```

15.2.4　日志和事务参数的设置

关系型数据库的日志与事务管理也是影响数据库性能的重要因素之一，尤其是以事务为主的系统，日志读写比较频繁，进行合理的设置可以降低系统 I/O 和其他资源开销，提升系统性能。

1. innodb_log_file_size 的设置

在 mysqld 中，对 innodb_log_file_size 参数的定义如下：

```
[root@localhost ~]# mysqld --verbose --help|grep "\-\-innodb_log_file_size" -A 1
  --innodb_log_file_size=#
                    Size of each log file in a log group.
```

该参数的作用是设置日志组中每个日志文件的大小。该参数在高写入负载尤其是大数据集的情

况下非常重要,这个值越大则性能相对越高。该参数系统默认值是 5MB。该参数的设置应该遵循以下原则:

(1)如果日志文件大小过小,日志切换比较频繁,会影响服务器性能;如果日志文件过大,当系统灾难时恢复时间会变大。

(2)在 MySQL 5.5 及其以前版本中的 InnODB 的 logfile 最大值设置为 4GB,MySQL 在 5.6 以后的版本中的 logfile 最大值可以设置为 512GB。一般控制在几个日志文件相加大小在 2GB 以内最佳,具体情况还需要以数据文件的大小为依据。

2. innodb_log_files_in_group 的设置

在 mysqld 中,对 innodb_log_files_in_group 参数的定义如下:

```
[root@localhost ~]# mysqld --verbose --help|grep "\-\-innodb_log_files_in_group" -A 3
--innodb_log_files_in_group=#
                      Number of log files in the log group. InnoDB writes to
                      the files in a circular fashion. Value 3 is recommended
                      here.
```

该参数用来指定数据库有几个日志组,默认为 2 个,因为有可能出现跨日志的大事务,所以一般来说,建议使用 3~4 个日志组。

3. innodb_log_buffer_size 的设置

在 mysqld 中,对 innodb_log_buffer_size 参数的定义如下:

```
[root@localhost ~]# mysqld --verbose --help|grep "\-\-innodb_log_buffer_size" -A2
--innodb_log_buffer_size=#
                      The size of the buffer which InnoDB uses to write log to
                      the log files on disk.
```

该参数的作用是设置日志缓存的大小,一旦提交事务,则将该缓存池中的内容写到磁盘的日志文件上。该参数的设置在中等强度写入负载以及较短事务的情况下,一般都可以满足服务器的性能要求。如果服务器负载较大,可以考虑加大该参数的值。缓存池中的内存每秒钟写到磁盘一次,所以设置该参数较大则会浪费内存空间,一般设置为 8~16MB 就足够了。

另外可以参考 Innodb_os_log_written 值,如果这个值增加过快,可以适当地增加该参数的值。

4. innodb_flush_log_at_trx_commit 的设置

在 mysqld 中,对 innodb_flush_log_at_trx_commit 参数的定义如下:

```
[root@localhost~]#mysqld --verbose --help|grep "\-\-innodb_flush_log_at_trx_commit" -A 3
--innodb_flush_log_at_trx_commit[=#]
                      Set to 0 (write and flush once per second), 1 (write and
                      flush at each commit) or 2 (write at commit, flush once
                      per second).
```

该参数用来控制缓冲区的数据写入到日志文件和将日志文件数据刷新到磁盘的操作时间。对该参数的设置可以在数据库性能与数据库安全之间进行折中。

(1)当该参数设置为 0 时,日志缓存每秒一次地被写入到日志文件中,并且将日志文件数据刷新到磁盘。

（2）当该参数设置为 1 时，InnoDB 的事务日志在每次提交后写入日志文件，并对日志做刷新到磁盘的操作。这个可以做到不丢失任何一个事务。

（3）当该参数设置为 2 时，在每个事务提交时，日志缓冲被写到日志文件，但不对日志文件做磁盘刷新的操作，其对日志文件的刷新也是每秒发生一次。但需要注意，由于进程调度方面的问题，并不能保证日志文件的刷新操作每秒一定会发生。

innodb_flush_log_at_trx_commit 参数只有 3 个值，默认为 1，这也是最安全的设置。处于性能考虑，可以设置为 0 或者 2，但会在数据库崩溃的时候丢失一秒钟的事务。为了保证事务的持久性和复制设置的一致性，建议将这个参数设置为 1。

15.2.5 存储和 I/O 相关参数的设置

数据库数据存储和 I/O 性能直接影响到数据库数据文件读写执行的效率，合理地配置存储和 I/O 方面的参数，将会提升数据库整体的性能。

1. innodb_open_files 的设置

在 mysqld 中，对 innodb_open_files 参数的定义如下：

```
[root@localhost ~]# mysqld --verbose --help|grep "\-\-innodb_open_files" -A 2
  --innodb_open_files=#
                      How many files at the maximum InnoDB keeps open at the
                      same time.
```

本参数的作用是限制 InnoDB 存储引擎在同一时间同时能打开表的数量，该参数默认值为 300，如果数据库里面的表特别多，可以考虑增加该参数的值。

2. innodb_flush_method 的设置

在 mysqld 中，对 innodb_flush_method 参数的定义如下：

```
[root@localhost ~]# mysqld --verbose --help|grep "\-\-innodb_flush_method" -A 1
  --innodb_flush_method=name
                      With which method to flush data.
```

该参数的作用是设置 InnoDB 引擎与操作系统进行 I/O 交互的模式，即刷新数据和日志的方法，该参数有三个可选项，含义如下：

（1）Fdatasync：默认值，InnoDB 使用 fsync()函数去更新日志和数据文件。

（2）O_DSYNC：InnoDB 使用 O_SYNC 模式打开并更新日志文件，使用 fsync()函数去更新数据文件。

（3）O_DIRECT：InnoDB 使用 O_DIRECT 模式打开数据文件，使用 fsync()函数去更新日志和数据文件。

3. innodb_max_dirty_pages_pct 的设置

在 mysqld 中，对 innodb_max_dirty_pages_pct 参数的定义如下：

```
[root@localhost ~]# mysqld --verbose --help|grep "\-\-innodb_max_dirty_pages_pct" -A 1
--innodb_max_dirty_pages_pct=#
                  Percentage of dirty pages allowed in bufferpool.
```

该参数的作用是控制 Innodb 的脏页在缓冲中的百分比,将脏页的比例控制在所设定的值百分比之下,如果该参数的值是 50,则脏页在缓冲中最多占 50%。建议该参数设置在 15~90,如果设置太大,缓存中每次更新需要置换的数据页太多;如果设置太小,可以存放脏数据页的缓冲区内存空间会很小,性能会受到一定的影响。

15.2.6 其他重要参数的设置

除了上面介绍的参数外,还有一些参数也会影响服务器的性能。

1. max_connect_errors 的设置

在 mysqld 中,对 max_connect_errors 参数的定义如下:

```
[root@localhost ~]# mysqld --verbose --help|grep '\-\-wait_timeout' -A 2
--wait_timeout=#    The number of seconds the server waits for activity on a
                    connection before closing it.
```

max_connect_errors 默认值为 10,表示 mysqld 线程没重新启动过,一台物理服务器只要连接异常中断累计超过 10 次,就再也无法连接上 mysqld 服务。建议大家设置此值至少大小等于 10。若异常中断累计超过参数设置的值,有两种解决方法:可以重启 mysqld 服务或者执行 FLUSH HOSTS 命令。

2. interactive_timeout 的设置

在 mysqld 中,对 interactive_timeout 参数的定义如下:

```
[root@localhost ~]# mysqld --verbose --help|grep '\-\-interactive_timeout' -A 2
--interactive_timeout=#
                    The number of seconds the server waits for activity on an
                    interactive connection before closing it.
```

该参数用于设置处于交互状态连接的活动被服务器端强制关闭后的等待时间。

3. wait_timeout

在 mysqld 中,对 wait_timeout 参数的定义如下:

```
[root@localhost ~]# mysqld --verbose --help|grep '\-\-wait_timeout' -A 2
--wait_timeout=#    The number of seconds the server waits for activity on a
                    connection before closing it.
```

该参数用于设置客户端与服务器在无交互状态连接,直到被服务器端强制关闭而等待的时间。此参数只有针对基于 TCP/IP 或基于 Socket 通信协议建立的连接才有效。

4. query_cache_type

在 mysqld 中,对 query_cache_type 参数的定义如下:

```
[root@localhost ~]# mysqld --verbose --help|grep '\-\-query_cache_type' -A 3
```

```
                --query_cache_type=#
                        0 = OFF = Don't cache or retrieve results. 1 = ON = Cache
                        all results except SELECT SQL_NO_CACHE ... queries. 2 =
                        DEMAND = Cache only SELECT SQL_CACHE ... queries.
```

该参数用于控制查询结果是否放到查询缓存中。查询缓存类型的可选值只能是 0、1、2，具体选项的含义如下：

（1）该参数选项设置为 0，表示禁止查询缓存的功能。

（2）该参数选项设置为 1，表示启用查询缓存的功能，缓存所有符合要求的查询结果集，除 SELECT SQL_NO_CACHE…，以及不符合查询缓存设置的结果集外。

（3）该参数选项设置为 2，表示仅仅缓存 SELECT SQL_CACHE…子句的查询结果集，除了不符合查询缓存设置的结果集之外。

5. query_cache_size

在 mysqld 中，对 query_cache_size 参数的定义如下：

```
[root@localhost ~]# mysqld --verbose --help|grep '\-\-query_cache_size' -A 1
  --query_cache_size=#
                        The memory allocated to store results from old queries.
```

该参数用来设置查询缓存的大小。需要从以下几个方面考虑如何设置该参数的大小：

（1）查询缓存区对 DDL 和 DML 语句的性能影响。
（2）查询缓存区的内部维护成本。
（3）查询缓存区的命中率以及内存使用率等因素。

15.3　MySQL 日志设置优化

MySQL 安装完成之后，需要通过对其各个参数的修改来进行优化数据库服务的性能。数据库日志对 MySQL 的 I/O 性能有很大的影响，本节将介绍通过对日志参数的设置来提升数据库性能的方法。

下面先来看一下二进制日志文件相关的参数，如下所示：

```
mysql> show variables like '%binlog%';
+-----------------------------------------+--------------+
| binlog_cache_size                       | 32768        |
| binlog_checksum                         | CRC32        |
| binlog_direct_non_transactional_updates | OFF          |
| binlog_encryption                       | OFF          |
| binlog_error_action                     | ABORT_SERVER |
| binlog_expire_logs_seconds              | 2592000      |
| binlog_format                           | ROW          |
| binlog_group_commit_sync_delay          | 0            |
| binlog_group_commit_sync_no_delay_count | 0            |
| binlog_gtid_simple_recovery             | ON           |
| binlog_max_flush_queue_time             | 0            |
```

```
| binlog_order_commits                                 | ON                   |
| binlog_rotate_encryption_master_key_at_startup       | OFF                  |
| binlog_row_event_max_size                            | 8192                 |
| binlog_row_image                                     | FULL                 |
| binlog_row_metadata                                  | MINIMAL              |
| binlog_row_value_options                             |                      |
| binlog_rows_query_log_events                         | OFF                  |
| binlog_stmt_cache_size                               | 32768                |
| binlog_transaction_compression                       | OFF                  |
| binlog_transaction_compression_level_zstd            | 3                    |
| binlog_transaction_dependency_history_size           | 25000                |
| binlog_transaction_dependency_tracking               | COMMIT_ORDER         |
| innodb_api_enable_binlog                             | OFF                  |
| log_statements_unsafe_for_binlog                     | ON                   |
| max_binlog_cache_size                                | 18446744073709547520 |
| max_binlog_size                                      | 1073741824           |
| max_binlog_stmt_cache_size                           | 18446744073709547520 |
| sync_binlog                                          | 1                    |
```

其中 binlog_cache_size 默认大小是 32768，即 32KB。该参数表示在事务中允许二进制日志缓存的大小。如果要设置该参数，首先需要了解下 binlog_cache_size 的使用情况，如下所示：

```
mysql> show status like '%binlog%';
+----------------------------+-------+
| Variable_name              | Value |
+----------------------------+-------+
| Binlog_cache_disk_use      | 16    |
| Binlog_cache_use           | 339   |
| Binlog_stmt_cache_disk_use | 0     |
| Binlog_stmt_cache_use      | 54    |
| Com_binlog                 | 0     |
| Com_show_binlog_events     | 0     |
| Com_show_binlogs           | 0     |
+----------------------------+-------+
```

其中影响 binlog_cashe_size 性能的参数主要如下：

（1）binlog_cache_use：表示使用二进制日志缓存事务的数量。

（2）binlog_cache_disk_use：该参数表示使用二进制日志缓存，并且值达到 binlog_cache_size 设置的值，此时用临时文件存储事务的数量。

max_binlog_cache_size 参数表示二进制日志所使用缓存的最大值，该参数的默认值是 18446744073709547520，该默认值已经足够大了。

max_binlog_size 参数代表二进制日志使用的最大值，如果系统中事务过多，而此参数值设置过小，则会报错，该参数默认值是 1073741824（代表 1GB），该参数一般设置为 512MB 或者 1GB。

sync_binlog 参数比较重要，不仅对数据完整性有影响，而且对数据库的性能也有影响。该参数的选项如下所示：

（1）sync_binlog=0：表示当事务提交之后，不做文件系统之类的磁盘同步指令刷新 binlog_cache 中的信息到磁盘，而会让文件系统自行决定同步，或者 cache 满了之后才同步到磁盘。

（2）sync_binlog=n：表示当事务提交 n 次之后，将进行一次同步，把 binlog_cache 中的数据写入到磁盘。

值得注意的是，sync_binlog=0 是数据库默认的设置，性能是最好的，不过也是最危险的，一旦系统崩溃，binlog_cache 中的数据就会丢失。sync_binlog=1 的时候，数据库消耗最大，不过安全性比较高，当事务提交后，数据库会将缓存中的数据同步到磁盘中。

通常在 MySQL 的复制过程中，实际上就是将客户端的日志通过 I/O 拷贝到服务器端，然后服务器端将日志中的数据解析出来应用到数据库中。

15.4 MySQL I/O 设置优化

目前大多数应用都是以 I/O 密集型为主，存储技术远没有计算机中其他系统发展迅速，尽管也有不少高端存储设备，不过考虑到成本因素，一般用户承受不起，目前大多数采用 SAS 盘结合应用不同的 RAID 组合来实现存储。

事实上，前面提到的 SQL 优化、数据库对象优化、数据库参数优化，以及应用程序优化等，都是想通过减少或延缓磁盘的读写来减轻磁盘 I/O 的压力。另外，管理员可以通过增强磁盘 I/O 的吞吐量以及增强磁盘 I/O 本身的性能，从而提高数据库的整体性能。

下面针对磁盘 I/O 进行不同类型的优化。

1. 选择合适的 RAID（磁盘阵列）级别

RAID 是 Redundant Array of Independent Disk 的缩写，就是将 N 台硬盘通过 RAID Controller 结合成虚拟的单台大容量硬盘使用。根据数据冗余和分布方式，RAID 分为不同的级别，表 15.1 所示的是常见的几种 RAID 级别。

表15.1 常见RAID级别的比较

RAID 级别	特 性	优 点	缺 点
RAID 0	也叫条带化（Stripe），按一定的条带大小将数据依次分布到各个磁盘，没有数据冗余	数据并发读写速度快，无额外磁盘空间开销	数据无冗余保护，可靠性差
RAID 1	也叫磁盘镜像（Mirror），两个磁盘一组，所有数据都同时写入两个磁盘，但读时从任一磁盘读都可以	可靠性高，并发读写性能优良	容量一定的话，需要 2 倍的磁盘，投资大
RAID 10	也叫 ARID 1+0，是 RAID 0 与 RAID 1 的组合体，集成了 RAID 0 快速和 RAID 1 的安全	可靠性能高，并发读写性能优良	RAID 10 会造成 50%的磁盘的浪费，投资比较大
RAID 3	RAID 3 是把数据分成多个块，按照一定的容错算法，存放在 N+1 个硬盘上，事实上空间是 N 个磁盘的空间总和，第 N+1 个磁盘存储的信息是校验容错信息	当 N+1 个硬盘中出现故障时，从其他 N 个硬盘中恢复原始数据	RAID 3 校验磁盘很容易成为整个系统的瓶颈。校验盘的负载如果很大，会导致整个 RAID 系统性能的下降

(续表)

RAID 级别	特 性	优 点	缺 点
RAID 4	跟 RAID 0 一样对磁盘组条带化，不同的是：需要额外增加一个磁盘，用来写各校验纠错数据	RAID 中一个磁盘损坏，其他数据可以通过校验纠错数据计算出来，读数据速度快	在失败恢复的时候，难度可要比 RAID 3 大，控制器的设计难度也要大许多，而且访问数据的效率不高
RAID 5	RAID 5 是一种存储性能、数据安全和存储成本兼顾的存储解决方案，它将每个条带的校验纠错数据块也分布到各个磁盘，而不是写到一个磁盘	跟 RAID 4 相似，只是其写性能和数据保护能力要更强一点	容错能力不及 RAID 1，在磁盘出现损坏时候，读写能会下降

了解了 RAID 级别的特性后，就可以根据数据读写的特点、可靠性要求，以及投资额度来选择合适的 RAID 级别了，比如：

（1）数据读写都比较频繁，最好选择 RAID 10。

（2）数据读很频繁，写相对少一点，对可靠性有一定的要求，可以考虑选择 RAID 5。

（3）数据读写比较频繁，但是对可靠性能要求不高，可以选择 RAID 10。

2. 使用 Symbolic Links 分布 I/O

使用操作系统的符号链接（Symbolic Links）将不同的数据库或表、索引指向不同的物理磁盘，从而达到分布磁盘 I/O 的目的。可以通过 SHOW VARIABLES LIKE 'have_symlink' 语句，检查系统是否支持符号链接：

```
mysql> SHOW VARIABLES LIKE 'have_symlink';
+---------------+-------+
| Variable_name | Value |
+---------------+-------+
| have_symlink  | YES   |
+---------------+-------+
1 row in set (0.00 sec)
```

在不使用 RAID 或者逻辑卷的情况下，要想达到分布式磁盘 I/O 的目的，可以使用操作系统的 Symbolic Links 将不同的数据库、表或索引链接到不同的物理磁盘上面。

（1）将一个数据库指向其他物理磁盘，首先在目标磁盘上创建目录，然后再创建从 MySQL 数据目录到目标目录的符号链接，命令如下所示：

```
[root@localhost ~]# mkdir /otherdisk/databases/test
[root@localhost ~]# ln -s /otherdisk/databases/test /path/to/datadir
```

（2）对于新建的 MyISAM 表，可以使用 DATA DIRECTORY 和 INDEX DIRECORY 来将表的数据文件或索引文件指向其他物理磁盘，命令如下所示：

```
CREATE TABLE test(id int primary key,name varchar(20))
Type = myisam
DATA DIRECOTORY ='/disk2/data'
INDEX DIRECTORY= '/disk3/index'
```

注意，对于已经存在的 MyISAM 表，可以将.MYD 或者.MYI 文件移到目标磁盘，然后再建立符号链接即可。

15.5　MySQL 并发设置优化

MySQL 数据库并发性能是影响数据整体性能的一个重要的参数，通过提高并发连接数量来提升并发效率，下面将讲解如何优化数据库连接的最大数。

首先查看一下并发连接的最大数和目前连接的最大数，命令如下：

```
mysql> show variables like 'max_connections';
+-----------------+-------+
| Variable_name   | Value |
+-----------------+-------+
| max_connections | 1000  |
+-----------------+-------+
1 row in set (0.00 sec)

mysql> show global status like 'max_used_connections';
+----------------------+-------+
| Variable_name        | Value |
+----------------------+-------+
| Max_used_connections | 125   |
+----------------------+-------+
1 row in set (0.00 sec)
```

参数 max_connections 表示允许同时连接 MySQL 数据库客户端的最大数量，max_used_connections 表示同时使用的最大连接数。max_used_connections 占 max_connections 的 80%左右比较合适。一般来讲，max_used_connections 参数在 MySQL 主机性能允许的范围内，设置 500~800 比较合适；而参数 max_used_connections 是用户正在连接最大的数量，可以尽量设置大一些。

参数 net_buffer_length 设置的只是消息缓冲区的初始化大小，系统默认值是 16KB，当然如果系统内存比较紧张的情况下，可以考虑将该参数降低到 8KB 左右比较合适。查看命令如下：

```
mysql> show variables like 'net%';
+-------------------+-------+
| Variable_name     | Value |
+-------------------+-------+
| net_buffer_length | 16384 |
| net_read_timeout  | 30    |
| net_retry_count   | 10    |
| net_write_timeout | 60    |
+-------------------+-------+
4 rows in set (0.00 sec)
```

参数 max_allowed_packet 表示在网络传输中，一次传递数据量的最大值。系统默认值为 1MB，最大值是 4GB，必须将该参数设置为 1024 的倍数，单位为字节。查看命令如下：

```
mysql> show variables like 'max_allowed_packet%';
+--------------------+-------+
| Variable_name      | Value |
```

```
| max_allowed_packet | 4194304 |
1 row in set (0.00 sec)
```

参数 back_log 表示在 MySQL 的连接请求等待队列中允许存放的最大连接请求数。系统默认的值是 80，最大可以设置为 65535。查看命令如下：

```
mysql> show variables like 'back_log%';
+---------------+-------+
| Variable_name | Value |
+---------------+-------+
| back_log      | 80    |
+---------------+-------+
1 row in set (0.00 sec)
```

15.6　服务器语句超时处理

在 MySQL 8.0 中可以设置服务器语句超时的限制，单位可以达到毫秒级别。

当中断的执行语句超过设置的超时毫秒数后，服务器将终止查询影响不大的事务或连接，然后将错误报给客户端。

设置服务器语句超时的限制，可以通过设置系统变量 max_execution_time 来实现。例如：

```
SET GLOBAL MAX_EXECUTION_TIME=2000;
```

默认情况下，MAX_EXECUTION_TIME 的值为 0，代表没有时间限制。通过上述设置后，如果 SELECT 语句执行超过 2000 毫秒，语句将会被终止执行。

设置服务器语句超时的限制，也可以通过设置系统变量 max_execution_time 来实现。该变量用于设置 SELECT 语句运行在一个特定的会话里，指定该会话的超时时间。例如：

```
SET SESSION MAX_EXECUTION_TIME=2000;
```

通过上述设置后，如果 SELECT 语句执行超过 2000 毫秒，会话将会被终止。

15.7　线程和临时表的优化

除了上面介绍的优化方法以外，还有一些因素也能提高 MySQL 的整体性能。例如，线程管理和临时表管理。

15.7.1　线程的优化

线程的优化主要是优化 MySQL 中与连接线程相关的系统参数以及状态变量。在 MySQL 中实现了一个线程池，将空闲的连接线程存放在线程池中，如果有新的请求，MySQL 会检查线程池中是否有空闲的连接线程，如果没有，则会创建新的线程；如果线程池中有空闲的线程，则会直接从线程池中取出空闲的线程。

查看与连接线程相关的系统参数，如下所示：

```
mysql> show variables like 'thread%';
+---------------------+---------------------------+
| Variable_name       | Value                     |
+---------------------+---------------------------+
| thread_cache_size   | 64                        |
| thread_concurrency  | 10                        |
| thread_handling     | one-thread-per-connection |
| thread_stack        | 196608                    |
+---------------------+---------------------------+
4 rows in set (0.04 sec)
```

参数 thread_cache_size 表示数据库线程池中应该存放的连接线程数。thread_cache_size 默认配置 thread_cache_size=8。建议根据物理内存来设置这个参数，1GB 内存可以设置为 8，2GB 内存可以设置为 16，3GB 内存可以考虑设置 32，大于 3GB 的内存可以设置成 64。

参数 thread_stack 表示每个连接线程被创建的时候，MySQL 分配内存的大小。系统默认值是 192KB，MySQL 的 max_connections * thread_stack 应小于可用内存。

15.7.2　临时表的优化

临时文件就是为了各种不同的目的而产生的中间文件。使用完毕后会被及时回收和清理。临时表也是如此，是 MySQL 在进行一些内部操作的时候生成的数据库表。这些操作主要包括：group by、distinct、一些 order by 查询语句、UNION、一些 from 语句中的子查询等。

可以使用 EXPLAIN 来分析查询语句，看看是否会用到临时表。EXPLAIN 输出中的 EXTRA 列会指明是否"Using temporary"。事实上，大多数用户都不会去关注临时表的产生、使用与消亡的过程，因为它对用户是透明的。但是对于数据库管理员或者其他关心性能的人员而言，就不得不关注临时表，因为如果设置不当或者是程序使用不当，可能会产生大量的磁盘临时表，对系统性能产生很大的影响。

什么是磁盘临时表呢？除了直接产生的磁盘临时表外，大量磁盘临时表是由内存临时表转化来的。临时表存在于内存中，由 MEMORY 引擎进行处理，速度较快。而磁盘临时表则是在磁盘上创建、使用、销毁的。磁盘是慢速访问设备，因此，磁盘临时表的操作效率要比临时表的操作差了几个数量级，具有较差的性能，需要尽量避免磁盘临时表的产生。

查看临时文件、临时表和磁盘临时表的命令如下：

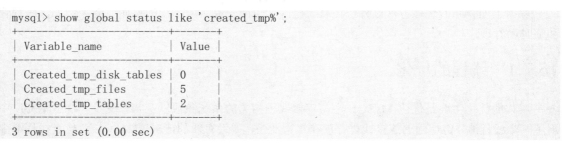

数据库系统每次创建临时表时，Created_tmp_tables 会增加，如果是在磁盘上创建临时表，

Created_tmp_disk_tables 也会增加，Created_tmp_files 表示 MySQL 数据库创建的临时文件的数量。

接下来看一下 MySQL 数据库对临时表的配置，如下所示：

```
mysql> show variables where
Variable_name in('max_heap_table_size','tmp_table_size');
+---------------------+-----------+
| Variable_name       | Value     |
+---------------------+-----------+
| max_heap_table_size | 16777216  |
| tmp_table_size      | 536870912 |
+---------------------+-----------+
2 rows in set (0.00 sec)
```

max_heap_table_size 参数表示 heap 数据表的最大长度，默认设置是 16MB。值得注意的是，存在 text 或者 blog 字段时，会导致不能用内存临时表而改用磁盘临时表。

tmp_table_size 参数规定了内存临时表的最大值（推荐该参数设置 200MB），如果内存临时表超过了最大值限制，数据库会将数据存储到磁盘临时表中，存储的目录默认是/tmp/，如下所示：

```
mysql> show variables like 'tmpdir';
+---------------+-------+
| Variable_name | Value |
+---------------+-------+
| tmpdir        | /tmp  |
+---------------+-------+
1 row in set (0.01 sec)
```

15.8 增加资源组

MySQL 8.0 新增了一个资源组功能，用于调控线程优先级以及绑定 CPU。MySQL 用户需要有 RESOURCE_GROUP_ADMIN 权限才能创建、修改、删除资源组。

注意，在 Linux 环境下，MySQL 进程需要有 CAP_SYS_NICE 权限才能使用资源组的完整功能。

```
[root@localhost~]# sudo setcap cap_sys_nice+ep /usr/local/mysql8.0/bin/mysqld
[root@localhost~]# getcap /usr/local/mysql8.0/bin/mysqld
/usr/local/mysql8.0/bin/mysqld = cap_sys_nice+ep
```

MySQL 8.0 默认提供两个资源组，分别是 USR_default 和 SYS_default。下面讲解一下资源组的常用操作。

创建名称为 my_resouce_group 的资源组，执行语句如下：

```
mysql> CREATE RESOURCE GROUP my_resouce_group type=USER vcpu=0,1 thread_priority=5;
Query OK, 0 rows affected (0.07 sec)
```

将当前线程加入资源组：

```
mysql> SET RESOURCE GROUP my_resouce_group;
Query OK, 0 rows affected (0.00 sec)
```

查看资源组 my_resouce_group 中包含的线程，执行语句如下（具体参看读者执行结果）：

```
mysql> SELECT * FROM Performance_Schema.threads WHERE RESOURCE_GROUP='my_resouce_group';
```

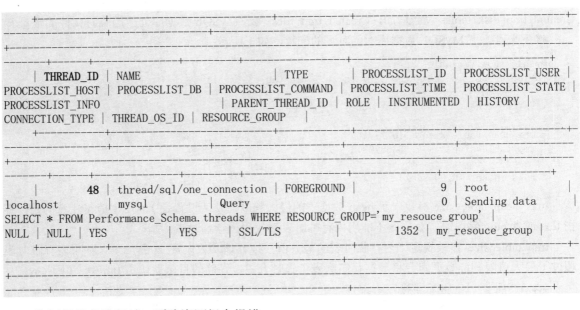

资源组里有线程时，删除资源组会报错：

```
mysql> DROP RESOURCE group my_resouce_group;
ERROR 3656 (HY000): Resource group test_resouce_group is busy.
```

修改资源组的语句如下：

```
mysql> alter resource group my_resouce_group vcpu = 2,3 THREAD_PRIORITY = 8;
Query OK, 0 rows affected (0.10 sec)
```

把资源组里的线程移出到默认资源组 USR_default 中，执行语句如下：

```
mysql> SET RESOURCE GROUP USR_default FOR 48;
Query OK, 0 rows affected (0.00 sec)
```

删除资源组的语句如下：

```
mysql> DROP RESOURCE group my_resouce_group;
Query OK, 0 rows affected (0.08 sec)
```

如果需要指定线程的 ID，可以使用以下语句：

```
SET RESOURCE GROUP my_resouce_group FOR thread_id;
```

这里的 thread_id 为线程的 ID。查询线程 ID 的方法如下：

```
SELECT thread_id FROM Performance_Schema.threads;
```

例如，将 ID 为 20 的线程加入到资源组 my_resouce_group 中，语句如下：

```
SET RESOURCE GROUP test_resouce_group FOR 20;
```

第 16 章

MySQL 性能监控

在 MySQL 数据库中，通常使用三方软件来监控数据库性能健康状况。本章将重点介绍监控工具——Nagios。使用 Nagios 监控工具不仅可以监控 MySQL 数据库性能，还可以对数据库库集群、数据复制等操作进行监控。因为数据库性能的瓶颈不仅仅和数据库本身的优化情况有关，还和操作系统本身的情况以及网络负载的状况有关。本章还将介绍如何监控 Linux 操作系统 I/O 和 CPU 的运行状况，通过本章的学习可以更好地对 MySQL 数据库及服务器的性能进行全方位的监控。

16.1 基本监控系统方法

在 Linux 操作系统中，用户可以使用一些分析系统性能的命令去分析数据库服务器的性能。本节主要介绍这些命令的使用方法。

16.1.1 ps 命令

ps 命令主要用来获取对于某个进程的一些信息。ps 命令的常用参数如下所示：

（1）-A：显示所有进程。
（2）-a：显示一个终端的所有进程，除了会话引线。
（3）-N：忽略选择。
（4）-d：显示所有进程，但省略所有的会话引线。
（5）-e：列出程序时，显示每个程序所使用的环境变量。
（6）-x：显示没有控制终端的进程，同时显示各个命令的具体路径。dx 参数不可合用。
（7）-p pid：进程使用 CPU 的时间。
（8）-u uid or username：选择有效的用户 id 或者用户名。
（9）-g gid or groupname：显示组的所有进程。

（10）-f：全部列出，通常和其他选项联用。

（11）-l：长格式（有 F、wchan、C 等字段）。

（12）-j：作业格式。

（13）-e：命令之后显示环境。

例如，运行一下 ps -aux 命令显示所有进程的信息，如下所示：

```
[root@localhost ~]# ps -aux
Warning: bad syntax, perhaps a bogus '-'? See /usr/share/doc/procps-3.2.7/FAQ
USER       PID %CPU %MEM    VSZ   RSS TTY      STAT START   TIME COMMAND
root         1  0.0  0.1   1948   740 ?        Ss   Nov19   0:03 /sbin/init
root         2  0.0  0.0      0     0 ?        S<   Nov19   0:00 [kthreadd]
...
nagios   20727  0.0  0.1  12952  1012 ?        Ssl  11:12   0:03 /var/www/nagios/bin/nagios -d /
nagios   21037  0.0  0.1   4960   908 ?        Ss   11:23   0:00 ./nrpe -c /var/www/nagios/etc/n
```

可以使用 ps –ef 命令查询所有进程及其环境变量信息，如下所示：

```
[root@localhost ~]# ps -ef
UID        PID  PPID  C STIME TTY          TIME CMD
root         1     0  0 Nov19 ?        00:00:03 /sbin/init
root         2     0  0 Nov19 ?        00:00:00 [kthreadd]
...
root     17334     1  0 Nov20 ?        00:00:00 /bin/sh /usr/local/mysql/bin/mysqld_safe --datadir
nagios   21037     1  0 11:23 ?        00:00:00 ./nrpe -c /var/www/nagios/etc/nrpe.cfg -d
nagios   23516 23515  0 14:52 ?        00:00:00 /var/www/nagios/libexec/check_ssh 127.0.0.1,192.16
root     23524 20118  3 14:52 pts/3    00:00:00 ps -ef
```

通常查看了进程信息后，如果需要终止某个进程，可以使用 kill 命令，如下所示：

```
[root@localhost ~]#kill -KILL [pid]
```

如果需要强行终止某个进程的话，可以使用 kill -9 [pid]命令，如下所示：

```
[root@localhost ~]#kill -9 [pid]
```

16.1.2　top 命令

top 命令是 Linux 操作系统下最常用的性能分析工具，能够实时地显示操作系统中各个进程消耗资源的情况。该命令可以显示 CUP 使用、内存使用和执行时间。下面执行 top 命令看一下执行的情况：

```
[root@localhost ~]# top
top - 15:12:58 up 3 days,  4:13,  4 users,  load average: 0.29, 0.27, 0.28
Tasks: 138 total,   3 running, 133 sleeping,   1 stopped,   1 zombie
Cpu(s): 24.4%us,  9.2%sy,  0.0%ni, 66.4%id,  0.0%wa,  0.0%hi,  0.0%si,  0.0%st
Mem:    673164k total,   650156k used,    23008k free,    90248k buffers
Swap:   524280k total,       52k used,   524228k free,   185692k cached

  PID USER      PR  NI  VIRT  RES  SHR S %CPU %MEM    TIME+  COMMAND
```

```
 2464 root      20   0 47108  20m 7888 R 15.9  3.1  39:50.58 Xorg
20115 root      20   0 90728  18m  11m S  4.0  2.8   0:07.86 gnome-terminal
...
   62 root      15  -5     0    0    0 S  0.0  0.0   0:00.00 kacpi_notify
  120 root      15  -5     0    0    0 S  0.0  0.0   0:00.00 cqueue
  122 root      15  -5     0    0    0 S  0.0  0.0   0:00.04 ksuspend_usbd
  127 root      15  -5     0    0    0 S  0.0  0.0   0:01.09 khubd
```

下面分析一下 top 命令的统计信息的含义：

```
top - 15:12:58 up 3 days,  4:13,  4 users,  load average: 0.29, 0.27, 0.28
```

先来分析 top 命令第一行信息，该行信息的具体含义如下：

（1）5:12:58 表示系统运行的当前时间。

（2）up 3 days 表示系统运行时间。

（3）4 users 表示登录用户的数量。

（4）load average: 0.29, 0.27, 0.28 表示系统负载，即任务队列的平均长度，三个数值分别为 1 分钟、5 分钟、15 分钟到现在的平均值。

接着分析 top 命令的第二行和第三行的具体含义，这两行分别给出进程和 CPU 性能的一些信息：

```
Tasks: 138 total,   3 running, 133 sleeping,   1 stopped,   1 zombie
Cpu(s): 24.4%us,  9.2%sy,  0.0%ni, 66.4%id,  0.0%wa,  0.0%hi,  0.0%si,  0.0%st
```

下面我们具体说明一下各个统计信息的含义：

（1）138 total 表示进程的总数。

（2）3 running 表示正在运行的进程数量。

（3）133 sleeping 表示睡眠的进程数量。

（4）1 stopped 表示停止的进程数量。

（5）1 zombie 表示僵尸进程数。

（6）24.4%us 表示用户空间占用 CPU 百分比。

（7）9.2%sy 表示内核空间占用 CPU 百分比。

（8）0.0%ni 表示用户进程空间内改变过优先级的进程占用 CPU 百分比。

（9）66.4%id 表示空闲 CPU 百分比。

（10）0.0%wa 表示等待输入输出的 CPU 百分比。

接着分析 top 命令的第四行和第五行的具体含义，这两行分别表示内存的性能分析信息，如下。

```
Mem:    673164k total,   650156k used,    23008k free,    90248k buffers
```

下面具体说明一下 Mem 各个统计信息的含义：

（1）673164k total 表示物理内存的总的大小。

（2）650156k used 表示使用的物理内存的大小。

（3）23008k free 表示空闲的物理内存的大小。

（4）90248k buffers 表示内核缓存内存空间的大小。

```
Swap:   524280k total,       52k used,   524228k free,   185692k cached
```

下面具体分析一下 Swap 各个统计信息的含义：

（1）524280k total 表示交换区的总量。
（2）52k used 表示使用的交换区总量。
（3）524228k free 表示空闲交换区总量。
（4）185692k cached 表示缓冲的交换区总量。内存中的内容被换出到交换区，而后又被换入到内存，但使用过的交换区尚未被覆盖，该数值即为这些内容已存在于内存中的交换区的大小。相应的内存再次被换出时，可不必再对交换区写入。

16.1.3　vmstat 命令

vmstat 命令可以用来显示 Linux 性能指标，该命令分别输出进程、内存、交互区、I/O、系统和 CPU 的情况。该命令执行情况如下所示：

```
[root@localhost ~]# vmstat
procs -----------memory---------- ---swap-- -----io---- --system-- -----cpu-----
 r  b   swpd   free   buff  cache   si   so    bi    bo   in   cs us sy id wa st
 2  0    404  12264 101280 164840    0    0     3    33    6  113  4  2 94  0  0
```

首先，分析一下进程（procs）的两列信息，如下所示：

（1）r 列表示可运行进程的数量。
（2）b 列表示阻塞进程的数量。

内存性能有 4 个报告虚拟内存如何使用的字段，具体意义如下：

（1）swpd 表示已经使用的交换空间的数量。
（2）free 表示自由 RAM 数量。
（3）buff 表示缓存使用的 RAM 数量。
（4）cache 表示文件系统缓存使用的 RAM 数量。

swap 交换字段说明如下：

（1）si 表示从磁盘分页到内存的数量。
（2）so 表示从内存分页到磁盘的数量。

io 字段说明如下：

（1）bi 表示从磁盘读入的块。
（2）bo 表示写入磁盘的块。

下面对系统字段和 CPU 字段进行说明，CPU 状态使用总 CPU 时间的百分比来表示：

（1）in 表示系统中断。
（2）cs 表示进程上下文开关。
（3）us 表示用户模式。
（4）sy 表示内核模式。

(5) wa 表示等待 I/O。

(6) id 表示空闲状态。

16.1.4 mytop 命令

mytop 是一个类似于 Linux 系统 top 命令风格的 MySQL 监控工具，可以监控当前用户正在执行的命令。

首先，使用源码安装 mytop-1.6.tar.gz，如下所示：

```
[root@localhost ~]# tar -zxvf mytop-1.6.tar.gz
[root@localhost ~]# cd mytop-1.6
[root@localhost mytop-1.6]# perl Makefile.PL
[root@localhost mytop-1.6]# make
[root@localhost mytop-1.6]# make test
[root@localhost mytop-1.6]# make install
[root@localhost mytop-1.6]#
```

直接输入 mytop 命令，结果发生错误，如下所示：

```
[root@localhost mytop-1.6]# mytop
Can't locate DBI.pm in @INC (@INC contains: /usr/lib/perl5/5.10.0/i386-linux-thread-multi /usr/lib/perl5/5.10.0 /usr/lib/perl5/site_perl/5.10.0/i386-linux-thread-multi /usr/lib/perl5/site_perl/5.10.0 /usr/lib/perl5/site_perl/5.8.8 /usr/lib/perl5/site_perl/5.8.7 /usr/lib/perl5/site_perl/5.8.6 /usr/lib/perl5/site_perl/5.8.5 /usr/lib/perl5/site_perl /usr/lib/perl5/vendor_perl/5.10.0/i386-linux-thread-multi /usr/lib/perl5/vendor_perl/5.10.0 /usr/lib/perl5/vendor_perl/5.8.8 /usr/lib/perl5/vendor_perl/5.8.7 /usr/lib/perl5/vendor_perl/5.8.6 /usr/lib/perl5/vendor_perl/5.8.5 /usr/lib/perl5/vendor_perl .) at /usr/bin/mytop line 22.
BEGIN failed--compilation aborted at /usr/bin/mytop line 22.
```

发现报错了，Can't locate DBI.pm in @INC 这种问题主要原因是系统没有安装 DBI 组件，可以安装 DBI、Data-ShowTable、DBD-mysql（假设已安装完 Perl 和 MySQL 数据库）。

下面可以直接到 ftp://ftp.funet.fi/pub/languages/perl/CPAN/modules/by-module 下载 3 个文件 DBI-1.601.tar.gz、Data-ShowTable-3.3.tar.gz、DBD-mysql-3.0007_1.tar.gz，分别放入 DBI、Data、DBD 目录下。

首先，先解压这 3 个压缩文件，如下所示：

```
[root@localhost ~]#tar zxvf DBI-1.601.tar.gz
[root@localhost ~]#tar zxvf Data-ShowTable-3.3.tar.gz
[root@localhost ~]#tar zxvf DBD-mysql-3.0007.tar.gz
```

下面安装 DBI，如下所示：

```
[root@localhost ~]# cd DBI-1.601
[root@localhost DBI-1.601]# perl ./Makefile.PL
[root@localhost DBI-1.601]# make
[root@localhost DBI-1.601]# make test
[root@localhost DBI-1.601]# make install
```

安装 Data-ShowTable，如下所示：

```
[root@localhost ~]# cd Data-ShowTable-3.3
[root@localhost Data-ShowTable-3.3]# perl ./Makefile.PL
```

```
[root@localhost Data-ShowTable-3.3]# make
[root@localhost Data-ShowTable-3.3]# make install     （注：无需 make test）
```

安装 DBD-mysql，如下所示：

```
[root@localhost src]# cd DBD-mysql-3.0007_1
[root@localhost DBD-mysql-3.0007_1]# perl ./Makefile.PL
--libs="-L/usr/local/mysql-6.0.9-alhpa/lib/mysql -lmysqlclient -lz -lrt -lcrypt -lnsl
-lm"   --cflags=" -I/usr/local/mysql-6.0.9-alpha/include/mysql -g -DUNIV_LINUX" --testuser=root
--testsocket=/home/cserken/mysql/tmp/mysql.sock
[root@localhost DBD-mysql-3.0007_1]# make
[root@localhost DBD-mysql-3.0007_1]# make test
[root@localhost DBD-mysql-3.0007_1]# make install
```

测试一下，结果问题还没有解决，发现没有安装 mytop 所需 TermReadKey，如下所示：

```
[root@localhost ~]# tar -xzvf TermReadKey-2.30.tar.gz
[root@localhost ~]# cd TermReadKey-2.30
[root@localhost TermReadKey-2.30]# perl ./Makefile.PL
[root@localhost TermReadKey-2.30]# make
[root@localhost TermReadKey-2.30]# make install
```

下面测试一下 mytop 命令，结果显示正常运行了，如下所示：

```
[root@localhost ~]# cd mytop-1.6
[root@localhost mytop-1.6]# mytop
MySQL on localhost (5.5.27-ndb-7.2.8-cluster-gpl-log)      up 4+11:10:38 [19:18:03]
   Queries: 3.0      qps:      0 Slow:     0.0       Se/In/Up/De(%):    00/00/00/00
[root@localhost ~]#
Key Efficiency: 75.0%  Bps in/out:   0.0/   0.0

      Id        User          Host/IP           DB        Time   Cmd Query or State
      --        ----          -------           --        ----   --- --------------
      3835      root          localhost         test         0   Query show full process
```

下面先了解一下 mytop 命令参数的含义：

（1）-u / --user <USERNAME> 指定 username，默认是 root。

（2）-p / --pass / --password <PASSWORD>：指定 password，默认是 none。

（3）-h / --host <HOSTNAME[:PORT]>：指定 MySQL Server 的 hostname，默认是 localhost。

（4）-P / --port <PORT>：指定连接 MySQL Server 的 port，默认是 3306。

（5）-s / --delay <SECONDS>：更新的秒数，默认是 5 秒。

（6）-d / --db / --database <DATABASE>：指定连接的数据库，默认是 test。

（7）-b / --batch / --batchmode：指定为 batch mode，每次更新不会清除旧的显示结果，会将更新资料显示在最上方，默认是 unset。

（8）-S / --socket <PATH_TO_SOCKET>：指定使用 MySQL Socket 直接连接，而不使用 TCP/IP 连线，默认是 none（当 mytop 和 MySQL 装在同一台计算机上时才能使用）。

（9） --header or -noheader：是否要显示表头，默认是 header。

（10）--color or --nocolor：是否要使用颜色，默认是 color。

（11）-i / -idle or -noidle：idle 的 thread 是否要出现在清单上，默认是 idle。

先在 MySQL 数据库中创建 xfimti 用户，如下所示：

```
mysql> grant all privileges on *.* to 'xfimti'@'localhost' identified by '123';
Query OK, 0 rows affected (0.01 sec)

mysql> flush privileges;
Query OK, 0 rows affected (0.02 sec)
```

输入 mytop 命令，将显示监控信息如下：

```
[root@localhost ~]# cd mytop-1.6
[root@localhost mytop-1.6]# ./mytop -u xfimti -p123 -d test -s /temp/mysql.socket

MySQL on localhost (5.5.27-ndb-7.2.8-cluster-gpl-log)    up 4+11:39:16 [19:46:41]
Queries: 3.0     qps:     0 Slow:     0.0       Se/In/Up/De(%):     00/00/00/00

Key Efficiency: 83.3%  Bps in/out:    0.0/  0.0

      Id      User         Host/IP         DB        Time   Cmd Query or State
      --      ----         -------         --        ----   --- --------------
    3875      xfimti       localhost       test         0   Query show full process
```

16.1.5 sysstat 工具

sysstat 工具包含检测系统性能及效率的一组工具。例如 CPU 的使用率、硬盘和网络吞吐数据，这些数据的收集和分析，有利于判断系统是否正常运行。如果是通过源码包安装，请到官方下载源码包，地址为 http://perso.wanadoo.fr/sebastien.godard。

Sysstat 软件包集成如下工具：

（1）iostat 工具：提供 CPU 使用率及硬盘吞吐效率的数据。
（2）mpstat 工具：提供单个处理器或多个处理器相关数据。
（3）sar 工具：负责收集、报告并存储系统活跃的信息。
（4）sa1 工具：负责收集并存储每天系统动态信息到一个二进制的文件中。
（5）sa2 工具：负责把每天的系统活跃信息写入总结性的报告中。
（6）sadc 工具：系统动态数据收集工具，收集的数据被写一个二进制的文件中。
（7）sadf 工具：显示被 sar 工具通过多种格式收集的数据。

下面开始安装 sysstat 工具包，如下所示：

```
[root@localhost ~]# tar zxvf sysstat-10.0.0.tar.gz
[root@localhost sysstat-10.0.0]# ./configure
[root@localhost sysstat-10.0.0]# make
[root@localhost sysstat-10.0.0]# make install
```

1. iostat 工具

iostat 工具用于输出 CPU 和磁盘 I/O 相关的统计信息，具体语法格式如下：

```
iostat [-c|-d] [-k] [-t] [-V] [-x] [-p device|ALL] [间隔描述] [检测次数]
```

上述参数的含义如下：

（1）-c：表示仅显示 CPU 的状态。

（2）-d：仅显示存储设备的状态，不允许和-c 一起使用。

（3）-k：默认显示读入读出的块信息。

（4）-t：显示搜集数据的时间。

（5）-V：显示版本号和帮助信息。

（6）-x：显示扩展信息。

（7）-p device|ALL：device 为某个设备或者某个分区，如果使用 ALL，就表示要显示所有分区和设备的信息。

直接运行 iostat 命令，结果如下所示：

```
[root@localhost ~]# iostat
Linux 2.6.25-14.fc9.i686 (localhost.localdomain)    11/23/2012    _i686_    (1 CPU)

avg-cpu:  %user   %nice %system %iowait  %steal   %idle
           3.60    0.04    1.53    0.49    0.00   94.34

Device:            tps   kB_read/s   kB_wrtn/s   kB_read    kB_wrtn
sda               1.66        2.47       32.28     800397   10474295
dm-0              8.34        2.46       32.28     797729   10473872
dm-1              0.00        0.00        0.00        592        404
sdb               0.00        0.01        0.00       4433          1
```

结果中关于 CPU 性能参数的含义如下：

（1）%user：在用户级别运行所使用的 CPU 百分比。

（2）%nice：nice 操作所使用的 CPU 百分比。

（3）%system：在系统级别（kernel）运行所使用的 CPU 百分比。

（4）%iowait：CPU 等待硬件 I/O 时，所占用的 CPU 百分比。

（5）%idle：表示 CPU 空闲时间所占比例。

结果中关于磁盘 I/O 性能参数的含义如下：

（1）Device：表示设备块的名字。

（2）tps：表示每秒钟发送到的 I/O 请求数。

（3）kB_read/s：表示从该设备每秒读取的数据块数量。

（4）kB_wrtn/s：表示从该设备每秒写入的数据块数量。

（5）kB_read：表示从该设备读取的数据块总数。

（6）kB_wrtn：表示从该设备写入的数据块总数。

使用-x 参数可以获得更多的统计信息，如下所示（请读者执行命令对照结果）：

```
[root@localhost init.d]# iostat -d -x -k 1 10
Linux 2.6.25-14.fc9.i686 (localhost.localdomain)    11/23/2012    _i686_    (1 CPU)

Device:         rrqm/s   wrqm/s     r/s     w/s    rkB/s    wkB/s avgrq-sz avgqu-sz   await
r_await w_await  svctm  %util
    sda           0.03     6.65    0.25    1.42     2.47    32.28    41.81     0.16   98.24
  28.60  110.40   6.41   1.07
```

```
      dm-0              0.00    0.00    0.27    8.07    2.46   32.28    8.33    1.35  161.51
 34.13  165.77   1.28    1.06
      dm-1              0.00    0.00    0.00    0.00    0.00    0.00    8.00    0.00   53.71
 52.81   55.04   7.02    0.00
       sdb              0.02    0.00    0.00    0.00    0.01    0.00   37.26    0.00   16.76
 15.00  225.50   9.53    0.00
```

结果中各个的参数的含义如下:

（1）rrqm/s：表示每秒这个设备相关的读取请求有多少被合并。

（2）wrqm/s：表示每秒这个设备相关的写入请求有多少被合并。

（3）r/s：表示每秒请求读该设备的数量。

（4）w/s：表示每秒请求写该设备的数量。

（5）await：表示每一个 IO 请求的处理的平均时间。

（6）%util：表示在统计时间内所有处理 I/O 时间，除以总共统计时间。

此外可以通过如下命令查询 CPU 的部分信息，如下所示：

```
[root@localhost ~]# iostat -c 1 10
avg-cpu:  %user   %nice    %sys %iowait   %idle
           1.98    0.00    0.35   11.45   86.22
avg-cpu:  %user   %nice    %sys %iowait   %idle
           1.62    0.00    0.25   34.46   63.67
```

也可以通过以下命令查询某个具体的设备块的信息，如下所示：

```
[root@localhost ~]# iostat -d -k 1 |grep sda10
Device:             tps    kB_read/s    kB_wrtn/s    kB_read    kB_wrtn
sda10             60.72        18.95        71.53  395637647 1493241908
sda10            299.02      4266.67       129.41       4352        132
sda10            483.84      4589.90      4117.17       4544       4076
sda10            218.00      3360.00       100.00       3360        100
sda10            546.00      8784.00       124.00       8784        124
sda10            827.00     13232.00       136.00      13232        136
```

2. mpstat 工具

mpstat 是系统实时监控工具，主要报告 CPU 的一些信息，该命令的语法如下：

```
mpstat [-P {|ALL}] [internal [count]]
```

下面分析一下具体参数的含义：

（1）-P {|ALL}：表示需要监控哪个 CPU。

（2）internal：表示相邻的两次采样的时间间隔。

（3）count：表示采样的次数。

下面通过示例来理解一下该命令的具体用法。

（1）显示所有 CPU 的信息，每秒钟执行一次。

```
[root@localhost ~]# mpstat -P ALL 1
Linux 2.6.25-14.fc9.i686 (localhost.localdomain)        11/23/2012        _i686_  (1 CPU)
```

```
07:49:21 AM  CPU   %usr   %nice  %sys   %iowait  %irq   %soft  %steal  %guest  %idle
07:49:22 AM  all   0.99   0.00   0.99   0.00     0.00   0.00   0.00    0.00    98.02
07:49:22 AM  0     0.99   0.00   0.99   0.00     0.00   0.00   0.00    0.00    98.02

07:49:22 AM  CPU   %usr   %nice  %sys   %iowait  %irq   %soft  %steal  %guest  %idle
07:49:23 AM  all   3.03   0.00   3.03   0.00     0.00   1.01   0.00    0.00    92.93
07:49:23 AM  0     3.03   0.00   3.03   0.00     0.00   1.01   0.00    0.00    92.93
```

（2）显示 ID 为 0 的 CPU 的信息，每秒钟执行一次。

```
[root@localhost ~]# mpstat -P 0 1
Linux 2.6.25-14.fc9.i686 (localhost.localdomain)   11/23/2012   _i686_   (1 CPU)

10:42:38 PM  CPU   %user  %nice  %system  %iowait  %irq   %soft  %idle   intr/s
10:42:43 PM  all   6.89   0.00   44.76    0.10     0.10   0.10   48.05   1121.60
10:42:43 PM  0     9.20   0.00   49.00    0.00     0.00   0.20   41.60   413.00
10:42:43 PM  1     4.60   0.00   40.60    0.00     0.20   0.20   54.60   708.40
10:42:43 PM  CPU   %user  %nice  %system  %iowait  %irq   %soft  %idle   intr/s
10:42:48 PM  all   7.60   0.00   45.30    0.30     0.00   0.10   46.70   1195.01
10:42:48 PM  0     4.19   0.00   2.20     0.40     0.00   0.00   93.21   1034.53
10:42:48 PM  1     10.78  0.00   88.22    0.40     0.00   0.00   0.20    160.48
Average:     CPU   %user  %nice  %system  %iowait  %irq   %soft  %idle   intr/s
Average:     all   7.25   0.00   45.03    0.20     0.05   0.10   47.38   1158.34
Average:     0     6.69   0.00   25.57    0.20     0.00   0.10   67.43   724.08
Average:     1     7.69   0.00   64.44    0.20     0.10   0.10   27.37   434.17
```

该结果的具体参数项的含义如下所示：

（1）%user：表示处理用户进程所使用 CPU 的百分比。

（2）%nice：表示使用 nice 命令对进程进行降级时 CPU 的百分比。

（3）%sys：表示内核进程使用 CPU 的百分比。

（4）%iowait：表示表示等待进行 I/O 所使用的 CPU 百分比。

（5）%irq：表示用于处理系统中断的 CPU 百分比。

（6）%soft：表示用于软件中断的 CPU 百分比。

（7）%idle：显示 CPU 的空闲时间。

（8）%intr/s：显示每秒 CPU 接收的中断总数。

3. sar 工具

sar 是目前 Linux 最为全面的系统性能分析工具之一，它可以从多方面对系统的活动进行报告，提供文件的读写情况、系统调用的使用情况、磁盘 I/O、CPU 效率、内存使用状况、进程活动及 IPC 有关的活动等信息。

sar 的语法格式如下：

```
sar [options] [-A] [-o file] t [n]
```

其中：t 为采样间隔，n 为采样次数，默认值是 1；-o file 表示将命令结果以二进制格式存放在文件中，file 是文件名。options 为命令行选项，选项的含义如下：

（1）-A：所有报告的总和。

（2）-u：输出 CPU 使用情况的统计信息。

（3）-v：输出 inode、文件和其他内核表的统计信息。
（4）-d：输出每一个块设备的活动信息。
（5）-r：输出内存和交换空间的统计信息。
（6）-b：显示 I/O 和传送速率的统计信息。
（7）-a：文件读写情况。
（8）-c：输出进程统计信息，每秒创建的进程数。
（9）-R：输出内存页面的统计信息。
（10）-y：终端设备活动情况。
（11）-w：输出系统交换活动信息。

下面通过一个示例来理解 sar 工具的使用方法，如下所示：

```
[root@localhost ~]# sar -u 1 5
Linux 2.6.25-14.fc9.i686 (localhost.localdomain)     11/24/2012      _i686_   (1 CPU)

02:33:56 AM       CPU     %user     %nice    %system   %iowait    %steal     %idle
02:33:57 AM       all      2.06      0.00       5.15      0.00      0.00     92.78
02:33:58 AM       all      3.00      0.00       6.00      0.00      0.00     91.00
02:33:59 AM       all     41.41      0.00       9.09      0.00      0.00     49.49
02:34:00 AM       all     96.77      0.00       3.23      0.00      0.00      0.00
```

结果中具体参数的含义如下所示。

（1）CPU：all 表示统计信息为所有 CPU 的平均值。
（2）%user：显示在用户级别（Application）运行使用 CPU 总时间的百分比。
（3）%nice：显示在用户级别，用于 nice 操作，所占用 CPU 总时间的百分比。
（4）%system：在核心级别（Kernel）运行所使用 CPU 总时间的百分比。
（5）%iowait：显示用于等待 I/O 操作占用 CPU 总时间的百分比。
（6）%steal：管理程序（Hypervisor）为另一个虚拟进程提供服务而等待虚拟 CPU 的百分比。
（7）%idle：显示 CPU 空闲时间占用 CPU 总时间的百分比。

提示：如果%iowait 的值过高，表示硬盘存在 I/O 瓶颈；如果%idle 的值高但系统响应慢时，有可能是 CPU 等待分配内存，此时应加大内存容量；若%idle 的值持续低于 10，则系统的 CPU 处理能力相对较低，表明系统中最需要解决的资源是 CPU。

16.2　开源监控利器 Nagios 实战

Nagios 是一个用来监控主机、服务和网络的开源软件。在实际工作中需要监控的对象主要是主机资源监控和网络服务监控。主机资源监控包括监控系统负载、当前 IP 连接数、磁盘空间使用情况、当前进程数以及自定义的资源监控等；网络服务监控包括主机存活检查、Web 服务监控、FTP 服务监控、数据库服务监控、自定义服务器监控等。

16.2.1　安装 Nagios 之前的准备工作

在安装 Nagios 之前，需要做以下必要的工作：

（1）使用 Fedora 操作系统，并且在安装 Nagios 时候拥有 root 使用权限。

（2）使用 root 用户登录系统安装 Apache 服务器，首先下载 httpd-2.2.20.tar.gz 安装包，解压复制到/user/src/目录下面。

```
[root@localhost ~]# mv httpd-2.2.0.tar.gz /usr/src
[root@localhost ~]# gunzip httpd-2.2.0.tar.gz
[root@localhost ~]# tar -xvf httpd-2.2.0.tar
[root@localhost ~]# cd /usr/src
[root@localhost src]# ls
httpd-2.2.0   httpd-2.2.0.tar   redhat
[root@localhost src]# cd httpd-2.2.0
[root@localhost httpd-2.2.0]# ls
ABOUT_APACHE        configure         LAYOUT            README
acinclude.m4        configure.in      libhttpd.dsp      README.platforms
Apache.dsw          docs              LICENSE           ROADMAP
apachenw.mcp.zip    emacs-style       Makefile.in       server
build               httpd.dsp         Makefile.win      srclib
BuildBin.dsp        httpd.spec        modules           support
buildconf           include           NOTICE            test
CHANGES             INSTALL           NWGNUmakefile     VERSIONING
config.layout       InstallBin.dsp    os

[root@localhost httpd-2.2.0]# ./configure --prefix=/usr/local/apache
[root@localhost httpd-2.2.0]# make
[root@localhost httpd-2.2.0]# make install
```

（3）执行/usr/local/apache/bin/apachetl -t 命令，检测 Apache 的配置文件语法是否正确，如下所示：

```
[root@localhost ~]# cd /usr/local/apache/bin/
[root@localhost bin]# ls
ab              apu-1-config    dbmmanage       htcacheclean    htpasswd        logresolve
apachectl       apxs            envvars         htdbm           httpd           rotatelogs
apr-1-config    checkgid        envvars-std     htdigest        htttxt2dbm
[root@localhost bin]# ./apachectl -t
Syntax OK
```

（4）修改/etc/profile 文件修改 PATH 路径，直接把/usr/local/apache/bin 的路径加入到环境变量 PATH 中，方便使用执行 Apache 的命令：

```
[root@localhost ~]# vi /etc/profile
```

（5）在该文件的最后直接添加下面两条语句，设置环境变量，如下所示：

```
PATH=$PATH:/usr/local/apache/bin
export PATH
```

这样可以直接运行 apachetl start 命令，从而启动 Apache 服务，如下所示：

```
[root@localhost ~]#apachetl start
```

接下来，可以在浏览器中输入 http://IP 地址，测试 Apache 服务是否成功启动。

（6）将 Apache 服务设置为系统启动服务，如下所示：

```
[root@localhost ~]#chkconfig httpd on
```

16.2.2　安装 Nagios 主程序

用户可以到 http://www.nagios.org 页面下载 Nagios 安装文件，然后进行安装。本示例以安装 nagios-3.4.1.tar.gz 为例进行讲解，操作步骤说明如下。

步骤01 解压 nagios-3.4.1.tar.gz，如下所示：

```
[root@localhost ~]# useradd nagios
[root@localhost ~]# mkdir /usr/local/nagios
[root@localhost ~]# chown nagios.nagios /usr/local/nagios
[root@localhost ~]# ls
anaconda-ks.cfg                      MySQL Cluster.txt
Desktop                              nagios
Documents                            nagios-3.4.1.tar.gz
Download                             network.txt
install.log                          node.txt
install.log.syslog                   Pictures
libaio-0.3.96-3.i386.rpm             Public
Music                                Templates
my.cnf.bak                           Unsaved Document 1
Mysql5.5                             Videos
mysql-cluster-gpl-7.2.8-linux2.6-i686.tar
[root@localhost ~]# tar zxvf nagios-3.4.1.tar.gz
```

步骤02 开始进行编译，如下所示：

```
[root@localhost ~]# cd nagios
[root@localhost nagios]# ./configure --prefix=/usr/local/nagios
[root@localhost nagios]# make all
```

步骤03 Nagios 主程序编译成功后，如果在编译过程中没有发生错误，接下来进行安装，如下所示：

```
[root@localhost nagios]# make install
[root@localhost nagios]# make install-config
[root@localhost nagios]# make install-init
[root@localhost nagios]# make install-commandmode
[root@localhost nagios]# ls /usr/local/nagios
bin  etc  libexec  sbin  share  var
```

Nagios 服务器安装完成后，在安装目录/usr/local/nagios 下生成下面的目录，如表 16.1 所示。

表16.1　Nagios服务器目录的具体作用

目录	作用
bin	Nagios 执行程序所在的目录，这个目录只有一个文件 nagios
etc	Nagios 配置文件位置，初始安装完后，只有几个*.cfg-sample 文件
sbin	Nagios CGI 文件所在目录，也就是执行外部命令所需文件的目录
share	Nagios 网页文件所在的目录
var	Nagios 日志文件、spid 等文件所在的目录

步骤04 将 Nagios 的启动脚本添加到系统服务中，如下所示：

```
[root@localhost ~]# vi /etc/profile
```

步骤05 在该文件的最后直接添加下面两句设置环境变量的语句，如下所示：

```
PATH=$PATH:/usr/local/nagios/bin
export PATH
[root@localhost ~]# chkconfig -add nagios
```

步骤06 执行以下命令，将 Nagios 服务添加到系统服务中去。

```
[root@localhost ~]# chkconfig nagios on
```

16.2.3 整合 Nagios 到 Apache 服务

Nagios 安装完成后，即可将其添加到 Apache 服务中，具体操作步骤如下：

步骤01 将 Apache 服务宿主用户加入到 nagios 组里面，这样做的目的是为了使 Apache 有适当的权限通过 CGI 命令对 Nagios 进行调用，例如通过浏览器界面关闭 Nagios 服务。

```
[root@localhost ~]# cd /usr/local/nagios/
[root@localhost nagios]# usermode -G nagios apache
[root@localhost nagios]# usermod -G nagios apache
```

步骤02 打开/etc/httpd/conf/httpd.conf 文件，Apache 的运行用户和运行组默认是 daemon，需要把运行用户和运行组改成"nagios"。

```
User nagios
Group nagios
```

步骤03 将 Nagios 的相关信息加入到 Apache 配置文件中，打开/etc/httpd/conf/httpd.conf 文件，在文件最后添加如下代码：

```
#setting for nagios
ScriptAlias /nagios/cig-bin /usr/local/nagios/sbin
<Directory "/usr/local/nagios/sbin">
    AuthType Basic
    Options ExecCGI
    AllowOverride None
    Order allow,deny
    Allow from all
    AuthName "Nagios Access"
    AuthUserFile /usr/local/nagios/etc/htpasswd
    Require valid-user
</Directory>

Alias /nagios /usr/local/nagios/share
<Directory "/usr/local/nagios/share">
    AuthType Basic
    Options None
    AllowOverride None
    Order allow,deny
    Allow from all
    AuthName "nagios Access"
```

```
        AuthUserFile /usr/local/nagios/etc/htpasswd
        Require valid-usr
</Directory>
```

上述文本的作用主要是用来对 Nagios 目录进行用户验证，只有合法的用户才能访问 Nagios 的页面，配置完成之后，可以执行 apachctl -t 检测 Apache 配置文件的语法是否存在错误。

```
[root@localhost ~]# apachctl -t
Syntax OK
```

步骤 04 创建 Web 接口，如下所示：

```
[root@localhost ~]# cd nagios
[root@localhost nagios]# make install-webconf
/usr/bin/install -c -m 644 sample-config/httpd.conf /etc/httpd/conf.d/nagios.conf

*** Nagios/Apache conf file installed ***
```

步骤 05 创建一个用于登录 nagios 的用户名和密码，如下所示：

```
[root@localhost ~]# /usr/bin/htpasswd -c /usr/local/nagios/etc/htpasswd.users nagios
New password:
Re-type new password:
Adding password for user nagios
```

步骤 06 重启 Apache 服务，重启 Nagios 服务，如下所示：

```
[root@localhost ~]# httpd -k restart
[root@localhost ~]# service nagios restart
Running configuration check...done.
Stopping nagios: done.
Starting nagios: done.
```

步骤 07 从另外一台计算机上登录 http://IP/nagios，此时会要求输入上面创建的用户名和密码，输入正确的用户名和密码之后，登录到 Nagios 服务器主页面，如图 16.1 所示。

图 16.1 Nagios 服务器主页面

配置 Nagios 的时候，经常会遇到提示 Internal Server Error 错误，内容如下：

```
Internal Server Error
The server encountered an internal error or misconfiguration and was unable to complete your
```

request.
   Please contact the server administrator, root@localhost and inform them of the time the error occurred, and anything you might have done that may have caused the error.
   More information about this error may be available in the server error log.

此时查看一下 Apache 服务器的错误日志信息，如下所示：

```
[root@localhost ~]# vi /var/log/httpd/error_log
   [Tue Nov 20 09:33:30 2012] [error] [client 192.168.0.208] (13)Permission denied: exec of '/usr/local/nagios/sbin/status.cgi' failed, referer: http://192.168.0.100/nagios/side.php
   [Tue Nov 20 09:33:30 2012] [error] [client 192.168.0.208] Premature end of script headers: status.cgi, referer: http://192.168.0.100/nagios/side.php
   [Tue Nov 20 09:34:49 2012] [error] [client 192.168.0.208] PHP Warning: MagpieRSS: Failed to fetch http://www.nagios.org/backend/feeds/frontpage/ and cache is off in /usr/local/nagios/share/includes/rss/rss_fetch.inc on line 238, referer: http://192.168.0.100/nagios/main.php
```

错误的含义是 Apache 开启了 suexec 的功能，对执行 CGI 的路径进行了限制。先用命令检测一下 Apache 是否打开了 suexec 功能，如下所示：

```
[root@localhost sysconfig]# httpd -V
Server version: Apache/2.2.8 (Unix)
Server built:   Feb 25 2008 07:05:32
Server's Module Magic Number: 20051115:11
Server loaded:  APR 1.2.12, APR-Util 1.2.12
Compiled using: APR 1.2.12, APR-Util 1.2.12
Architecture:   32-bit
Server MPM:     Prefork
  threaded:     no
    forked:     yes (variable process count)
Server compiled with....
 -D APACHE_MPM_DIR="server/mpm/prefork"
 -D APR_HAS_SENDFILE
 -D APR_HAS_MMAP
 -D APR_HAVE_IPV6 (IPv4-mapped addresses enabled)
 -D APR_USE_SYSVSEM_SERIALIZE
 -D APR_USE_PTHREAD_SERIALIZE
 -D SINGLE_LISTEN_UNSERIALIZED_ACCEPT
 -D APR_HAS_OTHER_CHILD
 -D AP_HAVE_RELIABLE_PIPED_LOGS
 -D DYNAMIC_MODULE_LIMIT=128
 -D HTTPD_ROOT="/etc/httpd"
 -D SUEXEC_BIN="/usr/sbin/suexec"
 -D DEFAULT_PIDLOG="logs/httpd.pid"
 -D DEFAULT_SCOREBOARD="logs/apache_runtime_status"
 -D DEFAULT_LOCKFILE="logs/accept.lock"
 -D DEFAULT_ERRORLOG="logs/error_log"
 -D AP_TYPES_CONFIG_FILE="conf/mime.types"
 -D SERVER_CONFIG_FILE="conf/httpd.conf"
```

从结果可以看出，Apache 已经启用了 suexec 功能。执行下列命令，可以获知 Apache 如何限制 CGI 执行的路径，如下所示：

```
[root@localhost sysconfig]# suexec -V
 -D AP_DOC_ROOT="/var/www"
 -D AP_GID_MIN=100
```

```
-D AP_HTTPD_USER="apache"
-D AP_LOG_EXEC="/var/log/httpd/suexec.log"
-D AP_SAFE_PATH="/usr/local/bin:/usr/bin:/bin"
-D AP_UID_MIN=500
-D AP_USERDIR_SUFFIX="public_html"
```

CGI 程序只能在/var/www/目录下面才能够正常运行，因为将 Nagios 安装在/usr/local/nagios 目录下面的，所以 CGI 程序无法执行。最简单的解决方法就是将 Nagios 安装到/var/www/nagios 就可以了。值得注意的是，编译的时候需要指定—prefix=/var/www/nagios，其他编译安装选项不用改变。

如果不想重新编译，可以将/usr/local目录下面的nagios移动到/var/www目录下面，然后修改相关的配置文件。值得注意的是，nagios目录下面所有cfg的路径都需要修改，所以建议读者还是重新编译安装，/etc/profile环境变量也要重新设置。笔者在这里重新安装一遍Nagios，上述问题才顺利解决。

16.2.4 安装 Nagios 插件包

安装 Nagios 插件包的具体操作步骤说明如下：

步骤 01 下载 Nagios 插件包，然后解压缩，如下所示：

```
[root@localhost ~]# tar -zxvf nagios-plugins-1.4.16.tar.gz
[root@localhost ~]# cd nagios-plugins-1.4.16
```

步骤 02 编译并且开始安装 Nagios 插件，如下所示：

```
[root@localhost nagios-plugins-1.4.16]# ./configure --with-nagios-user=nagios
--with-nagios-group=nagios  - prefix=/var/www/nagios
[root@localhost nagios-plugins-1.4.16]# make
[root@localhost nagios-plugins-1.4.16]# make install
```

步骤 03 验证 Nagios 的样例配置文件，如下所示：

```
[root@localhost nagios-plugins-1.4.16]# /var/www/nagios/bin/nagios -v
/var/www/nagios/etc/nagios.cfg

Nagios Core 3.4.1
Copyright (c) 2009-2011 Nagios Core Development Team and Community Contributors
Copyright (c) 1999-2009 Ethan Galstad
Last Modified: 05-11-2012
License: GPL

Website: http://www.nagios.org
Reading configuration data...
    Read main config file okay...
Processing object config file '/var/www/nagios/etc/objects/commands.cfg'...
Processing object config file '/var/www/nagios/etc/objects/contacts.cfg'...
Processing object config file '/var/www/nagios/etc/objects/timeperiods.cfg'...
Processing object config file '/var/www/nagios/etc/objects/templates.cfg'...
Processing object config file '/var/www/nagios/etc/objects/localhost.cfg'...
    Read object config files okay...

Running pre-flight check on configuration data...

Checking services...
```

```
        Checked 8 services.
Checking hosts...
        Checked 1 hosts.
Checking host groups...
        Checked 1 host groups.
Checking service groups...
        Checked 0 service groups.
Checking contacts...
        Checked 1 contacts.
Checking contact groups...
        Checked 1 contact groups.
Checking service escalations...
        Checked 0 service escalations.
Checking service dependencies...
        Checked 0 service dependencies.
Checking host escalations...
        Checked 0 host escalations.
Checking host dependencies...
        Checked 0 host dependencies.
Checking commands...
        Checked 24 commands.
Checking time periods...
        Checked 5 time periods.
Checking for circular paths between hosts...
Checking for circular host and service dependencies...
Checking global event handlers...
Checking obsessive compulsive processor commands...
Checking misc settings...

Total Warnings: 0
Total Errors:   0

Things look okay - No serious problems were detected during the pre-flight check
[root@localhost nagios-plugins-1.4.16]#
```

步骤04 重新启动 Nagios 服务和 Apache 服务，如下所示：

```
[root@localhost nagios-plugins-1.4.16]# service nagios restart
Running configuration check...done.
Stopping nagios: done.
Starting nagios: done.
[root@localhost nagios-plugins-1.4.16]# service httpd restart
Stopping httpd:                                            [  OK  ]
Starting httpd:                                            [  OK  ]
```

16.2.5 监控服务器的 CPU、负载、磁盘 I/O 使用情况

在监控服务器之前，需要在服务器上安装 nrpe、配置 Nagios 以及使用 htpasswd 创建浏览器验证账号。具体操作步骤说明如下：

步骤01 在 MySQL 服务器上安装 nrpe，如下所示：

```
[root@localhost ~]# tar -zxvf nrpe-2.12.tar.gz
[root@localhost ~]# tar -zxvf nrpe-2.12.tar.gz
```

```
[root@localhost nrpe-2.12]# ./configure --prefix=/var/www/nagios
[root@localhost nrpe-2.12]# make all
[root@localhost nrpe-2.12]# make install-plugin
[root@localhost nrpe-2.12]# make install-daemon
[root@localhost nrpe-2.12]# make install-daemon-config
[root@localhost nrpe-2.12]# make install-xinetd
```

步骤02 可以查看 nrpe 的一些系统信息，比如端口 5666，可以通过以下操作修改端口。

```
[root@localhost ~]# vi /etc/xinetd.d/nrpe
# description: NRPE (Nagios Remote Plugin Executor)
service nrpe
{
        flags           = REUSE
        socket_type     = stream
        port            = 5666
        wait            = no
        user            = nagios
        group           = nagios
        server          = /var/www/nagios/bin/nrpe
        server_args     = -c /var/www/nagios/etc/nrpe.cfg --inetd
        log_on_failure  += USERID
        disable         = no
        only_from       = 127.0.0.1
}
```

步骤03 在/var/www/nagios/etc/objects/command.cfg 命令定义文件中添加 nrpe 命令，如下所示：

```
[root@localhost ~]# vi /var/www/nagios/etc/objects/command.cfg
```

添加 NRPE 功能命令：

```
define command{
 command_name    check_nrpe
 command_line    $USER1$/check_nrpe -H $HOSTADDRESS$ -c $ARG1$
 }
```

这里有几点需要说明：

（1）该命令的名字叫作 check_nrpe。

（2）$USER1$/check_nrpe 会通过应用 resource.cfg 获得$USER1$变量的路径，从而可以获得/var/www/nagios/libexec/check_nrpe 这个绝对路径。

（3）-H $HOSTADDRESS$ 用来获得指定被检测主机的 IP 地址。

（4）-c $ARG1$用来指定被检测主机上 nrpe 守护进程运行着的 nrpe 命令名。

步骤04 修改客户端配置，如下所示：

```
vi /etc/nagios/nrpe.cfg
#下列配置表示允许 127.0.0.1，192.168.121.55（Server）这两台机器访问当前机器的信息。
allowed_hosts=127.0.0.1,192.168.121.55
```

步骤05 测试客户端配置情况，如果没有启动 nrpe 服务，会提示"Connection refused by host"信息。

```
#启动 nrpe (Nagios Remote Plugin Executor)
[root@localhost ~]# cd /var/www/nagios/bin
[root@localhost bin]# ./nrpe -c /var/www/nagios/etc/nrpe.cfg -d
```

```
[root@localhost bin]# netstat -an|grep 5666
tcp        0      0 0.0.0.0:5666              0.0.0.0:*                   LISTEN
unix  3    [ ]    STREAM    CONNECTED     15666   /tmp/orbit-root/linc-aad-0-6c80032044a1a
```

步骤06 接着测试一下是否可以监控客户端，如下所示：

```
[root@localhost ~]# cd /var/www/nagios/libexec
[root@localhost libexec]# ./check_nrpe -H localhost -c check_load
OK - load average: 0.06, 0.03, 0.12|load1=0.060;15.000;30.000;0; load5=0.030;10.000;25.000;0; load15=0.120;5.000;20.000;0;
```

步骤07 在 Nagios 监控服务器上配置服务端的信息，如下所示：

```
# 创建 client 配置（192.168.0.100 为 client ip）
[root@localhost ~]# vi  /var/www/nagios/etc/objects/192.168.0.100.cfg
define host{
        use linux-server
    host_name 127.0.0.1
alias client-1
        address 127.0.0.1
}

define service{
        use generic-service
host_name 127.0.0.1
    service_description check_disk2
    check_command check_nrpe!check_sda2
}

define service{
    use generic-service
    host_name 127.0.0.1
    service_description check_load
    check_command check_nrpe!check_load
}

define service{
    use generic-service
        host_name 127.0.0.1
    service_description check_swap
    check_command check_nrpe!check_swap
}
```

步骤08 将上面配置加到 **nagios.cfg** 文件中，如下所示：

```
[root@localhost ~]# vi /var/www/nagios/etc/nagios.cfg
cfg_file=/etc/nagios/objects/192.168.0.100.cfg
```

步骤09 重新启动 **Nagios** 服务和 **Apache** 服务，如下所示：

```
[root@localhost nagios-plugins-1.4.16]# service nagios restart
Running configuration check...done.
Stopping nagios: done.
Starting nagios: done.
[root@localhost nagios-plugins-1.4.16]# service httpd restart
Stopping httpd:                                            [  OK  ]
Starting httpd:                                            [  OK  ]
```

提示：如果 Nagios Web 界面提示如下错误信息：

It appears as though you do not have permission to view information for any of the services you requested...

可以打开 cgi.cfg 配置文件，里面有个参数：use_authentication=1。Nagios 设置了这个参数，用于保障系统的安全性，默认为 1，改为 0 即可解决上面的问题。

打开 Nagios Web 页面左侧的 Services，页面右侧中显示检测的所有服务项，如图 16.2 所示。

Host	Service	Status	Last Check	Duration	Attempt
192.168.0.100	check_disk2	CRITICAL	11-21-2021 16:50:54	0d 0h 2m 15s	2/3
	check_load	CRITICAL	11-21-2021 16:50:03	0d 0h 1m 6s	1/3
	check_swap	PENDING	N/A	0d 0h 3m 24s+	1/3
localhost	Currdefine service{ent Load	OK	11-21-2021 16:49:25	0d 3h 41m 44s	1/4
	Current Users	OK	11-21-2021 16:46:10	0d 16h 17m 27s	1/4
	HTTP	CRITICAL	11-21-2021 16:46:25	0d 15h 16m 2s	4/4
	PING	OK	11-21-2021 16:47:23	0d 16h 14m 20s	1/4
	Root Partition	CRITICAL	11-21-2021 16:47:20	1d 5h 55m 59s	4/4
	SSH	CRITICAL	11-21-2021 16:46:35	0d 15h 17m 57s	4/4
	Swap Usage	OK	11-21-2021 16:50:29	0d 16h 16m 57s	1/4
	Total Processes	OK	11-21-2021 16:50:13	0d 16h 15m 56s	1/4

图 16.2　检测到的所有的服务项

如果提示以下错误信息：CHECK_NRPE: Error - Could not complete SSL handshake，可以按下面的步骤解决。

步骤 01 检测 nrpe 机器上防火墙有没有启动，启动的话需要把 5666 端口打开，也可以直接关闭防火墙，如下所示：

```
[root@localhost ~]# service iptables stop
```

步骤 02 查看 nrpe 是否支持 ssl（nrpe 编译默认是支持 ssl），另外 openssl 和 openssl-devel 也都要装上，如下所示：

```
[root@localhost ~]# tar -zxvf openssl-0.9.81.tar.gz
[root@localhost openssl-0.9.81]# ./config --prefix=/usr/local/ssl-0.9.81 shared zlib-dynamic enable-camellia
[root@localhost openssl-0.9.81]# make depend
[root@localhost openssl-0.9.81]# make
[root@localhost openssl-0.9.81]# make test
[root@localhost openssl-0.9.81]# make install
```

步骤 03 建立连接，如下所示：

```
[root@localhost openssl-0.9.81]# cd /usr/local/
[root@localhost local]# ln -s ssl-0.9.81 ssl
[root@localhost local]# vi /etc/ld.so.conf
include ld.so.conf.d/*.conf
```

```
/usr/local/ssl/lib

[root@localhost local]# ldconfig
```

步骤04 配置环境变量，在/etc/profile 文件的末尾添加以下信息：

```
[root@localhost local]# vi /etc/profile
PATH=$PATH:/var/local/ssl/bin
export PATH
```

步骤05 测试 SSL 是否安装成功，如下所示：

```
[root@localhost ~]# cd /usr/local/
[root@localhost local]# ls
apache   doc   games    lib       man      mysql-cluster   share     ssl
bin      etc   include  libexec   mysql    sbin            src       ssl-0.9.81
[root@localhost local]# ldd /usr/local/ssl/bin/openssl
        linux-gate.so.1 =>  (0x00110000)
        libssl.so.0.9.8 => /usr/local/ssl-0.9.81/lib/libssl.so.0.9.8 (0x00111000)
        libcrypto.so.0.9.8 => /usr/local/ssl-0.9.81/lib/libcrypto.so.0.9.8 (0x00157000)
        libdl.so.2 => /lib/libdl.so.2 (0x0090c000)
        libc.so.6 => /lib/libc.so.6 (0x00776000)
        /lib/ld-linux.so.2 (0x00756000)
[root@localhost local]# which openssl
/usr/bin/openssl
```

步骤06 错误有可能是配置文件的问题，需要修改 nrpe.cfg 文件，把 allowed_hosts=127.0.0.1 修改为 allowed_hosts=127.0.0.1,192.168.0.100，然后重启 nrpe 服务。

```
[root@localhost lib]# ps -ef|grep nrpe
nagios    6591      1  0 Nov20 ?        00:00:03 ./nrpe -c /var/www/nagios/etc/nrpe.cfg -d
root     13152   1932  0 Nov21 pts/2    00:00:00 find / nrpe
root     20793  20118  0 11:14 pts/3    00:00:00 grep nrpe
[root@localhost lib]# kill -9 6591
[root@localhost lib]# cd
[root@localhost var]# cd /var/www/nagios/bin/
[root@localhost bin]# ./nrpe -c /var/www/nagios/etc/nrpe.cfg -d
```

步骤07 把之前在/var/www/nagios/etc/nrpe.cfg 文件中添加的两条 command 记录删除，问题即可解决。

如果还是错误，那么采用以下解决方法。

（1）确认 check_nrpe 和 nrpe daemon 的版本一定要一致。

（2）确认 check_nrpe 和 nrpe deamon 端同时启用或者禁用 ssl 支持。

（3）确认 nrep.cfg 可以被 nrpe 用户正常读取。

16.2.6 配置 Nagios 监控 MySQL 服务器

下面讲解通过配置 Nagios，实现监控 MySQL 服务器的方法，具体操作步骤说明如下。

步骤01 首先建立 nagios 数据库，创建 nagios 数据库用户，授予 nagios 用户权限，如下所示：

```
[root@localhost ~]# mysql -uroot -p
Enter password:
Welcome to the MySQL monitor.  Commands end with ; or \g.
```

```
Your MySQL connection id is 3
Server version: 5.5.27-ndb-7.2.8-cluster-gpl-log MySQL Cluster Community Server (GPL)

Copyright (c) 2000, 2011, Oracle and/or its affiliates. All rights reserved.

Oracle is a registered trademark of Oracle Corporation and/or its
affiliates. Other names may be trademarks of their respective
owners.

Type 'help;' or '\h' for help. Type '\c' to clear the current input statement.

mysql> create database nagios;
Query OK, 1 row affected (0.29 sec)

mysql> grant select on nagios.* to nagios@'%' identified by 'password';
Query OK, 0 rows affected (0.31 sec)
```

步骤02 为了方便本地测试 nagios 服务,授权用户 nagios@'localhost',如下所示:

```
mysql> grant select on nagios.* to nagios@'localhost' identified by 'password';
Query OK, 0 rows affected (0.00 sec)

mysql> select User,Password,HOst from mysql.user;
+--------+-------------------------------------------+-----------------------+
| User   | Password                                  | HOst                  |
+--------+-------------------------------------------+-----------------------+
| root   |                                           | localhost             |
| root   |                                           | localhost.localdomain |
| root   |                                           | 127.0.0.1             |
| root   |                                           | ::1                   |
|        |                                           | localhost             |
|        |                                           | localhost.localdomain |
| nagios | *2470C0C06DEE42FD1618BB99005ADCA2EC9D1E19 | %                     |
| nagios | *2470C0C06DEE42FD1618BB99005ADCA2EC9D1E19 | localhost             |
+--------+-------------------------------------------+-----------------------+
8 rows in set (0.00 sec)
```

步骤03 接下来执行 check_mysql 命令,检测 nagios 是否能够连接 MySQL 服务,如下所示:

```
[root@localhost ~]# cd /var/www/nagios/libexec/
[root@localhost libexec]# ./check_mysql -u nagios -d nagios -p password
./check_mysql: error while loading shared libraries: libmysqlclient.so.18: cannot open shared object file: No such file or directory
```

步骤04 此时,发现加载 libmysqlclient.so.18 发生错误,需要建立一个链接,如下所示:

```
[root@localhost usr]# cd /usr/local/mysql/lib/
[root@localhost lib]# ls
libmysqlclient.a            libmysqlclient.so.18         libndbclient.so.6.0.0
libmysqlclient_r.a          libmysqlclient.so.18.0.0     libndbclient_static.a
libmysqlclient_r.so         libmysqld.a                  libtcmalloc_minimal.so
libmysqlclient_r.so.18      libmysqld-debug.a            ndb_engine.so
libmysqlclient_r.so.18.0.0  libmysqlservices.a           plugin
libmysqlclient.so           libndbclient.so
[root@localhost lib]# ln -s /usr/local/mysql/lib/libmysqlclient.so.18 /usr/lib/libmysqlclient.so.18
```

步骤05 接着执行 check_mysql 命令检测 nagios 是否能够连接 MySQL 服务，如下所示：

```
[root@localhost ~]# cd /var/www/nagios/libexec/
[root@localhost libexec]# ./check_mysql -H 192.168.0.100 -u nagios -d nagios -p password
Uptime: 73815  Threads: 2  Questions: 16  Slow queries: 0  Opens: 33  Flush tables: 1  Open tables: 26  Queries per second avg: 0.000
```

步骤06 下面开始 Nagios 配置文件，如下所示：

```
[root@localhost ~]# cd /var/www/nagios/objects/
[root@localhost objects]# vi commands.cfg
```

步骤07 在 commands.cfg 文件的末尾加上如下文本：

```
#check_mysql
define command{
    command_name    check_mysql
    command_line    $USER1$/check_mysql -H $HOSTADDRESS$ -u nagios -d nagios -p password
}
```

步骤08 在 192.168.0.100.cfg 文件末尾追加 check_mysql 服务，如下所示：

```
[root@localhost ~]# vi    /var/www/nagios/etc/objects/192.168.0.100.cfg
define service{
    use generic-service
    host_name 192.168.0.100
    service_description MySQL
    normal_check_interval 1
    retry_check_interval 1
    check_command check_mysql
}
```

步骤09 打开 Nagios Web 左侧的页面的 Services，然后选择 MySQL，可以看到信息如图 16.3 所示。

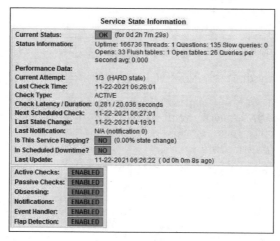

图 16.3　检测 MySQL 的状态信息

16.3 MySQL 监控利器 Cacti 实战

Cacti 是一套基于 PHP、MySQL、SNMP 以及 RRDTool 开发的网络流量监测分析工具，通过 snmpget 获得数据，使用 RRDTool 绘画图形。本节将讲解 Cacti 工具的使用方法。

16.3.1 Cacti 工具的安装

在安装 Cacti 工具之前，需要安装 MySQL 服务器和 Apache 服务，然后安装 PHP。安装 PHP 之前，还需要安装 zlib、freetype、libpng、jpegsrc 和 Fontconfig，从而使 PHP 支持 GD 库。GD 库的下载地址是 http://oss.oetiker.ch/rrdtool/pub/libs/。

（1）安装 zlib，如下所示：

```
[root@localhost ~]# tar zxvf zlib-1.2.3.tar.gz
[root@localhost ~]# cd zlib-1.2.3
[root@localhost zlib-1.2.3]# ./configure --prefix=/usr/local/zlib
[root@localhost zlib-1.2.3]# make
[root@localhost zlib-1.2.3]# make install
```

（2）安装 libpng，如下所示：

```
[root@localhost ~]# tar zxvf libpng-1.2.18.tar.gz
[root@localhost scripts]# mv makefile.linux ../makefile
[root@localhost scripts]# cd ..
[root@localhost libpng-1.2.18]# make
[root@localhost libpng-1.2.18]# make install
```

（3）安装 freetype，如下所示：

```
[root@localhost ~]# tar zxvf freetype-2.3.5.tar.gz
[root@localhost ~]# cd freetype-2.3.5
[root@localhost freetype-2.3.5]# ./configure --prefix=/usr/local/freetype
[root@localhost freetype-2.3.5]# make
[root@localhost freetype-2.3.5]# make install
```

（4）安装 jpegsrc，如下所示：

```
[root@localhost ~]# tar -zxvf jpegsrc.v6b.tar.gz
[root@localhost ~]# cd jpeg-6b/
[root@localhost jpeg-6b]# mkdir /usr/local/libjpeg
[root@localhost jpeg-6b]# mkdir /usr/local/libjpeg/include
[root@localhost jpeg-6b]# mkdir /usr/local/libjpeg/bin
[root@localhost jpeg-6b]# mkdir /usr/local/libjpeg/lib
[root@localhost jpeg-6b]# mkdir /usr/local/libjpeg/man
[root@localhost jpeg-6b]# mkdir /usr/local/libjpeg/man/man1
[root@localhost jpeg-6b]# ./configure --prefix=/usr/local/libjpeg --enable-shared --enable-static
[root@localhost jpeg-6b]# make
[root@localhost jpeg-6b]# make install
```

（5）安装 Fontconfig，如下所示：

```
[root@localhost ~]# tar -zxvf fontconfig-2.4.2.tar.gz
```

```
[root@localhost fontconfig-2.4.2]# ./configure --with-freetype-config=/usr/local/freetype
[root@localhost fontconfig-2.4.2]# make
[root@localhost fontconfig-2.4.2]# make install
```

（6）安装 GD，如下所示：

```
[root@localhost ~]#   tar -zxvf gd-2.0.35.tar.gz
[root@localhost ~]# cd gd-2.0.35
[root@localhost gd-2.0.35]# ./configure --prefix=/usr/local/libgd --with-png
--with-freetype=/usr/local/freetype --with-jpeg=/usr/local/libjpeg
[root@localhost gd-2.0.35]# make
[root@localhost gd-2.0.35]# make install
```

（7）安装 PHP，如下所示：

```
[root@localhost ~]# tar -zxvf   php-5.2.3.tar.gz
[root@localhost ~]# cd php-5.2.3
[root@localhost php-5.2.3]# ./configure --prefix=/usr/local/php
--with-apxs2=/usr/local/apache/bin/apxs --with-mysql=/usr/local/mysql --with-gd=/usr/local/libgd
--enable-gd-native-ttf --with-ttf --enable-gd-jis-conv --with-freetype-dir=/usr/local/freetype
--with-jpeg-dir=/usr/local/libjpeg --with-png-dir=/usr --with-zlib-dir=/usr/local/zlib
--enable-xml --enable-mbstring --enable-sockets
[root@localhost php-5.2.3]# make
[root@localhost php-5.2.3]# make install
[root@localhost php-5.2.3]# cp php.ini-recommended /usr/local/php/lib/php.ini
[root@localhost php-5.2.3]# ln -s /usr/local/php/bin/* /usr/local/bin/
```

下面需要修改 httpd.conf 文件，如下所示：

```
[root@localhost ~]# vi /usr/local/apache/conf/httpd.conf
查找 AddType application/x-compress .Z
AddType application/x-gzip .gz .tgz
在其下加入 AddType application/x-tar .tgz
AddType application/x-httpd-php .php
AddType image/x-icon .ico
修改 DirectoryIndex 行，添加 index.php
修改为 DirectoryIndex index.php index.html index.html.var
```

最后创建 test.php 测试是否成功安装 PHP，如下所示：

```
[root@localhost ~]# vi /usr/local/apache/htdocs/test.php
```

添加以下行，添加完成以后，wq 保存退出。

```
<?php
phpinfo();
?>
```

重启 Apache 服务器，如下所示：

```
[root@localhost ~]# /usr/local/apache/bin/httpd -k stop
[root@localhost ~]# /usr/local/apache/bin/httpd -k start
```

在浏览器中输入：http://192.168.0.100/test.php 进行 PHP 测试。

PHP 安装成功后，下面开始安装 RRDTool。安装 rrdtool-1.2.23 需要一些库文件支持，需要将 cgilib-0.5.tar.gz、zlib-1.2.3.tar.gz、libpng-1.2.18.tar.gz、freetype-2.3.5.tar.gz、libart_lgpl-2.3.17.tar.gz、 rrdtool-1.2.23.tar.gz 放到/root/rrdtool-1.2.23 目录下，将脚本保存为/root/rrdtool-1.2.23/rrdtoolinstall.sh，并给执行权限 chmod u+x /root/rrdtool-1.2.23/rrdtoolinstall.sh，rrdtoolinstall.sh 脚本信息如下所示：

```sh
#!/bin/sh
BUILD_DIR=/root/rrdtool-1.2.11
INSTALL_DIR=/usr/local/rrdtool
cd $BUILD_DIR
tar zxf cgilib-0.5.tar.gz
cd cgilib-0.5
make CC=gcc CFLAGS="-O3 -fPIC -I."
mkdir -p $BUILD_DIR/lb/include
cp *.h $BUILD_DIR/lb/include
mkdir -p $BUILD_DIR/lb/lib
cp libcgi* $BUILD_DIR/lb/lib
cd $BUILD_DIR
tar  zxf zlib-1.2.2.tar.gz
cd zlib-1.2.2
env CFLAGS="-O3 -fPIC" ./configure --prefix=$BUILD_DIR/lb
make
make install
cd $BUILD_DIR
tar zxvf libpng-1.2.8-config.tar.gz
cd libpng-1.2.8-config
env CPPFLAGS="-I$BUILD_DIR/lb/include" LDFLAGS="-L$BUILD_DIR/lb/lib" CFLAGS="-O3 -fPIC" \
   ./configure --disable-shared --prefix=$BUILD_DIR/lb
make
make install
cd $BUILD_DIR
tar zxvf freetype-2.1.9.tar.gz
cd freetype-2.1.9
env CPPFLAGS="-I$BUILD_DIR/lb/include" LDFLAGS="-L$BUILD_DIR/lb/lib" CFLAGS="-O3 -fPIC" \
   ./configure --disable-shared --prefix=$BUILD_DIR/lb
make
make install

cd $BUILD_DIR
tar zxvf libart_lgpl-2.3.17.tar.gz
cd libart_lgpl-2.3.17
env CFLAGS="-O3 -fPIC" ./configure --disable-shared --prefix=$BUILD_DIR/lb
make
make install

IR=-I$BUILD_DIR/lb/include
CPPFLAGS="$IR $IR/libart-2.0 $IR/freetype2 $IR/libpng"
LDFLAGS="-L$BUILD_DIR/lb/lib"
CFLAGS=-O3
export CPPFLAGS LDFLAGS CFLAGS

cd $BUILD_DIR
tar zxf rrdtool-1.2.11.tar.gz
cd rrdtool-1.2.11
./configure --prefix=$INSTALL_DIR --disable-python --disable-tcl && make && make install
```

直接运行 rrdtoolinstall.sh 脚本即可，如下所示：

```
[root@localhost ~]# tar -zxvf rrdtool-1.2.11.tar.gz
[root@localhost ~]# cd rrdtool-1.2.11
[root@localhost rrdtool-1.2.11]# ls
cgilib-0.5.tar.gz         libpng-1.2.8-config.tar.gz   zlib-1.2.2.tar.gz
freetype-2.1.9.tar.gz     rrdtool-1.2.11.tar.gz
```

安装 net-snmp,如下所示:

```
[root@localhost ~]# tar -zxvf net-snmp-5.2.4.tar.gz
[root@localhost ~]# cd net-snmp-5.2.4
[root@localhost net-snmp-5.2.4]# ./configure --prefix=/usr/local/net-snmp --enable-developer
[root@localhost ~]# make
[root@localhost ~]# make install
```

下面修改 snmpd.conf,然后启动 snmpd 服务。

```
[root@localhost net-snmp-5.2.4]# ln -s /usr/local/net-snmp/bin/* /usr/local/bin/
[root@localhost net-snmp-5.2.4]# cp EXAMPLE.conf /usr/local/net-snmp/share/snmp/snmp.conf
[root@localhost net-snmp-5.2.4]# /usr/local/net-snmp/sbin/snmpd
```

安装 Cacti,如下所示:

```
[root@localhost ~]# tar -zxvf cacti-0.8.6j.tar.gz
[root@localhost ~]# mv -r cacti-0.8.6j /usr/local/apache/htdocs/cacti
[root@localhost ~]# vi /usr/local/apache/htdocs/cacti/include/config.php
$database_type = "mysql";
$database_default = "cacti";
$database_hostname = "localhost";
$database_username = "xfimti";
$database_password = "123";
```

安装 Cactid,如下所示:

```
  [root@localhost ~]# tar -zxvf cacti-cactid-0.8.6i.tar.gz
[root@localhost ~]# cd cacti-cactid-0.8.6i
[root@localhost cacti-cactid-0.8.6i]# ./configure --with-mysql=/usr/local/mysql
--with-snmp=/usr/local/net-snmp
[root@localhost cacti-cactid-0.8.6i]# make
[root@localhost cacti-cactid-0.8.6i]# mkdir /usr/local/cactid
[root@localhost cacti-cactid-0.8.6i]# cp cactid cactid.conf /usr/local/cactid
[root@localhost ~]# vi   /usr/local/cactid/cactid.conf       //修改 cactid 配置文件
DB_Host              127.0.0.1
DB_Database          cacti
DB_User              xfimti
DB_Pass              123
```

配置 MySQL 数据库,如下所示:

```
mysql> create database cacti;
Query OK, 1 row affected (0.06 sec)
[root@localhost cacti-0.8.6j]# vi cacti.sql
[root@localhost cacti-0.8.6j]# pwd
/usr/local/apache/htdocs/cacti/cacti-0.8.6j
[root@localhost cacti-0.8.6j]# mysql -u xfimti -p cacti < cacti.sql
Enter password:
```

提示:cacti.sql 脚本直接导入数据库会发生错误,需要把建表语句 TYPE=MyISAM 改为 engine=myisam,就能解决问题。

最终完成 Cacti 的安装，在地址栏中输入：http://192.168.0.100/cacti，默认的用户名和密码是 admin，页面如图 16.4 所示。

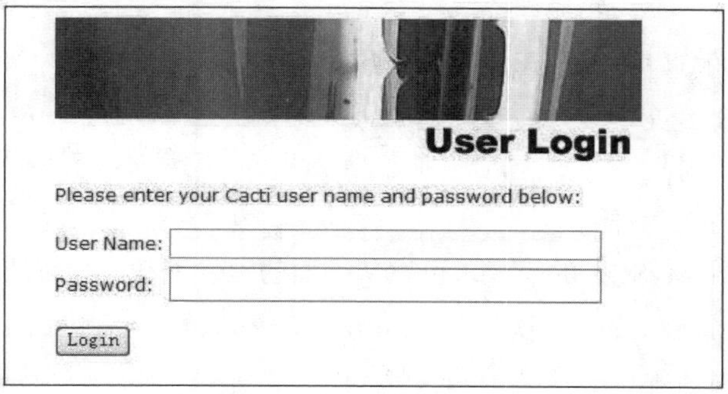

图 16.4 Cacti 登录页面

16.3.2 Cacti 监控 MySQL 服务器

登录 Cacti 后，主界面如图 16.5 所示，可以看到两个选项卡 console 和 graphs。console 表示控制台，可以进行所有的配置操作；graphs 表示用来查看所有的服务器性能。

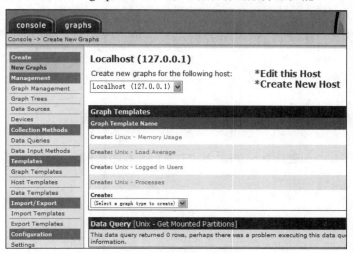

图 16.5 Cacti 登录后的界面

下面先了解下 console 菜单项的具体意义，如下所示：

（1）New Graphs：用来创建新图像的快捷方式；

（2）Graph Management：可以删除和复制图像，Cacti 会自动创建图像。

（3）Graph Trees：在 grahps 界面里，图像或 devices 是树状结果显示的，可以在此设置树的结构。

（4）Data Sources：用来管理 rrd 文件。一般无须修改，Cacti 会自己创建 rrd 文件。

（5）Devices：可以在此创建新的设备或修改其名称等信息。

（6）Data Queries 和 Data Input Methods：用来采集数据的方式。

（7）Graph Templates、Host Templates 和 Data Templates：分别表示图像模板、主机类型模板

和数据模板。

（8）Import Templates 和 Export Templates：是对模板（图像模板、主机类型模板和数据模板）的导入和导出操作。

（9）Setting：用来配置 Cacti 的主要配置菜单。

（10）System Utilities：用来显示 Cacti 系统的一些 cache 和 log 信息。

（11）User Management：用来对用户进行管理。可以在此添加、删除用户，并对每个用户设置详细的权限。

首先，需要下载监控 MySQL 的模板，可以从下面的地址下载，如下所示。

http://code.google.com/p/mysql-cacti-templates/downloads/detail?name=better-cacti-templates-1.1.8.tar.gz

该模板主要是用来监控 MySQL 服务器的性能，InnoDB 相关监控、MyISAM 相关监控、Nginx 状态监控。下面开始安装该模板，操作步骤说明如下。

步骤 01 安装 mysql-cacti-templates.tar.gz 文件，如下所示：

```
[root@localhost ~]# tar zxf better-cacti-templates-1.1.8.tar.gz
[root@localhost ~]# cd better-cacti-templates-1.1.8
[root@localhost better-cacti-templates-1.1.8]# cd scripts/
[root@localhost scripts]# ls
ss_get_by_ssh.php  ss_get_mysql_stats.php
[root@localhost scripts]# cp ss_get_mysql_stats.php /usr/local/apache/htdocs/cacti/scripts/
[root@localhost scripts]# cd /usr/local/apache/htdocs/cacti/scripts/
[root@localhost scripts]# vi ss_get_mysql_stats.php
```

步骤 02 编辑 ss_get_mysql_stats.php 文件，注意这三行：

```
$mysql_user = 'xfimti';
$mysql_pass = 'xfimti';
$cache_dir  = '/tmp';
```

步骤 03 需要在 MySQL 中创建具有 SUPER 和 PROCESS 权限的用户。如下所示：

```
mysql> grant super,process,select on *.* to xfimti@'%';
Query OK, 0 rows affected (0.00 sec)
```

步骤 04 导入模板。单击管理界面左侧"Import Templates"链接，然后将模板文件 cacti_host_template_x_mysql_server_ht_0.8.6i-server1.1.8.xml 导入进去，如图 16.6 所示。

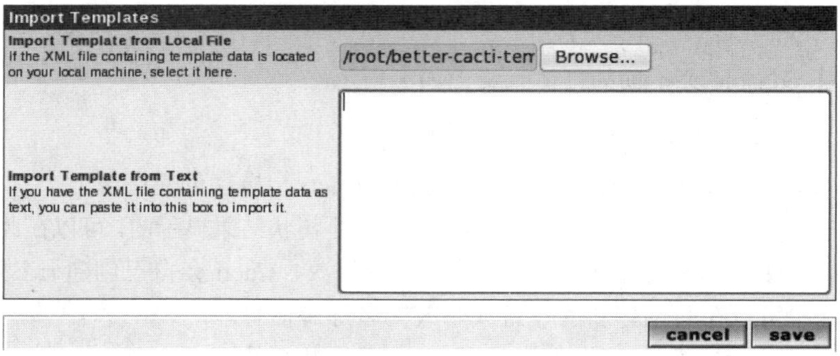

图 16.6　导入监控 MySQL 的模板

步骤 05 添加设备。在主页面中单击 Devices 链接，然后新建一个 Devices 或选择已有的 Devices，设置为需要监控的主机，如图 16.7 所示。

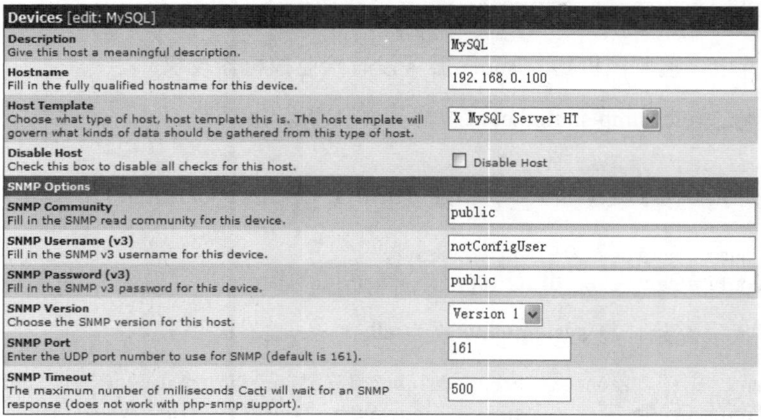

图 16.7 创建一个新的 Devices

步骤 06 设置完成之后保存，过一段时间就可以看见生成的图像了，如图 16.8 所示是部分监控对象。

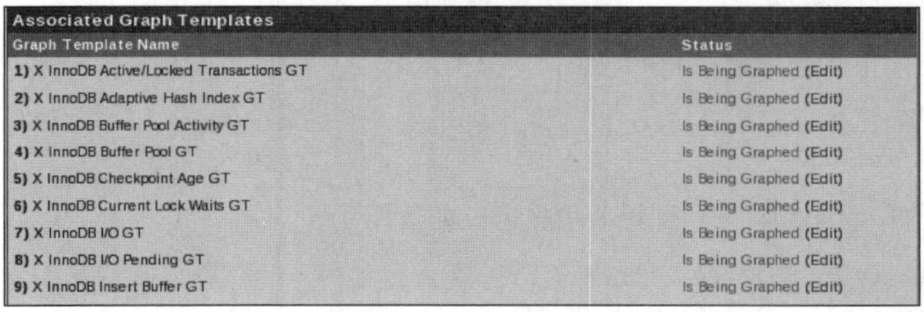

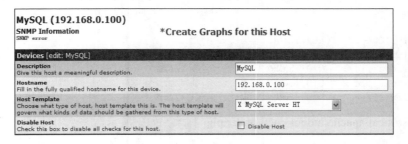

图 16.8 部分 Device 监控的对象

创建设备之后，如果出现 SNMP error 的问题，如图 16.9 所示。

图 16.9 SMNP error 出错界面

可以通过如下步骤来解决该问题：

步骤 01 确认被检测机器中 snmp 服务是否启动，如果没有启动，可以执行 service snmpd start 启动该服务，如下所示：

```
[root@localhost ~]# service snmpd status
snmpd (pid 29856) is running...
```

步骤02 snmp 服务启动成功后,在检测机器上执行如下命令,查看能否返回数据。

```
[root@localhost ~]# snmpwalk -v 2c -c public 192.168.0.100 if
Timeout: No Response from 192.168.0.100
```

从结果可以看出,被监控机器的 snmp 服务没有给监控机器授权。

步骤03 此时需要修改/etc/snmp/snmpd.conf 文件,如下所示:

```
####
# First, map the community name "public" into a "security name"

#              sec.name      source              community
com2sec notConfigUser      192.168.0.100         public
```

步骤04 找到如下代码部分,将 systemview 改为 all。

```
#         group            context sec.model sec.level prefix read   write   notif
access notConfigGroup      ""         any      noauth    exact  systemview none none
```

修改之后的代码如下所示:

```
#         group            context sec.model sec.level prefix read   write   notif
access notConfigGroup      ""         any      noauth    exact  all    none    none
```

步骤05 然后找到如下代码部分,将#去掉。

```
##               incl/excl subtree                         mask
#view all       included    .1                             80
```

步骤06 最后关闭掉 snmpd 服务,重新启动服务,如下所示:

```
[root@localhost ~]# kill -9 29856
[root@localhost ~]# service snmpd start
Starting snmpd:                                          [  OK  ]
```

步骤07 重新在检测机器上执行如下命令,检测能否返回数据。

```
[root@localhost ~]# snmpwalk -v 2c -c public 192.168.0.100 if
...
IF-MIB::ifOutQLen.2 = Gauge32: 0
IF-MIB::ifOutQLen.3 = Gauge32: 0
IF-MIB::ifSpecific.1 = OID: SNMPv2-SMI::zeroDotZero
IF-MIB::ifSpecific.2 = OID: SNMPv2-SMI::zeroDotZero
IF-MIB::ifSpecific.3 = OID: SNMPv2-SMI::zeroDotZero
```

步骤08 执行/usr/local/apache/htdocs/cacti/poller.php 脚本,如下所示:

```
[root@localhost bin]# pwd
/usr/local/php/bin
[root@localhost bin]# ls
pear  peardev  pecl  php  php-config  phpize
[root@localhost bin]# ./php /usr/local/apache/htdocs/cacti/poller.php
sh: /usr/local/bin/rrdtool: No such file or directory
12/05/2012 12:46:02 AM - POLLER: Poller[0] Maximum runtime of 292 seconds exceeded. Exiting.
12/05/2012 12:46:02 AM - SYSTEM STATS: Time:292.5174 Method:cmd.php Processes:1 Threads:N/A
Hosts:3 HostsPerProcess:3 DataSources:0 RRDsProcessed:0
  PHP Warning:  pclose(): 48 is not a valid stream resource in
```

/usr/local/apache/htdocs/cacti/lib/rrd.php on line 47

步骤09 在/usr/local/bin 目录下建立链接 rrdtool 指向 usr/local/rrdtool/bin/rrdtool。

```
[root@localhost bin]# ./php /usr/local/apache/htdocs/cacti/poller.php
12/05/2012 01:43:35 AM - POLLER: Poller[0] Maximum runtime of 292 seconds exceeded. Exiting.
12/05/2012 01:43:35 AM - SYSTEM STATS: Time:293.0108 Method:cmd.php Processes:1 Threads:N/A
Hosts:3 HostsPerProcess:3 DataSources:0 RRDsProcessed:0
    PHP Warning:  pclose(): 48 is not a valid stream resource in
/usr/local/apache/htdocs/cacti/lib/rrd.php on line 47
```

步骤10 下面重新单击 Devices 右侧的 MySQL 链接，如图 16.10 所示。

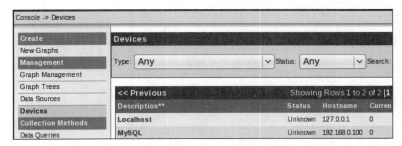

图 16.10　Devices 右侧的 MySQL 链接界面

步骤11 SNMP error 错误问题被解决了，最后结果如图 16.11 所示。

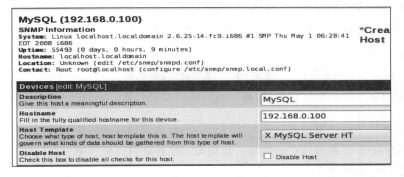

图 16.11　SNMP error 问题成功解决

第 17 章

数据备份与恢复

尽管一些管理措施可以用来保证数据库的安全,但是意外情况总是有可能发生,例如意外的停电、管理员操作失误等,都可能会造成数据的丢失。保证数据安全最重要的一个措施是确保定期对数据进行备份。如果数据库中的数据丢失或者出现错误,可以使用备份的数据进行恢复,这样就尽可能地降低了意外原因导致的损失。MySQL 提供了多种方法对数据进行备份和恢复。本章将介绍数据备份、数据恢复、数据迁移和数据导入导出的相关知识。

17.1 数据备份

数据备份是数据库管理员非常重要的工作之一。系统意外崩溃或者硬件损坏都可能导致数据库的丢失,因此 MySQL 管理员应该定期备份数据库,使得在意外情况发生时的损失尽可能降低。本节将介绍数据备份的 3 种方法。

17.1.1 使用 mysqldump 命令备份

mysqldump 是 MySQL 提供的一个非常有用的数据库备份工具。mysqldump 命令执行时,可以将数据库备份成一个文本文件,该文件中实际包含了多个 CREATE 和 INSERT 语句,使用这些语句可以重新创建表和插入数据。

mysqldump 备份数据库语句的基本语法格式如下:

```
mysqldump -u user -h host -ppassword dbname[tbname, [tbname...]]> filename.sql
```

user 表示用户名称;host 表示登录用户的主机名称;password 为登录密码;dbname 为需要备份的数据库名称;tbname 为 dbname 数据库中需要备份的数据表,可以指定多个需要备份的表;右箭头符号">"告诉 mysqldump 将备份数据表的定义和数据写入备份文件;filename.sql 为备份文件的名称。

1. 使用 mysqldump 备份单个数据库中的所有表

【例 17.1】使用 mysqldump 命令备份数据库中的所有表，执行过程如下：

为了更好地理解 mysqldump 工具是如何工作的，这里给出一个完整的数据库例子。首先登录 MySQL，按下面数据库结构创建 booksDB 数据库和各个表，并插入数据记录。数据库和表定义如下：

```sql
CREATE DATABASE booksDB;
use booksDB;

CREATE TABLE books
(
bk_id INT NOT NULL PRIMARY KEY,
bk_title VARCHAR(50) NOT NULL,
copyright YEAR NOT NULL
);
INSERT INTO books
VALUES (11078, 'Learning MySQL', 2010),
(11033, 'Study Html', 2011),
(11035, 'How to use php', 2003),
(11072, 'Teach yourself javascript', 2005),
(11028, 'Learning C++', 2005),
(11069, 'MySQL professional', 2009),
(11026, 'Guide to MySQL 8.0', 2008),
(11041, 'Inside VC++', 2011);

CREATE TABLE authors
(
auth_id INT NOT NULL PRIMARY KEY,
auth_name VARCHAR(20),
auth_gender CHAR(1)
);
INSERT INTO authors
VALUES (1001, 'WriterX' ,'f'),
(1002, 'WriterA' ,'f'),
(1003, 'WriterB' ,'m'),
(1004, 'WriterC' ,'f'),
(1011, 'WriterD' ,'f'),
(1012, 'WriterE' ,'m'),
(1013, 'WriterF' ,'m'),
(1014, 'WriterG' ,'f'),
(1015, 'WriterH' ,'f');

CREATE TABLE authorbook
(
auth_id INT NOT NULL,
bk_id INT NOT NULL,
PRIMARY KEY (auth_id, bk_id),
FOREIGN KEY (auth_id) REFERENCES authors (auth_id),
FOREIGN KEY (bk_id) REFERENCES books (bk_id)
);

INSERT INTO authorbook
VALUES (1001, 11033), (1002, 11035), (1003, 11072), (1004, 11028),
(1011, 11078), (1012, 11026), (1012, 11041), (1014, 11069);
```

完成数据插入后打开操作系统命令行输入窗口,输入备份命令如下:

```
C:\ >mysqldump -u root -p booksdb > C:/backup/booksdb_20220301.sql
Enter password: **
```

提示:这里要保证 C 盘下 backup 文件夹存在,否则将提示错误信息:系统找不到指定的路径。

输入密码之后,MySQL 便对数据库进行了备份,在 C:\backup 文件夹下面查看刚才备份过的文件,使用文本查看器打开文件可以看到其部分文件内容大致如下:

```sql
-- MySQL dump 10.13  Distrib 8.0.28, for Win64 (x86_64)
--
-- Host: localhost    Database: booksdb
-- ------------------------------------------------------
-- Server version  8.0.28

/*!40101 SET @OLD_CHARACTER_SET_CLIENT=@@CHARACTER_SET_CLIENT */;
/*!40101 SET @OLD_CHARACTER_SET_RESULTS=@@CHARACTER_SET_RESULTS */;
/*!40101 SET @OLD_COLLATION_CONNECTION=@@COLLATION_CONNECTION */;
 SET NAMES utf8mb4 ;
/*!40103 SET @OLD_TIME_ZONE=@@TIME_ZONE */;
/*!40103 SET TIME_ZONE='+00:00' */;
/*!40014 SET @OLD_UNIQUE_CHECKS=@@UNIQUE_CHECKS, UNIQUE_CHECKS=0 */;
/*!40014 SET @OLD_FOREIGN_KEY_CHECKS=@@FOREIGN_KEY_CHECKS, FOREIGN_KEY_CHECKS=0 */;
/*!40101 SET @OLD_SQL_MODE=@@SQL_MODE, SQL_MODE='NO_AUTO_VALUE_ON_ZERO' */;
/*!40111 SET @OLD_SQL_NOTES=@@SQL_NOTES, SQL_NOTES=0 */;

--
-- Table structure for table `authorbook`
--

DROP TABLE IF EXISTS `authorbook`;
/*!40101 SET @saved_cs_client     = @@character_set_client */;
 SET character_set_client = utf8mb4 ;
CREATE TABLE `authorbook` (
  `auth_id` int(11) NOT NULL,
  `bk_id` int(11) NOT NULL,
  PRIMARY KEY (`auth_id`,`bk_id`),
  KEY `bk_id` (`bk_id`),
  CONSTRAINT `authorbook_ibfk_1` FOREIGN KEY (`auth_id`) REFERENCES `authors` (`auth_id`),
  CONSTRAINT `authorbook_ibfk_2` FOREIGN KEY (`bk_id`) REFERENCES `books` (`bk_id`)
) ENGINE=InnoDB DEFAULT CHARSET=utf8mb4 COLLATE=utf8mb4_0900_ai_ci;
/*!40101 SET character_set_client = @saved_cs_client */;

--
-- Dumping data for table `authorbook`
--

LOCK TABLES `authorbook` WRITE;
/*!40000 ALTER TABLE `authorbook` DISABLE KEYS */;
INSERT INTO `authorbook` VALUES (1012,11026),(1004,11028),(1001,11033),(1002,11035),(1012,11041),(1014,11069),(1003,11072),(1011,11078);
/*!40000 ALTER TABLE `authorbook` ENABLE KEYS */;
UNLOCK TABLES;
```

```
--
-- Table structure for table `authors`
--

DROP TABLE IF EXISTS `authors`;
/*!40101 SET @saved_cs_client     = @@character_set_client */;
 SET character_set_client = utf8mb4 ;
CREATE TABLE `authors` (
 `auth_id` int(11) NOT NULL,
 `auth_name` varchar(20) DEFAULT NULL,
 `auth_gender` char(1) DEFAULT NULL,
 PRIMARY KEY (`auth_id`)
) ENGINE=InnoDB DEFAULT CHARSET=utf8mb4 COLLATE=utf8mb4_0900_ai_ci;
/*!40101 SET character_set_client = @saved_cs_client */;

--
-- Dumping data for table `authors`
--

LOCK TABLES `authors` WRITE;
/*!40000 ALTER TABLE `authors` DISABLE KEYS */;
INSERT INTO `authors` VALUES (1001,'WriterX','f'),(1002,'WriterA','f'),
(1003,'WriterB','m'),(1004,'WriterC','f'),(1011,'WriterD','f'),(1012,'WriterE','m'),(1013,'Writ
erF','m'),(1014,'WriterG','f'),(1015,'WriterH','f');
/*!40000 ALTER TABLE `authors` ENABLE KEYS */;
UNLOCK TABLES;

--
-- Table structure for table `books`
--

DROP TABLE IF EXISTS `books`;
/*!40101 SET @saved_cs_client     = @@character_set_client */;
 SET character_set_client = utf8mb4 ;
CREATE TABLE `books` (
 `bk_id` int(11) NOT NULL,
 `bk_title` varchar(50) NOT NULL,
 `copyright` year(4) NOT NULL,
 PRIMARY KEY (`bk_id`)
) ENGINE=InnoDB DEFAULT CHARSET=utf8mb4 COLLATE=utf8mb4_0900_ai_ci;
/*!40101 SET character_set_client = @saved_cs_client */;

--
-- Dumping data for table `books`
--

LOCK TABLES `books` WRITE;
/*!40000 ALTER TABLE `books` DISABLE KEYS */;
INSERT INTO `books` VALUES (11026,'Guide to MySQL 8.0',2008),(11028,'Learning
C++',2005),(11033,'Study Html',2011),(11035,'How to use php',2003),(11041,'Inside
VC++',2011),(11069,'MySQL professional',2009),(11072,'Teach yourself
javascript',2005),(11078,'Learning MySQL',2010);
/*!40000 ALTER TABLE `books` ENABLE KEYS */;
UNLOCK TABLES;
/*!40103 SET TIME_ZONE=@OLD_TIME_ZONE */;
```

```
/*!40101 SET SQL_MODE=@OLD_SQL_MODE */;
/*!40014 SET FOREIGN_KEY_CHECKS=@OLD_FOREIGN_KEY_CHECKS */;
/*!40014 SET UNIQUE_CHECKS=@OLD_UNIQUE_CHECKS */;
/*!40101 SET CHARACTER_SET_CLIENT=@OLD_CHARACTER_SET_CLIENT */;
/*!40101 SET CHARACTER_SET_RESULTS=@OLD_CHARACTER_SET_RESULTS */;
/*!40101 SET COLLATION_CONNECTION=@OLD_COLLATION_CONNECTION */;
/*!40111 SET SQL_NOTES=@OLD_SQL_NOTES */;

-- Dump completed on 2018-11-29 18:00:43
```

可以看到，备份文件包含了一些信息，文件开头首先表明了备份文件使用的 mysqldump 工具的版本号；然后是备份账户的名称和主机信息，以及备份的数据库的名称；最后是 MySQL 服务器的版本号，在这里为 8.0.28。

备份文件接下来的部分是一些 SET 语句，这些语句将一些系统变量值赋给用户定义变量，以确保被恢复的数据库的系统变量和原来备份时的变量相同，例如：

```
/*!40101 SET @OLD_CHARACTER_SET_CLIENT=@@CHARACTER_SET_CLIENT */;
```

该 SET 语句将当前系统变量 character_set_client 的值赋给用户定义变量@old_character_set_client。其他变量与此类似。

备份文件的最后几行 MySQL 使用 SET 语句恢复服务器系统变量原来的值，例如：

```
/*!40101 SET CHARACTER_SET_CLIENT=@OLD_CHARACTER_SET_CLIENT */;
```

该语句将用户定义的变量@old_character_set_client 中保存的值赋给实际的系统变量 character_set_client。

备份文件中以"--"字符开头的行为注释语句；以"/*!"开头、"*/"结尾的语句为可执行的 MySQL 注释，这些语句可以被 MySQL 执行，但在其他数据库管理系统中将被作为注释忽略，以提高数据库的可移植性。

另外，备份文件开始的一些语句以数字开头，代表的是 MySQL 版本号，这些语句只有在指定的 MySQL 版本或者比该版本高的情况下才能执行。例如，40101，表明这些语句只有在 MySQL 版本号为 4.01.01 或者更高的条件下才可以被执行。

2. 使用 mysqldump 备份数据库中的某个表

在前面 mysqldump 的语法中介绍过，mysqldump 还可以备份数据中的某个表，其语法格式为：

```
mysqldump -u user -h host -p dbname [tbname, [tbname...]] > filename.sql
```

tbname 表示数据库中的表名，多个表名之间用空格隔开。

备份表和备份数据库中所有表的语句中不同的地方在于，要在数据库名称 dbname 之后指定需要备份的表名称。

【例 17.2】备份 booksDB 数据库中的 books 表，输入语句如下：

```
mysqldump -u root -p booksDB books > C:/backup/books_20220301.sql
```

该语句创建名称为 books_20220301.sql 的备份文件，文件中包含了前面介绍的 SET 语句等内容，不同的是，该文件只包含 books 表的 CREATE 和 INSERT 语句。

3. 使用 mysqldump 备份多个数据库

如果要使用 mysqldump 备份多个数据库，就需要使用--databases 参数。备份多个数据库的语句格式如下：

```
mysqldump -u user -h host -p --databases [dbname, [dbname...]] > filename.sql
```

使用--databases 参数之后，必须指定至少一个数据库的名称，多个数据库名称之间用空格隔开。

【例 17.3】使用 mysqldump 备份 booksDB 和 test_db 数据库，输入语句如下：

```
mysqldump -u root -p --databases  booksDB test_db>C:\backup\books_testDB_20220301.sql
```

该语句创建名称为 books_testDB_20220301.sql 的备份文件，该文件中包含了创建两个数据库 booksDB 和 test_db 所必需的所有语句。

另外，使用--all-databases 参数可以备份系统中所有的数据库，语句如下：

```
mysqldump -u user -h host -p --all-databases > filename.sql
```

使用参数--all-databases 时，不需要指定数据库名称。

【例 17.4】使用 mysqldump 备份服务器中的所有数据库，输入语句如下：

```
mysqldump -u root -p --all-databases > C:/backup/alldbinMySQL.sql
```

该语句创建名称为alldbinMySQL.sql的备份文件,文件中包含了对系统中所有数据库的备份信息。

提示：如果在服务器上进行备份，并且表均为 MyISAM 表，就应该考虑使用 mysqlhotcopy，因为可以更快地进行备份和恢复。

17.1.2　直接复制整个数据库目录

因为 MySQL 表保存为文件方式，所以可以直接复制 MySQL 数据库的存储目录及文件进行备份。MySQL 的数据库目录位置不一定相同，在 Windows 平台下，MySQL 8.0 存放数据库的目录通常默认为"C:\Documents and Settings\All Users\Application Data\MySQL\MySQL Server 8.0\data"或者其他用户自定义的目录；在 Linux 平台下，数据库目录位置通常为/var/lib/mysql/，不同 Linux 版本下目录会有所不同，读者应在自己使用的平台下查找该目录。

这是一种简单、快速、有效的备份方式。要想保持备份的一致性,备份前需要对相关表执行 LOCK TABLES 操作，然后对表执行 FLUSH TABLES。这样当复制数据库目录中的文件时，允许其他客户继续查询表。开始备份前，需要执行 FLUSH TABLES 语句来确保将所有激活的索引页写入硬盘。当然，也可以停止 MySQL 服务再进行备份操作。

这种方法虽然简单，但并不是最好的方法。因为这种方法对 InnoDB 存储引擎的表不适用。使用这种方法备份的数据最好恢复到相同版本的服务器中，不同的版本可能不兼容。

提示：在 MySQL 版本号中，第一个数字表示主版本号，主版本号相同的 MySQL 数据库文件格式相同。

17.1.3 使用 mysqlhotcopy 工具快速备份

mysqlhotcopy 是一个 Perl 脚本，最初由 Tim Bunce 编写并提供。它使用 LOCK TABLES、FLUSH TABLES 和 cp 或 scp 来快速备份数据库。它是备份数据库或单个表最快的方法，但它只能运行在数据库目录所在的机器上，并且只能备份 MyISAM 类型的表。mysqlhotcopy 在 UNIX 系统中运行。

mysqlhotcopy 命令语法格式如下：

```
mysqlhotcopy db_name_1, ... db_name_n /path/to/new_directory
```

db_name_1,…,db_name_n 分别为需要备份的数据库的名称；/path/to/new_directory 指定备份文件目录。

【例 17.5】使用 mysqlhotcopy 备份 test_db 数据库到/usr/backup 目录下，输入语句如下：

```
mysqlhotcopy -u root -p test_db /usr/backup
```

要想执行 mysqlhotcopy，需要访问备份的表文件，并设置表的 SELECT 权限、RELOAD 权限（以便能够执行 FLUSH TABLES）和 LOCK TABLES 权限。

提示：mysqlhotcopy 只能将表所在的目录复制到另一个位置，且只能用于备份 MyISAM 和 ARCHIVE 表。备份 InnoDB 类型的数据表时会出现错误信息。由于它复制本地格式的文件，因此也不能移植到其他硬件或操作系统下。

17.2 数据恢复

管理人员操作的失误、计算机故障以及其他意外情况，都会导致数据的丢失和破坏。当数据丢失或意外破坏时，可以通过恢复已经备份的数据尽量减少数据丢失和破坏造成的损失。本节将介绍数据恢复的方法。

17.2.1 使用 MySQL 命令恢复

对于已经备份的包含 CREATE、INSERT 语句的文本文件，可以使用 MySQL 命令导入到数据库中。本小节将介绍 MySQL 命令导入 sql 文件的方法。

备份的 sql 文件中包含 CREATE、INSERT 语句（有时也会有 DROP 语句）。MySQL 命令可以直接执行文件中的这些语句。其语法如下：

```
mysql -u user -p [dbname] < filename.sql
```

user 是执行 backup.sql 中语句的用户名；-p 表示输入用户密码；dbname 是数据库名。如果 filename.sql 文件为 mysqldump 工具创建的、包含创建数据库语句的文件，执行的时候不需要指定数据库名。

【例 17.6】使用 MySQL 命令将 C:\backup\booksdb_20220301.sql 文件中的备份导入到数据库中，输入语句如下：

```
mysql -u root -p booksDB < C:/backup/booksdb_20220301.sql
```

执行该语句前，必须先在 MySQL 服务器中创建 booksDB 数据库，如果 booksDB 数据库不存在，恢复过程将会出错。命令执行成功之后 booksdb_20220301.sql 文件中的语句就会在指定的数据库中恢复以前的表。

如果已经登录 MySQL 服务器，还可以使用 source 命令导入 sql 文件。source 语句语法如下：

```
source filename
```

【例 17.7】使用 root 用户登录到服务器，然后使用 source 导入本地的备份文件 booksdb_20220301.sql，输入语句如下：

```
--选择要恢复到的数据库
mysql> use booksDB;
Database changed

--使用 source 命令导入备份文件
mysql> source C:\backup\booksdb_20220301.sql
```

命令执行后，会列出备份文件 booksdb_20220301.sql 中每一条语句的执行结果。source 命令执行成功后，booksdb_20220301.sql 中的语句会全部导入到现有数据库中。

提示：执行 source 命令前，必须使用 use 语句选择数据库。否则，恢复过程中会出现"ERROR 1046 (3D000): No database selected"的错误。

17.2.2 直接复制到数据库目录

如果数据库通过复制数据库文件备份，可以直接复制备份的文件到 MySQL 数据目录下实现恢复。通过这种方式恢复时，备份数据的数据库和待恢复的数据库服务器的主版本号必须相同。而且这种方式只对 MyISAM 引擎的表有效，对于 InnoDB 引擎的表不可用。

执行恢复以前关闭 MySQL 服务，将备份的文件或目录覆盖 MySQL 的 data 目录，再重新启动 MySQL 服务。对于 Linux/UNIX 操作系统来说，复制完文件需要将文件的用户和组更改为 MySQL 运行的用户和组，通常用户是 mysql，组也是 mysql。

17.2.3 mysqlhotcopy 快速恢复

mysqlhotcopy 备份后的文件也可以用来恢复数据库，在 MySQL 服务器停止运行时，将备份的数据库文件复制到 MySQL 存放数据的位置（MySQL 的 data 文件夹），重新启动 MySQL 服务即可。如果以根用户执行该操作，必须指定数据库文件的所有者，输入语句如下：

```
chown -R mysql.mysql /var/lib/mysql/dbname
```

【例 17.8】从 mysqlhotcopy 备份文件恢复数据库，输入语句如下：

```
cp -R /usr/backup/test usr/local/mysql/data
```

执行完该语句，重启服务器，MySQL 将恢复到备份状态。

提示：如果需要恢复的数据库已经存在，则在使用 DROP 语句删除已经存在的数据库之后，恢复才能成功。另外，MySQL 不同版本之间必须兼容，恢复之后的数据才可以使用。

17.3 数据库迁移

数据库迁移就是把数据从一个系统移动到另一个系统上。数据库迁移有以下原因：

（1）需要安装新的数据库服务器。
（2）MySQL 版本更新。
（3）数据库管理系统的变更（如从 Microsoft SQL Server 迁移到 MySQL）。

本节将讲解数据库迁移的方法。

17.3.1 相同版本的 MySQL 数据库之间的迁移

相同版本的 MySQL 数据库之间的迁移，就是在主版本号相同的 MySQL 数据库之间进行数据库移动。迁移过程其实就是在源数据库备份和目标数据库恢复过程的组合。

在讲解数据库备份和恢复时，已经知道最简单的方式是通过复制数据库文件目录，但是此种方法只适用于 MyISAM 引擎的表。而对于 InnoDB 表，不能用直接复制文件的方式备份数据库，因此最常用和最安全的方式是使用 mysqldump 命令导出数据，然后在目标数据库服务器上使用 MySQL 命令导入。

【例 17.9】将 www.abc.com 主机上的 MySQL 数据库全部迁移到 www.bcd.com 主机上。在 www.abc.com 主机上执行的命令如下：

```
mysqldump -h www.bac.com -uroot -ppassword dbname |
mysql -h www.bcd.com -uroot -ppassword
```

mysqldump 导入的数据直接通过管道符"|"传给 MySQL 命令，导入到主机 www.bcd.com 数据库中。dbname 为需要迁移的数据库名称，如果要迁移全部的数据库，可使用参数 --all-databases。

17.3.2 不同版本的 MySQL 数据库之间的迁移

因为数据库升级等原因，需要将较旧版本 MySQL 数据库中的数据迁移到较新版本的数据库中。MySQL 服务器升级时，需要先停止服务，然后卸载旧版本，并安装新版的 MySQL，这种更新方法很简单，如果想保留旧版本中的用户访问控制信息，就需要备份 MySQL 中的 MySQL 数据库，在新版本 MySQL 安装完成之后，重新读入 MySQL 备份文件中的信息。

旧版本与新版本的 MySQL 可能使用不同的默认字符集，例如 MySQL 8.0 版本之前，默认字符集为 latin1，而 MySQL 8.0 版本默认字符集为 **utf8mb4**。数据库中有中文数据时，迁移过程中需要对默认字符集进行修改，否则可能无法正常显示相关数据。

新版本会对旧版本有一定兼容性。从旧版本的 MySQL 向新版本的 MySQL 迁移时，对于 MyISAM 引擎的表，可以直接复制数据库文件，也可以使用 mysqlhotcopy 工具、mysqldump 工具。对于 InnoDB 引擎的表，一般只能使用 mysqldump 将数据导出。然后再使用 MySQL 命令导入到目标服务器上。从新版本向旧版本 MySQL 迁移数据时要特别小心，最好使用 mysqldump 命令导出，然后导入目标数据库中。

17.3.3　不同数据库之间的迁移

不同类型的数据库之间的迁移，是指把 MySQL 的数据库转移到其他类型的数据库，例如从 MySQL 迁移到 Oracle，从 Oracle 迁移到 MySQL，从 MySQL 迁移到 SQL Server 等。

迁移之前，需要了解不同数据库的架构，比较它们之间的差异。不同数据库中定义相同类型的数据的关键字可能会不同。例如，MySQL 中日期字段分为 DATE 和 TIME 两种，而 Oracle 日期字段只有 DATE。另外，数据库厂商并没有完全按照 SQL 标准来设计数据库系统，导致不同的数据库系统的 SQL 语句有差别。例如，MySQL 几乎完全支持标准 SQL 语言，而 Microsoft SQL Server 使用的是 T-SQL 语言，T-SQL 中有一些非标准的 SQL 语句，因此在迁移时必须对这些语句进行语句映射处理。

数据库迁移可以使用一些工具，例如在 Windows 系统下，可以使用 MyODBC 实现 MySQL 和 SQL Server 之间的迁移。MySQL 官方提供的工具 MySQL Migration Toolkit 也可以在不同数据库间进行数据迁移。

17.4　表的导出和导入

有时需要将 MySQL 数据库中的数据导出到外部存储文件中，MySQL 数据库中的数据可以导出成 sql 文本文件、xml 文件或者 html 文件。同样，这些导出文件也可以导入到 MySQL 数据库中。本节将介绍数据导出和导入的常用方法。

17.4.1　使用 SELECT…INTO OUTFILE 导出文本文件

MySQL 数据库导出数据时，允许使用包含导出定义的 SELECT 语句进行数据的导出操作。该文件被创建到服务器主机上，因此必须拥有文件写入权限（FILE 权限）才能使用此语法。使用 "SELECT...INTO OUTFILE 'filename'" 形式的 SELECT 语句，可以把被选择的行写入一个文件中，并且 filename 不能是一个已经存在的文件。SELECT...INTO OUTFILE 语句的基本格式如下：

```
SELECT columnlist  FROM table WHERE condition  INTO OUTFILE 'filename'  [OPTIONS]

--OPTIONS 选项
FIELDS   TERMINATED BY 'value'
FIELDS   [OPTIONALLY] ENCLOSED BY 'value'
FIELDS   ESCAPED BY 'value'
LINES    STARTING BY 'value'
LINES    TERMINATED BY 'value'
```

可以看到 SELECT columnlist FROM table WHERE condition 为一个查询语句，查询结果返回满足指定条件的一条或多条记录；INTO OUTFILE 语句的作用就是把前面 SELECT 语句查询出来的结果导出到名称为 filename 的外部文件中。[OPTIONS]为可选参数选项，OPTIONS 部分的语法包括 FIELDS 和 LINES 子句，其可能的取值有：

- FIELDS TERMINATED BY 'value'：设置字段之间的分隔字符，可以为单个或多个字符，默认情况下为制表符"\t"。
- FIELDS [OPTIONALLY] ENCLOSED BY 'value'：设置字段的包围字符，只能为单个字符，若使用了 OPTIONALLY，则只有 CHAR 和 VERCHAR 等字符数据字段被包括。
- FIELDS ESCAPED BY 'value'：设置如何写入或读取特殊字符，只能为单个字符，即设置转义字符，默认值为"\"。
- LINES STARTING BY 'value'：设置每行数据开头的字符，可以为单个或多个字符，默认情况下不使用任何字符。
- LINES TERMINATED BY 'value'：设置每行数据结尾的字符，可以为单个或多个字符，默认值为"\n"。

FIELDS 和 LINES 两个子句都是自选的，但是如果两个都被指定了，FIELDS 必须位于 LINES 的前面。

SELECT...INTO OUTFILE 语句可以非常快速地把一个表转储到服务器上。如果想要在服务器主机之外的部分客户主机上创建结果文件，不能使用 SELECT...INTO OUTFILE。在这种情况下，应该在客户主机上使用比如"MySQL –e "SELECT ..." > file_name"的命令来生成文件。

SELECT...INTO OUTFILE 是 LOAD DATA INFILE 的补语。用于语句的 OPTIONS 部分的语法包括部分 FIELDS 和 LINES 子句，这些子句与 LOAD DATA INFILE 语句同时使用。

【例 17.10】使用 SELECT...INTO OUTFILE 将 test_db 数据库中的 person 表的记录导出到文本文件，输入命令如下：

```
mysql> SELECT * FROM test_db.person INTO OUTFILE 'D:/person0.txt';
```

执行后报错信息如下：

```
ERROR 1290 (HY000): The MySQL server is running with the --secure-file-priv option so it cannot execute this statement
```

这是因为 MySQL 默认对导出的目录有权限限制，也就是说使用命令行进行导出的时候，需要指定目录进行操作。那么指定的目录是什么呢？

查询指定目录的命令如下：

```
show global variables like '%secure%';
```

执行结果如下所示：

```
+--------------------------+--------------------------------------------------+
| Variable_name            | Value                                            |
+--------------------------+--------------------------------------------------+
| require_secure_transport | OFF                                              |
| secure_file_priv         | C:\ProgramData\MySQL\MySQL Server 8.0\Uploads\   |
+--------------------------+--------------------------------------------------+
```

因为 secure_file_priv 配置的关系，所以必须导出到 C:\ProgramData\MySQL\MySQL Server 8.0\Uploads\目录下。如果想自定义导出路径，需要修改 my.ini 配置文件。打开路径 C:\ProgramData\MySQL\MySQL Server 8.0，用记事本打开 my.ini，然后搜索到以下代码：

```
secure-file-priv="C:/ProgramData/MySQL/MySQL Server 8.0/Uploads"
```

在上述代码前添加#，然后添加以下内容：

```
secure-file-priv="D:/"
```

结果如图 17.1 所示。

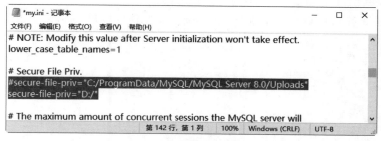

图 17.1　设置数据表的导出路径

重启 MySQL 服务器后，再次使用 SELECT…INTO OUTFILE 将 test_db 数据库中 person 表的记录导出到文本文件，输入如下命令：

```
mysql>SELECT * FROM test_db.person INTO OUTFILE 'D:/person0.txt';
Query OK, 1 row affected (0.01 sec)
```

由于指定了 INTO OUTFILE 子句，因此 SELECT 会将查询出来的 3 个字段值保存到 C:\person0.txt 文件中。打开文件，内容如下：

```
1    Green    21    Lawyer
2    Suse  22    dancer
3    Mary  24    Musician
4    Willam   20    sports man
5    Laura    25    \N
6    Evans    27    secretary
7    Dale  22    cook
8    Edison   28    singer
9    Harry    21    magician
10   Harriet  19    pianist
```

默认情况下，MySQL 使用制表符"\t"分隔不同的字段，字段没有被其他字符括起来。另外，第 5 行中有一个字段值为"\N"，表示该字段的值为 NULL。默认情况下，如果遇到 NULL 值，将会返回"\N"，代表空值，其中的反斜线"\"表示转义字符，如果使用 ESCAPED BY 选项，则 N 前面为指定的转义字符。

【例 17.11】使用 SELECT…INTO OUTFILE 将 test_db 数据库 person 表的记录导出到文本文件，使用 FIELDS 选项和 LINES 选项，要求字段之间使用","间隔，所有字段值用双引号括起来，定义转义字符为单引号"\'"，执行的命令如下：

```
SELECT * FROM  test_db.person INTO OUTFILE "D:/person1.txt"
FIELDS
TERMINATED BY ','
ENCLOSED BY '\"'
ESCAPED BY '\''
LINES
TERMINATED BY '\r\n';
```

该语句将把 person 表中所有记录导入到 D 盘目录下的 person1.txt 文本文件中。

FIELDS TERMINATED BY ','表示字段之间用逗号分隔；ENCLOSED BY '\"'表示每个字段用双引号括起来；ESCAPED BY '\''表示将系统默认的转义字符替换为单引号；LINES TERMINATED BY '\r\n'表示每行以回车换行符结尾，保证每一条记录占一行。

执行成功后，在目录 D 盘下生成一个 person1.txt 文件。打开文件，内容如下：

```
"1","Green","21","Lawyer"
"2","Suse","22","dancer"
"3","Mary","24","Musician"
"4","Willam","20","sports man"
"5","Laura","25",'N'
"6","Evans","27","secretary"
"7","Dale","22","cook"
"8","Edison","28","singer"
"9","Harry","21","magician"
"10","Harriet","19","pianist"
```

可以看到，所有的字段值都被双引号包括；第 5 条记录中空值的表示形式为"N"，即使用单引号替换了反斜线转义字符。

【例 17.12】使用 SELECT...INTO OUTFILE 将 test_db 数据库 person 表中的记录导出到文本文件，使用 LINES 选项，要求每行记录以字符串">"开始、以"<end>"字符串结尾，执行的命令如下：

```
SELECT * FROM  test_db.person INTO OUTFILE "D:/person2.txt"
LINES
STARTING BY '>'
TERMINATED BY '<end>';
```

执行成功后，在目录 D 盘下生成一个 person2.txt 文件。打开文件，内容如下：

```
> 1  Green    21  Lawyer <end>> 2  Suse  22  dancer <end>> 3  Mary  24  Musician <end>> 4  Willam   20   sports man <end>> 5  Laura   25   \N <end>> 6  Evans    27   secretary <end>> 7 Dale 22  cook <end>> 8 Edison  28  singer <end>> 9  Harry   21  magician <end>> 10  Harriet  19  pianist <end>
```

可以看到，虽然将所有的字段值导出到文本文件中，但是所有的记录没有分行区分，出现这种情况是因为 TERMINATED BY 选项替换了系统默认的"\n"换行符，如果希望换行显示，则需要修改导出语句：

```
SELECT * FROM  test_db.person INTO OUTFILE "D:/person3.txt"
LINES
STARTING BY '>'
TERMINATED BY '<end>\r\n';
```

执行完语句之后，换行显示每条记录，结果如下：

```
> 1  Green     21   Lawyer <end>
> 2  Suse      22   dancer <end>
> 3  Mary      24   Musician <end>
> 4  Willam    20   sports man <end>
> 5  Laura     25   \N <end>
> 6  Evans     27   secretary <end>
```

```
> 7  Dale      22   cook     <end>
> 8  Edison    28   singer   <end>
> 9  Harry     21   magician <end>
> 10 Harriet   19   pianist  <end>
```

17.4.2 使用 mysqldump 命令导出文本文件

除了使用 SELECT…INTO OUTFILE 语句导出文本文件之外，还可以使用 mysqldump。本章开始介绍了使用 mysqldump 备份数据库，该工具不仅可以将数据导出为包含 CREATE、INSERT 的 sql 文件，也可以导出为纯文本文件。

mysqldump 创建一个包含 CREATE TABLE 语句的 tablename.sql 文件和一个包含其数据的 tablename.txt 文件。mysqldump 导出文本文件的基本语法格式如下：

```
mysqldump -T path -u root -p dbname [tables] [OPTIONS]

--OPTIONS 选项
--fields-terminated-by=value
--fields-enclosed-by=value
--fields-optionally-enclosed-by=value
--fields-escaped-by=value
--lines-terminated-by=value
```

只有指定了 -T 参数才可以导出纯文本文件；path 表示导出数据的目录；tables 为指定要导出的表名称，如果不指定，将导出数据库 dbname 中所有的表；[OPTIONS]为可选参数选项，这些选项需要结合 -T 选项使用。使用 OPTIONS 常见的取值有：

- --fields-terminated-by=value：设置字段之间的分隔字符，可以为单个或多个字符，默认情况下为制表符 "\t"。
- --fields-enclosed-by=value：设置字段的包围字符。
- --fields-optionally-enclosed-by=value：设置字段的包围字符，只能为单个字符，只能包括 CHAR 和 VERCHAR 等字符数据字段。
- --fields-escaped-by=value：控制如何写入或读取特殊字符，只能为单个字符，即设置转义字符，默认值为反斜线 "\"。
- --lines-terminated-by=value：设置每行数据结尾的字符，可以为单个或多个字符，默认值为 "\n"。

提示：与 SELECT…INTO OUTFILE 语句中的 OPTIONS 各个参数设置不同，这里 OPTIONS 各个选项等号后面的 value 值不要用引号括起来。

【例 17.13】使用 mysqldump 将 test_db 数据库 person 表的记录导出到文本文件，执行的命令如下：

```
mysqldump -T D:\ test_db person -u root -p
```

语句执行成功，系统 D 盘目录下面将会有两个文件，分别为 person.sql 和 person.txt。person.sql 包含创建 person 表的 CREATE 语句，其内容如下：

```
-- MySQL dump 10.13  Distrib 8.0.28, for Win64 (x86_64)
```

```
--
-- Host: localhost    Database: test_db
-- ------------------------------------------------------
-- Server version  8.0.28

/*!40101 SET @OLD_CHARACTER_SET_CLIENT=@@CHARACTER_SET_CLIENT */;
/*!40101 SET @OLD_CHARACTER_SET_RESULTS=@@CHARACTER_SET_RESULTS */;
/*!40101 SET @OLD_COLLATION_CONNECTION=@@COLLATION_CONNECTION */;
 SET NAMES utf8mb4 ;
/*!40103 SET @OLD_TIME_ZONE=@@TIME_ZONE */;
/*!40103 SET TIME_ZONE='+00:00' */;
/*!40101 SET @OLD_SQL_MODE=@@SQL_MODE, SQL_MODE='' */;
/*!40111 SET @OLD_SQL_NOTES=@@SQL_NOTES, SQL_NOTES=0 */;

--
-- Table structure for table `person`
--

DROP TABLE IF EXISTS `person`;
/*!40101 SET @saved_cs_client     = @@character_set_client */;
/*!40101 SET character_set_client = utf8 */;
CREATE TABLE `person` (
 `id` int(10) unsigned NOT NULL AUTO_INCREMENT,
 `name` char(40) NOT NULL DEFAULT '',
 `age` int(11) NOT NULL DEFAULT '0',
 `info` char(50) DEFAULT NULL,
PRIMARY KEY (`id`)
) ENGINE=InnoDB AUTO_INCREMENT=11 DEFAULT CHARSET=utf8;
/*!40101 SET character_set_client = @saved_cs_client */;

/*!40103 SET TIME_ZONE=@OLD_TIME_ZONE */;

/*!40101 SET SQL_MODE=@OLD_SQL_MODE */;
/*!40101 SET CHARACTER_SET_CLIENT=@OLD_CHARACTER_SET_CLIENT */;
/*!40101 SET CHARACTER_SET_RESULTS=@OLD_CHARACTER_SET_RESULTS */;
/*!40101 SET COLLATION_CONNECTION=@OLD_COLLATION_CONNECTION */;
/*!40111 SET SQL_NOTES=@OLD_SQL_NOTES */;

-- Dump completed on 2022-03-03 16:40:55
```

备份文件中的信息介绍参看 17.1.1 小节。

person.txt 包含数据表中的数据，其内容如下：

```
1      Green       21      Lawyer
2      Suse        22      dancer
3      Mary        24      Musician
4      Willam      20      sports man
5      Laura       25      \N
6      Evans       27      secretary
7      Dale        22      cook
8      Edison      28      singer
9      Harry       21      magician
10     Harriet     19      pianist
```

【例 17.14】 使用 mysqldump 命令将 test_db 数据库 person 表的记录导出到文本文件，使用

FIELDS 选项，要求字段之间使用逗号","间隔，所有字符类型字段值用双引号括起来，定义转义字符为问号"?"，每行记录以回车换行符"\r\n"结尾，执行的命令如下：

```
mysqldump -T D:\ test_db person -u root -p --fields-terminated-by=,
--fields-optionally-enclosed-by=\" --fields-escaped-by=? --lines-terminated-by=\r\n
Enter password: ******
```

上面语句要在一行中输入，语句执行成功，系统 D 盘目录下面将会有两个文件，分别为 person.sql 和 person.txt。person.sql 包含创建 person 表的 CREATE 语句，其内容与前面例子中的相同，person.txt 文件的内容与上一个例子不同，显示如下：

```
1,"Green",21,"Lawyer"
2,"Suse",22,"dancer"
3,"Mary",24,"Musician"
4,"Willam",20,"sports man"
5,"Laura",25,?N
6,"Evans",27,"secretary"
7,"Dale",22,"cook"
8,"Edison",28,"singer"
9,"Harry",21,"magician"
10,"Harriet",19,"pianist"
```

可以看到，只有字符类型的值被双引号括了起来，而数值类型的值没有；第 5 行记录中的 NULL 值表示为"?N"，使用问号"?"替代了系统默认的反斜线转义字符"\"。

17.4.3 使用 MySQL 命令导出文本文件

MySQL 是一个功能丰富的工具命令，使用 MySQL 还可以在命令行模式下执行 SQL 指令，将查询结果导入到文本文件中。相比 mysqldump，mysql 工具导出的结果可读性更强。

如果 MySQL 服务器是单独的机器，用户是在一个 client 上进行操作，用户要把数据结果导入到 client 机器上。可以使用 mysql -e 语句。

使用 MySQL 命令导出数据文本文件语句的基本格式如下：

```
mysql -u root -p --execute= "SELECT 语句" dbname > filename.txt
```

该命令使用--execute 选项，表示执行该选项后面的语句并退出，后面的语句必须用双引号括起来，dbname 为要导出的数据库名称；导出的文件中不同列之间使用制表符分隔，第 1 行包含了各个字段的名称。

【例 17.15】使用 MySQL 语句，导出 test_db 数据库 person 表的记录到文本文件，输入语句如下：

```
mysql -u root -p --execute="SELECT * FROM person;" test_db > D:\person3.txt
```

语句执行完毕之后，D 盘目录下将会出现名称为 person3.txt 的文本文件，其内容如下：

```
id    name       age       info
1     Green      21        Lawyer
2     Suse       22        dancer
3     Mary       24        Musician
4     Willam     20        sports man
5     Laura      25        NULL
6     Evans      27        secretary
```

```
 7    Dale      22     cook
 8    Edison    28     singer
 9    Harry     21     magician
10    Harriet   19     pianist
```

可以看到，person3.txt 文件中包含了每个字段的名称和各条记录，该显示格式与 MySQL 命令行下 SELECT 查询结果的显示格式相同。

使用 MySQL 命令还可以指定查询结果的显示格式，如果某行记录字段很多，可能一行不能完全显示，可以使用--vartical 参数，将每条记录分为多行显示。

【例 17.16】使用 MySQL 命令导出 test_db 数据库 person 表的记录到文本文件，使用--vertical 参数显示结果，输入语句如下：

```
mysql -u root -p --vertical --execute="SELECT * FROM person;" test_db > D:\person4.txt
```

语句执行之后，D:\person4.txt 文件中的内容如下：

```
*** 1. row ***
  id: 1
name: Green
 age: 21
info: Lawyer
*** 2. row ***
  id: 2
name: Suse
 age: 22
info: dancer
*** 3. row ***
  id: 3
name: Mary
 age: 24
info: Musician
*** 4. row ***
  id: 4
name: Willam
 age: 20
info: sports man
*** 5. row ***
  id: 5
name: Laura
 age: 25
info: NULL
*** 6. row ***
  id: 6
name: Evans
 age: 27
info: secretary
*** 7. row ***
  id: 7
name: Dale
 age: 22
info: cook
*** 8. row ***
  id: 8
```

```
    name: Edison
     age: 28
    info: singer
*** 9. row ***
      id: 9
    name: Harry
     age: 21
    info: magician
*** 10. row ***
      id: 10
    name: Harriet
     age: 19
    info: pianist
```

可以看到，SELECT 的查询结果导出到文本文件之后，显示格式发生了变化，如果 person 表中记录内容很长，这样显示将会更加容易阅读。

MySQL 可以将查询结果导出到 html 文件中，使用--html 选项即可。

【例 17.17】使用 MySQL 命令导出 test_db 数据库 person 表的记录到 html 文件，输入语句如下：

```
mysql -u root -p --html --execute="SELECT * FROM person;" test_db > D:\person5.html
```

语句执行成功，将在 D 盘创建文件 person5.html，该文件在浏览器中的显示效果如图 17.2 所示。

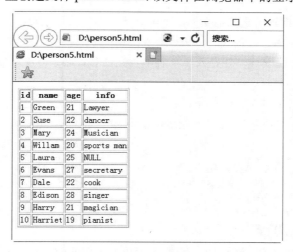

图 17.2　使用 MySQL 导出数据到 html 文件

如果要将表数据导出到 xml 文件中，可使用--xml 选项。

【例 17.18】使用 MySQL 命令导出 test_db 数据库 person 表的记录到 xml 文件，输入语句如下：

```
mysql -u root -p --xml --execute="SELECT * FROM person;" test_db >D:\person6.xml
```

语句执行成功，将在 D 盘创建文件 person6.xml，该文件在浏览器中的显示效果如图 17.3 所示。

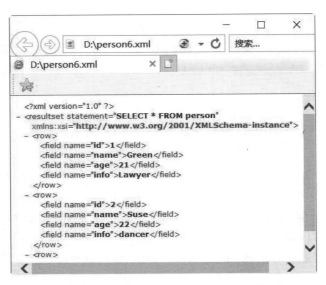

图 17.3　使用 MySQL 导出数据到 xml 文件

17.4.4　使用 LOAD DATA INFILE 方式导入文本文件

MySQL 允许将数据导出到外部文件，也可以从外部文件导入数据。MySQL 提供了一些导入数据的工具，包括 LOAD DATA 语句、source 命令和 MySQL 命令。LOAD DATA INFILE 语句用于高速地从一个文本文件中读取行，并导入一个表中。文件名称必须为文字字符串。本节将介绍 LOAD DATA 语句的用法。

LOAD DATA 语句的基本格式如下：

```
LOAD DATA  INFILE 'filename.txt' INTO TABLE tablename [OPTIONS] [IGNORE number LINES]

-- OPTIONS 选项
    FIELDS  TERMINATED BY 'value'
FIELDS  [OPTIONALLY] ENCLOSED BY 'value'
FIELDS  ESCAPED BY 'value'
LINES   STARTING BY 'value'
LINES   TERMINATED BY 'value'
```

可以看到 LOAD DATA 语句中，关键字 INFILE 后面的 filename 文件为导入数据的来源；tablename 表示待导入的数据表名称；[OPTIONS]为可选参数选项，OPTIONS 部分的语法包括 FIELDS 和 LINES 子句，其可能的取值有：

- FIELDS TERMINATED BY 'value'：设置字段之间的分隔字符，可以为单个或多个字符，默认情况下为制表符"\t"。
- FIELDS [OPTIONALLY] ENCLOSED BY 'value'：设置字段的包围字符，只能为单个字符。如果使用了 OPTIONALLY，则只有 CHAR 和 VERCHAR 等字符数据字段被包括。
- FIELDS ESCAPED BY 'value'：控制如何写入或读取特殊字符，只能为单个字符，即设置转义字符，默认值为"\"。
- LINES STARTING BY 'value'：设置每行数据开头的字符，可以为单个或多个字符，

- LINES TERMINATED BY 'value'：设置每行数据结尾的字符，可以为单个或多个字符，默认值为"\n"。

IGNORE number LINES 选项表示忽略文件开始处的行数，number 表示忽略的行数。执行 LOAD DATA 语句需要 FILE 权限。

【例 17.19】使用 LOAD DATA 命令将 D:\person0.txt 文件中的数据导入到 test_db 数据库的 person 表中，输入语句如下：

```
LOAD DATA  INFILE 'D:\person0.txt' INTO TABLE test_db.person;
```

恢复之前，要将 person 表中的数据全部删除。登录 MySQL，使用 DELETE 语句：

```
mysql> USE test_db;
Database changed;
mysql> DELETE FROM person;
Query OK, 10 rows affected (0.00 sec)
```

从 person0.txt 文件中恢复数据，语句如下：

```
mysql> LOAD DATA  INFILE 'D:\person0.txt' INTO TABLE test_db.person;
Query OK, 10 rows affected (0.00 sec)
Records: 10  Deleted: 0  Skipped: 0  Warnings: 0

mysql> SELECT * FROM person;
+----+---------+-----+------------+
| id | name    | age | info       |
+----+---------+-----+------------+
| 1  | Green   | 21  | Lawyer     |
| 2  | Suse    | 22  | dancer     |
| 3  | Mary    | 24  | Musician   |
| 4  | Willam  | 20  | sports man |
| 5  | Laura   | 25  | NULL       |
| 6  | Evans   | 27  | secretary  |
| 7  | Dale    | 22  | cook       |
| 8  | Edison  | 28  | singer     |
| 9  | Harry   | 21  | magician   |
| 10 | Harriet | 19  | pianist    |
+----+---------+-----+------------+
```

可以看到，语句执行成功之后，原来的数据重新恢复到了 person 表中。

【例 17.20】使用 LOAD DATA 命令将 D:\person1.txt 文件中的数据导入到 test_db 数据库的 person 表，使用 FIELDS 选项和 LINES 选项，要求字段之间使用逗号","间隔，所有字段值用双引号括起来，定义转义字符为单引号"\'"，每行记录以回车换行符"\r\n"结尾，输入语句如下：

```
LOAD DATA INFILE 'D:\person1.txt' INTO TABLE test_db.person
FIELDS
TERMINATED BY ','
ENCLOSED BY '\"'
ESCAPED BY '\''
LINES
TERMINATED BY '\r\n';
```

恢复之前，使用 DELETE 语句将 person 表中的数据全部删除，执行过程如下：

```
mysql> DELETE FROM person;
Query OK, 10 rows affected (0.00 sec)
```

从 person1.txt 文件中恢复数据，执行过程如下：

```
mysql> LOAD DATA  INFILE 'D:\person1.txt' INTO TABLE test_db .person
    -> FIELDS
    -> TERMINATED BY ','
    -> ENCLOSED BY '\"'
    -> ESCAPED BY '\'
    -> LINES
    -> TERMINATED BY '\r\n';
Query OK, 10 rows affected (0.00 sec)
Records: 10  Deleted: 0  Skipped: 0  Warnings: 0
```

语句执行成功，使用 SELECT 语句查看 person 表中的记录，结果与前一个例子相同。

17.4.5 使用 mysqlimport 命令导入文本文件

使用 mysqlimport 可以导入文本文件，并且不需要登录 MySQL 客户端。mysqlimport 命令提供许多与 LOAD DATA INFILE 语句相同的功能，大多数选项直接对应 LOAD DATA INFILE 子句。使用 mysqlimport 语句需要指定所需的选项、导入的数据库名称以及导入的数据文件路径和名称。mysqlimport 命令的基本语法格式如下：

```
mysqlimport -u root-p dbname filename.txt [OPTIONS]

--OPTIONS 选项
--fields-terminated-by=value
--fields-enclosed-by=value
--fields-optionally-enclosed-by=value
--fields-escaped-by=value
--lines-terminated-by=value
--ignore-lines=n
```

dbname 为导入的表所在的数据库名称。注意，mysqlimport 命令不指定导入数据库的表名称，数据表的名称由导入文件名称确定，即文件名作为表名，导入数据之前该表必须存在。[OPTIONS] 为可选参数选项，其常见的取值有：

- --fields-terminated-by='value'：设置字段之间的分隔字符，可以为单个或多个字符，默认情况下为制表符 "\t"。
- --fields-enclosed-by='value'：设置字段的包围字符。
- --fields-optionally-enclosed-by='value'：设置字段的包围字符，只能为单个字符，包括 CHAR 和 VERCHAR 等字符数据字段。
- --fields-escaped-by='value'：控制如何写入或读取特殊字符，只能为单个字符，即设置转义字符，默认值为反斜线 "\"。
- --lines-terminated-by='value'：设置每行数据结尾的字符，可以为单个或多个字符，默认值为 "\n"。

- --ignore-lines=n：忽视数据文件的前 n 行。

【例 17.21】使用 mysqlimport 命令将 D 盘目录下的 person.txt 文件内容导入到 test_db 数据库中，字段之间使用逗号","间隔，字符类型字段值用双引号括起来，将转义字符定义为问号"?"，每行记录以回车换行符"\r\n"结尾，执行的命令如下：

```
C:\ >mysqlimport -u root -p test_db D:\person.txt --fields-terminated-by=,
--fields-optionally-enclosed-by=\"--fields-escaped-by=?--lines-terminated-by=\r\n
```

上面的语句要在一行中输入，语句执行成功，将把 person.txt 中的数据导入到数据库。

除了前面介绍的几个选项之外，mysqlimport 支持许多选项，常见的选项有：

- --columns=column_list, -c column_list：采用逗号分隔的列名作为其值。列名的顺序指示如何匹配数据文件列和表列。
- --compress, -C：压缩在客户端和服务器之间发送的所有信息（如果二者均支持压缩）。
- -d，--delete：导入文本文件前清空表。
- --force, -f：忽视错误。例如，如果某个文本文件的表不存在，会继续处理其他文件。不使用--force，如果表不存在，则 mysqlimport 会退出。
- --host=host_name，-h host_name：将数据导入给定主机上的 MySQL 服务器。默认主机是 localhost。
- --ignore，-i：参见--replace 选项的描述。
- --ignore-lines=n：忽视数据文件的前 n 行。
- --local，-L：从本地客户端读入输入文件。
- --lock-tables，-l：处理文本文件前锁定所有表，以便写入。这样可以确保所有表在服务器上保持同步。
- --password[=password]，-p[password]：当连接服务器时使用的密码。如果使用短选项形式（-p），选项和密码之间不能有空格。如果在命令行中--password 或-p 选项后面没有密码值，则提示输入密码。
- --port=port_num，-P port_num：用于连接的 TCP/IP 端口号。
- --protocol={TCP | SOCKET | PIPE | MEMORY}：使用的连接协议。
- --replace，-r：--replace 和--ignore 选项用来控制导入时唯一键值已有记录的导入行的处理方式。如果指定--replace，新导入行替换原有相同的唯一键值的已有行；如果指定--ignore，已有唯一键值的记录行被跳过；如果不指定这两个选项，当发现一个已有键值记录行时会出现错误，并且忽视文本文件的剩余部分。
- --silent，-s：沉默模式。只有出现错误时才输出信息。
- --user=user_name，-u user_name：当连接服务器时 MySQL 使用的用户名。
- --verbose，-v：冗长模式。打印出程序操作的详细信息。
- --version，-V：显示版本信息并退出。

第 18 章

MySQL 日志

MySQL 日志记录了 MySQL 数据库日常操作和错误信息。MySQL 有不同类型的日志文件（各自存储了不同类型的日志），从日志当中可以查询到 MySQL 数据库的运行情况、用户操作、错误信息等，可以为 MySQL 管理和优化提供必要的参考。对于 MySQL 的管理工作而言，这些日志文件是不可缺少的。本章将介绍 MySQL 各种日志的作用以及日志的管理。

18.1 日志简介

MySQL 日志主要分为 4 类，使用这些日志文件，可以查看 MySQL 内部发生的事情。这 4 类日志分别是：

（1）错误日志：记录 MySQL 服务的启动、运行或停止 MySQL 服务时出现的问题。
（2）查询日志：记录建立的客户端连接和执行的语句。
（3）二进制日志：记录所有更改数据的语句，可以用于数据复制。
（4）慢查询日志：记录所有执行时间超过 long_query_time 的所有查询或不使用索引的查询。

默认情况下，所有日志创建于 MySQL 数据目录中。通过刷新日志，可以强制 MySQL 关闭和重新打开日志文件（或者在某些情况下切换到一个新的日志）。当执行一个 FLUSH LOGS 语句或执行 mysqladmin flush-logs 或 mysqladmin refresh 时，将刷新日志。

如果正使用 MySQL 复制功能，在复制服务器上可以维护更多的日志文件，这种日志称为接替日志。

启动日志功能会降低 MySQL 数据库的性能。例如，在查询非常频繁的 MySQL 数据库系统中，如果开启了通用查询日志和慢查询日志，MySQL 数据库会花费很多时间记录日志。同时，日志会占用大量的磁盘空间。

18.2 二进制日志

二进制日志主要记录 MySQL 数据库的变化。二进制日志以一种有效的格式并且是事务安全的方式包含更新日志中可用的所有信息。二进制日志包含了所有更新了数据或者已经潜在更新了数据（例如，没有匹配任何行的一个 DELETE）的语句。语句以"事件"的形式保存，描述数据更改。

二进制日志还包含关于每个更新数据库的语句的执行时间信息。它不包含没有修改任何数据的语句。如果想要记录所有语句（例如，为了识别有问题的查询），需要使用一般查询日志。使用二进制日志的主要目的是最大可能地恢复数据库，因为二进制日志包含备份后进行的所有更新。本节将介绍二进制日志的相关内容。

18.2.1 启动和设置二进制日志

默认情况下，二进制日志是开启的，可以通过修改 MySQL 的配置文件来启动和设置二进制日志。

my.ini 中[mysqld]组下面有关于二进制日志的设置：

```
log-bin [=path/ [filename] ]
expire_logs_days = 10
maxbinlogsize = 100M
```

log-bin 定义开启二进制日志；path 表明日志文件所在的目录路径；filename 指定了日志文件的名称，如文件的全名为 filename.000001、filename.000002 等，除了上述文件之外，还有一个名称为 filename.index 的文件，文件内容为所有日志的清单，可以使用记事本打开该文件。

expire_logs_days 定义了 MySQL 清除过期日志的时间，即二进制日志自动删除前的保留天数。默认值为 0，表示"没有自动删除"。当 MySQL 启动或刷新二进制日志时可能删除过期日志。

max_binlog_size 定义了单个文件的大小限制，如果二进制日志写入的内容大小超出给定值，日志就会发生滚动（关闭当前文件，重新打开一个新的日志文件）。不能将该变量设置为大于 1GB 或小于 4096B，默认值是 1GB。

如果正在使用大的事务，二进制日志文件大小还可能会超过 max_binlog_size 定义的大小。

在 my.ini 配置文件中的[MySQLd]组下，添加以下几个参数与参数值：

```
[mysqld]
log-bin
expire_logs_days = 10
max_binlog_size = 100M
```

添加完毕之后，关闭并重新启动 MySQL 服务进程，即可打开二进制日志，然后可以通过 SHOW VARIABLES 语句来查询日志设置。

【例 18.1】使用 SHOW VARIABLES 语句查询日志设置，执行的语句及结果如下：

```
mysql> SHOW VARIABLES LIKE 'log_%';
+-----------------------------+---------------------------------+
|Variable_name                | Value                           |
+-----------------------------+---------------------------------+
```

```
|log_bin                                  | ON                                                              |
|log_bin_basename         | C:\ProgramData\MySQL\MySQL Server 8.0\Data\X0NHUN07YDZVSSI-bin       |
|log_bin_index            | C:\ProgramData\MySQL\MySQL Server 8.0\Data\X0NHUN07YDZVSSI-bin.index |
|log_bin_trust_function_creators          |OFF                                                |
|log_bin_use_v1_row_events                | OFF                                               |
|log_error                                | .\X0NHUN07YDZVSSI.err                             |
|log_error_services                       |log_filter_internal;log_sink_internal              |
|log_error_suppression_list               |                                                   |
|log_error_verbosity                      |2                                                  |
|log_output                               | FILE                                              |
|log_queries_not_using_indexes            |OFF                                                |
|log_slave_updates                        |ON                                                 |
|log_slow_admin_statements                |OFF                                                |
|log_slow_slave_statements                |OFF                                                |
|log_statements_unsafe_for_binlog|ON                                                          |
|log_throttle_queries_not_using_indexes|0                                                      |
|log_timestamps                           | UTC                                               |
```

通过上面的查询结果可以看出，log_bin 变量的值为 ON，表明二进制日志已经打开。MySQL 重新启动之后，读者可以在自己机器上的 MySQL 数据文件夹下面看到新生成的文件后缀为.000001 和 .index 的两个文件，文件名称为默认主机名称。例如，在笔者的机器上的文件名称为 X0NHUN07YDZVSSI-bin.000001 和 X0NHUN07YDZVSSI-bin.index。

提示：数据库文件最好不要与日志文件放在同一个磁盘上，这样当数据库文件所在的磁盘发生故障时，可以使用日志文件恢复数据。

18.2.2 查看二进制日志

MySQL 二进制日志存储了所有的变更信息，MySQL 二进制日志是经常用到的。当 MySQL 创建二进制日志文件时，先创建一个以 filename 为名称、以.index 为后缀的文件，再创建一个以 filename 为名称、以.000001 为后缀的文件。MySQL 服务重新启动一次，以.000001 为后缀的文件会增加一个，并且后缀名加 1 递增；如果日志长度超过了 max_binlog_size 的上限（默认是 1GB），就会创建一个新的日志文件。

SHOW BINARY LOGS 语句可以查看当前的二进制日志文件个数及其文件名。MySQL 二进制日志并不能直接查看，如果要查看日志内容，可以通过 mysqlbinlog 命令查看。

【例 18.2】使用 SHOW BINARY LOGS 查看二进制日志文件个数及文件名，执行命令及结果如下：

```
mysql> SHOW BINARY LOGS;
+----------------------------+-----------+-----------+
| Log_name                   | File_size | Encrypted |
+----------------------------+-----------+-----------+
| X0NHUN07YDZVSSI-bin.000001 |    178    | No        |
+----------------------------+-----------+-----------+
1 row in set (0.00 sec)
```

可以看到，当前只有一个二进制日志文件。日志文件的个数与 MySQL 服务启动的次数相同。每启动一次 MySQL 服务，就会产生一个新的日志文件。

【例 18.3】 使用 mysqlbinlog 查看二进制日志，执行命令及结果如下：

```
C:\> mysqlbinlog D:/mysql/log/binlog.000001
/*!40019 SET @@session.max_insert_delayed_threads=0*/;
/*!50003 SET @OLD_COMPLETION_TYPE=@@COMPLETION_TYPE,COMPLETION_TYPE=0*/;
DELIMITER /*!*/;
# at 4
#190130 15:27:48 server id 1  end_log_pos 107    Start: binlog v 4, server v 8.0.28-log created 160330
# Warning: this binlog is either in use or was not closed properly.
ROLLBACK/*!*/;
BINLOG '
9JBcTg8BAAAAZwAAAGsAAAABAAQANS41LjEzLWxvZwAAAAAAAAAAAAAAAAAAAAAAAA
AAAAAAAAAA
AAAAAAAAAAAAAAAAAADOkFxOEzgNAAgAEgAEBAQEEgAAVAAEGggAAAAICAgCAA==
'/*!*/;
# at 107
#190330 15:34:17 server id 1   end_log_pos 175    Query    thread_id=2    exec_time=0    error_code=0
SET TIMESTAMP=1314689657/*!*/;
SET @@session.pseudo_thread_id=2/*!*/;
SET @@session.foreign_key_checks=1, @@session.sql_auto_is_null=0, @@session.unique_checks=1, @@session
SET @@session.sql_mode=0/*!*/;
SET @@session.auto_increment_increment=1, @@session.auto_increment_offset=1/*!*/;
/*!\C gb2312 *//*!*/;
SET @@session.character_set_client=24,@@session.collation_connection=24, @@session.collation_server=24/
SET @@session.lc_time_names=0/*!*/;
SET @@session.collation_database=DEFAULT/*!*/;
BEGIN
/*!*/;
# at 175
#190330 15:34:17 server id 1   end_log_pos 289    Query    thread_id=2    exec_time=0    error_code=0
use test/*!*/;
SET TIMESTAMP=1314689657/*!*/;
UPDATE fruits set f_price = 5.00  WHERE f_id = 'a1'
/*!*/;
# at 289
#190330 15:34:17 server id 1   end_log_pos 316    Xid = 14
COMMIT/*!*/;
DELIMITER ;
# End of log file
ROLLBACK /* added by mysqlbinlog */;
/*!50003 SET COMPLETION_TYPE=@OLD_COMPLETION_TYPE*/;
```

这是一个简单的日志文件，日志中记录了一些用户的操作。从文件内容中可以看到，用户对 fruits 表进行了更新操作，语句为 "UPDATE fruits set f_price = 5.00 WHERE f_id = 'a1';"。

18.2.3 删除二进制日志

MySQL 的二进制文件可以配置为自动删除，同时 MySQL 也提供了安全的手动删除二进制文件的方法：RESET MASTER 删除所有的二进制日志文件；PURGE MASTER LOGS 只删除部分二进制日志文件。本小节将介绍这两种二进制日志文件的删除方法。

1. 使用 RESET MASTER 语句删除所有二进制日志文件

RESTE MASTER 语法如下：

```
RESET MASTER;
```

执行完该语句后，所有二进制日志将被删除，MySQL 会重新创建二进制日志，新的日志文件扩展名将重新从 000001 开始编号。

2. 使用 PURGE MASTER LOGS 语句删除指定日志文件

PURGE MASTER LOGS 语法如下：

```
PURGE {MASTER | BINARY} LOGS TO 'log_name'
PURGE {MASTER | BINARY} LOGS BEFORE 'date'
```

第 1 种方法指定文件名，执行该命令将删除文件名编号比指定文件名编号小的所有日志文件。第 2 种方法指定日期，执行该命令将删除指定日期以前的所有日志文件。

【例 18.4】使用 PURGE MASTER LOGS 删除创建时间比 X0NHUNO7YDZVSSI-bin.000003 早的所有日志文件。

首先，为了演示这个语句的操作过程，需要准备多个日志文件，读者可以对 MySQL 服务进行多次重新启动。例如，这里有 3 个日志文件：

```
mysql> SHOW binary logs;
+-----------------------------+-----------+-----------+
| Log_name                    | File_size | Encrypted |
+-----------------------------+-----------+-----------+
| X0NHUNO7YDZVSSI-bin.000001  |       178 | No        |
| X0NHUNO7YDZVSSI-bin.000002  |       641 | No        |
| X0NHUNO7YDZVSSI-bin.000003  |       345 | No        |
+-----------------------------+-----------+-----------+
3 rows in set (0.00 sec)
```

执行删除命令：

```
mysql> PURGE MASTER LOGS TO " X0NHUNO7YDZVSSI-bin.000003";
Query OK, 0 rows affected (0.07 sec)
```

执行完成后，使用 SHOW binary logs 语句查看二进制日志：

```
mysql> SHOW binary logs;
+-----------------------------+-----------+
| Log_name                    | File_size |
+-----------------------------+-----------+
| X0NHUNO7YDZVSSI-bin.000003  |       345 |
+-----------------------------+-----------+
1 rows in set (0.00 sec)
```

可以看到，X0NHUNO7YDZVSSI-bin.000001、X0NHUNO7YDZVSSI-bin 000002 两个日志文件被删除了。

【例 18.5】使用 PURGE MASTER LOGS 删除 2022 年 1 月 30 日前创建的所有日志文件，执行命令及结果如下：

```
mysql> PURGE MASTER LOGS BEFORE '20220130';
Query OK, 0 rows affected (0.05 sec)
```

语句执行之后，2019 年 1 月 30 日之前创建的日志文件都将被删除，但 2022 年 1 月 30 日的日志会被保留（读者可根据自己机器中创建日志的时间修改命令参数）。使用 mysqlbinlog 可以查看指定日志的创建时间，如前面的例 18.3 所示，部分日志内容如下：

```
/*!40019 SET @@session.max_insert_delayed_threads=0*/;
/*!50003 SET @OLD_COMPLETION_TYPE=@@COMPLETION_TYPE,COMPLETION_TYPE=0*/;
DELIMITER /*!*/;
# at 4
#220130 15:27:48 server id 1   end_log_pos 107   Start: binlog v 4, server v 8.0.28-log created 160330
# Warning: this binlog is either in use or was not closed properly.
ROLLBACK/*!*/;
BINLOG '
```

其中，220130 为日志创建的时间，即 2022 年 1 月 30 日。

18.2.4 使用二进制日志恢复数据库

如果 MySQL 服务器启用了二进制日志，在数据库出现意外丢失数据时，可以使用 mysqlbinlog 工具从指定的时间点开始（例如，最后一次备份）直到现在，或另一个指定的时间点的日志中恢复数据。

要想从二进制日志恢复数据，需要知道当前二进制日志文件的路径和文件名，一般可以从配置文件（my.cnf 或者 my.ini，文件名取决于 MySQL 服务器的操作系统）中找到路径。

mysqlbinlog 恢复数据的语法如下：

```
mysqlbinlog [option] filename |mysql - uuser -ppass
```

option 是一些可选的选项，filename 是日志文件名。比较重要的两对 option 参数是--start-date、与--stop-date、--start-position 与--stop-position。--start-date 与--stop-date 可以指定恢复数据库的起始时间点和结束时间点。--start-position 与--stop-position 可以指定恢复数据的开始位置和结束位置。

【例 18.6】使用 mysqlbinlog 恢复 MySQL 数据库到 2022 年 1 月 30 日 15:27:48 时的状态，执行命令及结果如下：

```
mysqlbinlog --stop-date="2022-01-30 15:27:48" D:\mysql\log\binlog\ X0NHUNO7YDZVSSI-bin.000003 | mysql -uuser -ppass
```

该命令执行成功后，会根据 X0NHUNO7YDZVSSI-bin.000003 日志文件恢复 2022 年 01 月 30 日 15:27:48 以前的所有操作。这种方法对于意外操作非常有效，比如因操作不当误删了数据表的情况。

18.2.5 暂时停止二进制日志功能

如果在 MySQL 的配置文件中配置启动了二进制日志，MySQL 会一直记录二进制日志。修改配置文件，可以停止二进制日志，但是需要重启 MySQL 数据库。MySQL 提供了暂时停止二进制日志的功能。通过 SET SQL_LOG_BIN 语句可以使用 MySQL 暂停或者启动二进制日志。

SET SQL_LOG_BIN 的语法如下：

```
SET sql_log_bin = {0|1}
```

执行如下语句将暂停记录二进制日志：

```
mysql> SET sql_log_bin = 0;
Query OK, 0 rows affected (0.00 sec)
```

执行如下语句将恢复记录二进制日志：

```
mysql> SET sql_log_bin = 1;
Query OK, 0 rows affected (0.00 sec)
```

18.3 错误日志

错误日志文件包含了当 MySQLd 启动和停止时，以及服务器在运行过程中发生任何严重错误时的相关信息。在 MySQL 中，错误日志也是非常有用的，MySQL 会把启动和停止数据库信息以及一些错误信息记录到错误日志文件中。

18.3.1 启动和设置错误日志

在默认情况下，错误日志会记录到数据库的数据目录下。如果没有在配置文件中指定文件名，则文件名默认为 hostname.err。例如，MySQL 所在的服务器主机名为 MySQL-db，记录错误信息的文件名为 MySQL-db.err。如果执行了 FLUSH LOGS，错误日志文件会重新加载。

错误日志的启动和停止以及指定日志文件名都可以通过修改 my.ini（或者 my.cnf）来配置。错误日志的配置项是 log-error。在[mysqld]下配置 log-error，则启动错误日志。如果需要指定文件名，则配置项如下：

```
[mysqld]
log-error=[path / [file_name]]
```

path 为日志文件所在的目录路径，file_name 为日志文件名。修改配置项后，需要重启 MySQL 服务以生效。

18.3.2 查看错误日志

通过错误日志可以监视系统的运行状态，便于及时发现故障、修复故障。MySQL 错误日志是以文本文件的形式存储的，可以使用文本编辑器直接查看 MySQL 错误日志。

如果不知道日志文件的存储路径，可以使用 SHOW VARIABLES 语句查询错误日志的存储路径。SHOW VARIABLES 语句如下：

```
SHOW VARIABLES LIKE 'log_error';
```

【例 18.7】使用记事本查看 MySQL 错误日志。

首先，通过 SHOW VARIABLES 语句查询错误日志的存储路径和文件名：

```
mysql> SHOW VARIABLES LIKE 'log_error';
+---------------+------------------------+
| Variable_name | Value                  |
+---------------+------------------------+
| log_error     | .\X0NHUNO7YDZVSSI.err  |
+---------------+------------------------+
1 row in set (0.00 sec)
```

可以看到错误的文件是 X0NHUNO7YDZVSSI.err，位于 MySQL 默认的数据目录下。使用记事本打开该文件，可以看到 MySQL 的错误日志：

```
190130 16:45:14 [Note] Plugin 'FEDERATED' is disabled.
190130 16:45:14 InnoDB: The InnoDB memory heap is disabled
190130 16:45:14 InnoDB: Mutexes and rw_locks use Windows interlocked functions
190130 16:45:14 InnoDB: Compressed tables use zlib 1.2.3
190130 16:45:15 InnoDB: Initializing buffer pool, size = 46.0M
190130 16:45:15 InnoDB: Completed initialization of buffer pool
190130 16:45:15 InnoDB: highest supported file format is Barracuda.
190130 16:45:15 InnoDB: Waiting for the background threads to start
190130 16:45:16 InnoDB: 1.1.7 started; log sequence number 1679264
190130 16:45:16 [Note] Event Scheduler: Loaded 0 events
190130 16:45:16 [Note] C:\Program Files\MySQL\MySQL Server 8.0\bin\MySQLd: ready for connections.
Version: '8.0.28-log'  socket: ''  port: 3306  MySQL Community Server (GPL)
```

以上是错误日志文件的一部分，这里面记载了系统的一些错误。

18.3.3 删除错误日志

MySQL 的错误日志是以文本文件的形式存储在文件系统中的，可以直接删除。

对于 MySQL 5.5.7 以前的版本，flush logs 可以将错误日志文件重命名为 filename.err_old，并创建新的日志文件。但是从 MySQL 5.5.7 开始，flush logs 只是重新打开日志文件，并不做日志备份和创建的操作。如果日志文件不存在，MySQL 启动或者执行 flush logs 时会创建新的日志文件。

在运行状态下删除错误日志文件后，MySQL 并不会自动创建日志文件。flush logs 在重新加载日志的时候，如果文件不存在，则会自动创建。所以在删除错误日志之后，如果需要重建日志文件，需要在服务器端执行以下命令：

```
mysqladmin -u root -p flush-logs
```

或者在客户端登录 MySQL 数据库，执行 flush logs 语句：

```
mysql> flush logs;
Query OK, 0 rows affected (0.23 sec)
```

18.4 通用查询日志

通用查询日志记录 MySQL 的所有用户操作，包括启动和关闭服务、执行查询和更新语句等。本节将为读者介绍通用查询日志的启动、查看、删除等内容。

18.4.1 启动通用查询日志

MySQL 服务器默认情况下并没有开启通用查询日志。通过 "show variables like '%general%';" 语句可以查询当前通用查询日志的状态。

```
mysql> show variables like '%general%';
+------------------+----------------------+
| Variable_name    | Value                |
+------------------+----------------------+
| general_log      | OFF                  |
| general_log_file | X0NHUNO7YDZVSSI.log  |
+------------------+----------------------+
2 rows in set, 1 warning (0.11 sec)
```

从结果可以看出，通用查询日志的状态为 OFF，表示通用日志是关闭的。

开启通用日志的方法如下：

```
mysql> set @@global.general_log=1;
Query OK, 0 rows affected (0.04 sec)
```

再次查询通用查询日志的状态：

```
mysql> show variables like '%general%';
+------------------+----------------------+
| Variable_name    | Value                |
+------------------+----------------------+
| general_log      | ON                   |
| general_log_file | X0NHUNO7YDZVSSI.log  |
+------------------+----------------------+
2 rows in set, 1 warning (0.00 sec)
```

从结果可以看出，通用查询日志的状态为 ON，表示通用查询日志已经开启了。

如果想关闭通用查询日志，执行以下语句即可：

```
mysql> set @@global.general_log=0;
```

18.4.2 查看通用查询日志

通用查询日志中记录了用户的所有操作。通过查看通用查询日志，可以了解用户对 MySQL 进行的操作。通用查询日志以文本文件的形式存储在文件系统中，可以使用文本编辑器直接打开通用查询日志文件进行查看，Windows 下可以使用记事本，Linux 下可以使用 vim、gedit 等。

【例 18.8】使用记事本查看 MySQL 通用查询日志。

使用记事本打开 C:\ProgramData\MySQL\MySQL Server 8.0\Data\目录下的 X0NHUNO7YDZVSSI.log，

可以看到如下内容：

```
    C:\Program Files\MySQL\MySQL Server 8.0\bin\mysqld.exe, Version: 8.0.28 (MySQL Community Server
- GPL). started with:
    TCP Port: 3306, Named Pipe: MySQL
    Time                    Id Command    Argument
    2022-03-18 17:24:32         1 Connect    root@localhost on
          1 Query          select @@version_comment limit 1
    2022-03-18 17:24:36         1 Query         SELECT DATABASE()
          1 Init DB         test
    2022-03-18 17:24:53         1 Query          SELECT * FROM fruits
    2022-03-18 17:24:55         1 Quit
```

上面是笔者打开的通用查询日志的一部分，可以看到 MySQL 启动信息和用户 root 连接服务器与执行查询语句的记录。读者的文件内容可能与这里不同。

18.4.3 删除通用查询日志

通用查询日志记录用户的所有操作，因此在用户查询、更新频繁的情况下，通用查询日志会增长得很快。数据库管理员可以定期删除比较早的日志，以节省磁盘空间。本小节将介绍通用查询日志的删除方法。

可以用直接删除日志文件的方式删除通用查询日志。要重新建立新的日志文件，可使用语句 mysqladmin -flush logs。

【例 18.9】直接删除 MySQL 通用查询日志。执行步骤如下：

步骤 01 在数据目录中找到日志文件所在目录 C:\ProgramData\MySQL\MySQL Server 8.0\Data\，删除该后缀为 .log 的文件。

步骤 02 通过 mysqladmin –flush logs 命令建立新的日志文件，执行命令如下：

```
C:\> mysqladmin -u root -p flush-logs
```

步骤 03 执行完该命令，可以看到 C:\ProgramData\MySQL\MySQL Server 8.0\Data\ 目录中已经建立了新的日志文件。

18.5 慢查询日志

慢查询日志是记录查询时长超过指定时间的日志，它可以记录执行时间较长的查询语句。通过慢查询日志，可以找出执行时间较长、执行效率较低的语句，然后进行优化。本小节将讲解慢查询日志的相关内容。

18.5.1 启动和设置慢查询日志

MySQL 中慢查询日志默认是关闭的，可以通过配置文件 my.ini 或者 my.cnf 中的 log-slow-queries 选项打开，也可以在 MySQL 服务启动的时候，使用 --log-slow-queries[=file_name] 启动慢查询日志。

启动慢查询日志时,需要在 my.ini 或者 my.cnf 文件中配置 long_query_time 选项指定记录阈值,如果某条查询语句的查询时间超过了这个值,这个查询过程将被记录到慢查询日志文件中。

在 my.ini 或者 my.cnf 开启慢查询日志的配置如下:

```
[mysqld]
log-slow-queries[=path / [filename] ]
long_query_time=n
```

path 为日志文件所在目录路径,filename 为日志文件名。如果不指定目录和文件名称,默认存储在数据目录中,文件为 hostname-slow.log,hostname 是 MySQL 服务器的主机名。参数 n 是时间值,单位是秒。如果没有设置 long_query_time 选项,默认时间为 10 秒。

18.5.2 查看慢查询日志

MySQL 的慢查询日志是以文本形式存储的,可以直接使用文本编辑器查看。在慢查询日志中,记录着执行时间较长的查询语句,用户可以获取这些执行效率较低的查询语句,为查询优化提供重要的依据。

【例 18.10】查看慢查询日志。使用文本编辑器打开数据目录下的 Kevin-slow.log 文件,文件部分如下:

```
C:\Program Files\MySQL\MySQL Server 8.0\bin\mysqld.exe, Version: 8.0.28 (MySQL Community Server
- GPL). started with:
TCP Port: 3306, Named Pipe: MySQL
Time Id Command Argument
# Time: 181230 17:50:35
# User@Host: root[root] @ localhost [127.0.0.1]
# Query_time: 136.500000  Lock_time: 0.000000 Rows_sent: 1  Rows_examined: 0
SET timestamp=1314697835;
SELECT BENCHMARK(100000000, PASSWORD('newpwd'));
```

可以看到,这里记录了一条慢查询日志。执行该条查询语句的账户是 root@localhost,查询时间是 136.500000 秒,查询语句是 "SELECT BENCHMARK(100000000, PASSWORD('newpwd'));",该语句的查询时间大大超过了默认值 10 秒钟,因此被记录在慢查询日志文件中。

提示:借助慢查询日志分析工具,可以更加方便地分析慢查询语句。比较著名的慢查询工具有 MySQL Dump Slow、MySQL SLA、MySQL Log Filter、MyProfi。关于这些慢查询分析工具的用法,可以参考相关软件的帮助文档。

18.5.3 删除慢查询日志

和通用查询日志一样,慢查询日志也可以直接删除。删除后在不重启服务器的情况下,需要执行 mysqladmin -u root-p flush-logs 重新生成日志文件,或者在客户端登录到服务器执行 flush logs 语句重建日志文件。

第 19 章

MySQL 权限与安全管理

MySQL 是一个多用户数据库，具有功能强大的访问控制系统，可以为不同用户指定允许的权限。MySQL 用户可以分为普通用户和 root 用户。root 用户是超级管理员，拥有数据库的所有权限，包括创建用户、删除用户和修改用户的密码等管理权限；普通用户只拥有被授予的各种权限。用户管理包括管理用户账户、权限等内容。本章将向读者介绍 MySQL 用户管理中的相关知识点，包括权限表、账户管理和权限管理。

19.1 权限表

MySQL 服务器通过权限表来控制用户对数据库的访问，权限表存放在 MySQL 数据库中，由 MySQL_install_db 脚本初始化。存储账户权限信息的表主要有 user、db、host、tables_priv、columns_priv 和 procs_priv。本节将为读者介绍这些表的内容和作用。

19.1.1 user 表

user 表是 MySQL 中最重要的一个权限表，记录允许连接到服务器的账号信息，里面的权限是全局级的。例如，一个用户在 user 表中被授予了 DELETE 权限，则该用户可以删除 MySQL 服务器上所有数据库中的任何记录。MySQL 8.0 中 user 表有 42 个字段，如表 19.1 所示，这些字段可以分为 4 类，分别是用户列、权限列、安全列和资源控制列。本小节将介绍 user 表中各字段的含义。

表 19.1 user 表结构

字 段 名	数据类型	默 认 值
Host	char(60)	
User	char(16)	
authentication_string	text	

(续表)

字 段 名	数 据 类 型	默 认 值
Select_priv	enum('N','Y')	N
Insert_priv	enum('N','Y')	N
Update_priv	enum('N','Y')	N
Delete_priv	enum('N','Y')	N
Create_priv	enum('N','Y')	N
Drop_priv	enum('N','Y')	N
Reload_priv	enum('N','Y')	N
Shutdown_priv	enum('N','Y')	N
Process_priv	enum('N','Y')	N
File_priv	enum('N','Y')	N
Grant_priv	enum('N','Y')	N
References_priv	enum('N','Y')	N
Index_priv	enum('N','Y')	N
Alter_priv	enum('N','Y')	N
Show_db_priv	enum('N','Y')	N
Super_priv	enum('N','Y')	N
Create_tmp_table_priv	enum('N','Y')	N
Lock_tables_priv	enum('N','Y')	N
Execute_priv	enum('N','Y')	N
Repl_slave_priv	enum('N','Y')	N
Repl_client_priv	enum('N','Y')	N
Create_view_priv	enum('N','Y')	N
Show_view_priv	enum('N','Y')	N
Create_routine_priv	enum('N','Y')	N
Alter_routine_priv	enum('N','Y')	N
Create_user_priv	enum('N','Y')	N
Event_priv	enum('N','Y')	N
Trigger_priv	enum('N','Y')	N
Create_tablespace_priv	enum('N','Y')	N
ssl_type	enum('','ANY','X509','SPECIFIED')	
ssl_cipher	blob	NULL
x509_issuer	blob	NULL
x509_subject	blob	NULL
max_questions	int(11) unsigned	0
max_updates	int(11) unsigned	0
max_connections	int(11) unsigned	0
max_user_connections	int(11) unsigned	0
plugin	char(64)	
authentication_string	text	NULL

1. 用户列

user 表的用户列包括 Host、User、authentication_string，分别表示主机名、用户名和密码。其中 User 和 Host 为 User 表的联合主键。当用户与服务器之间建立连接时，输入的账户信息中的用户名称、主机名和密码必须匹配 User 表中对应的字段，只有 3 个值都匹配的时候，才允许建立连接。这 3 个字段的值就是创建账户时保存的账户信息。修改用户密码，实际上就是修改 user 表的 authentication_string 字段的值。

2. 权限列

权限列的字段决定了用户的权限，描述了在全局范围内允许对数据和数据库进行的操作。包括查询权限、修改权限等普通权限，还包括了关闭服务器、超级权限和加载用户等高级权限。普通权限用于操作数据库，高级权限用于数据库管理。

user 表中对应的权限是针对所有用户数据库的。这些字段值的类型为 ENUM，可以取的值只能为 Y 和 N，Y 表示该用户有对应的权限；N 表示用户没有对应的权限。查看 user 表的结构可以看到，这些字段的值默认都是 N。如果要修改权限，可以使用 GRANT 语句或 UPDATE 语句更改 user 表的这些字段来修改用户对应的权限。

3. 安全列

安全列只有 6 个字段，其中两个是 ssl 相关的，两个是 x509 相关的，另外两个是授权插件相关的。ssl 用于加密；x509 标准可用于标识用户；plugin 字段标识可以用于验证用户身份的插件，如果该字段为空，服务器使用内建授权验证机制来验证用户身份。读者可以通过 SHOW VARIABLES LIKE 'have_openssl'语句来查询服务器是否支持 ssl 功能。

4. 资源控制列

资源控制列的字段用来限制用户使用的资源，包含 4 个字段，分别为：

（1）max_questions：用户每小时允许执行的查询操作次数。
（2）max_updates：用户每小时允许执行的更新操作次数。
（3）max_connections：用户每小时允许执行的连接操作次数。
（4）max_user_connections：用户允许同时建立的连接次数。

一个小时内用户查询或者连接数量超过资源控制限制，用户将被锁定，直到下一个小时，才可以再次执行对应的操作。可以使用 GRANT 语句更新这些字段的值。

19.1.2 db 表

db 表是 MySQL 数据中非常重要的权限表。db 表中存储了用户对某个数据库的操作权限，决定用户能从哪个主机访问哪个数据库。db 表比较常用。db 表的结构如表 19.2 所示。

表 19.2　db 表结构

字 段 名	数据类型	默 认 值
Host	char(60)	
Db	char(64)	
User	char(32)	
Select_priv	enum('N','Y')	N
Insert_priv	enum('N','Y')	N
Update_priv	enum('N','Y')	N
Delete_priv	enum('N','Y')	N
Create_priv	enum('N','Y')	N
Drop_priv	enum('N','Y')	N
Grant_priv	enum('N','Y')	N
References_priv	enum('N','Y')	N
Index_priv	enum('N','Y')	N
Alter_priv	enum('N','Y')	N
Create_tmp_table_priv	enum('N','Y')	N
Lock_tables_priv	enum('N','Y')	N
Create_view_priv	enum('N','Y')	N
Show_view_priv	enum('N','Y')	N
Create_routine_priv	enum('N','Y')	N
Alter_routine_priv	enum('N','Y')	N
Execute_priv	enum('N','Y')	N
Event_priv	enum('N','Y')	N
Trigger_priv	enum('N','Y')	N

1. 用户列

db 表用户列有 3 个字段，分别是 Host、User、Db，标识某个用户从某个主机连接某个数据库的操作权限，这 3 个字段的组合构成了 db 表的主键。一般情况下 db 表就可以满足权限控制需求了。

2. 权限列

db 表中 create_routine_priv 和 alter_routine_priv 这两个字段表明用户是否有创建和修改存储过程的权限。

user 表中的权限是针对所有数据库的，如果希望用户只对某个数据库有操作权限，那么需要将 user 表中对应的权限设置为 N，然后在 db 表中设置对应数据库的操作权限。例如，有一个名称为 Zhangting 的用户分别从名称为 large.domain.com 和 small.domain.com 的两个主机连接到数据库，并需要操作 books 数据库。这时，可以将用户名称 Zhangting 添加到 db 表中，而 db 表中的 host 字段值为空，然后将两个主机地址分别作为两条记录的 host 字段值添加到 host 表中，并将两个表的数据库字段设置为相同的值 books。当有用户连接到 MySQL 服务器时，db 表中没有用户登录的主机名称，则 MySQL 会从 host 表中查找相匹配的值，并根据查询的结果决定用户的操作是否被允许。

19.1.3 tables_priv 表和 columns_priv 表

tables_priv 表用来对表设置操作权限，columns_priv 表用来对表的某一列设置权限。tables_priv 表和 columns_priv 表的结构分别如表 19.3 和表 19.4 所示。

表 19.3 tables_priv 表结构

字 段 名	数据类型	默 认 值
Host	char(60)	
Db	char(64)	
User	char(16)	
Table_name	char(64)	
Grantor	char(77)	
Timestamp	timestamp	CURRENT_TIMESTAMP
Table_priv	set('Select','Insert','Update','Delete','Create','Drop','Grant','References','Index','Alter','Create View','Show view','Trigger')	
Column_priv	set('Select','Insert','Update','References')	

表 19.4 columns_priv 表结构

字 段 名	数据类型	默 认 值
Host	char(60)	
Db	char(64)	
User	char(16)	
Table_name	char(64)	
Column_name	char(64)	
Timestamp	timestamp	CURRENT_TIMESTAMP
Column_priv	set('Select','Insert','Update','References')	

tables_priv 表有 8 个字段，分别是 Host、Db、User、Table_name、Grantor、Timestamp、Table_priv 和 Column_priv，各个字段说明如下：

（1）Host、Db、User 和 Table_name 4 个字段分表示主机名、数据库名、用户名和表名。
（2）Grantor 表示修改该记录的用户。
（3）Timestamp 字段表示修改该记录的时间。
（4）Table_priv 表示对表的操作权限，包括 Select、Insert、Update、Delete、Create、Drop、Grant、References、Index 和 Alter。
（5）Column_priv 字段表示对表中列的操作权限，包括 Select、Insert、Update 和 References。

columns_priv 表只有 7 个字段，分别是 Host、Db、User、Table_name、Column_name、Timestamp、Column_priv。其中，Column_name 用来指定对哪些数据列具有操作权限。

19.1.4 procs_priv 表

procs_priv 表可以对存储过程和存储函数设置操作权限。procs_priv 的表结构如表 19.5 所示。

表 19.5　procs_priv 表结构

字 段 名	数据类型	默 认 值
Host	char(60)	
Db	char(64)	
User	char(16)	
Routine_name	char(64)	
Routine_type	enum('FUNCTION','PROCEDURE')	NULL
Grantor	char(77)	
Proc_priv	set('Execute','Alter Routine','Grant')	
Timestamp	timestamp	CURRENT_TIMESTAMP

procs_priv 表包含 8 个字段，分别是 Host、Db、User、Routine_name、Routine_type、Grantor、Proc_priv 和 Timestamp，各个字段的说明如下：

（1）Host、Db 和 User 字段分别表示主机名、数据库名和用户名。Routine_name 表示存储过程或函数的名称。

（2）Routine_type 表示存储过程或函数的类型。Routine_type 字段有两个值，分别是 FUNCTION 和 PROCEDURE：FUNCTION 表示这是一个函数，PROCEDURE 表示这是一个存储过程。

（3）Grantor 表示插入或修改该记录的用户。

（4）Proc_priv 表示拥有的权限，包括 Execute、Alter Routine、Grant 3 种。

（5）Timestamp 表示记录更新时间。

19.2　账户管理

MySQL 提供了许多语句来管理用户账号，包括登录和退出 MySQL 服务器、创建用户、删除用户、密码管理和权限管理等内容。MySQL 数据库的安全性需要通过账户管理来保证。本节将介绍在 MySQL 中如何对账户进行管理。

19.2.1　登录和退出 MySQL 服务器

读者已经知道登录 MySQL 时，可以使用 MySQL 命令并在后面指定登录主机以及用户名和密码。本小节将详细介绍 MySQL 命令的常用参数以及登录、退出 MySQL 服务器的方法。

通过 mysql -help 命令可以查看 mysql 命令帮助信息。mysql 命令的常用参数如下：

（1）-h 主机名，可以使用该参数指定主机名或 IP，如果不指定，默认是 localhost。

（2）-u 用户名，可以使用该参数指定用户名。

（3）-p 密码，可以使用该参数指定登录密码。如果该参数后面有一段字段，则该段字符串将作为用户的密码直接登录。如果后面没有内容，则登录的时候会提示输入密码。注意：该参数后面的字符串和-p 之前不能有空格。

（4）-P 端口号，该参数后面接 MySQL 服务器的端口号，默认为 3306。

（5）数据库名，可以在命令的最后指定数据库名。

（6）-e 执行 SQL 语句，如果指定了该参数，将在登录后执行 -e 后面的命令或 SQL 语句并退出。

【例 19.1】使用 root 用户登录到本地 MySQL 服务器的 mysql 库中，命令如下：

```
mysql -h localhost -u root -p mysql
```

命令执行结果如下：

```
C:\Users\Administrator>mysql -h localhost -u root -p mysql
Enter password: ******
Welcome to the MySQL monitor.  Commands end with ; or \g.
Your MySQL connection id is 19
Server version: 8.0.28 MySQL Community Server - GPL

Copyright (c) 2000, 2022, Oracle and/or its affiliates. All rights reserved.

Oracle is a registered trademark of Oracle Corporation and/or its
affiliates. Other names may be trademarks of their respective
owners.

Type 'help;' or '\h' for help. Type '\c' to clear the current input statement.
```

执行命令时，会提示 Enter password，如果没有设置密码，可以直接按 Enter 键。密码正确就可以直接登录到服务器下面的 mysql 数据库中了。

【例 19.2】使用 root 用户登录到本地 MySQL 服务器的 test_db 数据库中，同时执行一条查询语句。命令如下：

```
mysql -h localhost -u root -p test_db -e "DESC person;"
```

命令执行结果如下：

```
C:\ > mysql -h localhost -u root -p test_db -e "DESC person;"
Enter password: **
+-------+--------------+------+-----+---------+----------------+
| Field | Type         | Null | Key | Default | Extra          |
+-------+--------------+------+-----+---------+----------------+
| id    | int unsigned | NO   | PRI | NULL    | auto_increment |
| name  | char(40)     | NO   |     |         |                |
| age   | int          | NO   |     | 0       |                |
| info  | char(50)     | YES  |     | NULL    |                |
+-------+--------------+------+-----+---------+----------------+
```

按照提示输入密码，命令执行完成后查询出 person 表的结构，查询返回之后会自动退出 MySQL。

19.2.2 新建普通用户

创建新用户，必须有相应的权限来执行创建操作。在 MySQL 数据库中，有两种方式创建新用户：一种是使用 CREATE USER 语句；另一种是直接操作 MySQL 授权表。下面分别介绍这两种创建新用户的方法。

1. 使用 CREATE USER 语句创建新用户

执行 CREATE USER 或 GRANT 语句时，服务器会修改相应的用户授权表，添加或者修改用户

及其权限。CREATE USER 语句的基本语法格式如下:

```
CREATE USER user_specification
    [, user_specification] ...

user_specification:
    user@host
    [
        IDENTIFIED BY [PASSWORD] 'password'
      | IDENTIFIED WITH auth_plugin [AS 'auth_string']
    ]
```

user 表示创建的用户的名称;host 表示允许登录的用户主机名称;IDENTIFIED BY 表示用来设置用户的密码;[PASSWORD]表示使用哈希值设置密码,该参数可选;'password'表示用户登录时使用的普通明文密码;IDENTIFIED WITH 语句为用户指定一个身份验证插件;auth_plugin 是插件的名称,插件的名称可以是一个带单引号的字符串或者带双引号的字符串;auth_string 是可选的字符串参数,该参数将传递给身份验证插件,由该插件解释该参数的意义。

CREATE USER 语句会添加一个新的 MySQL 账户。使用 CREATE USER 语句的用户,必须有全局的 CREATE USER 权限或 mysql 数据库的 INSERT 权限。每添加一个用户,CREATE USER 语句会在 mysql.user 表中添加一条新记录,但是新创建的账户没有任何权限。如果添加的账户已经存在,CREATE USER 语句会返回一个错误。

【例 19.3】使用 CREATE USER 创建一个用户,用户名是 jeffrey,密码是 mypass,主机名是 localhost,命令如下:

```
CREATE USER 'jeffrey'@'localhost' IDENTIFIED BY 'mypass';
```

如果只指定用户名部分'jeffrey',主机名部分则默认为'%'(对所有的主机开放权限)。

user_specification 告诉 MySQL 服务器当用户登录时怎么验证用户的登录授权。如果指定用户登录不需要密码,可以省略 IDENTIFIED BY 部分:

```
CREATE USER 'jeffrey'@'localhost';
```

此种情况,MySQL 服务端使用内建的身份验证机制,用户登录时不能指定密码。

如果要创建指定密码的用户,需要 IDENTIFIED BY 指定明文密码值:

```
CREATE USER 'jeffrey'@'localhost' IDENTIFIED BY 'mypass';
```

此种情况,MySQL 服务端使用内建的身份验证机制,用户登录时必须指定密码。

MySQL 的某些版本中会引入授权表的结构变化,添加新的特权或功能。每当更新 MySQL 到一个新的版本时,应该检查授权表,以确保它们有最新的结构,确认可以使用任何新功能。

2. 直接操作 MySQL 用户表

通过前面的介绍,使用 CREATE USER 创建新用户时,实际上都是在 user 表中添加一条新的记录。因此,可以使用 INSERT 语句向 user 表中直接插入一条记录来创建一个新的用户。使用 INSERT 语句,必须拥有对 mysql.user 表的 INSERT 权限。使用 INSERT 语句创建新用户的基本语法格式如下:

```
INSERT INTO MySQL.user(Host, User, authentication_string)
VALUES('host', 'username', MD5('password'));
```

Host、User、authentication_string 分别为 user 表中的主机、用户名称和密码字段；MD5()函数为密码加密函数。

【例 19.4】使用 INSERT 创建一个新账户，其用户名称为 customer1，主机名称为 localhost，密码为 aa123456，INSERT 语句如下：

```
INSERT INTO user (Host,User, authentication_string)
VALUES('localhost','customer1', MD5('aa123456'));
```

语句执行结果如下：

```
MySQL> use mysql;
MySQL> INSERT INTO user (Host,User, authentication_string)
    -> VALUES('localhost','customer1', MD5('aa123456'));
ERROR 1364 (HY000): Field 'ssl_cipher' doesn't have a default value
```

语句执行失败，查看警告信息如下：

```
MySQL> SHOW WARNINGS;
+-------+------+-----------------------------------------------------+
| Level | Code | Message                                             |
+-------+------+-----------------------------------------------------+
| Error | 1364 | Field 'ssl_cipher' doesn't have a default value     |
| Error | 1364 | Field 'x509_issuer' doesn't have a default value    |
| Error | 1364 | Field 'x509_subject' doesn't have a default value   |
+-------+------+-----------------------------------------------------+
```

因为 ssl_cipher、x509_issuer 和 x509_subject 3 个字段在 user 表定义中没有设置默认值，所以在这里提示错误信息，影响 INSERT 语句的执行。使用 SELECT 语句查看 user 表中的记录：

```
MySQL> SELECT host,user, authentication_string FROM user ;
+-----------+------------------+------------------------------------------------+
| host      | user             | authentication_string                          |
+-----------+------------------+------------------------------------------------+
| localhost | jeffrey          | *6C8989366EAF75BB670AD8EA7A7FC1176A95CEF4       |
| localhost | mysql.infoschema | $A$005$THISISACOMBINATIONOFIVALIDSALTANDP       |
| localhost | mysql.session    | $A$005$THISISACOMBINATIONOFINVALIDSALTANDPAS    |
| localhost | mysql.sys        | $A$005$THISISACOMBINATIONOFINVAUSTNEVERREUSED   |
| localhost | root             | *6BB4837EB74329105EE4568DDA7DC67ED2CA2AD9       |
+-----------+------------------+------------------------------------------------+
```

可以看到新用户 customer1 并没有添加到 user 表中，说明添加新用户失败。

19.2.3 删除普通用户

在 MySQL 数据库中，可以使用 DROP USER 语句删除用户，也可以直接通过 DELETE 从 mysql.user 表中删除对应的记录来删除用户。

1. 使用 DROP USER 语句删除用户

DROP USER 语句的语法如下：

```
DROP USER user [, user];
```

DROP USER 语句用于删除一个或多个 MySQL 账户。要使用 DROP USER，必须拥有 MySQL

数据库的全局 CREATE USER 权限或 DELETE 权限。使用与 GRANT 或 REVOKE 相同的格式为每个账户命名。例如，"'jeffrey'@'localhost'" 账户名称的用户和主机部分与用户表记录的 User 和 Host 列值相对应。

使用 DROP USER，可以删除一个用户账户及其权限，操作如下：

```
DROP USER 'user'@'localhost';
DROP USER;
```

第 1 条语句可以删除 user 在本地登录权限；第 2 条语句可以删除来自所有授权表的用户账户权限记录。

【例 19.5】使用 DROP USER 删除账户 "'jeffrey'@'localhost'"，DROP USER 语句如下：

```
DROP USER 'jeffrey'@'localhost';
```

执行过程如下：

```
MySQL> DROP USER 'jeffrey'@'localhost';
Query OK, 0 rows affected (0.00 sec)
```

可以看到语句执行成功，查看执行结果：

```
MySQL> SELECT host,user, authentication_string FROM user ;
+---------+------------------+-------------------------------------------+
| host    | user             | authentication_string                     |
+---------+------------------+-------------------------------------------+
|localhost|mysql.infoschema  |$A$005$THISISACOMBINATIONOFIVALIDSALTANDP  |
|localhost|mysql.session     |$A$005$THISISACOMBINATIONOFINVALIDSALTANDPAS|
|localhost|mysql.sys         |$A$005$THISISACOMBINATIONOFINVAUSTNEVERREUSED|
|localhost|root              |*6BB4837EB74329105EE4568DDA7DC67ED2CA2AD9  |
+---------+------------------+-------------------------------------------+
```

user 表中已经没有名称为 jeffrey、主机名为 localhost 的账户，即"'jeffrey'@'localhost'"的用户账号已经被删除。

提示：DROP USER 不能自动关闭任何打开的用户对话。而且，如果用户有打开的对话，此时取消用户，命令则不会生效，直到用户对话被关闭后才能生效。一旦对话被关闭，用户也被取消，此用户再次试图登录时将会失败。

2. 使用 DELETE 语句删除用户

DELETE 语句基本语法格式如下：

```
DELETE FROM MySQL.user WHERE host='hostname' and user='username'
```

host 和 user 为 user 表中的两个字段，两个字段的组合确定所要删除的用户账户记录。

【例 19.6】使用 DELETE 删除用户'customer1'@'localhost'。

首先创建用户 customer1，命令如下：

```
MySQL>CREATE USER 'customer1'@'localhost' IDENTIFIED BY 'my123';
Query OK, 0 rows affected (0.12 sec)
```

然后使用 DELETE 删除用户'customer1'@'localhost'，命令如下：

```
mysql> DELETE FROM MySQL.user WHERE host='localhost' and user='customer1';
Query OK, 1 row affected (0.01 sec)
```

可以看到语句执行成功,'customer1'@'localhost'的用户账号已经被删除。读者可以使用 SELECT 语句查询 user 表中的记录,确认删除操作是否成功。

19.2.4 root 用户修改普通用户密码

root 用户拥有很高的权限,可以修改其他用户的密码。root 用户登录 MySQL 服务器后,可以通过 SET 语句修改 mysql.user 表,通过 UPDATE 语句修改用户的密码。

创建用户 user,命令如下:

```
MySQL>CREATE USER 'user'@'localhost' IDENTIFIED BY 'my123';
Query OK, 0 rows affected (0.12 sec)
```

1. 使用 SET 语句修改普通用户的密码

使用 SET 语句修改普通用户密码的语法格式如下:

```
SET PASSWORD FOR 'user'@'localhost' = 'sa123';
```

【例 19.7】使用 SET 语句将 user 用户的密码修改为 "sa123"。

使用 root 用户登录到 MySQL 服务器后,执行如下语句:

```
MySQL> SET PASSWORD FOR 'user'@'localhost' = 'sa123';
Query OK, 0 rows affected (0.00 sec)
```

SET 语句执行成功,user 用户的密码被成功设置为 "sa123"。

2. 使用 UPDATE 语句修改普通用户的密码

使用 root 用户登录到 MySQL 服务器后,可以使用 UPDATE 语句修改 mysql 数据库 user 表的 authentication_string 字段,从而修改普通用户的密码。使用 UPDATA 语句修改用户密码的语法如下:

```
UPDATE mysql.user SET authentication_string=MD5("123456")
WHERE User="username" AND Host="hostname";
```

MD5()函数用来加密用户密码。执行 UPDATE 语句后,需要执行 FLUSH PRIVILEGES 语句重新加载用户权限。

【例 19.8】使用 UPDATE 语句将 user 用户的密码修改为 "sns123"。

使用 root 用户登录到 MySQL 服务器后,执行如下语句:

```
MySQL> UPDATE mysql.user SET authentication_string =MD5("sns123")
    -> WHERE User="user" AND Host="localhost";
Query OK, 1 row affected (0.00 sec)
Rows matched: 1  Changed: 1  Warnings: 0
MySQL> FLUSH PRIVILEGES;
Query OK, 0 rows affected (0.11 sec)
```

执行完 UPDATE 语句后,user 的密码被修改成了 "sns123"。使用 FLUSH PRIVILEGES 重新加载权限,user 用户就可以使用新密码登录系统了。

19.3 权限管理

权限管理主要是对登录到 MySQL 的用户进行权限验证。所有用户的权限都存储在 MySQL 的权限表中，不合理的权限规划会给 MySQL 服务器带来安全隐患。数据库管理员要对所有用户的权限进行合理规划和管理。MySQL 权限系统的主要功能是证实连接到一台给定主机的用户，并且赋予该用户在数据库上的 SELECT、INSERT、UPDATE 和 DELETE 权限。本节将为读者介绍 MySQL 权限管理的相关内容。

19.3.1 MySQL 的各种权限

账户权限信息被存储在 MySQL 数据库的 user、db、host、tables_priv、columns_priv 和 procs_priv 表中。在 MySQL 启动时，服务器将这些数据库表中权限信息读入内存。

GRANT 和 REVOKE 语句所涉及的权限名称如表 19.6 所示，其中还有在授权表中每个权限的列（字段）名称和每个权限有关的操作对象等。

表 19.6　GRANT 和 REVOKE 语句中可以使用的权限

权　　限	user 表中对应的列	权限的范围
CREATE	Create_priv	数据库、表或索引
DROP	Drop_priv	数据库、表或视图
GRANT OPTION	Grant_priv	数据库、表或存储过程
REFERENCES	References_priv	数据库或表
EVENT	Event_priv	数据库
ALTER	Alter_priv	数据库
DELETE	Delete_priv	表
INDEX	Index_priv	表
INSERT	Insert_priv	表
SELECT	Select_priv	表或列
UPDATE	Update_priv	表或列
CREATE TEMPORARY TABLES	Create_tmp_table_priv	表
LOCK TABLES	Lock_tables_priv	表
TRIGGER	Trigger_priv	表
CREATE VIEW	Create_view_priv	视图
SHOW VIEW	Show_view_priv	视图
ALTER ROUTINE	Alter_routine_priv	存储过程和函数
CREATE ROUTINE	Create_routine_priv	存储过程和函数
EXECUTE	Execute_priv	存储过程和函数
FILE	File_priv	访问服务器上的文件
CREATE TABLESPACE	Create_tablespace_priv	服务器管理
CREATE USER	Create_user_priv	服务器管理
PROCESS	Process_priv	存储过程和函数
RELOAD	Reload_priv	访问服务器上的文件
REPLICATION CLIENT	Repl_client_priv	服务器管理

（续表）

权　　限	user 表中对应的列	权限的范围
REPLICATION SLAVE	Repl_slave_priv	服务器管理
SHOW DATABASES	Show_db_priv	服务器管理
SHUTDOWN	Shutdown_priv	服务器管理
SUPER	Super_priv	服务器管理

（1）CREATE 和 DROP 权限，可以创建新数据库和表，或删除（移掉）已有数据库和表。如果将 MySQL 数据库的 DROP 权限授予某用户，用户可以删掉 MySQL 访问权限保存的数据库。

（2）SELECT、INSERT、UPDATE 和 DELETE 权限，允许在一个数据库现有的表上实施操作。

（3）SELECT 权限，只有在它们真正从一个表中检索行时才被用到。

（4）INDEX 权限，允许创建或删除索引，INDEX 适用已有表。如果具有某个表的 CREATE 权限，可以在 CREATE TABLE 语句中包括索引权限。

（5）ALTER 权限，可以使用 ALTER TABLE 来更改表的结构和重新命名表。

（6）CREATE ROUTINE 权限用来创建保存的程序（函数和程序），ALTER ROUTINE 权限用来更改和删除保存的程序，EXECUTE 权限用来执行保存的程序。

（7）GRANT 权限，允许授权给其他用户，可用于数据库、表和保存的程序。

（8）FILE 权限，给予用户使用 LOAD DATA INFILE 和 SELECT...INTO OUTFILE 语句读或写服务器上的文件，任何被授予 FILE 权限的用户，都能读或写 MySQL 服务器上的任何文件（说明用户可以读任何数据库目录下的文件，因为服务器可以访问这些文件）。FILE 权限允许用户在 MySQL 服务器具有写权限的目录下创建新文件，但不能覆盖已有文件。

其余的权限用于管理性操作，它们使用 mysqladmin 程序或 SQL 语句实施。表 19.7 显示每个权限允许执行的 mysqladmin 命令。

表 19.7　不同权限下可以使用的 mysqladmin 命令

权　　限	权限拥有者允许执行的命令
RELOAD	flush-hosts、flush-logs、flush-privileges、flush-status、flush-tables、flush-threads、refresh、reload
SHUTDOWN	shutdown
PROCESS	processlist
SUPER	kill

（1）reload 命令告诉服务器将授权表重新读入内存；flush-privileges 是 reload 的同义词；refresh 命令清空主机相关的缓存并关闭/打开日志文件；其他 flush-xxx 命令执行类似 refresh 的功能，但是范围更有限，并且在某些情况下可能更好用。例如，如果只是想清空日志文件，flush-logs 是比 refresh 更好的选择。

（2）shutdown 命令关掉服务器，只能从 mysqladmin 发出这个命令。

（3）processlist 命令显示在服务器内执行的线程的信息（其他账户相关的客户端执行的语句）。kill 命令杀死服务器线程。用户总是能显示或杀死自己的线程，但是需要 PROCESS 权限来显示或杀死其他用户以及 SUPER 权限启动的线程。

（4）kill 命令能用来终止其他用户或更改服务器的操作方式。

总的来说，权限只能授予需要它们的那些用户。

19.3.2 授权

授权就是为某个用户授予权限。合理的授权可以保证数据库的安全。MySQL 数据库使用 GRANT 语句为用户授予权限。

授予的权限可以分为多个层级。

1. 全局层级

全局权限适用于一个给定服务器中的所有数据库。这些权限存储在 mysql.user 表中。GRANT ALL ON *.*和 REVOKE ALL ON *.*语句可授予和撤销全局权限。

2. 数据库层级

数据库权限适用于一个给定数据库中的所有目标。这些权限存储在 mysql.db 和 mysql.host 表中。GRANT ALL ON db_name.和 REVOKE ALL ON db_name.*与语句可授予和撤销数据库权限。

3. 表层级

表权限适用于一个给定表中的所有列。这些权限存储在 mysql.talbes_priv 表中。GRANT ALL ON db_name.tbl_name 和 REVOKE ALL ON db_name.tbl_name 语句可授予和撤销表权限。

4. 列层级

列权限适用于一个给定表中的单一列。这些权限存储在 mysql.columns_priv 表中。当使用 REVOKE 语句时，必须指定与被授权列相同的列。

5. 子程序层级

CREATE ROUTINE、ALTER ROUTINE、EXECUTE 和 GRANT 权限适用于已存储的子程序。这些权限可以被授予为全局层级和数据库层级。而且，除了 CREATE ROUTINE 外，这些权限可以被授予子程序层级，并存储在 mysql.procs_priv 表中。

在 MySQL 中，必须是拥有 GRANT 权限的用户才能执行 GRANT 语句。

要使用 GRANT 或 REVOKE，必须拥有 GRANT OPTION 权限，并且必须用于正在授予或撤销的权限。GRANT 的语法如下：

```
GRANT priv_type [(columns)] [, priv_type [(columns)]] ...
ON [object_type]    table1, table2,..., tablen
TO user   [WITH GRANT OPTION]

object_type = TABLE  |  FUNCTION  |  PROCEDURE
```

其中，priv_type 参数表示权限类型；columns 参数表示权限作用于哪些列上，不指定该参数，表示作用于整个表；table1,table2,…,tablen 表示授予权限的列所在的表。object_type 指定授权作用的对象类型，包括 TABLE（表）、FUNCTION（函数）和 PROCEDURE（存储过程）；当从旧版本的 MySQL 升级时，要使用 object_tpye 子句，必须升级授权表。user 参数表示用户账户，由用户名和主机名构成，形式是 "'username'@'hostname'"。

WITH 关键字后可以跟一个或多个 with_option 参数。这个参数有 5 个选项，意义如下：

（1）GRANT OPTION：被授权的用户可以将这些权限赋予别的用户。

（2）MAX_QUERIES_PER_HOUR count：设置每小时可以执行 count 次查询。

（3）MAX_UPDATES_PER_HOUR count：设置每小时可以执行 count 次更新。

（4）MAX_CONNECTIONS_PER_HOUR count：设置每小时可以建立 count 个连接。

（5）MAX_USER_CONNECTIONS count：设置单个用户可以同时建立 count 个连接。

【例 19.9】创建一个新用户 grantUser。使用 GRANT 语句对用户 grantUser 赋予所有数据的查询、插入权限，并授于 GRANT 权限。GRANT 语句及其执行结果如下：

```
MySQL> CREATE USER 'grantUser'@'localhost' IDENTIFIED BY 'mypass';
MySQL> GRANT SELECT, INSERT ON *.* TO 'grantUser'@'localhost' WITH GRANT OPTION;
Query OK, 0 rows affected (0.03 sec)
```

结果显示执行成功。使用 SELECT 语句查询用户 grantUser 的权限：

```
MySQL> SELECT Host,User,Select_priv,Insert_priv, Grant_priv FROM mysql.user where user='grantUser';
+-----------+-----------+-------------+-------------+------------+
| Host      | User      | Select_priv | Insert_priv | Grant_priv |
+-----------+-----------+-------------+-------------+------------+
| localhost | grantUser | Y           | Y           | Y          |
+-----------+-----------+-------------+-------------+------------+
1 row in set (0.00 sec)
```

查询结果显示用户 grantUser 被创建成功，并被赋予 SELECT、INSERT 和 GRANT 权限，其相应字段值均为"Y"。

被授予 GRANT 权限的用户可以登录 MySQL 并创建其他用户账户，比如这里名称是 grantUser 的用户。

19.3.3 收回权限

收回权限就是取消已经赋予用户的某些权限。收回用户不必要的权限，可以在一定程度上保证系统的安全性。MySQL 使用 REVOKE 语句取消用户的某些权限。使用 REVOKE 收回权限之后，用户账户的记录将从 db、host、tables_priv 和 columns_priv 表中删除，但是用户账号记录仍然在 user 表中保存（删除 user 表中的账户记录，使用 DROP USER 语句，在 13.2.3 节已经介绍过）。

在将用户账户从 user 表删除之前，应该收回相应用户的所有权限。REVOKE 语句有两种语法格式。

（1）第一种语法是收回所有用户的所有权限，用于取消已授权的用户的所有全局层级、数据库层级、表层级和列层级的权限，具体如下：

```
REVOKE ALL PRIVILEGES, GRANT OPTION
FROM 'user'@'host' [, 'user'@'host' ...]
```

REVOKE 语句必须和 FROM 语句一起使用。FROM 语句指明需要收回权限的账户。

（2）另一种为长格式的 REVOKE 语句，基本语法如下：

```
REVOKE priv_type [(columns)] [, priv_type [(columns)]] ...
```

```
ON table1, table2,..., tablen
FROM 'user'@'host'[, 'user'@'host' ...]
```

该语法收回指定的权限。其中，priv_type 参数表示权限类型；columns 参数表示权限作用于哪些列上，如果不指定该参数，表示作用于整个表；table1,table2,…,tablen 表示从哪个表中收回权限；'user'@'host'参数表示用户账户，由用户名和主机名构成。

要使用 REVOKE 语句，必须拥有 MySQL 数据库的全局 CREATE USER 权限或 UPDATE 权限。

【例 19.10】使用 REVOKE 语句取消用户 grantUser 的查询权限。REVOKE 语句及其执行结果如下：

```
MySQL> REVOKE Select ON *.* FROM 'grantUser'@'localhost';
Query OK, 0 rows affected (0.00 sec)
```

结果显示执行成功。使用 SELECT 语句查询用户 grantUser 的权限：

```
mysql> SELECT Host,User,Select_priv,Insert_priv,Grant_priv FROM MySQL.user where user='grantUser';
+-----------+-----------+-------------+-------------+------------+
| Host      | User      | Select_priv | Insert_priv | Grant_priv |
+-----------+-----------+-------------+-------------+------------+
| localhost | grantUser | N           | Y           | Y          |
+-----------+-----------+-------------+-------------+------------+
```

查询结果显示用户 user 的 Select_priv 字段值为"N"，SELECT 权限已经被收回。

提示：当从旧版本的 MySQL 升级时，如果要使用 EXECUTE、CREATE VIEW、SHOW VIEW、CREATE USER、CREATE ROUTINE 和 ALTER ROUTINE 权限，必须首先升级授权表。

19.3.4 查看权限

SHOW GRANTS 语句可以显示指定用户的权限信息。使用 SHOW GRANT 查看账户信息的基本语法格式如下：

```
SHOW GRANTS FOR 'user'@'host';
```

其中，user 表示登录用户的名称，host 表示登录的主机名称或者 IP 地址。在使用该语句时，要确保指定的用户名和主机名都要用单引号括起来，并使用"@"符号将两个名字分隔开。

【例 19.11】使用 SHOW GRANTS 语句查询用户 grantUser 的权限信息。SHOW GRANTS 语句及其执行结果如下：

```
MySQL> SHOW GRANTS FOR 'grantUser'@'localhost';
+-----------------------------------------------------------------+
| Grants for grantUser@localhost                                  |
+-----------------------------------------------------------------+
| GRANT INSERT ON *.* TO `grantUser`@`localhost` WITH GRANT OPTION |
+-----------------------------------------------------------------+
```

返回的结果显示了表中 grantUser 的账户信息。接下来的行以"GRANT INSERT ON"关键字开头，表示用户被授予了 INSERT 权限，*.*表示 INSERT 权限作用于所有数据库的所有数据表。

在这里，只是定义了个别的用户权限，GRANT 可以显示更加详细的权限信息，包括全局级的和非全局级的权限，如果表层级或者列层级的权限被授予用户，那么它们也能在结果中显示出来。

在前面创建用户时，查看新建的账户时使用 SELECT 语句，也可以通过 SELECT 语句查看 user 表中的各个权限字段以确定用户的权限信息，其基本语法格式如下：

```
SELECT privileges_list FROM user WHERE user='username', host='hostname';
```

其中，privileges_list 为想要查看的权限字段，可以为 Select_priv、Insert_priv 等。读者可以根据需要选择想要查询的字段。

19.4 访问控制

正常情况下，并不希望每个用户都可以执行所有的数据库操作。当 MySQL 允许一个用户执行各种操作时，它将首先核实该用户向 MySQL 服务器发送的连接请求，然后确认用户的操作请求是否被允许。本节将向读者介绍 MySQL 中的访问控制过程。MySQL 的访问控制分为两个阶段：连接核实阶段和请求核实阶段。

19.4.1 连接核实阶段

当连接 MySQL 服务器时，服务器基于用户的身份以及用户是否能通过正确的密码验证来接受或拒绝连接，即客户端用户连接请求中会提供用户名称、主机地址名和密码。MySQL 使用 user 表中的 3 个字段（Host、User 和 authentication_string）执行身份检查，服务器只有在 user 表记录的 Host 和 User 字段匹配客户端主机名和用户名，并且提供正确的密码时才接受连接。如果连接检查没有通过，则服务器完全拒绝访问，否则服务器接受连接，然后进入阶段 2 等待用户请求。

19.4.2 请求核实阶段

建立了连接之后，服务器进入访问控制的阶段 2。对在此连接上的每个请求，服务器检查用户要执行的操作，然后检查是否有足够的权限来执行它。这正是在授权表中的权限列发挥作用的地方。这些权限可以来自 user、db、host、tables_priv 或 columns_priv 表。

确认权限时，MySQL 首先检查 user 表，如果指定的权限没有在 user 表中被授权，MySQL 将检查 db 表，db 表是下一安全层级，其中的权限限定于数据库层级，在该层级的 SELECT 权限允许用户查看指定数据库中所有表的数据；如果在该层级没有找到限定的权限，则 MySQL 继续检查 tables_priv 表以及 columns_priv 表；如果所有权限表都检查完毕，但还是没有找到允许的权限操作，MySQL 将返回错误信息，用户请求的操作不能执行，操作失败。

请求核实的过程如图 19.1 所示。

提示：MySQL 通过向下层级的顺序检查权限表（从 user 表到 columns_priv 表），但并不是所有的权限都要执行该过程。例如，一个用户登录到 MySQL 服务器之后只执行对 MySQL 的管理操作，此时，只涉及管理权限，因此 MySQL 只检查 user 表。另外，如果请求的权限操作不被允许，MySQL 也不会继续检查下一层级的表。

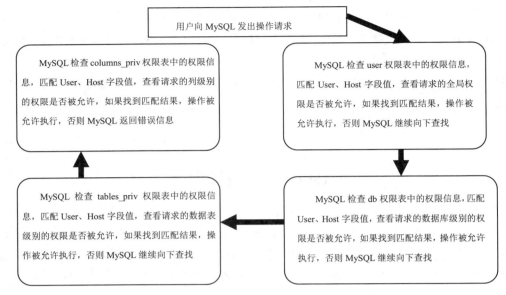

图 19.1 MySQL 请求核实过程

19.5 提升安全性

MySQL 8.0 可以采取一些措施进一步提升数据库的安全性，主要方法如下。

19.5.1 密码到期更换策略

MySQL 8.0 允许数据库管理员手动设置账户密码的过期时间。任何密码超期的账户想要连接服务端时都必须更改密码。通过设置 default_password_lifetime 参数，可以设置账户过期时间。

下面通过示例来学习。

首先查看系统中的账户过期时间。输入语句如下：

```
mysql> SELECT user,host, password_last_changed, password_lifetime, password_expired FROM mysql.user \G
*************************** 1. row ***************************
                 user: grantUser
                 host: localhost
password_last_changed: 2022-03-27 12:02:09
    password_lifetime: NULL
     password_expired: N
*************************** 2. row ***************************
                 user: jeffrey
                 host: localhost
password_last_changed: 2022-03-27 12:00:40
    password_lifetime: NULL
     password_expired: N
*************************** 3. row ***************************
```

```
                user: mysql.infoschema
                host: localhost
password_last_changed: 2022-03-27 11:43:30
    password_lifetime: NULL
     password_expired: N
*************************** 4. row ***************************
                user: mysql.session
                host: localhost
password_last_changed: 2022-03-27 11:43:30
    password_lifetime: NULL
     password_expired: N
*************************** 5. row ***************************
                user: mysql.sys
                host: localhost
password_last_changed: 2022-03-27 11:43:30
    password_lifetime: NULL
     password_expired: N
*************************** 6. row ***************************
                user: root
                host: localhost
password_last_changed: 2022-03-27 11:43:50
    password_lifetime: NULL
     password_expired: N
```

在这个结果中，password_lifetime: NULL 表示密码永不过期。

下面设置 root 用户的密码过期时间为 365 天，输入语句如下：

```
mysql> ALTER USER root@localhost PASSWORD EXPIRE INTERVAL 365 DAY;
Query OK, 0 rows affected (0.00 sec)
```

再次查看 root 用户的信息，输入语句如下：

```
mysql>SELECT user,host, password_last_changed, password_lifetime, password_expired FROM mysql.user WHERE user = 'root'\G
*************************** 1. row ***************************
                user: root
                host: localhost
password_last_changed: 2022-03-27 14:10:08
    password_lifetime: 365
     password_expired: N
1 row in set (0.00 sec)
```

从结果可以看出，root 用户的 password_lifetime 已经被修改为 365 天。

将 root 用户的密码过期重新设置为永不过期，输入语句如下：

```
mysql> ALTER USER root@localhost PASSWORD EXPIRE DEFAULT;
Query OK, 0 rows affected (0.00 sec)
```

再次查看 root 用户的信息，输入语句如下：

```
mysql> SELECT user,host, password_last_changed, password_lifetime, password_expired FROM mysql.user WHERE user = 'root'\G
*************************** 1. row ***************************
                user: root
                host: localhost
password_last_changed: 2022-03-27 11:43:50
```

```
        password_lifetime: NULL
        password_expired: N
1 row in set (0.00 sec)
```

从结果可以看出，password_lifetime 又被重新设置为 NULL，此时该用户的密码将永不过期。

19.5.2 安全模式安装

MySQL 8.0 新增了"安全模式"的安装形式，从而可以避免用户的数据被泄漏。用户可以通过以下方式来提升 MySQL 安装的安全性：

（1）为 root 账户设置密码。
（2）移除能从本地主机以外的地址访问数据库的 root 账户。
（3）移除匿名账户。
（4）移除 test 数据库，该数据库默认可被任意用户甚至匿名账户访问。

使用 mysqld -initialize 命令来安装 MySQL 实例默认是安全的，主要原因如下：

（1）在安装过程只创建一个 root 账户'root'@'localhost'，自动为这个账户生成一个随机密码并标记密码过期。
（2）数据库管理员必须用 root 账户及该随机密码登录并设置一个新密码后，才能对数据库进行正常操作。
（3）安装过程不创建任何匿名账户。
（4）安装过程不创建 test 数据库。

19.6 管理角色

在 MySQL 8.0 数据库中，角色可以看成是一些权限的集合，可以为用户赋予统一的角色，权限的修改直接通过角色来进行，无须为每个用户单独授权。

下面通过示例来学习如何管理角色。

创建角色，执行语句如下：

```
mysql> CREATE ROLE role_tt; # 创建角色
Query OK, 0 rows affected (0.02 sec)
```

给角色授予权限，执行语句如下：

```
mysql> GRANT SELECT ON db.* to 'role_tt'; # 给角色 role_tt 授予查询权限
Query OK, 0 rows affected (0.10 sec)
```

创建用户 myuser1，执行语句如下：

```
mysql> CREATE USER 'myuser1'@'%' identified by '123456';
Query OK, 0 rows affected (0.07 sec)
```

为用户 myuser1 赋予角色 role_tt，执行语句如下：

```
mysql> GRANT 'role_tt' TO 'myuser1'@'%';
Query OK, 0 rows affected (0.02 sec)
```

为角色 role_tt 增加 INSERT 权限，执行语句如下：

```
mysql> GRANT INSERT ON db.* to 'role_tt';
Query OK, 0 rows affected (0.08 sec)
```

为角色 role_tt 删除 INSERT 权限，执行语句如下：

```
mysql> REVOKE INSERT ON db.* FROM 'role_tt';
Query OK, 0 rows affected (0.10 sec)
```

查看角色与用户关系，执行语句如下：

```
mysql> SELECT * FROM mysql.role_edges;
+-----------+-----------+---------+---------+-------------------+
| FROM_HOST | FROM_USER | TO_HOST | TO_USER | WITH_ADMIN_OPTION |
+-----------+-----------+---------+---------+-------------------+
| %         | role_tt   | %       | myuser1 | N                 |
+-----------+-----------+---------+---------+-------------------+
1 row in set (0.00 sec)
```

删除角色 role_tt，执行语句如下：

```
mysql> DROP ROLE role_tt;
Query OK, 0 rows affected (0.04 sec)
```

第 20 章

MySQL 高可用架构

对于企业应用而言，MySQL 数据库的持续可用性和可访问性非常重要，尤其对于某些互联网企业来说，数据库提供持续可靠的可用性，才会给企业带来良好的效益，为其客户提供优质可靠的服务体验。因此，设计 MySQL 数据库架构初期，就需要考虑如何构建一套适合自身应用程序需要的高可用架构。本章将详细讲解 MySQL 几种常用的高可用架构配置方案。

20.1 MySQL 高可用简介

在单点访问服务中，客户访问应用系统，应用系统直接访问数据库，这种系统架构中所有存储数据和读写操作都发生在唯一的一台主机服务器上，数据库数据都存储在一个数据库系统中，往往一旦数据库系统发生故障，应用系统没有办法在短时间内恢复正常。

数据库系统是所有企业的核心系统，它存储着企业客户资料、生产数据和业务数据，一旦发生意外的停机，在没有及时恢复上线或者没有冗余方案的情况下，会造成最直接的经济损失。通常使用数据库高可用架构、同步读写分离、负载均衡等手段，达到减少系统的停机时间，提高系统服务的可靠性。

MySQL 数据库作为最流行的开源数据库产品，拥有许多成熟的高可用架构方案，其方案的可用性覆盖从 90%到 99.999%，能够适用对可用性级别多种不同的需求。其主要是利用复制技术，多个不同数据库主机之间进行复制，以保持数据的一致性，并通过一些第三方开源软件来实现负载均衡和统一的访问接口，既减少应用程序开发的复杂性，也降低了企业的运营成本。下面将具体讲解 MySQL 实现数据库高可用的几个常用架构。

20.2　MySQL 主从复制架构

MySQL 主从复制指单台 MySQL 服务器的数据复制到另一台 MySQL 服务器上。这种方案不需要复杂的配置，数据可以从单台 Master 主机复制到任意数量的 Slave 主机上。复制使用异步方式，从 MySQL 5.5 版本开始增加了半同步复制，有效地提高了复制的可靠性。

20.2.1　MySQL 主从架构设计

该架构由单台主服务器和多台从服务器构成，主服务器主要接受来自应用程序的写请求，从服务器接受读请求，在从服务器与应用层之间可以搭建负载均衡设备。例如：LVS、Haproxy、F5 等。该架构可将读写请求分离到不同的机器上执行，但需要应用程序实现不同的连接池，将读操作负载均衡到多个从服务器主机上，提高系统整体读操作的处理能力。但只有单台主服务器可以写数据，所以写操作无法负载均衡，也限制了其扩展性；并且当主服务器发生故障时，系统将无法写数据，存在单点故障。本架构通常适用于对可用性要求不高的场景，例如在线备份，将其中一台从服务器用于备份数据，避免了备份对主服务器产生的影响。如图 20.1 所示。

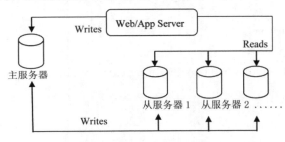

图 20.1　单主多从示意图

从图中可以看出应用程序可以访问多个从服务器主机，压力可以分布在不同的从服务器主机上，在应用层和从服务器中间可以添加 LVS，便可实现负载均衡，同时从服务器可以水平扩展来提高系统整体读性能。

20.2.2　配置环境

下面准备了 4 台机器，一台为主服务器、两台从服务器，一台用来做负载均衡。具体的设置如表 20.1 所示。

表 20.1　环境配置表

主机名	IP 地址	角色	备注
主服务器	192.168.1.20	主机	数据库写
从服务器 1	192.168.1.21	从机	数据库读
从服务器 2	192.168.1.22	从机	数据库读
LVS	192.168.1.23	LVS	负载均衡

20.2.3 服务器的安装配置

首先在主服务器、从服务器 1、从服务器 2 三台主机上安装 MySQL 数据库。把 MySQL 安装脚本上传到需要安装 MySQL 的服务器上，然后执行脚本操作。脚本有三个参数分别是：软件包、安装目录、配置文件，需要指定这三个参数才能顺利安装，如下所示：

```bash
#================================================================
#       FileName: mysql_install_local.sh
#================================================================

#!/usr/bin/env bash

fullname=$1
prefix=$2
mycnf=$3
fname=`basename $fullname`
realname=`echo "$fname" | awk -F ".tar" '{print $1}'`
if [ -z "$fullname" ]; then
    echo "Usage: $0 /home/xyz/mysql-xxx.tar.gz /path/to/install_prefix /path/to/my.cnf"
    exit 1;
fi
echo "Create MySQL user"
tt=`grep mysql /etc/passwd`
if [[ -z $tt ]]; then
    groupadd -f mysql
    useradd -g mysql -d /dev/null -s /sbin/nologin mysql
fi
echo "Clean old MySQL"
rm -rf $prefix/mysql
echo "Unpacking..."
if [[ ! -d $prefix ]]; then
  mkdir -p $prefix
fi
tar xfz $fullname -C $prefix 2>> /tmp/mysqlinstall.log
echo "Setting up symlink mysql->mysql-XYZ..."
ln -s $prefix/$realname $prefix/mysql 2>> /tmp/mysqlinstall.log
echo "export PATH=$PATH:$prefix/mysql/bin" >> /etc/profile
source /etc/profile
test -x $prefix/mysql/bin/mysqld
[ $? != 0 ] && exit 1
test -x $prefix/mysql/bin/mysql
[ $? != 0 ] && exit 1
if [[ -f /etc/my.cnf ]]; then
    mv /etc/my.cnf{,.old}
 fi
cp $mycnf /etc/my.cnf
echo "Initial MySQL Database"
if [[ $PWD != $prefix/mysql ]]; then
  cd $prefix/mysql/
fi
./scripts/mysql_install_db --user=mysql --defaults-file=/etc/my.cnf 2>> /tmp/mysqlinstall.log |grep -i 'ok'
    if [[ $? != 0 ]]; then
```

```
    echo "initial mysql dataabse failed see /tmp/mysqlinstall.log "; exit 1
fi
cp -f support-files/mysql.server /etc/init.d/mysql
echo "Installing of MySQL is complete"
echo "You can start MySQL server /etc/init.d/mysql start"
```

MySQL 安装完成后，即可对它们进行配置，具体操作步骤说明如下：

步骤01 在主服务器主机上开启 binlog 日志，保持 server-id 的唯一性，这一步最好安装 MySQL 之前就写在配置文件中。

步骤02 在主服务器上配置复制所需要的账户 repl，如下所示：

```
mysql> grant replication slave on *.* to repl@%;
Query OK, 0 rows affected (0.05 sec)

mysql>flush privileges;
Query OK, 0 rows affected (0.01 sec)
```

步骤03 接着，在从服务器 1 和从服务器 2 上分别配置主服务器信息，如下所示：

```
mysql>change master to master_host='192.168.1.20', master_user='repl', master_password='repl_password';
Query OK, 0 rows affected (0.05 sec)

mysql>slave start;
Query OK, 0 rows affected (0.01 sec)

mysql>show slave status\G
*************************** 1. row ***************************
             Slave_IO_State: Waiting for master to send event
                Master_Host: 192.168.1.20
                Master_User: repl
                Master_Port: 3306
              Connect_Retry: 3
            Master_Log_File: mysql-bin.001
        Read_Master_Log_Pos: 79
             Relay_Log_File: mysql-relay-bin.001
              Relay_Log_Pos: 548
      Relay_Master_Log_File: mysql-bin.001
           Slave_IO_Running: Yes
          Slave_SQL_Running: Yes
            Replicate_Do_DB:
        Replicate_Ignore_DB:
         Replicate_Do_Table:
     Replicate_Ignore_Table:
    Replicate_Wild_Do_Table:
Replicate_Wild_Ignore_Table:
                 Last_Errno: 0
                 Last_Error:
               Skip_Counter: 0
        Exec_Master_Log_Pos: 79
            Relay_Log_Space: 552
            Until_Condition: None
             Until_Log_File:
              Until_Log_Pos: 0
```

```
            Master_SSL_Allowed: No
            Master_SSL_CA_File:
            Master_SSL_CA_Path:
               Master_SSL_Cert:
             Master_SSL_Cipher:
                Master_SSL_Key:
         Seconds_Behind_Master: 8
1 row in set (0.00 sec)
```

下面了解一下其中主要的几个变量的含义：

（1）Master_Host：被复制的主机地址。
（2）Master_User：在主服务器上创建的用于复制的账户。
（3）Master_Port：主服务器上 MySQL 运行的端口。
（4）Slave_IO_Running：显示 I/O 线程运行状态。
（5）Slave_SQL_Running：显示 SQL 线程运行状态。
（6）Last_Error：当复制中断时，此处会显示错误原因，并由此来修复复制。
（7）Seconds_Behind_Master：显示从服务器中 I/O 线程与 SQL 线程之间的时间差。可以反映出主从服务器之间的网络快慢。

至此，MySQL 的主从服务器就配置好了。

20.2.4 LVS 的安装配置

接下来可以配置 LVS，用来做负载均衡，具体操作步骤说明如下：

步骤 01 在 LVS 主机上安装 LVS 相关软件，可以通过 yum 的方式，如下所示：

```
root@LVS#yum install ipvsadm piranha
```

步骤 02 安装完毕，开始配置 LVS，编辑 lvs 的配置文件。

```
root@LVS#vi /etc/sysconfig/ha/lvs.cf
serial_no = 26
primary = 192.168.1.10              #配置虚拟 IP
service = lvs
heartbeat = 1
heartbeat_port = 539
keepalive = 6
deadtime = 18
network = direct
debug_level = NONE
monitor_links = 1
syncdaemon = 0
virtual mysql {
     active = 1
     address = 192.168.1.23 eth0:1   #配置 LVS 的主机物理网口 IP
     vip_nmask = 255.255.255.0
     port = 3306
     use_regex = 0
```

```
            load_monitor = none
            scheduler = wrr                    #负载方式可选 wrr 加权轮询，或 wlc 加权最小连接
            protocol = tcp
            timeout = 6
            reentry = 15
            quiesce_server = 0
            server Slave-1 {
                address = 192.168.1.21         #Slave-1 的主机 IP
                active = 1
                port = 3306
                weight = 1
            }
            server Slave-2 {
                address = 192.168.1.22         #Slave-2 的主机 IP
                active = 1
                port = 3306
                weight = 1
            }
```

步骤 03 分别在从服务器 1 和从服务器 2 上安装 arptables_jf，进行 arp 设置，设置完成后启动 arptable_jf 服务，并设置其开机自动启动。（下面是从服务器 1 的配置，从服务器 2 请读者自行配置。）

```
root@Slave-1# yum install arptables_jf
root@Slave-1# arptables -A IN -d 192.168.1.10 -j DROP
root@Slave-1# arptables -A OUT -s 192.168.1.10 -j mangle --mangle-ip-s 192.168.1.21
root@Slave-1# echo "ip addr add  192.168.1.10    dev lo" >> /etc/rc.local
root@Slave-1# service arptables_jf start
root@Slave-1# chkconfig - level 2345 arptables_jf on
```

步骤 04 启动 LVS 服务，并设置其开机自动启动。

```
root@LVS#service pulse start
root@LVS#chkconfig - level 2345 pulse on
```

步骤 05 启动 LVS 服务后，可以通过 ipvsadm 命令来检查 LVS 状态是否正常。

```
root@LVS# ipvsadm -L -n
IP Virtual Server version 1.2.1 (size=4096)
Prot LocalAddress:Port Scheduler Flags
  -> RemoteAddress:Port           Forward Weight ActiveConn InActConn
TCP  192.168.1.10:3306 wrr
  -> 192.168.1.21:3306            Route    1      1          0
  -> 192.168.1.22:3306            Route    1      2          0
```

从结果中可以看到，虚拟 IP 为 192.168.1.10，转发端口是 3306，真实主机的 IP 是 192.168.1.21 和 192.168.1.22。当应用程序访问 192.168.1.10 的 3306 端口时，连接会自动被转发的 21 和 22 这两台机器上，此时再看 LVS 的状态，ActiveConn 的值就会有变化，说明 LVS 已经正常工作了。

到此为止，大家已经学会了如何配置一个最简单的 MySQL 高可用架构。当然这种架构存在单点故障问题，可以为主服务器准备一台备份机器来解决。接下来就在此架构基础上来进一步完善它的可用性。

20.3　MySQL+DRBD+HA 主备架构

一些成长性企业的业务发展非常迅速，对数据库的扩展性、可用性和性能的要求会更高，上面所讲的普通方案无法满足这种业务快速增长的场景的需求。越来越大的压力会使主从服务器之间的复制出现较大的延迟，从而影响了数据库数据的一致性，降低了它的可靠性。

有什么办法可以解决这样的问题呢？使用 DRBD 技术执行文件级的复制，主机之间能够保持较高的文件一致性，可靠性可以达到 99.99%，此方案更多地考虑到了扩展性，能很好地适应企业业务的快速增长给数据库带来的读写压力。

20.3.1　什么是 DRBD

DRBD（Distributed Replicated Block Device）指分布式复制块设备，是一种基于 Linux 系统的软件组件，它是由内核模块和相关程序组成，用以构建高可用性的集群。当数据写入到本地 DRBD 设备上的文件时，数据会同时被发送到网络中的另外一台主机之上。

DRBD 实现原理如图 20.2 所示，它通过网络来镜像整个设备。DRBD 复制接收数据，然后写入到本地磁盘，之后会把数据发送给另一台主机，另一台主机会将数据写入到本地磁盘，DRBD 通过对故障进行切换操作，从而实现集群的高可用性。

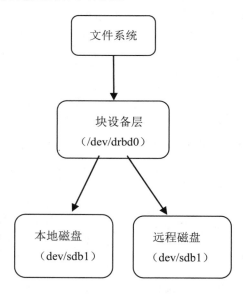

图 20.2　DRBD 工作原理图

20.3.2　MySQL+DRBD+HA 架构设计

该架构主要由 2 台机器组成。分别是 Active 和 Standby。它们之间通过 Heartbeat 进行连接，对外共用一个虚拟 IP 地址，正常情况下 Active 对外提供服务，当 Active 出现软硬件单点故障时，Heartbeat 自动将虚拟 IP 指向 Standby，由 Standby 作为新的 Active 对外提供服务。当原来 Active 的故障排除后，再将其接入系统作为新的 Standby 存在。

Active 和 Standby 之间通过 DRBD 技术实现数据的互为主从的复制，以减少故障切换的时间，架构设计如图 20.3 所示。

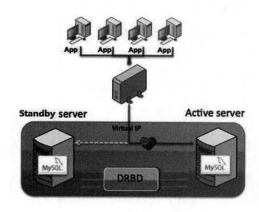

图 20.3　DRBD 方案架构

20.3.3　配置环境

首先我们需要准备两台数据库主机，具体配置信息如表 20.2 所示。

表20.2　环境配置表

主 机 名	IP 地址	心跳 IP	备　注
db-01	10.11.196.48	192.168.1.11	虚拟 IP：10.11.196.50
db-02	10.11.196.49	192.168.1.12	同上

在两台主机上安装好 MySQL 数据库，注意，MySQL 的数据目录需要设定在 DRBD 的块设备的挂载点，以/data 挂载点为例。

20.3.4　安装配置 Heartbeat

安装配置 Heartbeat 的具体操作步骤如下：

步骤01 分别在两台主机上安装 Hearbeat 心跳软件，可以通过 yum 的方式，如下所示：

```
root@db-01# yum install heartbeat
```

步骤02 分别在两台主机上配置 hosts 文件，把主机名与 IP 加入 hosts 文件。

```
root@db-01# cat /etc/hosts
127.0.0.1 localhost.localdomain localhost
::1       localhost6.localdomain6 localhost6
10.11.196.48　db-01
10.11.196.49　db-02
192.168.1.11　db-01
192.168.1.12　db-02
```

步骤03 分别在两台主机上配置网卡 IP 和心跳 IP。

```
root@db-01# cat /etc/sysconfig/network-scripts/ifcfg-eth0
DEVICE=eth0
BOOTPROTO=static
IPADDR=10.11.196.48          #另外一台改成 10.11.196.49
ONBOOT=yes

root@db-01# cat /etc/sysconfig/network-scripts/ifcfg-eth1
DEVICE=eth1
BOOTPROTO=static
BROADCAST=192.168.1.255
IPADDR=192.168.1.11          #另外一台改成 192.168.1.12
NETMASK=255.255.255.0
NETWORK=192.168.1.0
ONBOOT=yes
```

注意，两台数据库主机的系统时间要保持一致，如果内网环境中有 NTP 服务器，那么可以定期更新两台数据库服务器的时间；如果没有，可以从两台服务器中选一台配置，以保持两者的时间同步。

步骤04 以 db-01 为 NTP 服务器为例。首先安装 ntpd 软件包，可通过 yum 方式，然后配置其选项，如下所示：

```
root@db-01# cat /etc/ntp.conf
restrict default nomodify notrap noquery
server 127.0.0.1
restrict 192.168.1.0 mask 255.255.255.0 nomodify notrap
restrict 127.0.0.1
driftfile /var/lib/ntp/ntpd.drift
```

步骤05 启动 NTP 服务，设置其为开机自动启动，并在 db-02 服务器上设置定时任务来更新时间。

```
root@db-01# service ntpd start
root@db-01# chkconfig --level 2345 ntpd on

root@db-02# crontab -e
0-59/30 * * * * /usr/sbin/ntpdate 192.168.1.11
```

这些准备工作完成后，就可以开始配置 Heartbeat 了，在 Heartbeat 2.x 版本中相关的配置文件有三个，分别是 ha.cf、authkeys 和 haresources。

步骤06 首先配置 ha.cf 文件，这个是 Heartbeat 核心配置文件。

```
root@db-01# cat /etc/ha.d/ha.cf
debugfile /var/log/ha-debug
logfile /var/log/ha-log
logfacility local0
keepalive 1
deadtime 10
warntime 5
initdead 30
udpport 694
bcast eth1                   #此网口必须是心跳口，也可写成 ucast eth1 <ip>对方 ip
auto_failback off            #主节点故障恢复后，资源是否自动调回
node db-01 db-02             #节点名称与主机名（uname -n）保持一致
```

```
ping 10.11.196.254            #网关
respawn hacluster /usr/lib/heartbeat/ipfail
apiauth ipfail gid=haclient uid=hacluster
deadping 5                    #此时间要小于 deadtime
```

步骤07 接着配置 authkeys 文件并修改其权限设置，该文件主要用于设置通信的验证校验机制。

```
root@db-01# cat/etc/ha.d/authkeys
auth 1
1 crc
root@db-01# chmod 600 /etc/ha.d/authkeys
```

步骤08 配置 haresources 文件，该文件主要是设置挂载的文件系统类型和路径，以及服务名和虚拟 IP。

```
root@db-01# cat/etc/ha.d/haresources
db-01drbddisk Filesystem::/dev/drbd0::/data::ext3 mysql10.11.196.50
主机名 资源名 文件系统：设备名，挂载路径，类型 服务名 虚拟 IP
MailTo::youremail@address.com::DRBDFailure   #此行可选
```

这里需要注意，由于使用的是 DRBD，因此设备名是/dev/drbd0，服务名是 mysql，这个名称需要和/etc/init.d/mysql 这个脚本名保持一致，否则在发生故障切换时将无法启动 mysql。

Heartbeat 在两台数据库主机都需要配置，有些地方需要变动，例如主机名、IP 等。

20.3.5　安装配置 DRBD

首先下载源码包，地址为 http://oss.linbit.com/drbd/8.3/drbd-8.3.7.tar.gz，DRBD 分为内核模块和管理工具两部分，Linux 从 2.6.33 开始就将 DRBD 驱动模块合并到内核里，如果你的 Linux 系统内核大于 2.6.33，就只需要安装 DRBD 管理工具。这里以 DRBD 8.3.7 版本为例讲解一下配置过程。

步骤01 分别在两台机器上安装 DRBD，如下所示：

```
root@db-01# wget http://oss.linbit.com/drbd/8.3/drbd-8.3.7.tar.gz
root@db-01# tar zxf drbd-8.3.7.tar.gz
root@db-01# cd drbd-8.3.7
root@db-01# ./configure --with-utils --with-km
root@db-01# make && make install
root@db-01# cp drbd /etc/init.d/
root@db-01# chkconfig --add drbd
```

步骤02 安装完成后开始配置 DRBD，需要为 DRBD 分配一个块设备，以/dev/sdb1 为例：

```
root@db-01# cat/usr/local/etc/drbd.conf
global {
        usage-count yes;
}
common {
        protocol C;
        syncer {
                rate 100M;
        }
}
resource r0 {
        on db-01 {
```

```
            device    /dev/drbd0;
            disk      /dev/sdb1;
            address   192.168.1.11:7789;        #这里填对应主机的心跳地址
            meta-disk internal;
        }
        on db-02 {
            device    /dev/drbd0;
            disk      /dev/sdb1;
            address   192.168.1.12:7789;
            meta-disk internal;
        }
}
```

步骤03 配置完成后，就可以开始加载内核模块并进行初始化设置（两台机器都需要操作）。

```
加载 drbd 的内核模块：
root@db-01# depmod
root@db-01# modprobe drbd
创建 drbd 元数据：
root@db-01# drbdadm create-md r0
启动 drbd 相关进程：
root@db-01# drbdadm attach r0
root@db-01# drbdadm syncer r0
root@db-01# drbdadm connect r0
```

步骤04 开始创建主节点，并格式化文件系统，注意此步骤只需要在其中一台服务器上做，这里以 db-01 为例：

```
root@db-01# drbdadm -- --overwrite-data-of-peer primary r0
root@db-01# mkfs.ext3 /dev/drbd0
查看 DRBD 同步状态：
root@db-01# cat /proc/drbd
version: 8.3.7 (api:88/proto:86-91)
GIT-hash: ea9e28dbff98e331a62bcbcc63a6135808fe2917 build by root@db-01, 2012-04-08 16:57:43
 0: cs:Connected ro:Primary/Secondary ds:UpToDate/UpToDate C r----
    ns:265140 nr:33016 dw:298156 dr:4089 al:55 bm:15 lo:0 pe:0 ua:0 ap:0 ep:1 wo:d oos:0

root@db-02# cat /proc/drbd
version: 8.3.7 (api:88/proto:86-91)
GIT-hash: ea9e28dbff98e331a62bcbcc63a6135808fe2917 build by root@db-02, 2012-04-08 16:58:30
 0: cs:Connected ro:Secondary/Primary ds:UpToDate/UpToDate C r----
    ns:33016 nr:298512 dw:299164 dr:36705 al:8 bm:21 lo:0 pe:0 ua:0 ap:0 ep:1 wo:d oos:0
```

UpToDate 表示两台服务器的 DRBD 块设备数据已经同步，DRBD 配置成功。注意，DRBD 设备只有在 Primary 状态时才可以 mount，如上面的/proc/drbd 信息所示，当前 db-01 是 Primary，所以可以在 db-01 上把数据挂在/dev/drbd0：

```
root@db-01# mkdir /data
root@db-01# mount /dev/drbd0 /data
```

步骤05 启动 Heartbeat 服务，设置其为开机自动启动。

```
root@db-01# chkconfig - level 234 drbd on
root@db-01# service heartbeat start
```

现在整个配置就完成了。切记，在投入生产环境之前，一定要进行一些测试来验证各服务是否运行正常,模拟故障发生确认是否能自动切换主机,不会导致服务中断。

设想这样一种场景,两台 DRBD 主机一主一备,当两台机器之间的通信出现故障,备机无法和主机保持同步,此时 Heartbeat 识别到了连接故障,就会将备机切换成主机,但此时备机的数据并不是和主机完全一致,这就会导致数据的不一致性。如何应对这种情况?启用 Heartbeat 的 dopd 程序,当出现连接问题两台机器的数据不一致时,备机的元数据会被设置成 Outdated,当主机识别到后就不会切换到备机。这样一来就避免了两台机器的数据不一致,后面可以由人工去检查问题原因并修复。牺牲掉可用性,换来数据一致性也是值得的。下面简单介绍如何启用这个功能:

在 Heartbeat 的 ha.cf 配置文件中加入下面内容(两台主机都需要添加):

```
respawn hacluster /usr/lib/heartbeat/dopd
apiauth dopd gid=haclient uid=hacluster
```

重新加载 Heartbeat 配置,如下所示:

```
root@db-01# /etc/init.d/heartbeat reload
```

接着,修改 DRBD 配置文件 drbd.conf 的 common 部分,追加入下面内容:

```
common {
...
  handlers {
    outdate-peer "/usr/lib/heartbeat/drbd-peer-outdater";
  }
}
#在 resource 部分追加下面内容:
resource r0 {
...
  disk {
    fencing    resource-only;
  }
}
```

重新加载 DRBD 配置文件:

```
root@db-01# drbdadmin adjust all
```

此方案可以作为上一节方案扩展和增强,将原来的单台主服务器,升级为主备架构,避免了单点故障的发生,在实际的生产环境中往往需要灵活地设计架构,才能应对变化多样的业务需求。

20.4　LVS+Keepalived+MySQL 单点写入主主同步架构

目前在 MySQL 高可用方案中,LVS+Keepalived 是经常被使用到的,LVS 和 Keepalived 为高可用方案提供了负载均衡和故障转移。通常情况下 MySQL 高可用为了实现数据的一致性,会采用单点写入,本节将采用 Keepalived 中的 sorry_server 来实现写入数据库为单点的需求。下面先了解一下 LVS+Keepalived 单点写入主主同步架构,如图 20.4 所示。

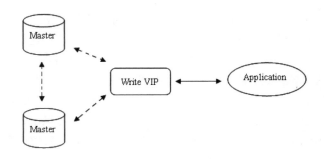

图 20.4　MM+LVS+Keepalived 单点写入主主同步架构

该架构安装配置非常简单，主服务器后面可以添加多个从服务器，并做到负载均衡。事实上，该方案并没有实现数据库的读写分离，只适用于只有两台数据库服务器的情况。

20.4.1　配置环境

下面先准备一下 LVS+Keepalived 高可用架构所需要的试验环境，如表 20.3 所示。

表20.3　LVS+Keepalived高可用方架构服务器列表

服务器名称	IP 地址	VIP	数据库服务器名称
Master	192.168.1.20	192.168.1.100	db1
Backup	192.168.1.30	192.168.1.100	db2

首先，准备两台机器并安装好 MySQL 服务，然后设置两台服务器主主同步，具体步骤如下：

步骤01 在 db1（192.168.1.20）服务器上配置 my.cnf 文件，如下所示：

```
# Replication Master Server (default)
# binary logging is required for replication
log-bin=mysql-bin

# binary logging format - mixed recommended
binlog_format=mixed

# required unique id between 1 and 2^32 - 1
# defaults to 1 if master-host is not set
# but will not function as a master if omitted
server-id= 1
replicate-do-db = test
slave-skip-errors = all
sync_binlog = 1
log-slave-updates

# set number of masters
auto_increment_increment = 2
auto_increment_offset = 1
```

值得注意的是，log-slave-updates 表示如果一个 Master 挂掉的话，另一个主机会接管。

步骤02 在 db2（192.168.1.30）服务器上配置 my.cnf 文件，如下所示：

```
# Replication Master Server (default)
# binary logging is required for replication
log-bin=mysql-bin

# binary logging format - mixed recommended
binlog_format=mixed

# required unique id between 1 and 2^32 - 1
# defaults to 1 if master-host is not set
# but will not function as a master if omitted
server-id= 2
replicate-do-db = test
slave-skip-errors = all
sync_binlog = 1
log-slave-updates

# set number of masters
auto_increment_increment = 2
auto_increment_offset = 2
```

注意，auto_increment_increment 的值表示整个结构中服务器的总数，本案例中使用到两台主机服务器，所以设置为 2。auto_increment_increment 和 auto_increment_offset 用于主主复制过程防止出现重复的值。

步骤 03 复制 db1（192.168.1.20）服务器上的数据库到 db2，以保持数据的一致性。首先将服务器上的表锁定起来，避免复制过程发生数据的变化，如下所示：

```
mysql> flush tables with read lock;
Query OK, 0 rows affected (0.00 sec)
```

步骤 04 然后在 db1 服务器上查询当前二进制文件的名称以及偏移位置。

```
mysql> show master status;
+------------------+----------+--------------+------------------+
| File             | Position | Binlog_Do_DB | Binlog_Ignore_DB |
+------------------+----------+--------------+------------------+
| mysql-bin.000022 |      107 |              |                  |
+------------------+----------+--------------+------------------+
1 row in set (0.00 sec)
```

步骤 05 接着将 db1 服务器的数据进行备份，然后在 db2 服务器上恢复，这样保持两个数据库一开始数据都是一致的，如下所示：

```
[root@localhost ~]# cd /usr/local/mysql/bin/
[root@localhost bin]# mysqldump -user=root -p test > /test.sql
Enter password:
```

下面在 db2 服务器上进行数据恢复，如下所示：

```
[root@localhost ~]# cd /usr/local/mysql/bin/
[root@localhost bin]# mysqldump -user=root -p test < /test.sql
Enter password:
```

步骤 06 值得注意的是，在实验的过程中，需要关闭系统自带的防火墙，如下所示：

```
[root@localhost ~]# service iptables stop
```

步骤 07 db1 和 db2 服务器相互通告二进制日志的位置并启动复制功能。在 db1 服务器上执行如下操作，启动复制功能：

```
mysql> stop slave;
Query OK, 0 rows affected (0.01 sec)

mysql> change master to
    -> master_host='192.168.1.30',
    -> master_user='replication',
    -> master_password='replication_password',
    -> master_log_file='mysql-bin.000022',
    -> master_log_pos=107;
Query OK, 0 rows affected (0.00 sec)

mysql> start slave;
Query OK, 0 rows affected (0.00 sec)

mysql> show variables like 'server%';
+---------------+-------+
| Variable_name | Value |
+---------------+-------+
| server_id     | 1     |
+---------------+-------+
1 row in set (0.00 sec)
```

步骤 08 在 db2 服务器上执行如下操作，启动复制功能：

```
mysql> stop slave;
Query OK, 0 rows affected (0.00 sec)

mysql> change master to
    -> master_host='192.168.1.20',
    -> master_user='replication',
    -> master_password='replication_password',
    -> master_log_file='mysql-bin.000022',
    -> master_log_pos=107;
Query OK, 0 rows affected (0.00 sec)

mysql> start slave;
Query OK, 0 rows affected (0.02 sec)

mysql> show variables like 'server_id';
+---------------+-------+
| Variable_name | Value |
+---------------+-------+
| server_id     | 2     |
+---------------+-------+
1 row in set (0.00 sec)
```

步骤 09 接下来，在 db1 服务器上测试一下，看看是否 db1 和 db2 是否实现同步，如下所示：

```
mysql> show slave status\G;
*************************** 1. row ***************************
               Slave_IO_State: Waiting for master to send event
                  Master_Host: 192.168.1.30
                  Master_User: replication
```

```
                Master_Port: 3306
              Connect_Retry: 60
            Master_Log_File: mysql-bin.000025
        Read_Master_Log_Pos: 107
             Relay_Log_File: localhost-relay-bin.000005
              Relay_Log_Pos: 253
      Relay_Master_Log_File: mysql-bin.000025
           Slave_IO_Running: Yes
          Slave_SQL_Running: Yes
            Replicate_Do_DB: test
        ...
             Last_SQL_Error:
  Replicate_Ignore_Server_Ids:
           Master_Server_Id: 2
1 row in set (0.00 sec)
```

步骤10 下面在 db2 服务器上测试一下，看看 db1 和 db2 是否实现同步，如下所示：

```
mysql> show slave status\G;
*************************** 1. row ***************************
             Slave_IO_State: Waiting for master to send event
                Master_Host: 192.168.1.20
                Master_User: replication
                Master_Port: 3306
              Connect_Retry: 60
            Master_Log_File: mysql-bin.000023
        Read_Master_Log_Pos: 107
             Relay_Log_File: localhost-relay-bin.000003
              Relay_Log_Pos: 253
      Relay_Master_Log_File: mysql-bin.000023
           Slave_IO_Running: Yes
          Slave_SQL_Running: Yes
            Replicate_Do_DB: test
        ...
             Last_SQL_Error:
  Replicate_Ignore_Server_Ids:
           Master_Server_Id: 1
1 row in set (0.00 sec)
```

步骤11 下面通过对数据的操作检验 db1 和 db2 是否同步。首先在 db1 服务器中新增加一个表，并添加一条数据，如下所示：

```
mysql> use test
Database changed
mysql> show tables;
+----------------+
| Tables_in_test |
+----------------+
| t3             |
+----------------+
1 row in set (0.06 sec)

mysql> create table t(data int);
Query OK, 0 rows affected (0.08 sec)
```

```
mysql> insert into t values(1);
Query OK, 1 row affected (0.01 sec)
```

步骤12 接下来，在 db2 服务器上查询 t 表是否添加成功，然后也在 db2 上添加一条数据。

```
mysql> use test;
Database changed
mysql> show tables;
+-----------------+
| Tables_in_test  |
+-----------------+
| t               |
| t3              |
+-----------------+
2 rows in set (0.03 sec)

mysql> select * from t;
+------+
| data |
+------+
|    1 |
+------+
1 row in set (0.03 sec)

mysql> insert into t values(2);
Query OK, 1 row affected (0.02 sec)
```

步骤13 接下来，回到 db1 服务器中测试，确认刚才 db2 中添加的数据是否同步过来了，如下所示：

```
mysql> select * from t;
+------+
| data |
+------+
|    1 |
|    2 |
+------+
2 rows in set (0.05 sec)
```

至此，主主同步配置完成，接下来开始安装 LVS 和 Keepalived。

20.4.2　LVS+Keepalived 的安装

需要在 db1 和 db2 两台服务器上都安装上 LVS，可以通过地址 http://www.linuxvirtualserver.org/software/kernel-2.6/ipvsadm-1.24.tar.gz 下载 LVS，具体安装步骤说明如下：

首先解压缩并安装 ipvsadm-1.24.tar.gz，如下所示：

```
wget http://www.linuxvirtualserver.org/software/kernel-2.6/ ipvsadm-1.24.tar.gz
ln -s /usr/src/kernels/2.6.18-164.el5-i68/ /usr/src/linux
tar zxvf ipvsadm-1.24.tar.gz
cd ipvsadm-1.24
make && make install
```

接下来，需要同时在 db1 和 db2 两台服务器上安装 Keepalived，Keepalived 下载地址为 http://www.keepalived.org/ software/keepalived-1.1.19.tar.gz，具体安装步骤如下：

```
wget http://www.keepalived.org/software/keepalived-1.1.19.tar.gz
cd keepalived-1.1.19
./configure --prefix=/usr/local/keepalived
make
make install

cp /usr/local/keepalived/sbin/keepalived /usr/sbin/
cp /usr/local/keepalived/etc/sysconfig/keepalived /etc/sysconfig/
cp /usr/local/keepalived/etc/rc.d/init.d/keepalived /etc/init.d/
mkdir /etc/keepalived
```

20.4.3　LVS+Keepalived 的配置

首先，需要对 Master 的 Keepalived 进行配置，如下所示：

```
vim /etc/keepalived/keepalived.conf
global_defs {
    notification_email {
        zhangxy@test.com
    }
    notification_email_from jiankong@test.com
    smtp_server mail.test.com
    smtp_connect_timeout 30
    router_id LVS1
}
vrrp_sync_group test {
    group {
        loadbalance
    }
}
vrrp_instance loadbalance {
    state MASTER
    interface eth0
    lvs_sync_daemon_inteface eth0
    virtual_router_id 51

    priority 180
    advert_int 1
    authentication {
        auth_type PASS
        auth_pass 1111
    }
    virtual_ipaddress {
        192.168.1.100 dev eth0 label eth0:1
    }
}
virtual_server 192.168.1.100 3306 {
    delay_loop 6
    lb_algo rr
    lb_kind DR
    persistence_timeout 20
    protocol TCP
    sorry_server 192.168.1.30 3306
    real_server 192.168.1.20 3306 {
```

```
            weight 3
            TCP_CHECK {
                connect_timeout 3
                nb_get_retry 3
                delay_before_retry 3
                connect_port 3306
            }
        }
    }
```

接下来，需要对 Backup 主机的 Keepalived 进行配置，如下所示：

```
vim /etc/keepalived/keepalived.conf
global_defs {
    notification_email {
        zhangxy@test.com
    }
    notification_email_from jiankong@test.com
    smtp_server mail.test.com
    smtp_connect_timeout 30
    router_id LVS1
}

vrrp_sync_group test {
    group {
        loadbalance
    }
}
vrrp_instance loadbalance {
    state BACKUP
    interface eth0
    lvs_sync_daemon_inteface eth0
    virtual_router_id 51
    priority 150
    advert_int 1
    authentication {
        auth_type PASS
        auth_pass 1111
    }
    virtual_ipaddress {
        192.168.1.100 dev eth0 label eth0:1
    }
}
virtual_server 192.168.1.100 3306 {
    delay_loop 6
    lb_algo rr
    lb_kind DR
    persistence_timeout 20
    protocol TCP
    sorry_server 192.168.1.30 3306
    real_server 192.168.1.20 3306 {
        weight 3
        TCP_CHECK {
            connect_timeout 3
            nb_get_retry 3
```

```
                delay_before_retry 3
                connect_port 3306
        }
    }
}
```

接下来，需要对 Master 和 Backup 的 realserver 进行配置，对于 realserver 的配置 Master 和 Backup 是一致的，如下所示：

```
vim /etc/rc.d/init.d/realserver.sh
#!/bin/bash
# description: Config realserver lo and apply noarp
SNS_VIP=192.168.1.100
/etc/rc.d/init.d/functions
case "$1" in
start)
    ifconfig lo:0 $SNS_VIP netmask 255.255.255.255 broadcast $SNS_VIP
    /sbin/route add -host $SNS_VIP dev lo:0
    echo "1" >/proc/sys/net/ipv4/conf/lo/arp_ignore
    echo "2" >/proc/sys/net/ipv4/conf/lo/arp_announce
    echo "1" >/proc/sys/net/ipv4/conf/all/arp_ignore
    echo "2" >/proc/sys/net/ipv4/conf/all/arp_announce
    sysctl -p >/dev/null 2>&1
    echo "RealServer Start OK"
    ;;
stop)
    ifconfig lo:0 down
    route del $SNS_VIP >/dev/null 2>&1
    echo "0" >/proc/sys/net/ipv4/conf/lo/arp_ignore
    echo "0" >/proc/sys/net/ipv4/conf/lo/arp_announce
    echo "0" >/proc/sys/net/ipv4/conf/all/arp_ignore
    echo "0" >/proc/sys/net/ipv4/conf/all/arp_announce
    echo "RealServer Stoped"
    ;;
*)
    echo "Usage: $0 {start|stop}"
    exit 1
esac
exit 0
```

20.4.4　Master 和 Backup 的启动

首先，需要先启动 Master 和 Backup 上的 MySQL 服务器，之后需要启动 keepalived 和 realserver 脚本，如下所示：

```
/etc/rc.d/init.d/realserver.sh start
/etc/rc.d/init.d/keepalived start
```

将 keepalived 和 realserver 启动脚本加入到操作系统 rc.local 自启动中：

```
echo "/etc/rc.d/init.d/realserver.sh start" >> /etc/rc.local
echo "/etc/rc.d/init.d/keepalived start" >> /etc/rc.local
```

20.5 MMM 高可用架构

MMM 即 MySQL 主主复制管理器（Master-Master Replication Manager for MySQL），它是 MySQL 主主复制配置监控、故障转移和管理的一套可伸缩的脚本套件。该套件也能对居于标准的主从配置的任意数量的从服务器进行读负载均衡，可以实现数据备份、节点之间重新同步功能。

20.5.1 MMM 高可用架构简介

在 MMM 高可用架构中，如果当前的服务器挂掉后，会将后端的从服务器自动转为新的主服务器进行同步复制，不需要手工更改同步配置。该架构是目前比较成熟的高可用解决方案。MySQL 本身没有提供 replication failover 的解决方案，通过 MMM 高可用架构方案能实现服务器的故障转移，从而实现 MySQL 高可用。MMM 高可用架构图如图 20.5 所示。

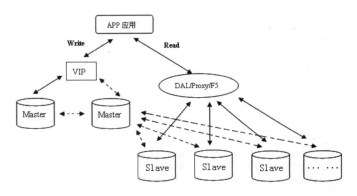

图 20.5　MMM 高可用架构图

MMM 高可用架构解决了很多问题，特别是在 read/write 比较高的 Web 2.0 应用中。该架构存在一些优点和缺点。先来看一下该架构的优点：

（1）安全性、稳定性比较高，可扩展性好，高可用。当主服务器挂掉以后，另一个主服务器立即接管，其他的从服务器能自动切换，不需要人工干预。

（2）写操作全部在主节点进行，并有从服务器数据库节点定时读取主服务器的二进制日志。

（3）将众多客户的读请求分散到更多的数据库节点，从而减轻了单点的压力。

该方案也在一些缺点，例如该方案中至少需要三个节点，对主机的数量有要求，需要实现读写分离，对程序来说难度系数较高。

MMM 高可用架构适合数据库访问量大，业务增长快，并且能够实现读写分离的场景。接下来我们进行 MMM 高可用架构的实战。

20.5.2 配置环境

MMM 高可用架构是对 MySQL Master-Slave Replication 的一个补充。MMM 有三个重要的部分，

包括 mmmd_mon、mmmd_agent 和 mmm_control。每一个 MySQL 服务器节点上都需要运行 mmmd_agent，同时需要使用一个机器运行 mmmd_mon，mmmd_mon 的作用主要用来监控其他数据库节点运行的状况，可以是独立的一台机器，也可以是和 App Server 共享同一个服务器。

MMM 利用了 VIP（虚拟 IP）的技术，1 个网卡可以同时使用多个 IP，当某个数据库节点出现故障时，mmmd_mon 检查不到 mmmd_agent 对应 MySQL 服务器的状态，此时 mmmd_mon 会下指令给某个正常的数据库节点的 mmmd_agent。

下面先准备一下 MMM 高可用架构所需要的实战环境，本环境需要 4 台服务器，设置如表 20.4 所示。

表20.4　MMM高可用架构服务器列表

服 务 器	IP 地址	Server ID	主机名称	操作系统
Monitoring host	192.168.0.5	-	mon	Fedora
Master1	192.168.0.2	1	db1	Fedora
Master2	192.168.0.3	2	db2	Fedora
Slave1	192.168.0.4	3	db3	Fedora

虚拟 IP 列表的角色和描述如表 20.5 所示。

表20.5　MMM高可用架构虚拟IP列表

VIP	角 色	描 述
192.168.0.12	Read	应用配置的读取IP，也可以在前段加LVS等，做负载均衡。三台数据库每台一个浮动 VIP
192.168.0.13	Read	
192.168.0.14	Read	
192.168.0.15	Write	应用配置的写入 VIP，单点写入

分别在各机器上安装 MySQL 服务，此时需要创建三个不同的用户，如表 20.6 所示。

表20.6　创建不同用户的权限和功能

用户功能	描 述	权 限
Monitor User	提供 Monitor Server 检查 MySQL 服务的状况	Replication Client
Agent User	MMM agent 用来切换 replication master	Super、replicationClient、Process
Replication	复制使用	Replication Slave

可以通过如下语句来创建三种不同功能的用户：

```
mysql> grant replication client on *.* to 'mmm_monitor'@'192.168.0.%' identified by 'monitor_password';
    Query OK, 0 rows affected (0.07 sec)

mysql> grant super,replication client,process on *.* to 'mmm_agent'@'192.168.0.%' identified by 'agent_password';
    Query OK, 0 rows affected (0.00 sec)

mysql> grant replication slave on *.* to 'replication'@'192.168.0.%' identified by 'replication_password';
    Query OK, 0 rows affected (0.00 sec)

mysql> flush privileges;
```

```
Query OK, 0 rows affected (0.00 sec)
```

值得注意的是，mmm_monitor 用户使用在 Monitor Server（192.168.0.5），另外 mmm_agent 用户和 replication 用户是其他机器中用到的账户。

在进行 MMM 配置之前，需要对 db1 和 db2 两台机器进行主主同步的相关配置，db1 和 db3 实现主从同步。主主同步配置（请参考本章的 20.4.1 节）完成后，接下来需要将 db1 和 db3 服务器设置成主从同步配置，db3 相关配置步骤如下：

步骤 01 首先，在 db3 中进行数据恢复，如下所示：

```
[root@localhost ~]# cd /usr/local/mysql/bin/
[root@localhost bin]# mysqldump -user=root -p test < /test.sql
Enter password:
```

步骤 02 接下来，需要配置 my.cnf 文件，如下所示：

```
server-id= 3
replicate-do-db = test
slave-skip-errors = all
sync_binlog = 1
```

步骤 03 设置主机的地址，实现 db1 和 d3 主从复制，如下所示：

```
mysql> stop slave;
Query OK, 0 rows affected (0.02 sec)

mysql> change master to
    -> master_host='192.168.0.2',
    -> master_user='replication',
    -> master_password='replication_password';
ipvsadm -L -n-> master_log_file='mysql-bin.000022';
-> master_log_pos=107;
Query OK, 0 rows affected (0.04 sec)

mysql> start slave;
Query OK, 0 rows affected (0.00 sec)
```

步骤 04 在 db3 上面测试一下，看看 db1 和 db3 是否实现了主从同步，如下所示：

```
mysql> show slave status\G;
*************************** 1. row ***************************
            Slave_IO_State: Waiting for master to send event
               Master_Host: 192.168.0.2
               Master_User: replication
               Master_Port: 3306
             Connect_Retry: 60
           Master_Log_File: mysql-bin.000023
       Read_Master_Log_Pos: 572
            Relay_Log_File: localhost-relay-bin.000003
             Relay_Log_Pos: 718
     Relay_Master_Log_File: mysql-bin.000023
          Slave_IO_Running: Yes
         Slave_SQL_Running: Yes
           Replicate_Do_DB: test
...
```

```
                Last_SQL_Errno: 0
                Last_SQL_Error:
    Replicate_Ignore_Server_Ids:
               Master_Server_Id: 1
1 row in set (0.00 sec)
```

步骤 05 在 db1 中新建一个表，然后添加一条数据，如下所示：

```
mysql> create table t2(data varchar(20));
Query OK, 0 rows affected (0.33 sec)

mysql> insert into t2 values('db1 insert a data');
Query OK, 1 row affected (0.08 sec)
```

步骤 06 在从服务器上面测试，查看该条记录是否同步成功，如下所示：

```
mysql> use test;
Database changed
mysql> show tables;
+----------------+
| Tables_in_test |
+----------------+
| t              |
| t2             |
| t3             |
+----------------+
3 rows in set (0.00 sec)

mysql> select *from t2;
+-------------------+
| data              |
+-------------------+
| db1 insert a data |
+-------------------+
1 row in set (0.01 sec)
```

至此，db1 和 db2 完成主主同步，db1 和 db3 完成主从同步，MMM 高可用方案的准备工作已经完成，接下来开始对 MMM 进行安装和配置。

20.5.3 MMM 的安装

MMM 软件是基于 Perl 的，所以在安装的过程中需要安装许多 Perl 模块，具体步骤如下所示：

步骤 01 下载并安装 MMM 软件，并在 db1、db2、db3 机器上配置 mysql-mmm-agent，monitor 机器上配置 mysql-mmm-monitor。下载地址：http://mysql-mmm.org/downloads，版本为 mysql-mmm-2.2.1.tar.gz。

步骤 02 下面在各个机器上开始安装，如下所示：

```
[root@localhost ~]# tar zxf mysql-mmm-2.2.1.tar.gz
[root@localhost ~]# cd mysql-mmm-2.2.1
[root@localhost mysql-mmm-2.2.1]# make install
```

步骤 03 MMM 软件安装可以根据官方的说明文档 http://mysql-mmm.org/mmm2:guide，严格按照版本号查找 rpm 包进行安装。下载网址是 http://pkgs.repoforge.org/，下载所需要的 rpm 包，然后进行安装即可。

20.5.4　Monitor 服务器的配置

MMM 的配置文件是在/etc/mysql-mmm 目录下面，monitor 服务器需要配置该目录下面的 mmm_common.conf、mmm_mon.conf 两个文件。mmm_common.conf 文件在 MMM 的各个节点都是一样的，因此配置完成之后可以直接复制到其他各个 DB 节点即可。具体操作步骤如下：

步骤 01 首先开始配置 mmm_conmm.conf 文件，如下所示：

```
[root@localhost ~]# vi /etc/mysql-mmm/mmm_common.conf
active_master_role      writer

<host default>
    cluster_interface       eth2

    pid_path                /var/run/mmm_agentd.pid
    bin_path                /usr/lib/mysql-mmm/

    replication_user        replication
    replication_password    replication_password

    agent_user              mmm_agent
    agent_password          agent_password
</host>

<host db1>
    ip                      192.168.0.2
    mode                    master
    peer                    db2
</host>

<host db2>
    ip                      192.168.0.3
    mode                    master
    peer                    db1
</host>

<host db3>
    ip                      192.168.0.4
    mode                    slave
</host>


<role writer>
    hosts                   db1, db2
    ips                     192.168.0.15
    mode                    exclusive
</role>

<role reader>
    hosts                   db1, db2, db3
```

```
        ips                      192.168.0.12, 192.168.0.13, 192.168.0.14
        mode                     balanced
</role>
```

步骤02 把配置后的 mmm_conmm.conf 文件复制到其他各个节点上。

步骤03 接下来，需要配置 monitor 机器上的 mmm_mon.conf 配置文件，如下所示：

```
[root@localhost ~]# vi /etc/mysql-mmm/mmm_mon.conf
include mmm_common.conf

<monitor>
    ip                       127.0.0.1
    pid_path                 /var/run/mmm_mond.pid
    bin_path                 /usr/lib/mysql-mmm/
    status_path              /var/lib/misc/mmm_mond.status
    ping_ips                 192.168.0.2, 192.168.0.3, 192.168.0.4
</monitor>

<host default>
    monitor_user             mmm_monitor
    monitor_password         monitor_password
</host>

debug 0
```

20.5.5　各个数据库服务器的配置

在 MMM 配置各个数据库服务器之前，需要保持各个服务器上 mmm_comm.conf 文件内容一致。下面对每个服务器上的 mmm_agent.conf 文件进行配置，具体步骤如下：

步骤01 配置 db1 上 agent 服务器的 mmm_agent.conf 配置文件，如下所示：

```
[root@localhost ~]# vi /etc/mysql-mmm/mmm_agent.conf
include mmm_common.conf
this db1
```

步骤02 配置 db2 上 agent 服务器的 mmm_agent.conf 配置文件，如下所示：

```
[root@localhost ~]# vi /etc/mysql-mmm/mmm_agent.conf
include mmm_common.conf
this db2
```

步骤03 配置 db3 上 agent 服务器的 mmm_agent.conf 配置文件，如下所示：

```
[root@localhost ~]# vi /etc/mysql-mmm/mmm_agent.conf
include mmm_common.conf
this db3
```

20.5.6　MMM 的管理

在启动 MMM 的时候，数据库机器需要启动 mmm-agent，命令如下：

```
[root@localhost ~]# /etc/init.d/mysql-mmm-agent start
```

monitor 机器上需要启动 mmm-monitor，命令如下：

```
[root@localhost ~]# /etc/init.d/mysql-mmm-monitor start
```

停止 MMM 只需要执行如下命令即可：

```
[root@localhost ~]# /etc/init.d/mysql-mmm-agent stop
[root@localhost ~]# /etc/init.d/mysql-mmm-monitor stop
```

此时在 monitor 机器上，可以通过如下命令查看服务器启动情况：

```
[root@localhost ~]# mmm_control show
```

如果需要将 db1 设置为 online 状态，此时会配置 db1 的 VIP，需要执行如下命令：

```
[root@localhost ~]# mmm_control set_online db1
```

在实现高可用方案中，系统各个层面都需要被监控起来，例如，监控 monitor 进程的状态、agent 服务器进程的状态。MySQL 可用性的监控，数据库同步状态的监控，推荐读者使用 Nagios 工具。

第 21 章

MySQL 复制

MySQL Replication 是 MySQL 的一个非常重要的功能，主要用于主服务器和从服务器之间的数据复制操作。本章将讲解 MySQL Replication 的基本概念、Windows 环境下的复制操作、Linux 环境下的复制操作、查看 Slave 的复制进度、日常管理和维护、切换主从服务器的方法等。

21.1 MySQL 复制概述

MySQL 从 3.25 版本开始提供数据库复制（replication）功能。MySQL 复制是指从一个 MySQL 主服务器（Master）将数据复制到另一台或多台 MySQL 从服务器（Slave）的过程，将主数据库的 DDL 和 DML 操作通过二进制日志传到从服务器上，然后在从服务器上对这些日志重新执行，从而使得主从服务器的数据保持同步。

在 MySQL 中，复制操作是异步进行的，Slave 服务器不需要持续地保持连接状态来接收 master 服务器的数据。

MySQL 支持一台主服务器同时向多台从服务器进行复制操作，从服务器同时可以作为其他从服务器的主服务器，如果 MySQL 主服务器访问量比较大，可以通过复制数据，然后在从服务器上进行查询操作，从而降低主服务器的访问压力，同时从服务器作为主服务器的备份，可以避免主服务器因为故障而造成数据丢失的问题。

MySQL 数据库复制操作大致可以分成 3 个步骤：

- 步骤 01 主服务器将数据的改变记录到二进制日志（binary log）中。
- 步骤 02 从服务器将主服务器的 binary log events 复制到它的中继日志（relay log）中。
- 步骤 03 从服务器重做中继日志中的事件，将修改数据与主服务器保持同步。

首先，主服务器会记录二进制日志，每个事务更新数据完成之前，主服务器将这些操作的信息记录在二进制日志里面，在事件写入二进制日志完成后，主服务器通知存储引擎提交事务。

Slave 上面的 I/O 进程连接上 Master，并发出日志请求，Master 接收到来自 Slave I/O 进程的请求后，根据请求信息添加位置信息，并返回给 Slave 的 I/O 进程。返回信息中除了日志所包含的信息之外，还包括 Master 端的 bin-log 文件的名称以及 bin-log 的位置信息。

Slave 的 I/O 进程接收到信息后，将接收到的日志内容依次添加到 Slave 端 relay-log 文件的最末端，并将读取到 Master 端的 bin-log 文件名和位置记录到 master-info 文件中。

Slave 的 SQL 进程检测到 relay-log 中新增加的内容后，会马上解析 relay-log 的内容，使之成为在 Master 端真实执行时的那些可执行内容，并在自身执行。

MySQL 复制环境 90%以上都是一个 Master 带一个或者多个 Slave 的架构模式。如果 Master 和 Slave 的压力不是太大的话，异步复制的延时一般都很小。尤其是 Slave 端的复制方式改成两个进程处理之后，更是减小了 Slave 端的延时。

提示：对于数据实时性要求不是特别严格的应用，只需要通过廉价的服务器来扩展 Slave 的数量，将读压力分散到多台 Slave 机器上面，即可解决数据库端的读压力瓶颈。这在很大程度上解决了目前很多中小型网站的数据库压力瓶颈问题，甚至有些大型网站也在使用类似的方案解决数据库瓶颈问题。

21.2　Windows 环境下的 MySQL 主从复制

本节将主要通过实验讲解 MySQL Replication 在 Windows 环境下如何配置主从复制的功能。

21.2.1　复制前的准备工作

在 Windows 环境下，如果想要实现主从复制功能，需要准备操作环境，本示例操作环境如表 21.1 所示。

表21.1　MySQL 主从复制所需的环境

角　色	IP	操作系统	MySQL 版本
Master	192.168.1.208	Windows	MySQL-installer-community-8.0
Slave	192.168.1.206	Windows	MySQL-installer-community-8.0

提示：读者在做实验的过程中，如果没有富余的计算机，可以使用 VirtualBox、VMware 虚拟机来模拟实现，操作系统可以选 Windows。这里 Master 可以由本机担任，Salve 再装一台虚拟机担任；IP 地址可以自行按网络情况做相应的修改。

21.2.2　Windows 环境下实现主从复制

准备好两台安装 MySQL 8.0 的计算机后，即可对主从复制备份进行配置，操作步骤如下：

步骤01 在 Windows 操作系统下安装好两台主机的 MySQL 服务器，一台 Master 主机，一台 Slave 从机，并配置好两台主机的 IP 地址，实现两台计算机可以网络连通。

步骤02 配置 Master 主机的相关信息，在 Master 主机上开启 binlog 日志。首先，确认一下 datadir 的具体路径。

```
mysql> show variables like '%datadir%';
+---------------+-------------------------------------------+
| Variable_name | Value                                     |
+---------------+-------------------------------------------+
| datadir       | C:\ProgramData\MySQL\MySQL Server 8.0\Data\|
+---------------+-------------------------------------------+
1 row in set (0.00 sec)
```

步骤03 此时到 C:\Documents and Settings\All Users\Application Data\MySQL\MySQL Server 8.0 目录下，打开配置文件 my.ini，添加如下代码，开启 binlog 功能：

```
[mysqld]
log_bin="D:/mysqllog/binlog"
expire_logs_days = 10
max_binlog_size = 100M
```

我们在 D 盘下创建 mysqllog 文件夹，binlog 日志记录在该文件夹里面，该配置中其他参数的含义如下所示。

- expire_logs_days：表示二进制日志文件删除的天数。
- max_binlog_size：表示二进制日志文件最大的大小。

步骤04 登录 MySQL 之后，可以执行 show variables like '%log_bin%'命令来测试 log_bin 是否成功开启：

```
mysql> show variables like '%log_bin%';
+---------------------------------+--------------------------+
| Variable_name                   | Value                    |
+---------------------------------+--------------------------+
| log_bin                         | ON                       |
| log_bin_basename                | D:\MySQLlog\binlog       |
| log_bin_index                   | D:\MySQLlog\binlog.index |
| log_bin_trust_function_creators | OFF                      |
| log_bin_use_v1_row_events       | OFF                      |
| sql_log_bin                     | ON                       |
+---------------------------------+--------------------------+
6 rows in set (0.00 sec)
```

如果 log_bin 参数的值为 ON，那么表示二进制日志文件已经成功开启；如果为 OFF，那么表示二进制日志文件开启失败。

步骤05 在 Master 主机上配置复制所需要的账户，这里创建一个名为 repl 的用户，%表示任何远程地址的 repl 用户都可以连接 Master 主机，语句执行如下所示：

```
mysql> grant replication slave on *.* to repl@'%' identified by '123';
Query OK, 0 rows affected (0.06 sec)

mysql> flush privileges;
Query OK, 0 rows affected (0.09 sec)
```

步骤06 在 my.ini 配置文件中配置 Master 主机的相关信息，如下所示：

```
[mysqld]
```

```
log_bin="D:/MySQLlog/binlog"
expire_logs_days = 10
max_binlog_size = 100M

server-id = 1
binlog-do-db = test
binlog-ignore-db = mysql
```

这些配置语句的含义如下:

- server-id: 表示服务器标识 id 号,Master 和 Slave 上的 server-id 不能一样。
- binlog-do-db: 表示需要复制的数据库,这里以 test 数据库为例。
- binlog-ignore-db: 表示不需要复制的数据库。

步骤 07 重启 Master 主机的 MySQL 服务,然后输入 show master status 命令查询 Master 主机的信息。

```
mysql> show master status \G;
*** 1. row ***
            File: binlog.000003
        Position: 120
    Binlog_Do_DB: test
Binlog_Ignore_DB: mysql
Executed_Gtid_Set:
1 row in set (0.00 sec)
```

步骤 08 将 Master 主机的数据备份出来,然后导入到 Slave 从机中去,具体执行语句如下:

```
C:\Program Files\MySQL\MySQL Server 8.0\bin>mysqldump -u root -p -h localhost test >c:\a.txt
Enter password:
```

将 c:/a.txt 复制到 Slave 从机上面去,然后执行以下操作:

```
C:\Program Files\MySQL\MySQL Server 8.0\bin>mysqldump -u root -p root -h localhost test <c:\a.txt
Warning: Using a password on the command line interface can be insecure.
-- MySQL dump 10.13  Distrib 8.0.28, for Win32 (x86)
--
-- Host: localhost    Database: test
-- ------------------------------------------------------
-- Server version       8.0.28-log

/*!40101 SET @OLD_CHARACTER_SET_CLIENT=@@CHARACTER_SET_CLIENT */;
/*!40101 SET @OLD_CHARACTER_SET_RESULTS=@@CHARACTER_SET_RESULTS */;
/*!40101 SET @OLD_COLLATION_CONNECTION=@@COLLATION_CONNECTION */;
/*!40101 SET NAMES utf8 */;
/*!40103 SET @OLD_TIME_ZONE=@@TIME_ZONE */;
/*!40103 SET TIME_ZONE='+00:00' */;
/*!40014 SET @OLD_UNIQUE_CHECKS=@@UNIQUE_CHECKS, UNIQUE_CHECKS=0 */;
/*!40014 SET @OLD_FOREIGN_KEY_CHECKS=@@FOREIGN_KEY_CHECKS, FOREIGN_KEY_CHECKS=0 */;
/*!40101 SET @OLD_SQL_MODE=@@SQL_MODE, SQL_MODE='NO_AUTO_VALUE_ON_ZERO' */;
/*!40111 SET @OLD_SQL_NOTES=@@SQL_NOTES, SQL_NOTES=0 */;
/*!40103 SET TIME_ZONE=@OLD_TIME_ZONE */;

/*!40101 SET SQL_MODE=@OLD_SQL_MODE */;
/*!40014 SET FOREIGN_KEY_CHECKS=@OLD_FOREIGN_KEY_CHECKS */;
```

```
/*!40014 SET UNIQUE_CHECKS=@OLD_UNIQUE_CHECKS */;
/*!40101 SET CHARACTER_SET_CLIENT=@OLD_CHARACTER_SET_CLIENT */;
/*!40101 SET CHARACTER_SET_RESULTS=@OLD_CHARACTER_SET_RESULTS */;
/*!40101 SET COLLATION_CONNECTION=@OLD_COLLATION_CONNECTION */;
/*!40111 SET SQL_NOTES=@OLD_SQL_NOTES */;

-- Dump completed on  2022-04-03 17:25:17
```

步骤09 在 Slave 从机（192.168.1.206）的 C:\Documents and Settings\All Users\Application Data\MySQL\MySQL Server 8.0 目录下，配置 my.ini 配置文件，具体配置信息如下所示：

```
[mysql]
default-character-set=utf8
log_bin="D:/MySQLlog/binlog"
expire_logs_days=10
max_binlog_size = 100M

[mysqld]
server-id = 2
```

提示：配置 Slave 从机上的 my.ini 文件时，需要将 server-id=2 写到 [mysqld] 后面。另外，如果配置文件中还有 log_bin 的配置，可以将它注释掉。例如：

```
# Binary Logging.
# log-bin
# log_bin = "D:/MySQLlog/mysql-bin.log"
```

步骤10 重启 Slave 从机（192.168.1.206），在 Slave 从机的 MySQL 数据库中执行如下命令，关闭 slave 服务：

```
mysql> stop slave;
Query OK, 0 rows affected (0.05 sec)
```

步骤11 在 Slave 从机上设置实现复制功能的相关信息，命令如下：

```
mysql> change master to
    -> master_host='192.168.1.208',
    -> master_user='repl',
    -> master_password='123',
    -> master_log_file='binlog.000003',
    -> master_log_pos=120;
Query OK, 0 rows affected, 2 warnings (0.34 sec)
```

各个参数所代表的具体含义如下：

- master_host：表示实现复制的主机的 IP 地址。
- master_user：表示实现复制的登录远程主机的用户。
- master_password：表示实现复制的登录远程主机的密码。
- master_log_file：表示实现复制的 binlog 日志文件。
- master_log_pos：表示实现复制的 binlog 日志文件的偏移量。

步骤12 继续执行操作，显示 Slave 从机的状况，如下所示：

```
mysql> start slave;
Query OK, 0 rows affected (0.11 sec)
```

```
mysql> show slave status \G;
*** 1. row ***
               Slave_IO_State:
                  Master_Host: 192.168.1.208
                  Master_User: repl
                  Master_Port: 3306
                Connect_Retry: 60
              Master_Log_File: binglog.000003
          Read_Master_Log_Pos: 120
               Relay_Log_File: 2022-20190220JX-relay-bin.000001
                Relay_Log_Pos: 4
        Relay_Master_Log_File: binglog.000003
             Slave_IO_Running: No
            Slave_SQL_Running: Yes
              Replicate_Do_DB:
          Replicate_Ignore_DB:
           Replicate_Do_Table:
       Replicate_Ignore_Table:
      Replicate_Wild_Do_Table:
  Replicate_Wild_Ignore_Table:
                   Last_Errno: 0
                   Last_Error:
                 Skip_Counter: 0
          Exec_Master_Log_Pos: 120
              Relay_Log_Space: 120
              Until_Condition: None
               Until_Log_File:
                Until_Log_Pos: 0
           Master_SSL_Allowed: No
           Master_SSL_CA_File:
           Master_SSL_CA_Path:
              Master_SSL_Cert:
            Master_SSL_Cipher:
               Master_SSL_Key:
        Seconds_Behind_Master: NULL
Master_SSL_Verify_Server_Cert: No
                Last_IO_Errno: 1236
                Last_IO_Error: Got fatal error 1236 from master when reading dat
a from binary log: 'Could not find first log file name in binary log index file'
               Last_SQL_Errno: 0
               Last_SQL_Error:
  Replicate_Ignore_Server_Ids:
             Master_Server_Id: 1
                  Master_UUID: a6bd1fa8-8d6b-11e2-97b4-001f3ca9bc3a
             Master_Info_File:C:\Documents and Settings\All Users\Application D
ata\MySQL\MySQL Server 8.0\data\master.info
                    SQL_Delay: 0
          SQL_Remaining_Delay: NULL
      Slave_SQL_Running_State: Slave has read all relay log; waiting for the sla
ve I/O thread to update it
           Master_Retry_Count: 86400
                  Master_Bind:
      Last_IO_Error_Timestamp: 190403 17:53:15
```

```
              Last_SQL_Error_Timestamp:
                       Master_SSL_Crl:
                   Master_SSL_Crlpath:
                   Retrieved_Gtid_Set:
                    Executed_Gtid_Set:
                        Auto_Position: 0
1 row in set (0.00 sec)
```

在上述执行 show slave status \G 命令中很显然存在一些问题，具体如下：

```
Last_IO_Error: Got fatal error 1236 from master when reading dat
a from binary log: 'Could not find first log file name in binary log index file'
```

下面给出解决该问题的方法。

步骤01 重启 Master（192.168.1.208）主机，执行 show master status \G 命令，记下 File 和 Position 的值，后面 Slave 从机会用到。命令执行如下：

```
mysql> show master status \G;
*** 1. row ***
             File: binlog.000004
         Position: 120
     Binlog_Do_DB: test
 Binlog_Ignore_DB: mysql
Executed_Gtid_Set:
1 row in set (0.00 sec)
```

步骤02 在 Slave（192.168.1.206）从机上重新设置信息，命令执行如下所示：

```
mysql> stop slave;
Query OK, 0 rows affected (0.01 sec)

mysql> change master to
    -> master_log_file='binlog.000004',
    -> master_log_pos = 120;
Query OK, 0 rows affected (0.16 sec)

mysql> start slave;
Query OK, 0 rows affected (0.05 sec)

mysql> show slave status\G;
*** 1. row ***
               Slave_IO_State: Waiting for master to send event
                  Master_Host: 192.168.1.208
                  Master_User: repl
                  Master_Port: 3306
                Connect_Retry: 60
              Master_Log_File: binlog.000004
          Read_Master_Log_Pos: 120
               Relay_Log_File: 2022-20190220JX-relay-bin.000002
                Relay_Log_Pos: 280
        Relay_Master_Log_File: binlog.000004
             Slave_IO_Running: Yes
            Slave_SQL_Running: Yes
              Replicate_Do_DB:
          Replicate_Ignore_DB:
```

```
                Replicate_Do_Table:
            Replicate_Ignore_Table:
           Replicate_Wild_Do_Table:
       Replicate_Wild_Ignore_Table:
                          Last_Errno: 0
                          Last_Error:
                        Skip_Counter: 0
                 Exec_Master_Log_Pos: 120
                     Relay_Log_Space: 463
                     Until_Condition: None
                      Until_Log_File:
                       Until_Log_Pos: 0
                  Master_SSL_Allowed: No
                  Master_SSL_CA_File:
                  Master_SSL_CA_Path:
                     Master_SSL_Cert:
                   Master_SSL_Cipher:
                      Master_SSL_Key:
               Seconds_Behind_Master: 0
       Master_SSL_Verify_Server_Cert: No
                       Last_IO_Errno: 0
                       Last_IO_Error:
                      Last_SQL_Errno: 0
                      Last_SQL_Error:
         Replicate_Ignore_Server_Ids:
                    Master_Server_Id: 1
                         Master_UUID: a6bd1fa8-8d6b-11e2-97b4-001f3ca9bc3a
                    Master_Info_File:C:\Documents and Settings\All Users\Application D
ata\MySQL\MySQL Server 8.0\data\master.info
                           SQL_Delay: 0
                 SQL_Remaining_Delay: NULL
             Slave_SQL_Running_State: Slave has read all relay log; waiting for the sla
ve I/O thread to update it
                  Master_Retry_Count: 86400
                         Master_Bind:
             Last_IO_Error_Timestamp:
            Last_SQL_Error_Timestamp:
                      Master_SSL_Crl:
                  Master_SSL_Crlpath:
                  Retrieved_Gtid_Set:
                   Executed_Gtid_Set:
                       Auto_Position: 0
1 row in set (0.00 sec)
```

由此可见，问题完全解决，接下来可以进行 Window 环境下主从复制的测试。

21.2.3　Windows 环境下主从复制测试

在 Windows 环境中测试主从复制操作，具体操作步骤如下：

步骤 01 在 Master 主机的 MySQL 环境下，执行如下命令：

```
mysql> use test;
Database changed
mysql> create table rep_test(
```

```
        -> data integer
        -> );
Query OK, 0 rows affected (0.02 sec)

mysql> insert into rep_test values(2);
Query OK, 1 row affected (0.06 sec)
```

步骤02 在 Slave 从机的 MySQL 环境下，查看主机刚才添加的表和数据是否成功同步过来，命令执行如下所示：

```
mysql> use test;
Database changed
mysql> show tables;
+----------------+
| Tables_in_test |
+----------------+
| t2             |
+----------------+
1 row in set (0.00 sec)

mysql> use test;
Database changed
mysql> show tables;
+----------------+
| Tables_in_test |
+----------------+
| rep_test       |
| t2             |
+----------------+
2 rows in set (0.00 sec)

mysql> select *from rep_test;
+------+
| data |
+------+
|    2 |
+------+
1 row in set (0.02 sec)
```

测试表明数据已经成功同步到 Slave 从机上了，本节试验只是一主一从的同步，在实际生产环境中 MySQL 架构可能会用到一主多从的架构，其配置方法类似，这里不再赘述。

21.3　Linux 环境下的 MySQL 复制

在现实的生产环境中单机实现的主从复制比较少，通常会使用一主多从的体系架构。为了读者朋友更好地实现本机主从复制，需要在 Linux 环境下通过 mysqld_multi 实现单机的主从复制。注意，本节使用的是 Fedora 操作系统。

21.3.1 下载并安装 MySQL 8.0

很多熟悉 MySQL 的用户都喜欢使用源码包来进行安装，因为在安装源码的过程中可以非常方便地进行性能优化。下面就源码安装过程中涉及的优化项进行简单介绍。

步骤01 下载 mysql-8.0.28.tar.gz 源文件。可以在下载页面 http://dev.mysql.com/downloads/mysql/ 中选择【Source Code】平台，然后选择下载 mysql-8.0.28.tar.gz 源码，如图 21.1 所示。

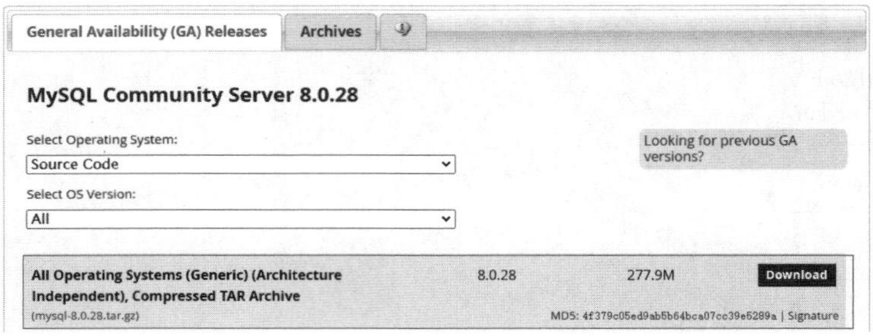

图 21.1　MySQL 源码下载

步骤02 下载完 mysql-8.0.28.tar.gz 后，创建 mysql 用户和组：

```
[root@localhost ~]# groupadd mysql
[root@localhost ~]# useradd -r -g mysql mysql
```

步骤03 解压缩 MySQL 源代码，这里使用 CMake 2.8.4 来编译 MySQL 源代码：

```
[root@localhost ~]# tar -zxvf mysql-8.0.28.tar.gz
[root@localhost ~]# cd mysql-8.0.28
[root@localhost mysql-8.0.28]# cmake .
[root@localhost mysql-8.0.28]# make && make install
```

步骤04 创建 MySQL 安装程序的目录和数据文件的目录：

```
[root@localhost ~]# mkdir -p /usr/local/mysql
[root@localhost ~]# mkdir -p /usr/local/mysql/data
[root@localhost ~]# chown -R mysql.mysql /usr/local/mysql
```

步骤05 安装 MySQL 8.0 的源码：

```
[root@localhost ~]# cd /usr/local/mysql/scripts
[root@localhost scripts]# ./mysql_install_db --user=mysql --basedir=/usr/local/mysql --datadir=/usr/local/mysql/data
[root@localhost ~]# cd /usr/local/mysql/support-files
[root@localhost support-files]# cp mysql.server /etc/rc.d/init.d/mysql
[root@localhost support-files]# cp my-default.cnf /etc/my.cnf
[root@localhost ~]# chkconfig --add mysql
[root@localhost ~]# chkconfig mysql on
```

步骤06 启动 MySQL 8.0 服务：

```
[root@localhost ~]# service mysql start
Starting MySQL...                                          [OK]
```

步骤07 在 Fedora 操作系统中登录 MySQL 8.0，默认用户 root，密码为空，命令如下所示：

```
[root@localhost ~]# mysql
Welcome to the mysql monitor.  Commands end with ; or \g.
Your MySQL connection id is 1
Server version: 8.0.28 Source distribution

Copyright (c) 2000, 2022, Oracle and/or its affiliates. All rights reserved.

Oracle is a registered trademark of Oracle Corporation and/or its
affiliates. Other names may be trademarks of their respective
owners.

Type 'help;' or '\h' for help. Type '\c' to clear the current input statement.

mysql>
```

21.3.2 单机主从复制前的准备工作

MySQL 服务器可以采用主从机制进行备份。一对一进行备份对于生成环境而言比较浪费资源，主服务器把数据变化记录到主日志，然后从服务器通过 I/O 线程读取主服务器的日志，并将它写入到从服务器的中继日志中，接着 SQL 线程读取中继日志，并且在从服务器上重放，从而实现 MySQL 复制，具体如图 21.2 所示。

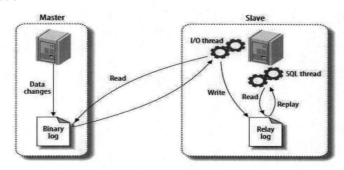

图 21.2　MySQL 复制

MySQL 具有可以运行多个实例的功能，这个功能是通过 mysqld_multi 来实现的。当一台机器上需要运行多个 MySQL 服务器时，mysqld_multi 是管理多个 mysqld 的服务进程，用不同的 Unix Socket 或是监听不同的端口，通过命令来启动、关闭和报告所管理的服务器的状态。

下面介绍如何在一台服务器上使用 Mysqld_multi 管理多个 MySQL 服务进程，具体操作步骤如下：

步骤01 初始化多实例数据库时，首先要停止 MySQL 服务器：

```
[root@localhost ~]# service mysql stop
Shutting down MySQL.                                           [  OK  ]
```

提示：此时可以采用 netstat 命令查看 3306 端口关闭了没有，如果没有查询出结果，就说明 MySQL 服务器已经成功关闭。

步骤02 把常用到的工具添加到/usr/bin 目录：

```
[root@localhost~]#ln -s /usr/local/mysql/bin/mysqld_multi /usr/bin/mysqld_multi
```

```
[root@localhost~]#ln -s /usr/local/mysql/scripts/mysql_install_db
/usr/bin/mysql_install_db
```

步骤 03 初始化 3 个数据目录并安装 3 个 MySQL 服务：

```
[root@localhost ~]# cd /usr/local/mysql/
[root@localhost mysql]# mkdir -p /usr/local/var/mysql1
[root@localhost mysql]# mkdir -p /usr/local/var/mysql2
[root@localhost mysql]# mkdir -p /usr/local/var/mysql3
[root@localhost mysql]# ./scripts/mysql_install_db --datadir=/usr/local/
var/mysql1 --user=mysql
[root@localhost mysql]# ./scripts/mysql_install_db --datadir=/usr/local/var/
mysql2 --user=mysql
[root@localhost mysql]# ./scripts/mysql_install_db --datadir=/usr/local/var/
mysql3 --user=mysql
```

步骤 04 从 MySQL 的源码中把 mysqld_multi.server 复制到/etc/init.d/目录下：

```
[root@localhost ~]# cd /usr/local/mysql/support-files/
[root@localhost support-files]# cp ./mysqld_multi.server /etc/init.d/mysql_multi.server
```

步骤 05 配置数据库文件。在配置文件/etc/my.cnf 中修改相应的属性：

```
# The MySQL server
[mysqld_multi]
mysqld          = /usr/local/mysql/bin/mysqld_safe
mysqladmin      = /usr/local/mysql/bin/mysqladmin
user            = root

[mysqld1]
port        = 3306

[mysqld2]
port        = 3307
socket      = /temp/mysql2.sock
datadir     = /usr/local/var/mysql2

[mysqld3]
port        = 3308
socket      = /temp/mysql3.sock
datadir     = /usr/local/var/mysql3

[mysqld]
```

步骤 06 查看数据库的状态：

```
[root@localhost ~]# mysqld_multi --defaults-extra-file=/etc/my.cnf report
Reporting MySQL servers
MySQL server from group: mysqld1 is not running
MySQL server from group: mysqld2 is not running
MySQL server from group: mysqld3 is not running
```

此时，发现 MySQL 服务器不能打开所需的文件，程序发生错误。

步骤 07 使用 mysqld_multi 启动 MySQL 服务器：

```
[root@localhost mysql1]# mysqld_multi --defaults-extra-file=/etc/my.cnf stop
```

```
[root@localhost mysql1]# mysqld_multi --defaults-extra-file=/etc/my.cnf start
[root@localhost mysql1]# mysqld_multi --defaults-extra-file=/etc/my.cnf report
Reporting MySQL servers
MySQL server from group: mysqld1 is running
MySQL server from group: mysqld2 is running
MySQL server from group: mysqld3 is running
```

步骤08 测试 MySQL 服务器的状态：

```
[root@localhost ~]# netstat -an|grep 330
```

此时发现端口同时开启 3306、3307、3308 端口，在进程里面可以发现同时开启了两个 mysql_safe 进程。

步骤09 登录查看 MySQL 数据库：

```
[root@localhost data]# MySQL -u root -p -P 3306
[root@localhost data]# MySQL -u root -p -P 3307
[root@localhost data]# MySQL -u root -p -P 3308
```

此时可以顺利登录到数据库，通过 ps 命令可以发现后台产生了 3 个 mysqld 进程的实例。

步骤10 直接登录 MySQL 服务器，执行 show variable 命令，发现 3 个 MySQL 服务器的 pid_file、socket 参数都一样，命令执行如下：

```
mysql> show variables like 'socket';
+---------------+-----------------+
| Variable_name | Value           |
+---------------+-----------------+
| socket        | /tmp/mysql.sock |
+---------------+-----------------+
1 row in set (0.00 sec)

mysql> show variables like 'pid%';
+---------------+-------------------------------+
| Variable_name | Value                         |
+---------------+-------------------------------+
| pid_file      | /usr/local/var/mysql1/mysql1.pid |
+---------------+-------------------------------+
1 row in set (0.00 sec)
```

此时，可以通过登录 MySQL 服务器自带参数解决以上的问题，命令执行如下所示：

```
[root@localhost ~]# mysql -u root -S /tmp/mysql2.sock
Welcome to the mysql monitor.  Commands end with ; or \g.
Your mysql connection id is 1
Server version: 8.0.28 mysql Community Server (GPL)

Type 'help;' or '\h' for help. Type '\c' to clear the current input statement.

mysql> show databases;
+--------------------+
| Database           |
+--------------------+
| information_schema |
| mysql              |
```

```
| test                |
+---------------------+
3 rows in set (0.04 sec)

mysql> show variables like 'pid%';
+---------------+--------------------------------+
| Variable_name | Value                          |
+---------------+--------------------------------+
| pid_file      | /usr/local/var/mysql2/mysql2.pid |
+---------------+--------------------------------+
1 row in set (0.00 sec)
```

由测试结果可知，问题已经解决了。接下来启动3个数据库，可以直接使用了：

```
[root@localhost ~]# mysqld_multi --defaults-extra-file=/etc/my.cnf stop 1-3
[root@localhost ~]# mysqld_multi --defaults-extra-file=/etc/my.cnf star 1-3
```

21.3.3 mysqld_multi 实现单机主从复制

MySQL 的复制至少需要两个 MySQL 服务，这些 MySQL 服务可以分布在不同的服务器上，也可以在一台服务器上启动多个服务。

MySQL 的复制（Replication）是一个异步的复制过程，从 Master 复制到 Slave。在 Master 和 Slave 中的整个复制过程由3个线程完成，其中两个线程（SQL 线程和 I/O 线程）在 Slave 端，另一个线程（I/O 线程）在 Master 端。要实现复制过程，Master 必须打开 Binary Log 功能，复制过程其实就是 Slave 端从 Master 端获取 bin 日志，然后在自己服务器上完全顺序执行日志中所记录的各种操作。

在 Fedora 操作系统中使用 mysqld_multi 单机实现主从复制的具体配置，如表 21.2 所示。

表21.2 mysqld_multi单机实现主从复制的具体配置环境

角色	IP	操作系统	MySQL 版本	端口
Master	192.168.1.208	Fedora	MySQL-8.0.28.tar.gz	3306
Slave	192.168.1.208	Fedora	MySQL-8.0.28.tar.gz	3307
Slave	192.168.1.208	Fedora	MySQL-8.0.28.tar.gz	3308

注意，这里只用一台 Fedora 虚拟机，下面采用 mysqld_multi 实现单机 MySQL 服务器主从复制。

步骤01 使用 mysqld_multi 开启上一节已经设定好的3个 MySQL 服务：

```
[root@localhost ~]# mysqld_multi --defaults-extra-file=/etc/my.cnf start 1-3
[root@localhost ~]# netstat -an|grep 330
tcp    0    0 0.0.0.0:3306           0.0.0.0:*               LISTEN
tcp    0    0 0.0.0.0:3307           0.0.0.0:*               LISTEN
tcp    0    0 0.0.0.0:3308           0.0.0.0:*               LISTEN
```

步骤02 登录 Master 主服务器，设置一个复制使用的账户，并授予 REPLICATION SLAVE 权限。这里创建一个复制用户 rep1。

```
[root@localhost ~]# mysql -u root -p -P 3306
Enter password:

mysql> grant replication slave on *.* to 'rep1'@'localhost' identified by '123';
Query OK, 0 rows affected (0.04 sec)
```

```
mysql> grant replication slave on *.* to 'repl'@'%' identified by '123';
Query OK, 0 rows affected (0.04 sec)
```

步骤03 修改 Master 主数据库服务器的配置文件 my.cnf，开启 BINLOG，并设置 server-id 的值。需要重启服务器之后才生效。

```
[root@localhost ~]#vi /etc/my.cnf
[mysqld1]
port       = 3306
log-bin = /usr/local/var/mysql1/mysql-bin
server-id = 1

[root@localhost ~]# mysqld_multi --defaults-extra-file=/etc/my.cnf stop 1-3
[root@localhost ~]# mysqld_multi --defaults-extra-file=/etc/my.cnf start 1-3
```

步骤04 在 Master 主服务器上，设置锁定有效。这个操作是为了确保没有数据库操作，以便获得一致性的快照。

```
[root@localhost ~]# mysql -u root -P 3306 -S /tmp/mysql.sock
Welcome to the MySQL monitor.  Commands end with ; or \g.
Your MySQL connection id is 2
Server version: 8.0.28-log Source distribution

Copyright (c) 2000, 2022, Oracle and/or its affiliates. All rights reserved.

Oracle is a registered trademark of Oracle Corporation and/or its
affiliates. Other names may be trademarks of their respective
owners.

Type 'help;' or '\h' for help. Type '\c' to clear the current input statement.

mysql> flush tables with read lock;
Query OK, 0 rows affected (0.01 sec)
```

步骤05 用 show master status 命令查看日志情况，查询得到主服务器上当前的二进制日志名和偏移量值。这个操作的目的是为了在从数据库启动以后，从这个时间点开始进行数据的恢复。

```
mysql> show master status \G;
*** 1. row ***
            File: mysql-bin.000001
        Position: 120
    Binlog_Do_DB:
Binlog_Ignore_DB:
Executed_Gtid_Set:
1 row in set (0.00 sec)
```

步骤06 主数据库服务此时可以做一个备份，在服务器停止的情况下直接使用系统复制命令。

```
[root@localhost mysql1]#tar -cvf data.tar data
```

步骤07 主数据库备份完成后，主数据库恢复写操作。

```
mysql> unlock tables;
Query OK, 0 rows affected (0.00 sec)
```

提示：Master 主服务器的配置已经成功，如果 my.cnf 的 mysqld 选项设置 server-id 参数，而从服务器没有设置 server-id，那么启动从服务器会发生如下错误：

```
mysql> start slave;
ERROR 1200 (HY000): The server is not configured as slave; fix in config file or with CHANGE MASTER TO
```

步骤08 接下来继续编辑/etc/my.cnf 文件，具体配置项如下：

```
# The MySQL server
[mysqld_multi]
mysqld          = /usr/local/mysql/bin/mysqld_safe
mysqladmin      = /usr/local/mysql/bin/mysqladmin
user            = root

[mysqld1]
port      = 3306
log-bin   = /usr/local/var/mysql1/mysql-bin
server-id = 1

[mysqld2]
port      = 3307
socket    = /temp/mysql2.sock
datadir   = /usr/local/var/mysql2
log-bin   = /usr/local/var/mysql2/mysql-bin
server-id = 2

[mysqld3]
port      = 3308
socket    = /temp/mysql3.sock
datadir   = /usr/local/var/mysql3
log-bin   = /usr/local/var/mysql3/mysql-bin
server-id = 3

[mysqld]
```

步骤09 重启 Master 主服务器。

```
[root@localhost ~]# mysqld_multi --defaults-extra-file=/etc/my.cnf stop 1-3
[root@localhost ~]# mysqld_multi --defaults-extra-file=/etc/my.cnf start 1-3
[root@localhost ~]# mysqld_multi --defaults-extra-file=/etc/my.cnf report
Reporting MySQL servers
MySQL server from group: mysqld1 is running
MySQL server from group: mysqld2 is running
MySQL server from group: mysqld3 is running
[root@localhost ~]#
```

步骤10 对从数据库服务器做相应的设置，此时需要指定复制使用的用户、主数据服务器的 IP 地址、端口，以及开始复制的日志文件和位置等，具体设置如下：

```
[root@localhost ~]# mysql -uroot -p -P 3307 -S /temp/mysql2.sock
Enter password:
Welcome to the MySQL monitor.  Commands end with ; or \g.
Your MySQL connection id is 1
Server version: 8.0.28-log Source distribution
```

```
Copyright (c) 2000, 2022, Oracle and/or its affiliates. All rights reserved.

Oracle is a registered trademark of Oracle Corporation and/or its
affiliates. Other names may be trademarks of their respective
owners.

Type 'help;' or '\h' for help. Type '\c' to clear the current input statement.

mysql> show variables like '%log_bin%';
+---------------------------------+-------------------------------------+
| Variable_name                   | Value                               |
+---------------------------------+-------------------------------------+
| log_bin                         | ON                                  |
| log_bin_basename                | /usr/local/var/mysql2/mysql-bin     |
| log_bin_index                   | /usr/local/var/mysql2/mysql-bin.index|
| log_bin_trust_function_creators | OFF                                 |
| log_bin_use_v1_row_events       | OFF                                 |
| sql_log_bin                     | ON                                  |
+---------------------------------+-------------------------------------+
6 rows in set (0.01 sec)

mysql> stop slave;
Query OK, 0 rows affected, 1 warning (0.00 sec)

mysql> change master to
    -> master_host='127.0.0.1',
    -> master_user='repl',
    -> master_password='123',
    -> master_log_file='mysql-bin.000001',
    -> master_log_pos=120;
Query OK, 0 rows affected, 2 warnings (0.10 sec)
```

步骤⑪ 在从服务器上执行 show slave status\G 命令，查询从服务器的状态。

```
mysql> start slave;
Query OK, 0 rows affected (0.12 sec)
mysql> show slave status \G;
*** 1. row ***
              Slave_IO_State: Waiting for master to send event
                 Master_Host: 127.0.0.1
                 Master_User: repl
                 Master_Port: 3306
               Connect_Retry: 60
             Master_Log_File: mysql-bin.000002
         Read_Master_Log_Pos: 120
              Relay_Log_File: localhost-relay-bin.000003
               Relay_Log_Pos: 283
       Relay_Master_Log_File: mysql-bin.000002
            Slave_IO_Running: Yes
           Slave_SQL_Running: Yes
             Replicate_Do_DB:
         Replicate_Ignore_DB:
          Replicate_Do_Table:
      Replicate_Ignore_Table:
     Replicate_Wild_Do_Table:
```

```
         Replicate_Wild_Ignore_Table:
                      Last_Errno: 0
                      Last_Error:
                    Skip_Counter: 0
             Exec_Master_Log_Pos: 120
                 Relay_Log_Space: 623
                 Until_Condition: None
                  Until_Log_File:
                   Until_Log_Pos: 0
              Master_SSL_Allowed: No
              Master_SSL_CA_File:
              Master_SSL_CA_Path:
                 Master_SSL_Cert:
               Master_SSL_Cipher:
                  Master_SSL_Key:
           Seconds_Behind_Master: 0
 Master_SSL_Verify_Server_Cert: No
                   Last_IO_Errno: 0
                   Last_IO_Error:
                  Last_SQL_Errno: 0
                  Last_SQL_Error:
       Replicate_Ignore_Server_Ids:
                Master_Server_Id: 1
                     Master_UUID: 5c2dfa9a-8327-11e2-94c2-000c296d88f8
                Master_Info_File: /usr/local/var/mysql2/master.info
                       SQL_Delay: 0
             SQL_Remaining_Delay: NULL
         Slave_SQL_Running_State: Slave has read all relay log; waiting for the slave I/O thread
 to update it
              Master_Retry_Count: 86400
                     Master_Bind:
         Last_IO_Error_Timestamp:
        Last_SQL_Error_Timestamp:
                  Master_SSL_Crl:
              Master_SSL_Crlpath:
              Retrieved_Gtid_Set:
               Executed_Gtid_Set:
                   Auto_Position: 0
1 row in set (0.00 sec)
```

步骤12 此时发现从服务器已经成功设置，同时也可以执行 show processlist \G 命令，查询从服务器的进程状态。

```
mysql> show processlist \G;
*** 1. row ***
     Id: 7
   User: root
   Host: localhost
     db: NULL
Command: Query
   Time: 0
  State: init
   Info: show processlist
*** 2. row ***
     Id: 10
```

```
         User: repl
         Host: localhost:45056
           db: NULL
      Command: Binlog Dump
         Time: 202
        State: Master has sent all binlog to slave; waiting for binlog to be updated
         Info: NULL
2 rows in set (0.00 sec)
```

结果表明 Slave 端已经连接上 Master 端，开始接受并执行日志。

步骤 13 此时可以测试复制服务的正确性。在 Master 主数据库上执行一个更新操作，观察是否在从服务器上做了同步。下面在主数据库的 test 库上创建一个测试表，然后插入数据：

```
[root@localhost ~]# mysql -uroot -p -P 3306 -S /tmp/mysql.sock
Enter password:
Welcome to the MySQL monitor.  Commands end with ; or \g.
Your MySQL connection id is 11
Server version: 8.0.28-log Source distribution

Copyright (c) 2000, 2022, Oracle and/or its affiliates. All rights reserved.

Oracle is a registered trademark of Oracle Corporation and/or its
affiliates. Other names may be trademarks of their respective
owners.

Type 'help;' or '\h' for help. Type '\c' to clear the current input statement.

mysql> use test;
Database changed
mysql> show tables;
Empty set (0.00 sec)

mysql> create table repl_test(id int);
Query OK, 0 rows affected (0.18 sec)

mysql> insert into repl_test values(1),(2);
Query OK, 2 rows affected (0.06 sec)
Records: 2  Duplicates: 0  Warnings: 0
```

步骤 14 在从服务器上检测新的表是否被创建，以及数据是否同步。

```
[root@localhost ~]# mysql -uroot -P 3307 -S /temp/mysql2.sock
Welcome to the MySQL monitor.  Commands end with ; or \g.
Your MySQL connection id is 6
Server version: 8.0.28-log Source distribution

Copyright (c) 2000, 2022, Oracle and/or its affiliates. All rights reserved.

Oracle is a registered trademark of Oracle Corporation and/or its
affiliates. Other names may be trademarks of their respective
owners.

Type 'help;' or '\h' for help. Type '\c' to clear the current input statement.

mysql> use test;
```

```
Database changed
mysql> show tables;
+----------------+
| Tables_in_test |
+----------------+
| repl_test      |
+----------------+
1 row in set (0.00 sec)

mysql> select * from repl_test;
+------+
| id   |
+------+
|    1 |
|    2 |
+------+
2 rows in set (0.01 sec)
```

从结果可以看出，端口为 3306 的 Master 主机上的数据已经可以正确地同步到端口为 3307 的 Slave 从机的数据库上，复制服务配置成功完成。另外一个端口为 3308 的从机的配置跟端口为 3307 从机的一样操作，这里不再重复叙述。

21.3.4 不同服务器之间实现主从复制

在大多数情况下，采用不同的 MySQL 主从复制比较常见。不同 IP 地址的服务器上的 MySQL 数据库实现一对一复制跟上一节比较相似，具体的配置步骤如下：

步骤01 准备两台 Fedora 虚拟机，确保主从服务器上安装了相同版本的数据库，设定主服务器的 IP 是 192.168.1.100，从服务器的 IP 是 192.168.1.101。

步骤02 登录主服务器，设置一个复制使用的账户，并授予 REPLICATION SLAVE 权限。这里创建一个复制用户 rep1。

```
mysql> grant replication slave on *.* to 'rep1'@'192.168.1.101' identified by '123';
Query OK, 0 rows affected (0.00 sec)
```

步骤03 修改主数据库服务器的配置文件 my.cnf，开启 BINLOG，并设置 server-id 的值。需要重启服务器之后才能生效。

在 my.cnf 中修改配置项如下：

```
[mysqld]
log-bin = /usr/local/var/mysql1/mysql-bin
server-id = 1
```

步骤04 在主服务器上，设置锁定有效。这个操作是为了确保没有数据库操作，以便获得一致性的快照。

```
mysql> flush tables with read lock;
Query OK, 0 rows affected (0.00 sec)
```

步骤05 查询主服务器上当前的二进制日志名和偏移值。这个操作的目的是为了在从数据库启动以后，从这个时间点进行数据库的恢复。

```
mysql> show master status;
+------------------+----------+--------------+------------------+
| File             | Position | Binlog_Do_DB | Binlog_Ignore_DB |
+------------------+----------+--------------+------------------+
| mysql-bin.000029 |      109 |              |                  |
+------------------+----------+--------------+------------------+
1 row in set (0.00 sec)
```

步骤 06 主数据库停止更新操作，生成数据库的备份。可以通过 mysqldump 导出数据或者使用 ibbackup 工具进行数据库的备份。如果主数据库停止，那么可以直接使用 cp 命令将数据文件复制到从数据库服务器上。

主数据库备份完成后，主数据库恢复写操作，命令执行如下：

```
mysql> unlock tables;
Query OK, 0 rows affected (0.00 sec)
```

步骤 07 修改从数据库的配置文件 my.cnf，增加 server-id 参数。server-id 的值是唯一的，不能和主数据库的配置相同，如果有多个从数据库，那么每个从数据库都必须有自己唯一的 server-id 值。

```
my.cnf
[mysqld]
Server-id = 2
```

步骤 08 在从服务器上，使用 --skip-slave-start 选项启动从数据库，这样不会立即启动从数据服务器上的复制进程，方便我们对从数据库的服务进行进一步的配置。

```
[root@localhost ~]# mysqld_safe --skip-slave-start &
```

步骤 09 对从数据库服务器做相应的设置，指定复制使用的用户、主数据库服务器的 IP 和端口，以及开始执行复制的日志文件名和位置。

```
mysql> stop slave;
Query OK, 0 rows affected, 1 warning (0.00 sec)

mysql> change master to
    -> master_host='192.168.1.100',
    -> master_user='repl',
    -> master_password='123',
    -> master_log_file='mysql-bin.000029',
    -> master_log_pos=109;
Query OK, 0 rows affected (0.00 sec)
```

步骤 10 在从服务器上启动 slave 线程。

```
mysql> start slave;
Query OK, 0 rows affected (0.00 sec)
```

步骤 11 在从服务器上执行 show slave status\G 命令，查询从服务器的状态。

```
mysql> show slave status\G;
```

此时也可以执行 show processlist \G 命令，查询从服务器的进程状态。

```
mysql> show processlist \G;
```

接下来可以测试复制服务的正确性，在主数据库上执行一个更新操作，观察是否在从数据库上做了同步。具体方法与上一节相似，这里不再重复讲解。

21.3.5 MySQL 主从复制启动选项

MySQL 安装配置的时候，已经介绍了几个启动时的常用参数，其中包括 MASTER_HOST、MASTER_PORT、MASTER_USER、MASTER_PASSWORD、MASTER_LOG_FILE 和 MASTER_LOG_POS。这几个参数需要在从服务器上配置。下面介绍几个常用的启动选项，如 log-slave-updates、master-connect-retry、read-only 和 slave-skip-errors 等。

（1）log-slave-updates

log-slave-updates 参数主要用来配置从服务器的更新是否写入二进制日志，该选项默认是不打开的。如果这个从服务器同时也作为其他服务器的主服务器，搭建一个链式的复制，那么就需要开启这个选项，这样从服务器才能获取它的二进制日志进行同步操作。

（2）master-connect-retry

master-connect-retry 参数是用来设置在和主服务器连接丢失的时候重试的时间间隔，默认是 60 秒。

（3）read-only

read-only 用来限制普通用户对从数据库的更新操作，以确保从数据库的安全性，不过如果是超级用户，依然可以对从数据库进行更新操作。如果主数据库创建了一个普通用户，在默认情况下，该用户可以更新从数据库中的数据；使用 read-only 选项启动从数据库以后，该用户对从数据库的更新会提示错误。

使用 read-only 选项启动的语法如下：

```
[root@localhost ~]#mysqld_safe - read-only&
```

（4）slave-skip-errors

在复制的过程中，从服务器可能会执行 BINLOG 中错误的 SQL 语句，此时如果不忽略错误，从服务器将会停止复制进程，等待用户处理错误。这种错误如果不能及时发现，将会对应用或者备份产生影响。slave-skip-errors 的作用就是用来定义复制过程中从服务器可以自动跳过的错误号。设置该参数后，MySQL 会自动跳过所配置的一系列错误，直接执行后面的 SQL 语句。该参数可以定义多个错误号，如果设置成 all，就表示跳过所有的错误，具体语法如下：

```
vi /etc/my.cnf
slave-skip-errors=1007,1051,1062
```

如果从数据库主要是作为主数据库的备份，那么不应该使用这个启动参数，设置不当的话很可能会造成主从数据库的数据不同步。如果从数据库仅仅是为了分担主数据库的查询压力，并且对数据的完整性要求不是很严格，那么这个选项可以减轻数据库管理员维护从数据库的工作量。

21.3.6 指定复制的数据库或者表

MySQL 数据库可以指定需要复制到从数据库上的数据库或者表。有时候，用户只需要将主数据库中的某些关键表复制到从服务器上，或者将某些提供查询的表复制到主数据库上即可，可以使用配置参数 replicate-do-db、replicate-do-table、replicate-ignore-db、replicate-ignore-table 或 replicate-while-do-table 来指定需要复制的数据或者表。

1. replicate-do-table 和 replicate-ignore-table 的用法

步骤 01 启动主从数据库，首先在主数据库 test 库中创建两个表 rep_t1 和 rep_t2。

```
[root@localhost ~]# mysqld_multi --defaults-extra-file=/etc/my.cnf start 1-3
[root@localhost ~]# mysql -uroot -p -P 3306 -S /tmp/mysql.sock
Enter password:
Welcome to the MySQL monitor.  Commands end with ; or \g.
Your MySQL connection id is 11
Server version: 8.0.28-log Source distribution

Copyright (c) 2000, 2022, Oracle and/or its affiliates. All rights reserved.

Oracle is a registered trademark of Oracle Corporation and/or its
affiliates. Other names may be trademarks of their respective
owners.

Type 'help;' or '\h' for help. Type '\c' to clear the current input statement.

mysql> show databases;
+--------------------+
| Database           |
+--------------------+
| information_schema |
| cc                 |
| mysql              |
| test               |
| tt                 |
+--------------------+
5 rows in set (0.11 sec)

mysql> show tables;
+----------------+
| Tables_in_test |
+----------------+
| rep_t1         |
| rep_t2         |
+----------------+
2 rows in set (0.00 sec)

mysql> select * from rep_t1;
+------+
| data |
+------+
|    1 |
|    2 |
```

```
|   3 |
+-----+
3 rows in set (0.05 sec)

mysql> select * from rep_t2;
Empty set (0.04 sec)
```

步骤 02 关闭数据库服务器，编辑从数据库配置参数，replicate-do-table=test.rep_t1 指定 test 数据库中的 rep_t1 表被复制，replicate-ignore-table=test.rep_t2 指定 test 库中的 rep_t2 表不会被复制。

```
[root@localhost ~]# mysqld_multi --defaults-extra-file=/etc/my.cnf stop 1-3

Vi /etc/my.cnf
[mysqld2]
...
port       = 3307
socket     = /temp/mysql2.sock
server-id  = 2
replicate-do-table=test.rep_t1
replicate-ignore-table=test.rep_t2
```

步骤 03 启动从服务器线程。

```
mysql> start slave;
Query OK, 0 rows affected, 1 warning (0.00 sec)

mysql> show slave status\G;
*** 1. row ***
               Slave_IO_State: Waiting for master to send event
                  Master_Host: 127.0.0.1
                  Master_User: rep1
                  Master_Port: 3306
                Connect_Retry: 60
              Master_Log_File: mysql-bin.000013
          Read_Master_Log_Pos: 98
               Relay_Log_File: mysql2-relay-bin.000035
                Relay_Log_Pos: 235
        Relay_Master_Log_File: mysql-bin.000013
             Slave_IO_Running: Yes
            Slave_SQL_Running: Yes
              Replicate_Do_DB:
          Replicate_Ignore_DB:
           Replicate_Do_Table: test.rep_t1,test.rep_t1
       Replicate_Ignore_Table: test.rep_t2,test.rep_t2
      Replicate_Wild_Do_Table:
  Replicate_Wild_Ignore_Table:
                   Last_Errno: 0
                   Last_Error:
                 Skip_Counter: 0
          Exec_Master_Log_Pos: 98
              Relay_Log_Space: 235
              Until_Condition: None
               Until_Log_File:
                Until_Log_Pos: 0
           Master_SSL_Allowed: No
```

```
            Master_SSL_CA_File:
            Master_SSL_CA_Path:
               Master_SSL_Cert:
             Master_SSL_Cipher:
                Master_SSL_Key:
         Seconds_Behind_Master: 0
1 row in set (0.00 sec)
```

步骤 04 主从服务器都成功启动后,开始更新主数据库 test 库中的 rep_t1 表和 rep_t2 表,具体数据如下:

```
mysql> insert into rep_t1 values(888);
Query OK, 1 row affected (0.00 sec)

mysql> insert into rep_t2 values(888);
Query OK, 1 row affected (0.01 sec)
```

步骤 05 登录从数据库,查询 test 库中的表 rep_t1 和 rep_t2 的数据更新情况,具体查询语句如下:

```
[root@localhost ~]# mysql -u root -P 3307 -S /temp/mysql2.sock
Enter password:
Welcome to the MySQL monitor.  Commands end with ; or \g.
Your MySQL connection id is 11
Server version: 8.0.28-log Source distribution

Copyright (c) 2000, 2022, Oracle and/or its affiliates. All rights reserved.

Oracle is a registered trademark of Oracle Corporation and/or its
affiliates. Other names may be trademarks of their respective
owners.

Type 'help;' or '\h' for help. Type '\c' to clear the current input statement.

mysql> use test;
Database changed
mysql> show tables;
+----------------+
| Tables_in_test |
+----------------+
| rep_t1         |
| rep_t2         |
+----------------+
2 rows in set (0.00 sec)

mysql> select * from rep_t1;
+------+
| data |
+------+
|    1 |
|    2 |
|    3 |
|  888 |
+------+
4 rows in set (0.00 sec)
```

```
mysql> select * from rep_t2;
Empty set (0.00 sec)
```

从测试的结果可以看到，主表中的 rep_t1 数据已经复制到从服务器上了，而 rep_t2 中的数据没有被复制。

2. replicate-do-db 和 replicate-ignore-db 的用法

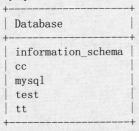启动主从数据库服务器，查询主数据库中主要有哪些数据库。

```
[root@localhost ~]# mysqld_multi --defaults-extra-file=/etc/my.cnf start 1-3

[root@localhost ~]# mysql -u root -P 3306 -S /tmp/mysql.sock
Enter password:
Welcome to the MySQL monitor.  Commands end with ; or \g.
Your MySQL connection id is 11
Server version: 8.0.28-log Source distribution

Copyright (c) 2000, 2022, Oracle and/or its affiliates. All rights reserved.

Oracle is a registered trademark of Oracle Corporation and/or its
affiliates. Other names may be trademarks of their respective
owners.

Type 'help;' or '\h' for help. Type '\c' to clear the current input statement.


mysql> show databases;
+--------------------+
| Database           |
+--------------------+
| information_schema |
| cc                 |
| mysql              |
| test               |
| tt                 |
+--------------------+
5 rows in set (0.00 sec)

mysql>
```

步骤02 使用 mysqldump 工具，将主数据库中的所有信息导出到 all.sql 脚本文件中。

```
[root@localhost ~]# mysqldump -u root -P 3306 -S /tmp/mysql.sock --all-databases >all.sql
```

步骤03 登录从数据库，导入 all.sql 中的数据，保持从服务器与主数据库数据一致。

```
[root@localhost ~]# mysql -u root -P 3307 -S /temp/mysql2.sock
Enter password:
Welcome to the MySQL monitor.  Commands end with ; or \g.
Your MySQL connection id is 11
Server version: 8.0.28-log Source distribution

Copyright (c) 2000, 2022, Oracle and/or its affiliates. All rights reserved.

Oracle is a registered trademark of Oracle Corporation and/or its
```

```
affiliates. Other names may be trademarks of their respective
owners.

Type 'help;' or '\h' for help. Type '\c' to clear the current input statement.

mysql> show databases;
+--------------------+
| Database           |
+--------------------+
| information_schema |
| mysql              |
| test               |
+--------------------+
3 rows in set (0.00 sec)

mysql> source ./all.sql
Query OK, 0 rows affected (0.05 sec)
Query OK, 0 rows affected (0.00 sec)
...

mysql> show databases;
+--------------------+
| Database           |
+--------------------+
| information_schema |
| cc                 |
| mysql              |
| test               |
| tt                 |
+--------------------+
5 rows in set (0.00 sec)
```

步骤 04 关闭从数据库,然后编辑数据库的配置文件。replicate-do-db 表示从服务器可以复制的数据库的名字,如果有多个数据库,那么可以重复写多个 replicate-do-db 配置。replicate-ignore-db 表示在从服务器复制过程中忽略复制该配置设置的数据库。

```
[root@localhost ~]# mysqld_multi --defaults-extra-file=/etc/my.cnf stop 1-3

vi /etc/my.cnf
[mysqld2]
...
port = 3307
socket = /tmp/mysql2.sock
server-id = 2
#replicate-do-table=test.rep_t1
#replicate-ignore-table=test.rep_t2
replicate-do-db=test
replicate-do-db=cc
replicate-ignore-db=tt
```

步骤 05 启动主从数据库,然后在主数据库 cc 库中增加表 cc_t1 表,在 tt 库中增加表 tt_t1 表。

```
[root@localhost ~]# mysqld_multi --defaults-extra-file=/etc/my.cnf start 1-3

mysql> use cc;
```

```
Database changed

mysql> create table cc_t1(data int);
Query OK, 0 rows affected (0.01 sec)

mysql> use tt;
Database changed
mysql> create table tt_t1(data int);
Query OK, 0 rows affected (0.01 sec)
```

步骤 06 登录从数据库，查询数据库 cc 库和 tt 库相应的数据是否更新。

```
[root@localhost ~]# mysql -uroot -P 3307 -S /tmp/mysql2.sock
Enter password:
Welcome to the MySQL monitor.  Commands end with ; or \g.
Your MySQL connection id is 11
Server version: 8.0.28-log Source distribution

Copyright (c) 2000, 2022, Oracle and/or its affiliates. All rights reserved.

Oracle is a registered trademark of Oracle Corporation and/or its
affiliates. Other names may be trademarks of their respective
owners.

Type 'help;' or '\h' for help. Type '\c' to clear the current input statement.

mysql> show databases;
+--------------------+
| Database           |
+--------------------+
| information_schema |
| cc                 |
| mysql              |
| test               |
| tt                 |
+--------------------+
5 rows in set (0.00 sec)

mysql> use cc;
Database changed
mysql> show tables;
+---------------+
| Tables_in_cc  |
+---------------+
| cc1           |
| cc_t1         |
+---------------+
2 rows in set (0.00 sec)

mysql> use tt;
Database changed
mysql> show tables;
Empty set (0.00 sec)
```

21.4 查看从服务器的复制进度

很多情况下，用户想知道从服务器复制的进度，从而判断从服务器上复制数据的完整性，以及是否需要手工来做主从的同步工作。事实上，用户可以通过 SHOW PROCESSLIST 列表中的 Slave_SQL_Running 线程的 Time 值得知，它记录了从服务器当前执行的 SQL 时间戳与系统时间之间的差距。下面通过例子测试一下这个时间的准确性。

步骤01 在主服务器上插入一个包含当前时间戳的记录。

```
mysql> alter table rep_t3 add column createtime datetime;
Query OK, 0 rows affected (0.10 sec)
Records: 0  Duplicates: 0  Warnings: 0

mysql> insert into rep_t3 values(1,now());
Query OK, 1 row affected (0.00 sec)

mysql> select * from rep_t3;
+------+---------------------+
| data | createtime          |
+------+---------------------+
|    1 | 2022-02-25 15:10:51 |
+------+---------------------+
1 row in set (0.01 sec)
```

步骤02 从服务器的 I/O 线程停止，使得从数据库服务器暂时不写中继日志，停止时执行的 SQL 就是最后执行的 SQL，命令执行如下：

```
mysql> stop slave;
Query OK, 0 rows affected (0.00 sec)

mysql> select * from rep_t3;
+------+---------------------+
| data | createtime          |
+------+---------------------+
|    1 | 2022-02-25 15:10:51 |
+------+---------------------+
1 row in set (0.00 sec)

mysql> select now();
+---------------------+
| now()               |
+---------------------+
| 2022-02-25 15:13:42 |
+---------------------+
1 row in set (0.00 sec)
```

步骤03 在从数据库服务器上执行 show processlist，查看 SQL 线程的时间。这个时间说明了主服务器最后执行的更新操作大概是主服务器 46 秒前的更新操作。

```
mysql> stop slave io_thread;
Query OK, 0 rows affected (0.00 sec)
```

```
mysql> show processlist \G;
*** 1. row ***
     Id: 6
   User: root
   Host: localhost
     db: test
Command: Query
   Time: 0
  State: NULL
   Info: show processlist
*** 2. row ***
     Id: 8
   User: system user
   Host:
     db: NULL
Command: Connect
   Time: 46
  State: Has read all relay log; waiting for the slave I/O thread to update it
   Info: NULL
2 rows in set (0.00 sec)
```

21.5 复制环境的监控和维护

数据复制环境配置完成后,数据库管理员需要进行日常的监控和管理维护工作,以便能够及时发现问题和解决问题,从而保证主从数据库能够正常工作。比如,有时候会因为主服务器的更新过于频繁,造成从服务器更新速度较慢。当然,问题是多种多样的,原因有可能是网络搭建的结构不好或者硬件的性能较差,从而使得主从服务器之间的差距越来越大,最终对某些应用产生了影响,在这种情况下,用户需要定期检查主从服务器的数据同步操作。

21.5.1 了解服务器的状态

一般使用 show slave status 命令来检查从服务器的状态,如下所示:

```
mysql> show slave status\G;
*** 1. row ***
             Slave_IO_State: Waiting for master to send event
                Master_Host: 127.0.0.1
                Master_User: repl
                Master_Port: 3306
              Connect_Retry: 60
            Master_Log_File: mysql-bin.000013
        Read_Master_Log_Pos: 98
             Relay_Log_File: mysql2-relay-bin.000035
              Relay_Log_Pos: 235
      Relay_Master_Log_File: mysql-bin.000013
           Slave_IO_Running: Yes
          Slave_SQL_Running: Yes
            Replicate_Do_DB:
        Replicate_Ignore_DB:
```

```
              Replicate_Do_Table: test.rep_t1,test.rep_t1
          Replicate_Ignore_Table: test.rep_t2,test.rep_t2
         Replicate_Wild_Do_Table:
     Replicate_Wild_Ignore_Table:
                      Last_Errno: 0
                      Last_Error:
                    Skip_Counter: 0
             Exec_Master_Log_Pos: 98
                 Relay_Log_Space: 235
                 Until_Condition: None
                  Until_Log_File:
                   Until_Log_Pos: 0
              Master_SSL_Allowed: No
              Master_SSL_CA_File:
              Master_SSL_CA_Path:
                 Master_SSL_Cert:
               Master_SSL_Cipher:
                  Master_SSL_Key:
           Seconds_Behind_Master: 0
1 row in set (0.00 sec)
```

在从服务器信息中，首先要查看 Slave_IO_Running 和 Slave_SQL_Running 这两个进程状态是否是"yes"。Slave_IO_Running 表明此进程是否能够由从服务器到主服务器上正确地读取 BINLOG 日志，并写入到从服务器的中继日志中。Slave_SQL_Running 表明能否读取并执行中继日志中的 BINLOG 信息。

21.5.2 服务器复制出错的原因

在某些时候，会出现从服务器更新失败的情况，此时首先需要确定是否是主从服务器的表不同造成的。如果是表结构不同而导致的，就修改从服务器上的表与主服务器上的表一致，然后重新执行 START SLAVE 命令。服务器复制出错的常见问题如下：

（1）问题一：出现"log event entry exceeded max_allowed_pack"错误。

如果在应用中使用大的 BLOG 列或者长字符串，那么在从服务器上回复时可能会出现"log event entry exceeded max_allowed_pack"错误，这是因为含有大文本的记录无法通过网络进行传输，解决方法是在主从服务器上添加 max_allowed_packet 参数（默认设置是 1MB），具体如下：

```
mysql> SHOW VARIABLES LIKE 'MAX_ALLOWED_PACKET';
+--------------------+---------+
| Variable_name      | Value   |
+--------------------+---------+
| max_allowed_packet | 1048576 |
+--------------------+---------+
1 row in set (0.00 sec)

mysql> SET @@global.max_allowed_packet=16777216;
Query OK, 0 rows affected (0.00 sec)
```

同时，在 my.cnf 里设置 max_allowed_packet=16MB，数据库重新启动之后该参数将生效。

（2）问题二：多主复制时的自增长变量冲突问题。

大多数情况下使用一台主服务器对一台或者多台从服务器，但是在某些情况下可能会将多个服务器配置为复制主服务器，所以使用 auto_increment 时应采取特殊步骤以防止键值冲突，否则插入行时多个主服务器会试图使用相同的 auto_increment 值。

服务器变量 auto_increment_increment 和 auto_increment_offset 可以协调多主服务器复制和 auto_increment 列。在多主服务器复制到从服务器过程中，迟早会发生主键冲突，为了解决这种问题，可以重新设置不同主服务器的这两个参数，比如在 A 数据库服务器上设置 auto_increment_increment=1、auto_increment_offset=1，在 B 数据库服务器上设置 auto_increment_increment=1、auto_increment_offset=0。

下面的例子演示修改这两个参数后的效果。

步骤 01 创建表 auto_t，系统默认的 auto_increment_increment 和 auto_increment_offset 参数都是 1，增加数据默认的也是增加幅度为 1，命令执行如下：

```
mysql> create table auto_t( data int primary key auto_increment )engine=myisam default charset=gbk;
Query OK, 0 rows affected (0.05 sec)

mysql> show variables like 'auto_inc%';
+--------------------------+-------+
| Variable_name            | Value |
+--------------------------+-------+
| auto_increment_increment | 1     |
| auto_increment_offset    | 1     |
+--------------------------+-------+
2 rows in set (0.00 sec)

mysql> insert into auto_t values(null),(null),(null);
Query OK, 3 rows affected (0.01 sec)
Records: 3  Duplicates: 0  Warnings: 0

mysql> select * from auto_t;
+------+
| data |
+------+
|    1 |
|    2 |
|    3 |
+------+
3 rows in set (0.00 sec)
```

步骤 02 重新设置参数 auto_increment_increment 的值为 10，然后插入数据。

```
mysql> set @@auto_increment_increment=10;
Query OK, 0 rows affected (0.00 sec)

mysql> show variables like 'auto_inc%';
+--------------------------+-------+
| Variable_name            | Value |
+--------------------------+-------+
```

```
| auto_increment_increment | 10 |
| auto_increment_offset    | 1  |
+--------------------------+----+
2 rows in set (0.00 sec)

mysql> insert into auto_t values(null),(null),(null);
Query OK, 3 rows affected (0.00 sec)
Records: 3  Duplicates: 0  Warnings: 0

mysql> select * from auto_t;
+------+
| data |
+------+
|    1 |
|    2 |
|    3 |
|   11 |
|   21 |
|   31 |
+------+
6 rows in set (0.00 sec)
```

从测试效果看，每次递增值是 10。下面看参数 auto_increment_offset 的用法：

步骤 03 重新设置参数 auto_increment_offset 的值为 5，再插入数据。

```
mysql> set @@auto_increment_offset=5;
Query OK, 0 rows affected (0.00 sec)

mysql> insert into auto_t values(null),(null),(null);
Query OK, 3 rows affected (0.00 sec)
Records: 3  Duplicates: 0  Warnings: 0

mysql> select * from auto_t;
+------+
| data |
+------+
|    1 |
|    2 |
|    3 |
|   11 |
|   21 |
|   31 |
|   35 |
|   45 |
|   55 |
+------+
9 rows in set (0.00 sec)
```

从插入的记录可以看出，auto_increment_increment 参数是每次增加的量，而参数 auto_increment_offset 参数设置的是每次增加后的偏移量，也就是每次按照 10 累加后，还需要增加 5 个偏移量。

21.6 切换主从服务器

在实际工作环境中,有时候会遇到这样的问题:在一个工作环境中,有一个主数据库服务器 A,两个从数据库服务器 B、C 同时指向主数据库服务器 A,当主数据库服务器 A 发生故障时,需要将其中的一个从数据库服务器 B 切换成主数据库服务器,同时修改数据库服务器 C 的配置,使其指向新的主数据库服务器 B。

下面介绍一下切换主从服务器的具体操作步骤:

步骤01 首先要确保所有的从数据库都已经执行了 relay log 中的全部更新,查看从数据库的状态是否是 Has read all relay log(是否更新都已经执行完成)。

```
mysql> stop slave IO_THREAD;
Query OK, 0 rows affected (0.00 sec)

mysql> SHOW PROCESSLIST \G;
*** 1. row ***
     Id: 2
   User: system user
   Host:
     db: NULL
Command: Connect
   Time: 39
  State: Has read all relay log; waiting for the slave I/O thread to update it
   Info: NULL
*** 2. row ***
     Id: 3
   User: root
   Host: localhost
     db: NULL
Command: Query
   Time: 0
  State: NULL
   Info: SHOW PROCESSLIST
2 rows in set (0.00 sec)
```

步骤02 在从数据库 B 上停止 slave 服务,然后执行 reset master,重置成主数据库。

```
mysql> stop slave;
Query OK, 0 rows affected (0.00 sec)

mysql> reset master;
ERROR 1186 (HY000): Binlog closed, cannot RESET MASTER
```

此时报 Binlog 错误(表示没有设置),不能够执行 reset master 命令。下面关闭数据库服务,然后修改/etc/my.cnf,在[mysqld2]后面的配置选项中添加 log-bin 选项:

```
[mysqld2]
...
log-bin = /usr/local/var/mysql2/mysql-bin
```

配置完成后,重启数据库服务,登录数据库 B,然后执行如下命令开启主数据库功能。

```
mysql> stop slave;
Query OK, 0 rows affected (0.00 sec)

mysql> reset master;
Query OK, 0 rows affected (0.04 sec)
```

此时从数据库 B 已经成功切换成为主数据库，下面接着设置从数据库。

步骤 03 在从数据库服务器 B 上添加具有 replication 权限的用户 rep1，查询主数据库状态。

```
mysql> grant replication slave on *.* to 'rep1'@'localhost' identified by '123';
Query OK, 0 rows affected (0.00 sec)

mysql> show master status;
+------------------+----------+--------------+------------------+
| File             | Position | Binlog_Do_DB | Binlog_Ignore_DB |
+------------------+----------+--------------+------------------+
| mysql-bin.000001 |      229 |              |                  |
+------------------+----------+--------------+------------------+
1 row in set (0.00 sec)
```

步骤 04 在从数据库服务器 C 上配置复制的参数。

```
mysql> change master to
    -> master_host='127.0.0.1',
    -> master_user='rep1',
    -> master_password='123',
    -> master_port=3307,
    -> master_log_file='mysql-bin.000002',
    -> master_log_pos=98;
Query OK, 0 rows affected (0.01 sec)

mysql> start slave;
Query OK, 0 rows affected (0.01 sec)
```

步骤 05 在从数据库服务器 C 上执行 show slave status 命令，查看从数据库服务是否成功启动。

```
mysql> show slave status \G;
*** 1. row ***
             Slave_IO_State: Waiting for master to send event
                Master_Host: 127.0.0.1
                Master_User: rep1
                Master_Port: 3307
              Connect_Retry: 60
            Master_Log_File: mysql-bin.000002
        Read_Master_Log_Pos: 98
             Relay_Log_File: mysql3-relay-bin.000002
              Relay_Log_Pos: 235
      Relay_Master_Log_File: mysql-bin.000002
           Slave_IO_Running: Yes
          Slave_SQL_Running: Yes
            Replicate_Do_DB:
        Replicate_Ignore_DB:
         Replicate_Do_Table:
     Replicate_Ignore_Table:
    Replicate_Wild_Do_Table:
```

```
            Replicate_Wild_Ignore_Table:
                        Last_Errno: 0
                        Last_Error:
                      Skip_Counter: 0
               Exec_Master_Log_Pos: 98
                   Relay_Log_Space: 235
                   Until_Condition: None
                    Until_Log_File:
                     Until_Log_Pos: 0
                Master_SSL_Allowed: No
                Master_SSL_CA_File:
                Master_SSL_CA_Path:
                   Master_SSL_Cert:
                 Master_SSL_Cipher:
                    Master_SSL_Key:
             Seconds_Behind_Master: 0
1 row in set (0.00 sec)
```

步骤 06 在主数据库服务器 B 和从数据库服务器 C 上面测试数据库是否成功设置复制功能。首先，查看主数据库服务器 B 中 test 库中表的情况：

```
mysql> use test;
Database changed
mysql> show tables;
+-----------------+
| Tables_in_test  |
+-----------------+
| rep_t1          |
| rep_t2          |
+-----------------+
2 rows in set (0.00 sec)
```

然后，查询从数据库服务器 C 中 test 库中表的情况：

```
mysql> use test;
Database changed
mysql> show tables;
Empty set (0.01 sec)
```

步骤 07 在主数据库服务器 B 中增加表 rep_t3。

```
mysql> create table rep_t3(data int);
Query OK, 0 rows affected (0.01 sec)
```

步骤 08 在从数据库服务器 C 中查询，看看表 rep_t3 是否从主数据库服务器 B 上成功复制过来。

```
mysql> show tables;
+-----------------+
| Tables_in_test  |
+-----------------+
| rep_t3          |
+-----------------+
1 row in set (0.00 sec)
```

至此，主从数据库成功地做了切换。最后，如果原来的主数据库服务器 A 可以修复，可以考虑采用以上方法将数据库服务器 A 配置成为数据库服务器 B 的从数据库。

21.7 多源复制的改进

在早期的版本中，MySQL 服务器的复制只能在一个主服务器和多个从服务器之间实现。图 21.3 所示为一个主服务器和多个从服务器的复制情况。

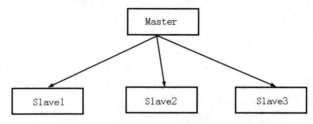

图 21.3　一个主服务器和多个从服务器

MySQL 8.0 添加了多源复制功能，可以实现多主服务器和一从服务器的复制。图 21.4 所示为一个从服务器和多个主服务器的复制情况。

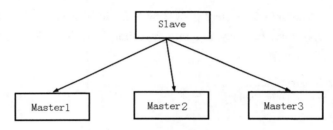

图 21.4　一个从服务器和多个主服务器

多源复制功能的优势如下：

（1）如果在主服务器进行了分库分表的操作，可以在从服务器进行数据汇总。为了实现后期的一些数据统计功能，往往需要把数据汇总在一起再统计。

（2）在从服务器上对主服务器的数据进行备份，在 MySQL 8.0 之前每一个主服务器都需要一个从服务器，很容易造成资源浪费，同时也加大了数据库管理员的维护成本；MySQL 8.0 则引入了多源复制，可以把多个主服务器的数据同步到一个从服务器进行备份。

下面举例说明。

首先，配置两个主服务器和一个从服务器：

- 主服务器 Master1：192.168.1.101。
- 主服务器 Master2：192.168.1.102。
- 从服务器 Slave：192.168.1.103。

步骤 01 在 Master1 上导出需要同步的数据库：

```
[root@Master1 mysql]# mysqldump -u root -p 123456 --master-data=2 --single-transaction --databases --add-drop-database mytest> mytest.sql
```

步骤02 在 Master2 上导出需要同步的数据库：

```
[root@Master2 mysql]# mysqldump -u root -p123456 --master-data=2 --single-transaction --databases --add-drop-database mytest2 > mytest2.sql
```

步骤03 分别在 Master1 和 Master2 上把备份文件复制到 Slave 上：

```
[root@Master1 mysql]# scp -P22 mytest.sql 192.168.1.103:/data/service/mysql/
[root@Master2 mysql]# scp -P22 mytest2.sql 192.168.1.103:/data/service/mysql/
```

步骤04 在 Master1 上创建复制账号：

```
<Master1>[(none)]>grant replication slave on *.* to 'myuser'@'192.168.1.103' identified by '123456';
Query OK, 0 rows affected, 1 warning (0.00 sec)
```

步骤05 在 Master2 上创建复制账号：

```
<Master1>[(none)]>grant replication slave on *.* to 'myuser'@'192.168.1.103' identified by '123456';
Query OK, 0 rows affected, 1 warning (0.00 sec)
```

步骤06 后续操作将把 Master1 和 Master2 的数据导入 Slave 服务器。在导入前先修改 MySQL 存储 master_info_repository 和 relay_log_info_repository 的方式，即从文件存储改为表存储，在 my.cnf 里添加以下设置：

```
master_info_repository=TABLE
relay_log_info_repository=TABLE
```

步骤07 在 Slave 上进行数据导入：

```
[root@Slave mysql]# mysql -u root -p 123456 <./mytest.sql
[root@Slave mysql]# mysql -u root -p 123456 <./mytest2.sql
```

步骤08 分别找出 Master1 和 Master2 的 binlog 位置和 Pos 位置：

```
[root@Slave mysql]# cat mytest.sql |grep " CHANGE MASTER"
-- CHANGE MASTER TO MASTER_LOG_FILE='Master1-bin.000001', MASTER_LOG_POS=1539;
[root@Slave mysql]# cat mytest2.sql |grep " CHANGE MASTER"
-- CHANGE MASTER TO MASTER_LOG_FILE='Master2-bin.000003', MASTER_LOG_POS=630;
[root@Slave mysql]#
```

步骤09 登录 Slave 进行同步操作，分别执行 CHANGE MASTER 到两台 Master 服务器：

```
<Slave> [(none)]> CHANGE MASTER TO MASTER_HOST='192.168.1.101',MASTER_USER='myuser',MASTER_PASSWORD='123456',MASTER_LOG_FILE='Master1-bin.000001',MASTER_LOG_POS=1539 FOR CHANNEL 'Master1';
Query OK, 0 rows affected, 2 warnings (0.05 sec)
<Slave> [(none)]> CHANGE MASTER TO MASTER_HOST='192.168.1.102',MASTER_USER='myuser',MASTER_PASSWORD='123456',MASTER_LOG_FILE='Master2-bin.000001',MASTER_LOG_POS=1630 FOR CHANNEL 'Master2';
Query OK, 0 rows affected, 2 warnings (0.04 sec)
```

步骤10 通过 start slave 命令的方式去启动所有的复制：

```
<Slave> [(none)]> start slave for CHANNEL 'Master1';
Query OK, 0 rows affected (0.01 sec)
<Slave> [(none)]> start slave for CHANNEL 'Master2';
Query OK, 0 rows affected (0.02 sec)
```

正常启动后,可以查看复制源 Master1 和 Master2 的同步状态,命令如下:

```
SHOW SLAVE STATUS FOR CHANNEL 'Master1'\G
SHOW SLAVE STATUS FOR CHANNEL 'Master2'\G
```

MySQL 8.0 的多源复制能有效地解决分库分表的数据统计问题,同时也可以实现在一台从服务器对多台主服务器的数据备份的功能。

第 22 章

MySQL Utilities

作为一款非常流行的开源数据库，MySQL 支持的工具越来越多，对用户来说，选择一个好的工具，对于提高工作效率有很大的帮助。本章将介绍一款利器——MySQL Utilities。MySQL Utilities 是官方提供的 MySQL 管理工具，功能非常强大，对 MySQL 数据库管理员提供诸多方便，减轻维护工作量和难度。通过本章的学习，读者可以熟练使用 MySQL Utilities 来管理 MySQL 数据库。

22.1 MySQL Utilities 概述

MySQL Utilities 是官方提供的 MySQL 管理工具，功能比较完善，通过提供一组命令行工具来维护和管理 MySQL 服务器，主要包含以下几方面的工具：

（1）管理工具：主要功能为克隆、复制、比较、导入导出数据。
（2）一般工具：监控磁盘使用情况、检查冗余索引和搜索元数据。
（3）高可用工具：支持主从复制、故障转移和主从服务器同步数据功能。

22.2 安装与配置

安装 MySQL Utilities 之前，首先需要安装 Python 2.6 或以上版本，这是 MySQL Utilities 必须依赖的环境。另外，还需要连接驱动 MySQL Connector/Python 1.0.8 或以上版本，其中 MySQL Connector/Python 的下载地址为 http://dev.mysql.com/downloads/connector/python/。用户可以下载后自行安装。

22.2.1 下载与安装 MySQL Utilities

下载 MySQL Utilities 的方法是：打开浏览器，在地址栏中输入网址"http://dev.mysql.com/downloads/utilities/"，打开 MySQL Utilities 1.6.5 下载页面，选择 Microsoft Windows 平台，然后根据本机的系统平台选择 32 位或者 64 位安装包，在这里选择 64 位，单击右侧的 Download 按钮开始下载，如图 22.1 所示。

图 22.1　MySQL Utilities 下载页面

MySQL Utilities 下载完成后，找到下载文件，双击进行安装，具体操作步骤如下：

步骤 01 双击下载的 mysql-utilities-1.6.5-win64.msi 文件，如图 22.2 所示。

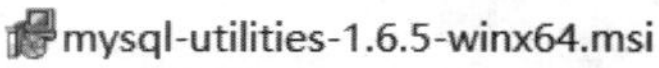

图 22.2　MySQL Utilities 安装文件名称

步骤 02 打开欢迎安装窗口，单击 Next（下一步）按钮，如图 22.3 所示。

步骤 03 打开 Destination Folder（安装路径文件夹）窗口，单击 Change（修改）按钮，可以修改安装路径，这里采用默认的安装路径，接着单击 Next 按钮，如图 22.4 所示。

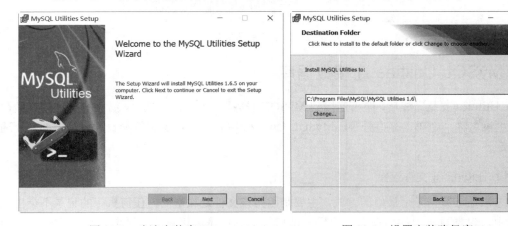

图 22.3　欢迎安装窗口　　　　　　　　图 22.4　设置安装路径窗口

步骤 04 打开准备安装窗口，确认无误后，单击 Install（安装）按钮，如图 22.5 所示。

步骤 05 开始安装 MySQL Utilities 文件，并显示安装的进度，如图 22.6 所示。

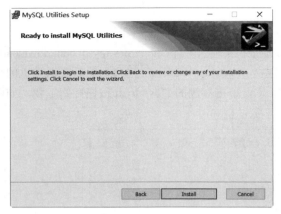
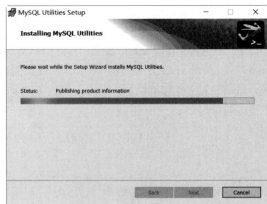

图 22.5　准备安装窗口　　　　　　图 22.6　开始安装 MySQL Utilities

步骤 06 安装完成后，打开安装完成窗口，单击 Finish（完成）按钮即可，如图 22.7 所示。

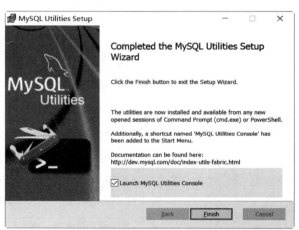

图 22.7　安装完成窗口

22.2.2　MySQL Utilities 连接数据库

MySQL Utilities 安装完成后，即可连接到 MySQL 服务器。连接服务器时，必须指定连接参数，如用户名、主机名称、密码、端口号等。MySQL Utilities 中有 3 种提供这些参数的方法，都需要通过命令行指定。

1．使用 .mylogin.cnf 文件连接

使用该方法连接数据库是最好的方式。主要是因为该文件是加密的，任何执行的记录不会出现在连接信息中，包括日志中的用户名、密码、端口等信息都是不可见的。因此，这种方法是使用 MySQL Utilities 工具连接数据库的首选方法。

连接数据库的字符串格式为 login-path-name[:port][:socket]。其中，port、socket 是可选的参数。

使用 mysql_config_editor 工具添加如下连接信息：

```
mysql_config_editor set --login-path=instance_3306 --host=localhost --user=root --port=3306 --password Enter password:
```

```
Enter password:
```
此时会创建一个隐藏的加密文件.mylogin.cnf，接着只需要指定.mylogin.cnf 文件中的服务段进行连接即可。例如，在上面的示例中创建了"instance_3306"服务段，因此可以使用 --server=instance_3306 进行连接：

```
mysqlserverinfo --server=instance_3306 --format=vertical
```

2. 使用配置文件连接

MySQL Utilities 也可以使用配置文件 my.cnf 连接数据库，但是由于该文件是文本文件，因此只要能访问到该文件的用户，都可以查看连接的具体信息。

使用配置文件连接数据库，具体如下：

```
Mysqlserverinfo --server=/my.cnf[client] --format=vertical
```

3. 使用命令行连接数据库

通过命令行参数指定连接数据库服务器，这种方式是最不安全的，因为数据在命令行可见，在日志文件中也是可见的。

这种方式下，指定参数的顺序为<user>[:<passwd>]@<host>[:<port>][:<socket>]，其中，[]内的参数是可选的。

22.3 管理与维护

数据库连接完成后，即可使用 MySQL Utilities 对数据库进行管理和维护操作。MySQL Utilities 提供了一系列 MySQL 服务器和数据库的管理工具，完全支持 MySQL 5.1 及以上版本；注意它也兼容 MySQL 5.0 版本，不过有些特性不支持。

22.3.1 使用 mysqldbcompare 比较数据

通过 mysqldbcompare 工具可以实现以下功能：

- 比较两个服务器或同一个服务器上的数据库。
- 比较定义文件和数据。
- 产生差异报告，生成差异性的转换 SQL 语句。

mysqldbcompare 从两个数据库比较对象和数据的不同。在比较过程中，数据不可以改变，否则会出现错误。数据库中的对象包括表、视图、触发器、存储过程、函数和事件。通过一系列步骤检查进行测试，默认情况下，一旦测试失败就终止检测。

可以指定--run-all-tests 选项来进行所有的测试。如果想忽略检测数据库的对象，可以使用 --skip-object-compare 跳过该测试；如果想忽略对象定义的比较，可以使用--skip-diff 跳过该测试；如果想忽略检测表的行数，可以使用 --skip-row-count 选项跳过这一步。

使用下面的命令比较本地服务器上的 test1 和 test2 数据库，进行所有的测试，不管是否失败：

```
mysqldbcompare --server1=root@localhost test1:test2 --run-all-tests #
```

22.3.2 使用 mysqldbcopy 复制数据

管理员使用 mysqldbcopy 可以从源服务器上复制一个数据库到另一个目标服务器上。源服务器和目标服务器可以是同一服务器，也可以是不同服务器。数据库名字可以相同或不相同。如果源服务器和目标服务器是同一台服务器，那么数据库名字必须不一样，也就是同一个实例下不能有相同的数据库名。

mysqldbcopy 接受一个或多个数据库对，格式为 db_name:new_db_name，分别表示源和目标。例如，下面的命令即为复制数据的例子：

```
mysqldbcopy --source=instance_3306 --destination=instance_3307 test:test_copy --rpl=master
--rpl-user=root -rrr123 --drop-first
```

其中，--source 为源服务器；--destination 为目标服务器；test 为需要复制的数据库；test_copy 为复制后的数据库；--rpl=master 表示创建并执行 CHANGE MASTER 语句，将目标服务器作为--source 选项指定的服务器的从服务器，在复制数据之前，执行 STOP SLAVE 语句，在复制完成后执行 CHANGE MASTER 和 START SLAVE 语句；--rpl-user=root -rrr123 表示指定执行复制操作的用户名和密码；--drop-first 表示如果目标数据库上存在与被复制数据库重名的数据库，就将其删除。

默认情况下，复制所有对象（如表、视图、触发器、事件、存储过程、函数和数据库级别权限）和数据到目标服务器中。管理员可以有选择性地复制，例如只复制部分对象、不复制数据等。如果想要针对性地复制，可以使用--exclude 选项来排除。

默认情况下，目标服务器上使用的存储引擎与源服务器相同。如果目标服务器上使用另一种存储引擎，可以使用--new-storage-engine 选项来指定。如果目标服务器支持指定的引擎，那么所有表都使用该引擎。如果目标服务器不支持源服务器所用的存储引擎，可以使用--default-storage-engine 选项来指定默认使用的存储引擎。--new-storage-engine 选项的优先级高于--default-storage-engine。如果这两个选项都指定，首先判断指定的存储引擎是否被支持：如果被支持，就使用指定的存储引擎；如果不被支持，就使用默认的存储引擎。

22.3.3 使用 mysqldbexport 导出数据

使用 mysqldbexport 工具的主要作用是从一个或多个数据库导出对象定义的元数据和数据。和 mysqldump 类似，最大的区别是 mysqldbexport 工具支持的格式很多，包括 SQL、CSV、TAB、Grid 和 Vertical，这样使数据更容易提取和转移。

如果管理员想指定导出特定的对象，可以使用-exclude 选项来排除。

（1）只导出定义语句的命令如下：

```
mysqldbexport --server=instance_3306 --format=sql test --export=definitions
```

其中，--server 指定导出数据的服务器；--format 指定导出数据的格式；--export 指定导出的内容为只导出定义语句。

（2）只导出数据，且批量插入语句，命令如下：

```
mysqldbexport --server=instance_3306 --format=sql test --export=data --bulk-insert
```

（3）为当前的数据库创建一个从服务器，命令如下：

```
mysqldbexport --server=instance_3306 --format=sql test --export=both --rpl-user=root --rpl=master
```

22.3.4 使用 mysqldbimport 导入数据

mysqldbimport 工具可以将导出的数据导入到指定的目标服务器上，如果目标服务器上已经存在一个同名的对象，那么先将其删除再导入。mysqldbimport 支持导入的文件格式包括 SQL、CSV、TAB、Grid 和 Vertical。

（1）导入指定 test.sql 文件的定义语句，命令如下：

```
Mysqldbimport --server=root@localhost --import=definitions \ --format=sql test.sql
```

其中，--server 指定导入的服务器；--import 指定导入的内容为只导入定义语句；--format 指定导出数据的格式；test.sql 为导入的文件。

（2）通过批量插入语句的方式导入数据，命令如下：

```
mysqldbimport --server=root@localhost --import=data\ --bulk-insert --format=sql test.csv
```

其中，--import=data 表示导入数据；--bulk-insert 表示通过批量插入语句的方式。

（3）通过批量插入语句的方式导入定义语句和数据，命令如下：

```
mysqldbimport --server=root@localhost --import=both --bulk-insert --format=sql test.sql
```

22.3.5 使用 mysqldiff 比较对象的定义

在前面的小节中讲解了如何比较两个对象的数据差异，这一小节继续学习如何比较对象的定义是否相关并显示不同之处。mysqldiff 工具的主要功能是比较对象的定义，然后输出差异的报告。

例如，比较服务器 server1 上数据库 test 和服务器 server2 上 test 数据库的定义是否相同，命令如下：

```
mysqldiff --server1=instance_3306 --server2=instance_3308 test:test
```

其中，server1 的数据库 test 为参照数据库。

若 server1 中出现的对象在 server2 中不存在，则报错信息如下：

```
WARNING: Objects in server2.ttlsa_com but not in server1.ttlsa_com: # TABLE: t Compare failed. One or more differences found.
```

如果要比较特定对象，可以使用 db.obj 格式。db1.obj1:db2 和 db1:db2.obj2 这些类型的格式是非法的，报错信息如下：

```
mysqldiff: error: Incorrect object compare argument, one specific object is missing. Please verify that both object are correctly specified. No object has been specified for db1 'test', while object 't' was specified for db2 'test'. Format should be: db1[.object1]:db2[.object2].
```

第 23 章

MySQL Proxy

MySQL Proxy 是一个位于客户端和 MySQL 服务器端之间的程序,通过它可以实现监听和管理客户端与 MySQL 服务器端之间的通信,最大的作用是实现数据库的读写分离,从而达到负载均衡的目的。本章将讲解 MySQL Proxy 的使用方法和技巧。

23.1 概　述

MySQL Proxy 的常用用途包括负载平衡、故障分析、查询分析、查询过滤和修改等。通俗地说,作为一个中间层代理,它就是一个连接池,负责将前台应用的连接请求转发给后台的数据库,并且通过使用 lua 脚本,可以实现复杂的连接控制和过滤,从而实现读写分离和负载平衡。对于应用程序来说,MySQL Proxy 是完全透明的,应用程序只需要连接到 MySQL Proxy 的监听端口即可。

MySQL Proxy 最强大的功能是实现"读写分离",基本原理是让主数据库处理事务性查询,让从数据库处理 SELECT 查询,最后通过数据库的复制功能,把事务性查询导致的数据变更同步到集群中的从数据库中。MySQL Proxy 实现读写分离的过程如图 23.1 所示。

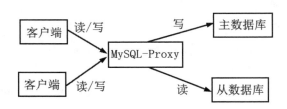

图 23.1　MySQL Proxy 实现读写分离的过程

提示:如果此时只有一个 MySQL Proxy 机器,可能会出现单点失效的问题,解决办法是使用多个 Proxy 机器作为冗余,在应用服务器的连接池配置中配置多个 Proxy 的连接参数即可。

MySQL Proxy 通过 mysql-proxy 来指定配置的参数。下面了解一下该命令常用参数的含义：

- --proxy-backend-addresses：该参数用来指定 MySQL 服务器的 IP 地址和端口号，如果代理多个服务器，可以用逗号分隔。
- --proxy-read-only-backend-addresses：该参数用来指定只读服务器的 IP 地址和端口号，如果代理多个服务器，可以用逗号分隔。
- --proxy-skip-profiling：该参数用来设置是否禁用查询性能分析。
- --proxy-lua-script：该参数用来指定 lua 脚本文件。
- --daemon：采用 daemon 方式启动。
- --admin-address：指定 MySQL Proxy 的管理端口。
- --proxy-address=：指定 MySQL Proxy 的监听端口。

读者也可以通过 mysql-proxy --help-all 命令查看完整的参数含义。

23.2 安装与配置

MySQL Proxy 支持的系统平台很多，包括 Linux、OS X、FreeBSD、IBM AIX、Sun Solaris 和微软 Windows，这里以 Windows 10 为例进行讲解。

23.2.1 下载与安装 MySQL Proxy

安装 MySQL Proxy 之前，需要下载官方的 MySQL Proxy 压缩包。打开浏览器，在地址栏中输入网址"http://downloads.mysql.com/archives/proxy/"，打开 MySQL Proxy 下载页面，选择 Microsoft Windows 平台，然后根据平台选择 32 位或者 64 位安装包，在这里选择 32 位，单击右侧的 Download 按钮开始下载，如图 23.2 所示。

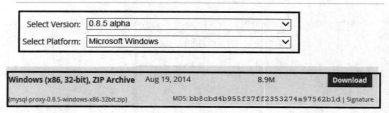

图 23.2　MySQL Proxy 下载页面

MySQL Proxy 压缩包下载完成后，找到下载压缩包文件，双击进行解压操作，然后将文件夹的名称修改为"mysql—proxy—0.8.5"，最后将该文件夹移动到 MySQL 安装的根目录下，如图 23.3 所示，即可完成 MySQL Proxy 的安装操作。

图 23.3　MySQL Proxy 文件夹放置位置

23.2.2　配置 MySQL Proxy 参数

安装完 MySQL Proxy，即可使用 mysql-proxy 命令进行代理参数的配置操作，具体步骤如下：

步骤 01 单击开始菜单，在搜索框中输入 cmd，按 Enter 键确认，如图 23.4 所示。

图 23.4　运行对话框

步骤 02 打开 DOS 窗口，输入以下命令并按 Enter 键确认，如图 23.5 所示。

```
sc create "Proxy" DisplayName= "MySQL Proxy" start= "auto" binPath= "C:\Program Files\MySQL\mysql-proxy-0.8.5\bin\mysql-proxy-svc.exe --proxy-backend-addresses=127.0.0.1:3306"
```

图 23.5　DOS 窗口

提示：这里使用 sc 命令创建一个 MySQL Proxy 代理服务，启动方式为自动启动。

步骤 03 在 DOS 窗口中可以通过启动命令启动 MySQL Proxy 代理服务，如图 23.6 所示。

```
net start proxy
```

第 23 章　MySQL Proxy ｜ 497

图 23.6　启动 MySQL Proxy 代理服务

步骤 04 在运行 mysql-proxy 命令之前，需要先进入 MySQL Proxy 解压包的 bin 文件夹下，运行如下 cd 命令即可，如图 23.7 所示。

```
cd C:\Program Files\MySQL\mysql-proxy-0.8.5\bin
```

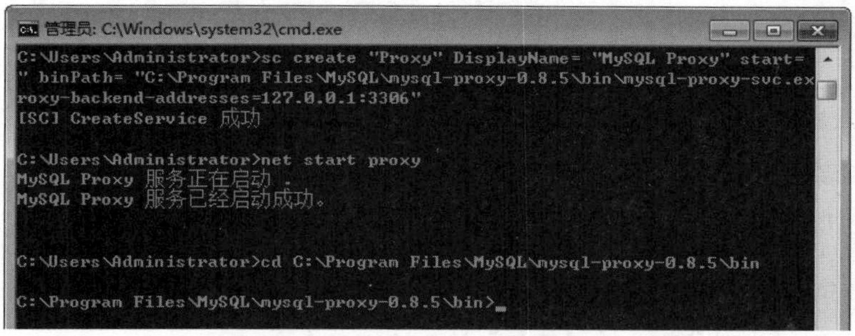

图 23.7　进入 MySQL Proxy 解压包的 bin 文件夹下

步骤 05 接着即可使用 mysql-proxy 命令配置 MySQL Proxy 代理参数，如图 23.8 所示。

```
mysql-proxy --proxy-address=localhost:49710
            --proxy-backend-addresses=201.13.100.41:8008
```

图 23.8　配置 MySQL Proxy 代理参数

提示：上面的配置命令含义是使用本地服务器的 49710 端口代理服务器 201.13.100.41:8008，如果应用程序完成本地服务器 49710 端口的连接，就等同于连接上了 201.13.100.41:8008 的数据库。

在前面配置 MySQL Proxy 代理参数的时候，不能直接输入 mysql-proxy 命令，是因为没有把 MySQL Proxy 的 bin 目录添加到系统的环境变量里面，所以不能直接使用 mysql-proxy 命令。如果每

次登录都输入 "cd C:\Program Files\MySQL\mysql-proxy-0.8.5\bin" 才能使用 MySQL Proxy，会比较麻烦。可以手动配置 PATH 变量，把 mysql-proxy 命令所在的目录加入 PATH 变量中，由于操作比较简单，此处省略。

23.3 使用 MySQL Proxy 实现读写分离

本节将通过案例来学习 MySQL Proxy 是如何实现读写分离的。本案例需要准备 4 台主机，配置如表 23.1 所示。

表23.1 MySQL Proxy读写分离测试环境配置

主机名称	IP 地址	功 能
主机 1	192.168.1.101	测试客户端主机
主机 2	192.168.1.102	安装 MySQL Proxy 服务主机
主机 3	192.168.1.103	MySQL 数据库
主机 4	192.168.1.104	MySQL 数据库

在主机 2 上安装 MySQL Proxy 并创建服务，命令如下：

```
sc create "Proxy" DisplayName= "MySQL Proxy" start= "auto" binPath= "C:\Program Files\MySQL\mysql-proxy-0.8.5\bin\mysql-proxy-svc.exe --proxy-backend-addresses=192.168.1.102:3306"
```

在主机 2 上配置代理参数，设置其与主机 3 连接，命令如下：

```
mysql-proxy --proxy-backend-addresses=192.168.1.103:3306
```

在主机 2 上配置代理参数，设置其与主机 4 连接，实现只读操作，命令如下：

```
mysql-proxy --proxy-read-only-backend-addresses=192.168.1.104:3306
```

分别在主机 3 和主机 4 上创建管理账号 zhangsan，密码为 zhang123456，并有读写操作数据库 test 的权限，命令如下：

```
GRANT SELECT,INSERT ON test TO 'zhangsan'@'localhost'
    IDENTIFIED BY 'zhang123456'
```

用主机 1 连接主机 2，命令如下：

```
mysql -h 192.168.1.102 -u zhangsan -p 123456 -P 4040
```

连接成功后，创建数据表：

```
create table ceshi(id int,name char(6));
```

然后到主机 3 和主机 4 上查询数据库的情况，发现主机 3 上存在刚刚写入的数据表 ceshi，而主机 4 上不存在数据表 ceshi。可见，此时已经实现了数据库的读写分离操作。

第 24 章

新闻发布系统数据库设计

MySQL 数据库的使用非常广泛，很多网站和应用系统使用 MySQL 数据库存储数据。本章将讲解新闻发布系统的数据库设计过程。通过本章的学习，读者可以学会如何使用 MySQL 数据库完成 Web 应用系统的数据库设计。

24.1 系统概述

本章介绍的是一个小型 Web 新闻发布系统，管理员可以通过该系统发布新闻信息、管理新闻信息。一个典型的新闻发布系统网站至少应该包含新闻信息管理、新闻信息显示和新闻信息查询 3 种功能。

新闻发布系统所要实现的功能具体包括新闻信息添加、新闻信息修改、新闻信息删除、显示全部新闻信息、按类别显示新闻信息、按关键字查询新闻信息、按关键字进行站内查询。

本站为一个简单的新闻信息发布系统，该系统具有以下特点。

- 实用：系统实现了一个完整的信息发布过程。
- 简单易用：为了使用户尽快掌握和使用整个系统，系统需要设计得结构简单且功能齐全，简洁的页面设计使用户操作起来非常方便。

本系统主要用于发布新闻信息、管理用户、管理权限、管理评论等功能。这些信息的录入、查询、修改和删除等操作都是该系统重点解决的问题。

本系统的主要功能包括以下几点：

（1）具有用户注册及个人信息管理功能。
（2）具有管理员注册和管理员信息管理功能。
（3）用户注册后可以对新闻进行评论。
（4）管理员可以发布新闻、管理新闻信息。
（5）管理员可以对评论进行审核和删除。

24.2 系统功能

新闻发布系统分为 5 个管理部分，即用户管理、管理员管理、权限管理、新闻管理和评论管理。本系统的功能模块如图 24.1 所示。

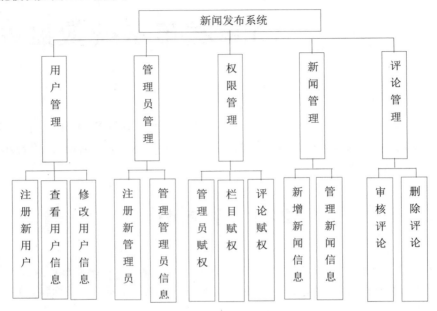

图 24.1　系统功能模块图

图 24.1 中模块的详细介绍如下：

（1）用户管理模块：实现新增用户、查看和修改用户信息的功能。

（2）管理员管理模块：实现新增管理员，查看、修改和删除管理员信息的功能。

（3）权限管理模块：实现对管理员、对管理的栏目和评论赋权的功能。

（4）新闻管理模块：实现有相关权限的管理员对新闻的增加、查看、修改和删除的功能。

（5）评论管理模块：实现有相关权限的管理员对评论的审核和删除的功能。

通过本节的介绍，读者对这个新闻发布系统的主要功能有了一定的了解，下一节将介绍本系统所需要的数据库和表的设计。

24.3 数据库设计和实现

数据库设计是开发管理系统最重要的一个步骤。如果数据库设计得不够合理，就会给后续的开发工作带来很大的麻烦。本节将讲解新闻发布系统的数据库设计过程。

数据库设计时要确定设计哪些表、表中包含哪些字段、字段的数据类型和长度。通过本节的学习，读者可以对 MySQL 数据库的设计有一个全面的了解。

24.3.1 设计表

本系统所有的表都放在 webnews 数据库下。创建和选择 webnews 数据库的 SQL 代码如下：

```
CREATE DATABASE webnews;
USE webnews;
```

在这个数据库下总共存放 9 张表，分别是 user、admin、roles、news、category、comment、admin_Roles、news_Comment 和 users_Comment。

1. user 表

用户表（user）用于存储用户 ID、用户名、密码和用户 Email 地址，所以 user 表设计了 5 个字段。user 表每个字段的信息如表 24.1 所示。

表 24.1 user 表

列 名	数据类型	允许 NULL 值	说 明
userID	INT	否	用户编号
userName	VARCHAR(20)	否	用户名称
userPassword	VARCHAR(20)	否	用户密码
sex	VARCHAR(10)	否	用户性别
userEmail	VARCHAR(20)	否	用户 Email

根据表 24.1 的内容创建 user 表，SQL 语句如下：

```
CREATE TABLE user(
userID INT PRIMARY KEY UNIQUE NOT NULL,
userName VARCHAR(20) NOT NULL,
userPassword VARCHAR(20) NOT NULL,
sex varchar(10) NOT NULL,
userEmail VARCHAR(20) NOT NULL
);
```

创建完成后，可以使用 DESC 语句查看 user 表的基本结构，也可以通过 SHOW CREATE TABLE 语句查看 user 表的详细信息。

2. admin 表

管理员表（admin）主要用来存放用户账号信息，如表 24.2 所示。

表 24.2 admin 表

列 名	数据类型	允许 NULL 值	说 明
adminID	INT	否	管理员编号
adminName	VARCHAR(20)	否	管理员名称
adminPassword	VARCHAR(20)	否	管理员密码

根据表 24.2 的内容创建 admin 表。创建 admin 表的 SQL 语句如下：

```
CREATE TABLE admin(
adminID INT PRIMARY KEY UNIQUE NOT NULL,
adminName VARCHAR(20) NOT NULL,
```

```
adminPassword VARCHAR(20) NOT NULL
);
```

创建完成后，可以使用 DESC 语句查看 admin 表的基本结构，也可以通过 SHOW CREATE TABLE 语句查看 admin 表的详细信息。

3. roles 表

权限表（roles）主要用来存放权限信息，如表 24.3 所示。

表 24.3 roles 权限表

列 名	数据类型	允许 NULL 值	说 明
roleID	INT	否	权限编号
roleName	VARCHAR(20)	否	权限名称

根据表 24.3 的内容创建 roles 表，SQL 语句如下：

```
CREATE TABLE roles(
roleID INT PRIMARY KEY UNIQUE NOT NULL,
roleName VARCHAR(20) NOT NULL
);
```

创建完成后，可以使用 DESC 语句查看 roles 表的基本结构，也可以通过 SHOW CREATE TABLE 语句查看 roles 表的详细信息。

4. news 表

新闻表（news）主要用来存放新闻信息，如表 24.4 所示。

表 24.4 news 新闻表

列 名	数据类型	允许 NULL 值	说 明
newsID	INT	否	新闻编号
newsTitle	VARCHAR(50)	否	新闻标题
newsContent	TEXT	否	新闻内容
newsDate	TIMESTAMP	是	发布时间
newsDesc	VARCHAR(50)	否	新闻描述
newsImagePath	VARCHAR(50)	是	新闻图片路径
newsRate	INT	否	新闻级别
newsIsCheck	BIT	否	新闻是否检验
newsIsTop	BIT	否	新闻是否置顶

根据表 24.4 的内容创建 news 表，SQL 语句如下：

```
CREATE TABLE news(
newsID INT PRIMARY KEY UNIQUE NOT NULL,
newsTitle VARCHAR(50) NOT NULL,
newsContent TEXT NOT NULL,
newsDate TIMESTAMP,
newsDesc VARCHAR(50) NOT NULL,
newsImagePath VARCHAR(50),
newsRate INT,
```

```
newsIsCheck BIT,
newsIsTop BIT
);
```

创建完成后,可以使用 DESC 语句查看 news 表的基本结构,也可以通过 SHOW CREATE TABLE 语句查看 news 表的详细信息。

5. category 表

栏目表(categroy)主要用来存放新闻栏目信息,如表 24.5 所示。

表 24.5 category 栏目表

列 名	数据类型	允许 NULL 值	说 明
categroyID	INT	否	栏目编号
categroyName	VARCHAR(50)	否	栏目名称
categroyDesc	VARCHAR(50)	否	栏目描述

根据表 24.5 的内容创建 categroy 表。创建 categroy 表的 SQL 语句如下:

```
CREATE TABLE categroy (
categoryID INT PRIMARY KEY UNIQUE NOT NULL,
categoryName VARCHAR(50) NOT NULL,
categoryDesc VARCHAR(50) NOT NULL
);
```

创建完成后,可以使用 DESC 语句查看 categroy 表的基本结构,也可以通过 SHOW CREATE TABLE 语句查看 categroy 表的详细信息。

6. comment 表

评论表(comment)主要用来存放新闻评论信息,如表 24.6 所示。

表 24.6 comment 评论表

列 名	数据类型	允许 NULL 值	说 明
commentID	INT	否	评论编号
commentTitle	VARCHAR(50)	否	评论名称
commentContent	TEXT	否	评论内容
commentDate	DATETIME	否	评论时间

根据表 24.6 的内容创建 comment 表,SQL 语句如下:

```
CREATE TABLE comment (
commentID INT PRIMARY KEY UNIQUE NOT NULL,
commentTitle VARCHAR(50) NOT NULL,
commentContent TEXT NOT NULL,
commentDate DATETIME
);
```

创建完成后,可以使用 DESC 语句查看 comment 表的基本结构,也可以通过 SHOW CREATE TABLE 语句查看 comment 表的详细信息。

7. admin_Roles 表

管理员权限表（admin_Roles）主要用来存放管理员和权限的关系，如表 24.7 所示。

表 24.7 admin_Roles 管理员权限表

列 名	数据类型	允许 NULL 值	说 明
aRID	INT	否	管理员权限编号
adminID	INT	否	管理员编号
roleID	INT	否	权限编号

根据表 24.7 的内容创建 admin_Roles 表，SQL 语句如下：

```
CREATE TABLE admin_Roles (
aRID INT PRIMARY KEY UNIQUE NOT NULL,
adminID INT NOT NULL,
roleID INT NOT NULL
);
```

创建完成后，可以使用 DESC 语句查看 admin_Roles 表的基本结构，也可以通过 SHOW CREATE TABLE 语句查看 admin_Roles 表的详细信息。

8. news_Comment 表

新闻评论表（news_Comment）主要用来存放新闻和评论的关系，如表 24.8 所示。

表 24.8 news_Comment 新闻评论表

列 名	数据类型	允许 NULL 值	说 明
nCommentID	INT	否	新闻评论编号
newsID	INT	否	新闻编号
commentID	INT	否	评论编号

根据表 24.8 的内容创建 news_Comment 表，SQL 语句如下：

```
CREATE TABLE news_Comment (
nCommentID INT PRIMARY KEY UNIQUE NOT NULL,
newsID INT NOT NULL,
commentID INT NOT NULL
);
```

创建完成后，可以使用 DESC 语句查看 news_Comment 表的基本结构，也可以通过 SHOW CREATE TABLE 语句查看 news_Comment 表的详细信息。

9. users_Comment 表

用户评论表（users_Comment）主要用来存放用户和评论的关系，如表 24.9 所示。

表 24.9 users_Comment 用户评论表

列 名	数据类型	允许 NULL 值	说 明
uCID	INT	否	用户评论编号
userID	INT	否	用户编号
commentID	INT	否	评论编号

根据表 24.9 的内容创建 users_Comment 表，SQL 语句如下：

```
CREATE TABLE news_Comment (
uCID    INT PRIMARY KEY UNIQUE NOT NULL,
userID  INT NOT NULL,
commentID   INT NOT NULL
);
```

创建完成后，可以使用 DESC 语句查看 users_Comment 表的基本结构，也可以通过 SHOW CREATE TABLE 语句查看 users_Comment 表的详细信息。

24.3.2 设计索引

索引是创建在表上的，是对数据库中一列或者多列的值进行排序的一种结构。索引可以提高查询的速度。新闻发布系统需要查询新闻的信息，这就需要在某些特定字段上建立索引，以便提高新闻查询速度。

1. 在 news 表上建立索引

新闻发布系统中需要按照 newsTitle 字段、newsDate 字段和 newsRate 字段查询新闻信息。在本书前面有关索引的章节中介绍了几种创建索引的方法。这里将使用 CREATE INDEX 语句和 ALTER TABLE 语句创建索引。

下面使用 CREATE INDEX 语句在 newsTitle 字段上创建名为 index_new_title 的索引。SQL 语句如下：

```
CREATE INDEX index_new_title ON news(newsTitle);
```

然后，使用 CREATE INDEX 语句在 newsDate 字段上创建名为 index_new_date 的索引。SQL 语句如下：

```
CREATE INDEX index_new_date  ON news(newsDate);
```

最后，使用 ALTER TABLE 语句在 newsRate 字段上创建名为 index_new_rate 的索引。SQL 语句如下：

```
ALTER TABLE news  ADD INDEX index_new_rate (newsRate);
```

2. 在 categroy 表上建立索引

在新闻发布系统中，需要通过栏目名称查询该栏目下的新闻，因此需要在这个字段上创建索引。创建索引的语句如下：

```
CREATE INDEX index_categroy_name ON categroy (categroyName);
```

代码执行完成后，读者可以使用 SHOW CREATE TABLE 语句查看 categroy 表的详细信息。

3. 在 comment 表上建立索引

新闻发布系统需要通过 commentTitle 字段和 commentDate 字段查询评论内容，因此可以在这两个字段上创建索引。创建索引的语句如下：

```
CREATE INDEX index_comment_title  ON comment (commentTitle);
```

```sql
CREATE INDEX index_comment_date ON comment (commentDate);
```

代码执行完成后，读者可以通过 SHOW CREATE TABLE 语句查看 comment 表的结构。

24.3.3 设计视图

视图是由数据库中一个表或者多个表导出的虚拟表，作用是方便用户对数据的操作。在这个新闻发布系统中，也设计了一个视图改善查询操作。

在新闻发布系统中，如果直接查询 news_Comment 表，显示信息时会显示新闻编号和评论编号。这种显示不直观，为了以后查询方便，可以建立一个视图 news_view。这个视图显示评论编号、新闻编号、新闻级别、新闻标题、新闻内容和新闻发布时间。创建视图 news_view 的 SQL 代码如下：

```sql
CREATE VIEW news_view
AS SELECT c.commentID, n.newsID, n.newsRate, n.newsTitle, n.newsContent, n.newsDate
FROM news_Comment c, news n
WHERE news_Comment.newsID=news.newsID;
```

news_Comment 表的别名为 c，news 表的别名为 n，news_view 视图从这两张表中取出相应的字段。视图创建完成后，可以使用 SHOW CREATE VIEW 语句查看 news_view 视图的详细信息。

24.3.4 设计触发器

触发器由 INSERT、UPDATE 和 DELETE 等事件来触发某种特定的操作。在满足触发器的触发条件时，数据库系统就会执行触发器中定义的程序语句。这样做可以保证某些操作之间的一致性。为了使新闻发布系统的数据更新更加快速和合理，可以在数据库中设计几个触发器。

1. 设计 UPDATE 触发器

在设计表时，news 表和 news_Comment 表的 newsID 字段的值是一样的。如果 news 表中的 newsID 字段的值更新了，那么 news_Comment 表中的 newsID 字段的值也必须同时更新。这可以通过一个 UPDATE 触发器来实现。创建 UPDATE 触发器 update_newsID 的 SQL 代码如下：

```sql
DELIMITER &&
CREATE TRIGGER update_newsID AFTER UPDATE
ON news FOR EACH ROW
BEGIN
    UPDATE news_Comment SET newsID=NEW.newsID
END
&&
DELIMITER ;
```

其中，NEW.newsID 表示 news 表中更新的记录的 newsID 值。

2. 设计 DELETE 触发器

如果从 user 表中删除一个用户的信息，那么这个用户在 users_Comment 表中的信息也必须同时删除。这也可以通过触发器来实现。在 user 表上创建 delete_user 触发器，只要执行 DELETE 操作，就会同时删除 users_Comment 表中相应的记录。创建 delete_user 触发器的 SQL 语句如下：

```
DELIMITER &&
CREATE TRIGGER delete_user AFTER DELETE
ON user FOR EACH ROW
BEGIN
    DELETE FROM users_Comment WHERE userID=OLD.userID
END
&&
DELIMITER ;
```

其中，OLD.userID 表示所删除用户的 userID 值。

第 25 章

论坛管理系统数据库设计

在论坛系统中，用户可以注册成为论坛会员，取得发表言论的资格，同时管理员可以对用户、版块、帖子进行管理。通过这样的系统，可以做到规范用户言论、促进用户交流、增强论坛对用户的凝聚力。为了实现论坛系统运行和管理规范有序，需要数据库的设计非常合理。本章将讲解论坛管理系统数据库的设计方法。

25.1 系统概述

论坛系统又名 BBS（Bulletin Board System，电子公告板）或者 Bulletin Board Service（公告板服务）。它是 Internet 上的一种电子信息服务系统，像一块电子公告牌，每个用户都可以在上面发布信息或发表见解。

现实生活中的交流存在时间和空间上的局限性——交流人群范围狭小以及间断交流，不能保证信息的准确性和可获得性。因此，用户需要通过网上论坛来扩大交流面，以获得自己需要的信息。另外，网上论坛信息传播速度更快，用户更容易迅速、准确地获得相关信息。

论坛系统能为分散在五湖四海的人提供一个共同交流、学习、倾吐心声的平台，实现来自不同地方用户的极强的信息互动性，用户在获得自己所需要的信息的同时，也可以广交朋友，拓展自己的视野和扩大自己的社交面。

论坛系统是一种交互性强、内容丰富且信息实时发布的电子信息服务系统。用户可以在 BBS 上获得各种信息服务、发布信息、进行讨论、聊天等。论坛系统按不同的主题分为许多版块，版面的设立依据是大多数用户的要求和喜好，用户可以阅读别人关于某个主题的看法，也可以将自己的想法毫无保留地发布到论坛中。随着互联网技术的不断发展，论坛系统的功能越来越强大。

本章将要讲解的论坛管理系统，其基本功能包括以下几点：

（1）用户和管理员的管理。
（2）版块信息的管理。
（3）主帖和回复帖的管理。

25.2 系统功能

　　论坛管理系统的主要功能是管理论坛系统的几个参与要素，包括用户、管理员、版块、主帖和回复帖。通过本管理系统，可以提高论坛管理员的工作效率。本节将详细介绍论坛管理系统的功能。

　　论坛管理系统主要分为 5 个管理部分，包括用户管理、管理员管理、版块管理、主帖管理和回复帖管理。本系统的功能模块图如图 25.1 所示。

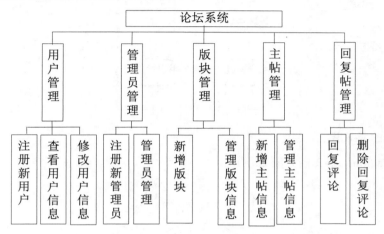

图 25.1　系统功能模块图

图 25.1 中各模块的详细介绍如下：

（1）用户管理模块：实现新增用户、查看和修改用户信息的功能。
（2）管理员管理模块：实现新增管理员，查看、修改和删除管理员信息的功能。
（3）版块管理模块：实现新增版块、管理版块的功能。
（4）主帖管理模块：实现对主帖的增加、查看、修改和删除功能。
（5）回复帖管理模块：实现有相关权限的管理员对回复帖的审核和删除功能。

　　通过本节的介绍，读者应该对这个论坛管理系统的主要功能有了一定的了解，下一节将介绍本系统所需要的数据库和表的设计。

25.3　数据库设计和实现

　　数据库设计时要确定设计哪些表、表中包含哪些字段、字段的数据类型和长度。本节将主要讲解论坛管理系统数据库的设计。

25.3.1　设计方案图表

　　在设计表之前，我们可以先设计出方案图表。

1. 用户表的 E-R 图

用户表为 user，E-R 图如图 25.2 所示。

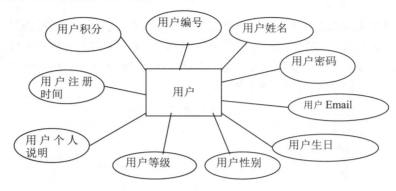

图 25.2　用户 user 表的 E-R 图

2. 管理员表的 E-R 图

管理员表为 admin，E-R 图如图 25.3 所示。

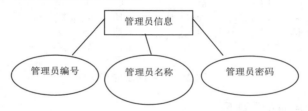

图 25.3　admin 管理员信息表的 E-R 图

3. 版块表的 E-R 图

版块表为 section，E-R 图如图 25.4 所示。

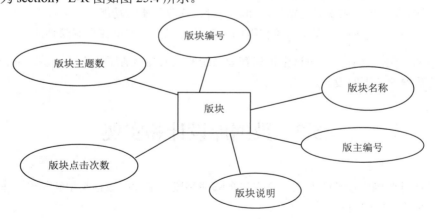

图 25.4　section 版块信息表的 E-R 图

4. 主帖表的 E-R 图

主帖表为 topic，E-R 图如图 25.5 所示。

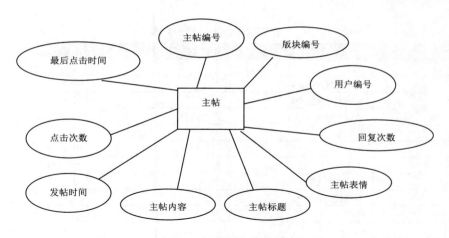

图 25.5　topic 主帖表的 E-R 图

5. 回复帖表的 E-R 图

回复帖表为 reply，E-R 图如图 25.6 所示。

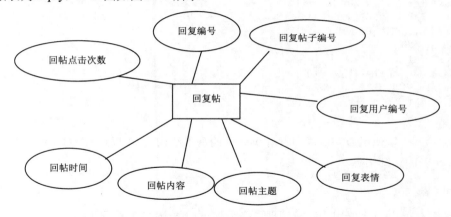

图 25.6　reply 回复帖表的 E-R 图

25.3.2　设计表

本系统所有的表都放在 bbs 数据库下。创建和选择 bbs 数据库的 SQL 代码如下：

```
CREATE DATABASE bbs;
USE bbs;
```

在这个数据库下总共存放 5 张表，分别是 user、admin、section、topic 和 reply。

1. user 表

user 表主要用于存储用户 ID、用户名、密码和用户 Email 地址等，所以 user 表设计了 10 个字段。user 表每个字段的信息如表 25.1 所示。

表 25.1 user（用户）表

列 名	数据类型	允许 NULL 值	说 明
uID	INT	否	用户编号
userName	VARCHAR(20)	否	用户姓名
userPassword	VARCHAR(20)	否	用户密码
userEmail	VARCHAR(20)	否	用户 Email
userBirthday	DATE	否	用户生日
userSex	BIT	否	用户性别
userClass	INT	否	用户等级
userStatement	VARCHAR(150)	否	用户个人说明
userRegDate	TIMESTAMP	否	用户注册时间
userPoint	INT	否	用户积分

根据表 25.1 的内容创建 user 表，SQL 语句如下：

```
CREATE TABLE user(
uID INT PRIMARY KEY UNIQUE NOT NULL,
userName VARCHAR(20) NOT NULL,
userPassword VARCHAR(20) NOT NULL,
userEmail VARCHAR(20) NOT NULL,
userBirthday DATE NOT NULL,
userSex BIT NOT NULL,
userClass INT NOT NULL,
userStatement VARCHAR(150) NOT NULL,
userRegDate TIMESTAMP NOT NULL,
userPoint INT NOT NULL
);
```

创建完成后，可以使用 DESC 语句查看 user 表的基本结构，也可以通过 SHOW CREATE TABLE 语句查看 user 表的详细信息。

2. admin 表

管理员表（admin）主要用来存放管理员账号信息，如表 25.2 所示。

表 25.2 admin（管理员）表

列 名	数据类型	允许 NULL 值	说 明
adminID	INT	否	管理员编号
adminName	VARCHAR(20)	否	管理员名称
adminPassword	VARCHAR(20)	否	管理员密码

根据表 25.2 的内容创建 admin 表，SQL 语句如下：

```
CREATE TABLE admin(
adminID INT PRIMARY KEY UNIQUE NOT NULL,
adminName VARCHAR(20) NOT NULL,
adminPassword VARCHAR(20) NOT NULL
);
```

创建完成后，可以使用 DESC 语句查看 admin 表的基本结构，也可以通过 SHOW CREATE TABLE 语句查看 admin 表的详细信息。

3. section 表

版块表（section）主要用来存放版块信息，如表 25.3 所示。

表 25.3　section（版块）表

列　名	数据类型	允许 NULL 值	说　明
sID	INT	否	版块编号
sName	VARCHAR(20)	否	版块名称
sMasterID	INT	否	版主编号
sStatement	VARCHAR(200)	否	版块说明
sClickCount	INT	否	版块点击次数
sTopicCount	INT	否	版块主题数

根据表 25.3 的内容创建 section 表，SQL 语句如下：

```
CREATE TABLE section (
sID INT PRIMARY KEY UNIQUE NOT NULL,
sName VARCHAR(20) NOT NULL,
sMasterID INT NOT NULL,
sStatement VARCHAR(200) NOT NULL,
sClickCount INT NOT NULL,
sTopicCount INT NOT NULL
);
```

创建完成后，可以使用 DESC 语句查看 section 表的基本结构，也可以通过 SHOW CREATE TABLE 语句查看 section 表的详细信息。

4. topic 表

主帖表（topic）主要用来存放主帖信息，如表 25.4 所示。

表 25.4　topic（主帖）表

列　名	数据类型	允许 NULL 值	说　明
tID	INT	否	主帖编号
tsID	INT	否	版块编号
tuid	INT	否	用户编号
tReplyCount	INT	否	回复次数
tEmotion	VARCHAR(20)	否	主帖表情
tTopic	VARCHAR(20)	否	主帖标题
tContents	TEXT	否	主帖内容
tTime	TIMESTAMP	否	发帖时间
tClickCount	INT	否	点击次数
tLastClickT	TIMESTAMP	否	最后点击时间

根据表 25.4 的内容创建 topic 表。创建 topic 表的 SQL 语句如下：

```
CREATE TABLE topic (
tID INT PRIMARY KEY UNIQUE NOT NULL,
tSID INT NOT NULL,
tuid INT NOT NULL,
tReplyCount INT NOT NULL,
tEmotion VARCHAR(20) NOT NULL,
```

```
tTopic VARCHAR(20) NOT NULL,
tContents TEXT NOT NULL,
tTime TIMESTAMP NOT NULL,
tClickCount INT NOT NULL,
tLastClickT TIMESTAMP NOT NULL
);
```

创建完成后，可以使用 DESC 语句查看 topic 表的基本结构，也可以通过 SHOW CREATE TABLE 语句查看 topic 表的详细信息。

5. reply 表

回复帖表（reply）主要用来存放回复帖的信息，如表 25.5 所示。

表 25.5 reply（回复帖）表

列 名	数据类型	允许 NULL 值	说 明
rID	INT	否	回复编号
tID	INT	否	回复帖子编号
uID	INT	否	回复用户编号
rEmotion	CHAR	否	回复表情
rTopic	VARCHAR（20）	否	回帖主题
rContents	TEXT	否	回帖内容
rTime	TIMESTAMP	否	回帖时间
rClickCount	INT	否	回帖点击次数

根据表 25.5 的内容创建 reply 表，SQL 语句如下：

```
CREATE TABLE reply (
rID INT PRIMARY KEY UNIQUE NOT NULL,
tID INT NOT NULL,
uID INT NOT NULL,
rEmotion CHAR NOT NULL,
rTopic VARCHAR(20) NOT NULL,
rContents TEXT NOT NULL,
rTime TIMESTAMP NOT NULL,
rClickCount INT NOT NULL
);
```

创建完成后，可以使用 DESC 语句查看 reply 表的基本结构，也可以通过 SHOW CREATE TABLE 语句查看 reply 表的详细信息。

25.3.3 设计索引

索引创建在表上，是对数据库中一列或者多列的值进行排序的一种结构。索引可以提高查询速度。论坛系统需要查询论坛的信息，这就需要在某些特定字段上建立索引，以便提高帖子的查询速度。

1．在 topic 表上建立索引

新闻发布系统中需要按照 tTopic 字段、tTime 字段和 tContents 字段查询新闻信息。在本书前面有关索引的章节中介绍了几种创建索引的方法。这里将使用 CREATE INDEX 语句和 ALTER TABLE 语句创建索引。

下面使用 CREATE INDEX 语句在 tTopic 字段上创建名为 index_topic_title 的索引。SQL 语句如下：

```
CREATE INDEX index_topic_title ON topic(tTopic);
```

然后，使用 CREATE INDEX 语句在 tTime 字段上创建名为 index_topic_time 的索引。SQL 语句如下：

```
CREATE INDEX index_topic_date  ON topic(tTime);
```

最后，使用 ALTER TABLE 语句在 tContents 字段上创建名为 index_topic_contents 的索引。SQL 语句如下：

```
ALTER TABLE topic  ADD INDEX index_new_contents (contents);
```

2. 在 section 表上建立索引

在论坛系统中需要通过版块名称查询该版块下的帖子信息，因此需要在这个字段上创建索引。创建索引的语句如下：

```
CREATE INDEX index_section_name ON section (sName);
```

代码执行完成后，读者可以使用 SHOW CREATE TABLE 语句查看 section 表的详细信息。

3. 在 reply 表上建立索引

论坛系统需要通过 rTime 字段、rTopic 字段和 tID 字段查询回复帖子的内容，因此可以在这 3 个字段上创建索引。创建索引的语句如下：

```
CREATE INDEX index_reply_rtime  ON reply (rTime);
CREATE INDEX index_reply_rtopic  ON reply (rTopic);
CREATE INDEX index_reply_rid  ON reply (tID);
```

代码执行完成后，读者可以通过 SHOW CREATE TABLE 语句查看 reply 表的结构。

25.3.4　设计视图

在论坛系统中，如果直接查询 section 表，显示信息时会显示版块编号和版块名称等信息。这种显示不直观显示主帖的标题和发布时间，为了以后查询方便，可以建立一个视图 topic_view。这个视图显示版块的编号、版块的名称、同一版块下主帖的标题、主贴的内容和主帖的发布时间。创建视图 topic_view 的 SQL 代码如下：

```
CREATE VIEW topic_view
AS SELECT s.sID,s.sName,t.tTopic,t.tContents,t.tTime
FROM section s,topic t
WHERE s.sID=t.sID;
```

上面 SQL 语句中给每个表都取了别名，section 表的别名为 s；topic 表的别名为 t，topic_view 视图从这两个表中取出相应的字段。视图创建完成后，可以使用 SHOW CREATE VIEW 语句查看 topic_view 视图的详细信息。

25.3.5 设计触发器

触发器由 INSERT、UPDATE 和 DELETE 等事件来触发某种特定的操作。满足触发器的触发条件时，数据库系统就会执行触发器中定义的程序语句。这样做可以保证某些操作之间的一致性。为了使论坛系统的数据更新更加快速和合理，可以在数据库中设计几个触发器。

1. 设计 INSERT 触发器

如果向 section 表插入记录，说明版块的主题数目要相应地增加。这可以通过触发器来完成。在 section 表上创建名为 section_count 的触发器，其 SQL 语句如下：

```
DELIMITER &&
CREATE TRIGGER section_count AFTER UPDATE
ON section FOR EACH ROW
BEGIN
    UPDATE section SET sTopicCount= sTopicCount+1
    WHERE sID=NEW.sID;
END
&&
DELIMITER ;
```

其中，NEW.sID 表示 section 表中增加的记录 sID 值。

2. 设计 UPDATE 触发器

在设计表时，user 表和 reply 表的 uID 字段的值是一样的。如果 user 表中的 uID 字段的值更新了，那么 reply 表中的 uID 字段的值也必须同时更新。这可以通过一个 UPDATE 触发器来实现。创建 UPDATE 触发器 update_userID 的 SQL 代码如下：

```
DELIMITER &&
CREATE TRIGGER update_userID AFTER UPDATE
ON user FOR EACH ROW
BEGIN
    UPDATE reply SET uID=NEW.uID;
END
&&
DELIMITER ;
```

其中，NEW.uID 表示 user 表中更新的记录的 uID 值。

3. 设计 DELETE 触发器

如果从 user 表中删除一个用户的信息，那么这个用户在 topic 表中的信息也必须同时删除。这也可以通过触发器来实现。在 user 表上创建 delete_user 触发器，只要执行 DELETE 操作，就删除 topic 表中用户相关的记录。创建 delete_user 触发器的 SQL 语句如下：

```
DELIMITER &&
CREATE TRIGGER delete_user AFTER DELETE
ON user FOR EACH ROW
BEGIN
    DELETE FROM top WHERE uID=OLD.uID;
END
&&
DELIMITER ;
```

其中，OLD.uID 表示所删除用户的 uID 值。